T5-AOI-136

R00069 59891

CHICAGO PUBLIC LIBRARY
HAROLD WASHINGTON LIBRARY CENTER

R0006959891

REF

QH
541.5
.S3   Early marine ecology
E2
Cop.1

DATE			

REF
QH
541.5
.S3
E2
Cop.1

FORM 125 M

Business/Science/Technology
Division

The Chicago Public Library

OCT 20 1978

Received

© THE BAKER & TAYLOR CO.

EARLY MARINE ECOLOGY

This is a volume in the Arno Press collection

HISTORY OF ECOLOGY

Advisory Editor
Frank N. Egerton III

Editorial Board
John F. Lussenhop
Robert P. McIntosh

*See last pages of this volume for a
complete list of titles.*

EARLY MARINE ECOLOGY

ARNO PRESS
A New York Times Company
New York / 1977

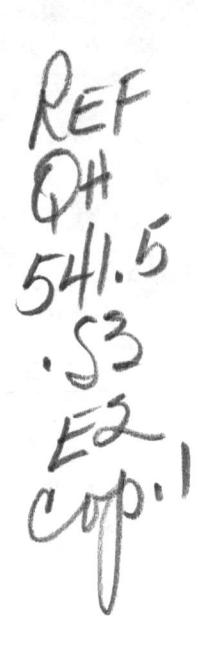

Editorial Supervision: LUCILLE MAIORCA

Reprint Edition 1977 by Arno Press Inc.

Copyright © 1977 by Arno Press Inc.

"Victor Hensen and the Development of
Sampling Methods in Ecology" by
John Lussenhop is reprinted from the
Journal of the History of Biology by
permission of D. Reidel Publishing Company.

"The Sea Bottom and Its Production of Fish
Food" by C. G. J. Peterson is reprinted from
Reports of the Danish Biological Station by
permission of the Danmarks Fiskeri-Og Havundersøgelser.

HISTORY OF ECOLOGY
ISBN for complete set: 0-405-10369-7
See last pages of this volume for titles.

Manufactured in the United States of America

Library of Congress Cataloging in Publication Data

Main entry under title:

Early marine ecology.

   (History of ecology)
    1. Marine ecology--Addresses, essays, lectures.
2. Mariculture--Addresses, essays, lectures.
QH541.5.S3E2   574.5'2636   77-74216
ISBN 0-405-10386-7 (set)

CONTENTS

Forbes, Edward
ON THE LIGHT THROWN ON GEOLOGY BY SUBMARINE RESEARCHES (Reprinted from *Edinburgh New Philosophical Journal*, Volume 36) Edinburgh, 1844

Haeckel, Ernst
GENERAL ECOLOGY AND DISTRIBUTION (OF RADIOLARIA) (Reprinted from *Report on the Scientific Results of the Voyage of H.M.S. Challenger during the Years 1873-1876*, Edited by C. Wyville Thompson and John Murray, Volume 18), London, 1887

Haeckel, Ernst
PLANKTONIC STUDIES: A Comparative Investigation of the Importance and Constitution of the Pelagic Fauna and Flora. Translated by George W. Field (Reprinted from *U.S. Commission of Fish and Fisheries Report*), Washington, D.C., 1891

Lussenhop, John
VICTOR HENSEN AND THE DEVELOPMENT OF SAMPLING METHODS IN ECOLOGY (Reprinted from *Journal of the History of Biology*, Volume 7), Dordrecht, Holland, 1974

Hensen, Victor
UEBER DIE BEFISCHUNG DER DEUTSCHEN KÜSTEN (ON THE FISHERIES OF THE GERMAN COAST) (Reprinted from *Wissenschaftlichen Meeresuntersuchungen der deutschen Kommission zur Wissenschaftlichen Untersuchungen der deutschen Meere*, Jahresbericht II-III), 1875

Dakin, W.J.
METHODS OF PLANKTON RESEARCH (Reprinted from *Proceedings and Transactions of the Liverpool Biological Society*, Volume XXII), Liverpool, 1908

Möbius, Karl
THE OYSTER AND OYSTER-CULTURE. Translated by H.J. Rice (Reprinted from *United States Commission of Fish and Fisheries, Report of the Commissioner for 1880*, Part VIII), Washington, D.C., 1883

Peterson, C.G.J.
THE SEA BOTTOM AND ITS PRODUCTION OF FISH FOOD: A Survey of the Work Done in Connection with Valuation of the Danish Waters from 1883-1917 (Reprinted from *Reports of the Danish Biological Station*, No. 25), Copenhagen, 1918

ON THE LIGHT THROWN ON GEOLOGY BY SUBMARINE RESEARCHES

Edward Forbes

*On the Light thrown on Geology by Submarine Researches;
being the Substance of a Communication made to the Royal
Institution of Great Britain, Friday Evening, the* 23d *February* 1844. By EDWARD FORBES, F.L.S., M.W.S., &c. Prof.
Bot. King's College, London. Communicated by the
Author.

About the middle of the last century, certain Italian naturalists[*] sought to explain the arrangement and disposition of organic remains in the strata of their country, by an examination of the distribution of living beings on the bed of the Adriatic Sea. They sought in the bed of the present sea for an explanation of the phenomena presented by the upheaved beds of former seas. The instrument, by means of which they conducted their researches, was the common oyster-dredge. The results they obtained bore importantly on Geology; but since their time, little has been done in the same line of research,—the geologist has been fully occupied above water, and the naturalist has pursued his studies with far too little reference to their bearing on geological questions, and on the history of animals and plants *in time*. The dredge, when used, has been almost entirely restricted to the search after rare animals, by the more adventurous among zoologists.

Convinced that inquiries of the kind referred to, if conducted with equal reference to all the natural history sciences,

---

[*] Marsili and Donati, and after them Soldani.

and to their mutual connection, must lead to results still more important than those which have been obtained, I have, for several years, conducted submarine researches by means of the dredge. In the present communication, I shall give a brief account of some of the more remarkable facts and conclusions to which they have led, and as briefly point out their bearings on the science of geology.

I. *Living beings are not distributed indifferently on the bed of the sea, but certain species live in certain parts, according to the depth, so that the sea-bed presents a series of zones or regions, each peopled by its peculiar inhabitants.*—Every person who has walked between high and low water-marks on the British coasts, when the tide was out, must have observed, that the animals and plants which inhabit that space, do not live on all parts of it alike, but that particular kinds reach only to certain distances from its extremities. Thus the species of *Auricula* are met with only at the very margin of high water mark, along with *Littorina cærulescens*, and *saxatilis*, *Velutina otis*, *Kellia rubra*, *Balani*, &c.; and among the plants, the yellow *Chondrus crispus* (*Carrigeen*, or Iceland moss of the shops), and *Corallina officinalis*. These are succeeded by other forms of animals and plants, such as *Littorina littorea*, *Purpura lapillus*, *Trochi*, *Actineæ*, *Porphyra laciniata* (Laver, Sloke), and *Ulvæ*. Towards the margin of low water, *Lottia testudinaria*, *Solen siliqua*, and the Dulse, *Rhodomenia palmata*, with numerous Zoophytes and Ascidian molluscs, indicate a third belt of life, connected, however, with the two others, by certain species common to all three, such as *Patella vulgata*, and *Mytilus edulis*. These sub-divisions of the sea-bed, exposed at ebb-tide, have long attracted attention on the coasts of our own country, and on those of France, where they have been observed by Audouin and Milne Edwards, and of Norway, where that admirable observer Sars has defined them with great accuracy.

Now this subdivision of the tract between tide-marks into zones of animal life, is a representation in miniature of the entire bed of the sea. The result of my observations, first in the British seas,* and more lately in the Ægean, has been to define a series of zones or regions in depth, and to ascertain *specifically* the animal and vegetable inhabitants of each. Regarding the tract between tide-marks as one region, which I have termed the *Littoral Zone*, we find a series of equivalent regions, succeeding it in depth. In the British seas, the littoral zone is succeeded by the region of Laminariæ, filled by forests of broad-leaved Fuci, among which live some of the most brilliantly coloured and elegant inhabitants of the ocean. This is the chosen habitat of *Lacunæ*, of *Rissoæ*, and of *Nudibranchous mollusca*. A belt generally of mud or gravel, in which numerous bivalve mollusca live, intervenes between the laminarian zone (in which the Flora of the sea appears to have its maximum), and the region of Corallines, which, ranging from a depth of from 20 to 40 fathoms, abounds in beautiful flexible zoophytes and in numerous species of Mollusca and Crustacea, to be procured only by means of the dredge. The great banks of Monomyarious Mollusca, which occur in many districts of the Northern Seas, are for the most part included in this region, and afford the zoologist his richest treasures. Deeper still is a region as yet but little explored, from which we draw up the more massy corals found on our shores, accompanied by shellfish of the class *Brachiopoda*. In the Eastern Mediterranean (where, through the invaluable assistance afforded by Captain Graves, and the Mediterranean Survey, I have been enabled to define the regions in depth, to an extent, and with a precision which, without similar aid, cannot be hoped for in the British seas), between the surface and the depth of 230 fathoms, the lowest point I had an opportunity of examining, there are eight well-defined zones, corresponding in part, and presenting similar characters with those which I have enumerated as presented by the sea-bed in the North. The details of these will be given in the forthcoming volume of the Transactions of the

---

* The first notice of these was published in the Edinburgh Academic Annual for 1840.

British Association, to which body I had the honour of presenting a report on the subject, at the last meeting.

When we examine the distribution and association of organic remains, in the upheaved beds of tertiary seas, we find the zones of depth as evident as they are in the present ocean. I have proved this to my own satisfaction, by a minute comparison of the newer Pliocene strata of Rhodes, where that formation attains a great thickness, with the present state of the neighbouring sea, and carrying on the comparison through the more recent tertiaries with the more ancient, have found indubitable evidences of the same phenomena. The strata of the cretaceous system yield similar evidences, and doubtless, in all time, the element of depth exercised a most important influence in regulating the distribution of animal life in the sea. If so, as our researches extend, we may hope eventually to ascertain the probable depth, or, at any rate, the region of depth, in which a given stratum containing organic remains was deposited. Every geologist will at once admit, that such a result would contribute materially to the history of sedimentary formations, and to the progress of geological science.

II. *The number of species is much less in the lower zones than in the upper. Vegetables disappear below a certain depth, and the diminution in the number of animal species indicates a zero not far distant.*—This conclusion is founded on my Ægean researches. Vegetables become fewer and fewer in the lower zones; and dwindle to a single species,—a *nullipora*, at the depth of 100 fathoms. Although the lower zones have a much greater vertical range than the higher, the number of animal species is infinitely greater in the latter. The lowest region (the 8th) in the Mediterranean, exceeds in extent all the other regions together; yet its fauna is comparatively small, and at the lowest portion explored, the number of species of testacea found was only eight. In the littoral zone, there were above 150 species. We may fairly infer, then, that as there is a zero of vegetable life, so is there one of animal life. In the sea, the vertical range of animals is greater than that of vegetables;—on the land, the reverse is the case. The geological application of this fact, of a zero of life in the ocean, is evident. All deposits formed below that zero, will be void, or almost void, of organic contents. The greater part of the sea is far deeper than the point zero; consequently, the greater part of deposits forming, will be void of organic remains. Hence we have no right to infer that any sedimentary formation, in which we find few or no traces of animal life, was formed either before animals were created, or at a time when the sea was less prolific in life than it now is. *It might have been formed in a very deep sea.* And that such was the case in regard to some of our older rocks, such as the great slates, is rendered the more probable, seeing that the few fossils we find in them, belong to tribes which, at present, have their maximum in the lowest regions of animal life, such as the Brachiopoda, and Pteropoda, of which, though free swimmers in the ocean, the remains accumulate only in very deep deposits. The uppermost deposits, those in which organic remains would be most abundant, are those most liable to disappear, in consequence of the destroying action of denudation. The great and almost nonfossiliferous strata of Scaglia, which form so large a part of the south of Europe and of Western Asia, were probably, for the most part, formed below the zero of life. The few fossils they contain, chiefly nummulites, correspond to the foraminifera which now abound mostly in the lowest region of animals. There is no occasion to attribute to metamorphic action the absence of traces of living beings in such rocks.

III. *The number of northern forms of animals and plants is not the same in all the zones of depth, but increases either positively, or by representation, as we descend.*—The association of species in the littoral zone is that most characteristic of the geographical region we are exploring; but the lower zones have their faunas and floras modified by the presence of species which, in more northern seas, are characteristic of the littoral zones. Of course, this remark applies only to the northern hemisphere; though, from analogy, we may expect to find such *inversely* the case also in the southern. The law, put in the abstract, appears to be, that *parallels in depth are equivalent to parallels in latitude*, corresponding to a well-known law in the distribution of terrestrial organic beings, viz. that *parallels in elevation are equivalent to parallels in latitude :* for example, as we ascend mountains in tropical countries, we find the successive belts of vegetation more and more northern or southern (according to the hemisphere) in character, either by identity of species, or by representation of forms by similar forms ; so in the sea, as we descend, we find a similar representation of climates in parallels of latitude in depth. The possibility of such a representation has been hypothetically anticipated in regard to marine animals by Sir Henry De La Beche,* and to marine plants by Lamouroux. To me it has been a great pleasure to confirm the felicitous speculations of those distinguished observers. The fact of such a representation has an important geological application. It warns us that all climatal inferences drawn from the number of northern forms in strata containing assemblages of organic remains, are fallacious, unless the element of depth be taken into consideration. But the influence of that element once ascertained (and I have already shewn the possibility of doing so), our inferences assume a value to which they could not otherwise pretend. In this way, I have no doubt, the per-centage test of Mr Lyell will become one of the most important aids in geology and natural history generally ; and, in fact, the most valuable conclusions to which I arrived by the reduction of my observations in the Ægean, were attained through the employment of Mr Lyell's method.

IV. *All varieties of sea-bottom are not equally capable of sustaining animal and vegetable life.*—In all the zones of depth there are occasionally more or less desert tracts, usually of sand or mud. The few animals which frequent such tracts are mostly soft and unpreservable. In some muddy and sandy districts, however, worms are very numerous, and to such places many fishes resort for food. The scarcity of remains of testacea in sandstones, the tracks of worms on ripple-marked sandstones, which had evidently been deposited in a

---

* Ten years ago, in his " Researches in Theoretical Geology."

shallow sea, and the fish remains often found in such rocks, are explained, in a great measure, by these facts.

V. *Beds of marine animals do not increase to an indefinite extent. Each species is adapted to live on certain sorts of sea-bottom only. It may die out in consequence of its own increase changing the ground.*—Thus, a bed of scallops, *Pecten opercularis*, for example, or of oysters having increased to such an extent that the ground is completely changed, in consequence of the accumulation of the remains of dead scallops or oysters, becomes unfitted for the further sustenance of the tribe. The young cease to be developed there, and the race dies out, and becomes silted up or imbedded in sediment, when the ground being renewed, it may be succeeded either by a fresh colony of scallops, or by some other species or assemblage of species. This " rotation of crops," as it were, is continually going on in the bed of the sea, and affords a very simple explanation of the alternation of fossiliferous and nonfossiliferous strata; organic remains in rocks being very rarely scattered through their substance, but arranged in layers of various thickness, interstratified with layers containing few or no fossils. Such interstratification may, in certain cases, be caused in another way, to-wit, by the elevation or subsidence of the sea-bottom, and the consequent destruction of the inhabitants of one region of depth, and the substitution of those of another. It is by such effects of oscillation of level, we may account for the repetition, at intervals, in certain formations of strata indicating the same region of depth.

VI. *Animals having the greatest ranges in depth have usually a great geographical, or else a great geological range, or both.*—I found that such of the Mediterranean testacea as occur both in the existing sea, and in the neighbouring tertiaries, were such as had the power of living in several of the zones in depth, or else had a wide geographical distribution, frequently both. The same holds true of the testacea in the tertiary strata of Great Britain. The cause is obvious: such species as had the widest horizontal and vertical ranges in space, are exactly such as would live longest in time, since they would be much more likely to be independent of catastrophes and destroying influences, than such as had a more limited distribution. In the cretaceous system, also, we find that such species as lived through several epochs of that era, are the few which are common to the cretaceous rocks of Europe, Asia, and America. Count D'Archiac and M. De Verneuil, in their excellent remarks on the fauna of the Palæozoic rocks, appended to Mr Murchison and Professor Sedgwick's valuable memoir on the Rhenish Provinces, have come to the conclusion that the fossils common to the most distant localities, are such as have the greatest vertical range. My observations on the existing testacea and their fossil analogues, lead to the same inference. It is very interesting thus to find a general truth coming out, as it were, in the same shape, from independent inquiries at the two ends of time.

VII. *Mollusca migrate in their larva state, but cease to exist at a certain period of their metamorphosis, if they do not meet with favourable conditions for their development;* i. e. *if they do not reach the particular zone of depth in which they are adapted to live as perfect animals.*

This proposition, which, as far as I am aware, is now put forward for the first time, includes two or three assertions which require explanation and proof, before I can expect the whole to be received. First, *that mollusca migrate.* In the fourth volume of the Annals of Natural History (1840), I gave a zoo-geological account of a shell-bank in the Irish Sea, being a brief summary of the results of seven years' observations at a particular season of the year. In that paper, I made known the appearance, after a time, of certain mollusca on the coasts of the Isle of Man, which had not previously inhabited those shores. They were species of limpet, about which there could be no mistake, and one was a littoral species. At that time, I could not account for their appearance. Many similar facts have since come to my knowledge, and fishermen are familiar with what they call "shifting" of shell-beds, which they erroneously attribute to the moving away and swimming off of a whole body of shell-fish, such as mussels and oysters. Even the *Pectens*, much less the testacea just named, have very little power of progressing to any distance, when fully developed. The "shifting" or migration is accomplished by the young animals when in a larva state. This brings me to a second point, which needs explanation. *All mollusca undergo a metamorphosis* either in the egg, or out of the egg, but, for the most part, among the marine species out of the egg. The relations of the metamorphoses of the several tribes are not yet fully made out; but sufficient is now known to warrant the generalization. In one great class of mollusca, the *Gasteropoda*, all appear to commence life under the same form, both of shell and animal, viz. a very simple, spiral, helicoid shell, and an animal furnished with two ciliated wings or lobes, by which it can swim freely through the fluid in which it is contained. *At this stage of the animal's existence, it is in a state corresponding to the permanent state of a Pteropod,** and the form is alike whether it be afterwards a shelled or shell-less species. (This the observations of Dalyell, Sars, Alder and Hancock, Allman, and others prove, and I have seen it myself.) It is in this form that most species migrate, swimming with ease through the sea. Part of the journey may be performed sometimes by the strings of eggs which fill the sea at certain seasons, and are wafted by currents. My friend, Lieut. Spratt, R.N., has lately forwarded me a drawing of a chain of eggs of mollusca, taken eighty miles from shore, and which, on being hatched, produced shelled larvæ of the forms which I have described. If they reach the region and ground, of which the perfect animal is a member, then they develope and flourish; but if the period of their development arrives before they have reached their destination, they perish, and their fragile shells sink into the depths of the sea. Millions and millions must thus perish, and every handful of the fine mud brought up from the eighth zone of depth in the Mediterranean, is literally filled with hundreds of these curious exuviæ of the larvæ of mollusca.†

---

* It is not improbable that the form of the larva of the Pteropod, when it shall be known, will be found to be that of an Ascidian polype, even as the larva of the Tunicata presents us with the representation of a hydroid polype.

† The nucleus of the shells of the Cephalopoda is a spiral-univalve

Were it not for the law which permits of the development of these larvæ only in the region of which the adult is a true native, the zones of depth would long ago have been confounded with each other, and the very existence of the zones of depth is the strongest proof of the existence of the law. Our confidence in their fixity, which the knowledge of the fact *that mollusca migrate* might at first shake, is thus restored, and with it our confidence in the inferences applicable to geology which we draw from submarine researches.

Some of the facts advanced in this communication are new, some of them have been stated before: but all, for which no authority is given, whether new or old, are put forth as the results of personal observation.

GENERAL ECOLOGY AND DISTRIBUTION OF RADIOLARIA

Ernst Haeckel

PHYSIOLOGICAL SECTION.

Chapter VII.—VEGETATIVE FUNCTIONS.

(§§ 201-217.)

201. *Mechanism of the Functions.*—The vital phenomena of the Radiolaria are dependent upon the mechanical functions of their unicellular body, and like those of all other organisms, are to be referred to physical and chemical natural laws. All processes which appear in the life of the Radiolaria are, therefore, ultimately to be explained by the attraction and repulsion of the smallest particles, which compose the different portions of their unicellular body; and the sensation of pleasure or the opposite is in its turn the exciting cause of these elementary movements. Many adaptive arrangements in the Radiolarian organism may produce the appearance of being the premeditated result of causes working towards an end ("zweckthätig," *causæ finales*), but as opposed to this deceptive appearance it must here be expressly stated that these may be recognised in accordance with the developmental theory as the necessary consequence of mechanical causes (*causæ efficientes*).

Our *physiological* acquaintance with the Radiolaria has by no means progressed so far as our *morphological*, so that the incomplete communications which are placed here for the sake of completeness must be regarded merely as preliminary fragments, not as fully elaborated results. Since my recent investigations have been mainly in the direction of morphology, I can add but little to the physiological conclusions, which I stated at length in my monograph twenty-four years ago (L. N. 16, pp. 127–165). Recently the vegetative physiology of the Radiolaria has been much advanced by the recognition of the symbiosis with the Xanthellæ (§ 205, L. N. 22, 39, 42). In addition Karl Brandt has recently (1885) published several important contributions to the physiology of the Polycyttaria or Sphaerozoea (L. N. 52).

202. *Distribution of Functions.*—The distribution of the functions among the various parts of the unicellular organism of the Radiolaria corresponds directly to their anatomical composition, so that physiologically as well as morphologically the central capsule and the extracapsulum appear as the two coordinated main components. On the one hand the *central capsule* with its endoplasm and enclosed nucleus is the central organ of the "cell-soul" (Zellseele), the unit regulating its animal and vegetative functions, and the special organ of reproduction and inheritance. The *extracapsulum* forms, on the other hand, by its calymma the protective envelope of the central

capsule, the support of the soft pseudopodia and the substratum of the skeleton; the calymma acts also as a hydrostatic apparatus, whilst the radiating pseudopodia are of the greatest importance both as organs of nutrition and adaptation, as well as of motion and sensation (§ 15). If, however, the vital functions as a whole be divided in accordance with the usual convention into the two great groups of *vegetative* (nutrition and reproduction) and *animal* (motion and sensation), then the central capsule would be mainly the organ of reproduction and sensation, and the extracapsulum the organ of nutrition and motion.

The numerous separate vital phenomena, which by accurate physiological investigation may be distinguished in the unicellular Radiolarian organism, may be distributed in the above indicated conventional fashion into a few larger and several smaller groups; it must always be borne in mind, however, that these overlap in many respects, and that the division of labour among the different organs in these Protista is somewhat complicated, notwithstanding the apparent simplicity of their unicellular organisation. A general classification of the groups of functions is difficult, because each individual organ discharges several different functions. Thus the central capsule is pre-eminently the organ of reproduction and inheritance, but not less (though less conspicuous) is its importance as the psychical central organ, the unit regulating the processes of sensation, motion, and also nutrition. In this last respect it is comparable to the nerve-centres of the Metazoa, whilst the peripheral nervous system of the latter (including the organs of sense and the muscles) are in the present instance represented by the pseudopodia, which are at the same time the most important organs of nutrition and adaptation. In the calymma also in similar fashion several different physiological functions are united.

203. *Metastasis.*—The functions of metastasis and nutrition have in all Radiolaria a purely animal character, so that these Rhizopoda from the physiological standpoint are to be regarded as truly *unicellular animals*, or Protozoa ("Urthiere"). Since they do not possess, like plants, the power of forming synthetically the compounds (protoplasm, carbohydrates, &c.) necessary for their sustenance, they are compelled to obtain them ready-formed from other organisms. Like other true animals they evolve carbon dioxide by the partial oxidation of those products, and hence they successively take up the oxygen necessary to their existence from their environment.

The question whether the Radiolaria are to be regarded as true animals I discussed fully from various points of view in 1862, and finally answered in the affirmative (L. N. 16, pp. 159–165). Afterwards, when in my Generelle Morphologie (1866) I sought to establish the kingdom Protista, I removed the Radiolaria along with the other Rhizopoda from the animal kingdom proper and placed them in the kingdom Protista (Bd. i. pp. 215–220; Bd. ii. p. xxix). Compare also my Protistenreich (L. N. 32) and my Natürliche Schöpfungsgeschichte (vii. Aufl., 1879, p. 364). Both these steps appear fully justified when considered in the light of our present increased knowledge. From the *physiological* standpoint the Radiolaria appear as unicellular *animals*, for in this respect the animal character of their metastasis (that proper to an oxidising organism) furnishes the sole criterion. On the other hand, from the morphological standpoint, they are to be classed as neutral Protista, for in this respect their unicellular character is the prominent feature, and distinguishes them from all true multicellular animals (Metazoa). Compare my Gastræa Theorie (1873, Jena. Zeitschr. für Naturwiss., Bd. viii. pp. 29, 53).

204. *Nutrition.*—The nutritive materials which the Radiolaria require for their sustenance, especially albuminates (plasma) and carbohydrates (starch, &c.), they obtain partly from foreign organisms which they capture and digest, and partly directly from the Xanthellæ or Philozoa, the unicellular Algæ, with which they live in symbiosis (§ 205). *Zooxanthella intracapsularis,* found in the ACANTHARIA (§ 76), is probably of the same significance in this respect as *Zooxanthella extracapsularis* of the SPUMELLARIA and NASSELLARIA (§ 90); and perhaps the same is true also of *Phæodella extracapsularis* (or *Zoochlorella phæodaris ?*) of the PHÆODARIA (§ 89). The considerable quantity of starch or amyloid bodies, elaborated by these inquiline symbiontes, as well as their protoplasm and nucleus, are available, on their death, for the nutrition of the Radiolaria which harbour them. Nutrition by means of other particles obtained by the pseudopodia from the surrounding medium is by no means excluded; indeed it may be regarded as certain that numerous Radiolaria (especially such as contain no symbiotic Algoid cells) are nourished for the most part or exclusively by this means. Diatoms, Infusoria, Thalamophora (Foraminifera) as well as decaying particles of animal and vegetable tissues can be seized directly by the pseudopodia and conveyed either to the sarcodictyum (on the surface of the calymma) or to the sarcomatrix (on the surface of the central capsule) in order to undergo digestion there. The indigestible constituents (siliceous shells of Diatoms and Tintinnoidea, calcareous shells of small Monothalamia and Polythalamia, &c.) are here collected often in large numbers and removed by the streaming of the protoplasm.

The inception and digestion of nutriment, as it usually appears to take place by the pseudopodia, has already been so fully treated in my Monograph (L. N. 16, pp. 135–140), and since then in my paper on the sarcode body of the Rhizopoda (L. N. 19, p. 342), that I have nothing of importance to add. Quite recently Karl Brandt has expressed a doubt as to whether the taking up of formed particles by the pseudopodia and their aggregation in the calymma be really connected with the process of nutrition. He is disposed rather to believe that these foreign bodies are usually only accidentally and mechanically brought into the calymma, and that the nourishment of the Radiolaria is derived exclusively or pre-eminently from the symbiotic Xanthellæ (L. N. 52, pp. 88–93). I must, however, maintain my former opinion, which I have only modified insomuch that I now regard the sarcodictyum (on the outer surface of the calymma, § 94) rather than the sarcomatrix (on the outer surface of the central capsule, § 92) as the principal seat of true digestion and assimilation. From the sarcodictyum the dissolved and assimilated nutritive matters may pass by the intracalymmar pseudopodia (or sarcoplegma, § 93) into the sarcomatrix, and hence may reach the endoplasm through the openings in the central capsule. To what an extent the Radiolaria are capable of taking up even large formed bodies into the calymma, is shown by the striking instance of *Thalassicolla sanguinolenta*, which becomes so deformed by the inception of numerous coccospheres and coccoliths, that I described it as a special genus under the name *Myxobrachia* (compare pp. 23, 30; also L. N. 21, p. 519, Taf. xviii., and L. N. 33, p. 37).

205. *Symbiosis*.—Very many Radiolaria, but by no means all members of this class, live in a definite commensal relation with yellow unicellular Algæ of the group Xanthellæ. In the ACANTHARIA they live within the central capsule (*Zooxanthella intracapsularis*, § 76), in the SPUMELLARIA and NASSELLARIA, on the other hand, within the calymma but outside the central capsule (*Zooxanthella extracapsularis*, § 90); in the PHÆODARIA a special form of these symbiotic unicellular Algæ appears to inhabit the phæodium in the extracapsulum, and to compose a considerable portion of the phæodellæ (*Zooxanthella phæodaris*, § 90, or better perhaps *Zoochlorella phæodaris*, § 89). Undoubtedly this commensal life is in very many cases of the greatest physiological significance for both the symbiontes, for the animal Radiolarian cells furnish the inquiline Xanthellæ not only with shelter and protection, but also with carbon dioxide and other products of decomposition for their nutriment; whilst on the other hand the vegetable cells of the Xanthellæ yield the Radiolarian host its most important supply of nutriment, protoplasm and starch, as well as oxygen for respiration. Hence it is not only theoretically possible, but has been experimentally proved, that Radiolaria which contain numerous Xanthellæ can exist without extraneous nutriment for a long period in closed vessels of filtered sea-water, kept exposed to the sunlight; the two symbiontes furnish each other mutually with nourishment, and are physiologically supplementary to each other by reason of the opposite nature of their metastasis. This symbiosis is not necessary, however, for the existence of the Radiolaria; for in many species the number of Xanthellæ is very variable and in many others they are entirely wanting.

The symbiosis of the Radiolaria and Xanthellæ, or "yellow cells" (§§ 76, 90) was first discovered by Cienkowski in 1871 (L. N. 22). Ten years later this important and often doubted fact was established by extended observations and experiments almost simultaneously by Karl Brandt (L. N. 38, 39) and Patrick Geddes (L. N. 42, 43). This commensal life may be compared with that of the lichens, in which an organism with vegetable metastasis (the Algoid gonidia) and an organism with animal metastasis (the Fungoid hyphæ) are intimately united for mutual benefit. But the symbiosis of the Xanthellæ and Radiolaria is not as in the lichens a phenomenon essential for their development, but has more or less the character of an accidental association. The number of the inquiline Xanthellæ is so variable even in one and the same species of Radiolaria, that they do not appear to be exactly essential to its welfare; and in many species they are entirely wanting. Their significance is questionable in the case of those numerous deep-sea Radiolaria which live in complete darkness, and in which, therefore, the Xanthellæ, even if present, could excrete no oxygen on account of the want of light. Nevertheless it is possible that the phæodellæ of the PHÆODARIA (usually green, olive, or brown in colour), which are true cells, represent vegetable symbiontes, which in the absence of sunlight are able to evolve oxygen by the aid of the phosphoresence of other abyssal animals. Since the PHÆODARIA are, for the most part, dwellers in the deep-sea, and since the voluminous phæodium must be of great physiological importance, a positive solution of this hypothetical question would be of no small interest (compare § 89).

206. *Respiration.*—The respiration of the Radiolaria is animal in nature, since all Protista of this class, like all other true Rhizopoda, take in oxygen and give off carbon dioxide. Probably this process goes on continuously and is tolerably active, as may be inferred from the fact that Radiolaria cannot be kept for long in small vessels of sea-water unless either they contain numerous Xanthellæ or the water is well aërated. The oxygen is obtained from two sources, either from the surrounding water or from the enclosed Xanthellæ, which in sunlight evolve considerable quantities of this gas. Correspondingly, the carbon dioxide which is formed during the process of oxidation of the Radiolaria is either given up to the surrounding water or to the inquiline Xanthellæ, which utilise it for their own sustenance (§§ 204, 205).

The significance of the symbiotic Xanthellæ for the respiration of the enclosing Radiolaria may be shown experimentally in the following way. If two Polycyttarian colonies of equal size, both of which contain numerous Xanthellæ, be placed in equal quantities of filtered sea-water in sealed glass tubes, and if one tube be placed in the dark the other in the light, the colony in the former rapidly perishes, but not that in the latter; the Xanthellæ excrete only under the influence of sunlight the oxygen necessary for the life of the Radiolarian (compare Patrick Geddes, L. N. 42, p. 304).

207. *Circulation.*—In the protoplasm of all Radiolaria, both inside and outside the central capsule, slow currents may be recognised which fall under the general term circulation, and have already been compared to the cyclosis in the interior of animal and vegetable cells, as well as to the sarcode streams in the body of other Rhizopoda. These plasmatic currents or "plasmorrheumata" probably continue throughout the whole life of the Radiolaria, and are of fundamental importance for the performance of their vital functions. They depend upon slow displacements of the molecules of the plasma (plastidules or micellæ) and cause a uniform distribution of the absorbed nutriment and a certain equalisation of the metastasis. Furthermore they are of great importance also in the inception of nutriment, the formation of the skeleton, locomotion, &c. Sometimes the circulation is directly perceptible in the plasma itself; but usually it is only visible owing to the presence of granules (sarcogranula), which are suspended in the plasma in larger or smaller numbers. The movements of these granules are usually regarded as passive, due to the active displacement of the molecules of the plasma. Although the intracapsular protoplasm is in communication with the extracapsular through the openings in the capsule membrane, nevertheless the currents exhibit certain differences

in the two portions of the malacoma. It is sometimes possible, however, to recognise the direct connection between them and to observe how the granules pass through the openings in the capsule-membrane.

208. *Currents in the Endoplasm.*—Intracapsular circulation or a certain slow flowing of the plasma within the central capsule is probably just as common in the Radiolaria as without it, but it is not so easy to observe in the former case as in the latter. A more satisfactory proof of these endoplasmatic currents is furnished by the arrangement of the protoplasm within the central capsule, since this is (at all events in part) an effect produced by them. In this respect the two main divisions of the class show characteristic differences. In the Porulosa (the SPUMELLARIA, § 77, and the ACANTHARIA, § 78) the endoplasm is in general distinguished by a more or less distinct radial structure, which is to be regarded as the effect of alternating centripetal and centrifugal radial streams. In the Osculosa, on the other hand, this radial structure is absent and the intracapsular plasmatic streams converge or diverge towards the osculum or main-opening in the central capsule which lies at the basal pole of its main axis, and through which the mass of the endoplasm issues into the calymma. The two legions of the Osculosa, however, present differences in this respect. In the NASSELLARIA (§ 79) the endoplasmatic currents appear to unite in an axial main stream at the apex of the monaxon central capsule, and this apical stream seems to split into a conical bundle, the individual threads of which pass diverging between the myophane fibrillæ of the podoconus towards the basis of the central capsule, and issue through the pores of the porochora. In the PHÆODARIA (§ 80), on the other hand, meridional currents of endoplasm are probably present on the inner surface of the capsule, which flow from the aboral pole of the vertical main axis to its basal pole, and return in the reverse direction.

209. *Currents in the Exoplasm.*—Extracapsular circulation, or a distinct flowing of the plasma outside the central capsule, may be readily observed in all Radiolaria which are examined alive; this is most readily seen in the astropodia, or those free pseudopodia which radiate from the sarcodictyum on the surface of the calymma into the surrounding water. The granular movement is often quite as clear in the sarcodictyum itself, and may be recognised in the collopodia, which compose the irregular plasmatic network within the calymma. More rarely it is possible to follow the granular stream thence through the sarcomatrix, and further into the interior of the central capsule. In general the direction of the extracapsular protoplasmic streams is radial, and it is frequently possible, even in a single free astropodium, to observe two streams opposite in direction, the granules on one side of the radial sarcode thread moving centripetally, those on the other side centrifugally. If the threads branch, and neighbouring ones

become united by connecting threads, the circulation of the granules may proceed quite irregularly in the network thus formed. The rapidity and character of the extracapsular currents are subject to great variations.

The different forms of extracapsular sarcode currents have been already very fully described in my Monograph (L. N. 16, pp. 89–126), and in my critical essay on the sarcode body of the Rhizopoda (L. N. 19, p. 357, Taf. XXVI.).

210. *Secretion.*—Under the name *secretions*, in the strict sense, all the skeletal formations of the Radiolaria may be included. They may be divided according to their chemical composition into three different groups : pure silica in the SPUMELLARIA and NASSELLARIA, a silicate of carbon in the PHÆODARIA, and acanthin in the ACANTHARIA (compare § 102). It may indeed be assumed that these skeletons arise directly by a chemical metamorphosis (silicification, acanthinosis, &c.) of the pseudopodia and protoplasmic network; and this view seems especially justified in the case of the Astroid skeleton of the ACANTHARIA (§ 114), the Spongoid skeleton of the SPUMELLARIA (§ 126), the Plectoid skeleton of the NASSELLARIA (§ 125), the Cannoid skeleton of the PHÆODARIA (§ 127), and several other types. On closer investigation, however, it appears yet more probable that the skeleton does not arise by direct chemical metamorphosis of the protoplasm, but by secretion from it; for when the dissolved skeletal material (silica, acanthin) passes from the fluid into the solid state, it does not appear as imbedded in the plasma, but as deposited from it. However, it must be borne in mind that a hard line of demarcation can scarcely, if at all, be drawn between these two processes. In the ACANTHARIA the intracapsular sarcode is the original organ of secretion of the skeleton; in the other three legions, on the other hand, the extracapsulum performs this function (§§ 106, 107). In addition to the skeleton, we may regard as secretions (or excretions) the intracapsular crystals (§ 75) and concretions (§ 75A), and perhaps certain pigment-bodies (§§ 74, 88); and further, the calymma (§ 82) may be considered to be a gelatinous secretion of the central capsule, and perhaps also the capsule-membrane, in so far as it represents only a secondary excretory product of the unicellular organism.

211. *Adaptation.*—The innumerable and very various adaptive phenomena which we meet with in the morphology of the Radiolaria, and especially in that of their skeleton, are like other phenomena of the same kind, to be ultimately referred to altered nutritional relations. These may be caused directly either by the influence of external conditions of existence (nutrition, light, temperature, &c.), or by the proper activity of the unicellular organism (use or disuse of its organs, &c.), or, finally, by the combined action of both causes in the struggle for existence. In very many cases the cause to which the origin of a particular form of Radiolaria is due may be directly perceived or at least guessed at with considerable probability; thus, for example, the lattice-shells

may be explained as protective coverings, the radial spines as defensive weapons, and the anchor-hooks and spathillæ as organs of prehension, which are of advantage to their possessors in the struggle for existence; the regular arrangement of the radial spines in the Radiolaria may also be explained on hydrostatic grounds, it being advantageous that the body should be maintained in a definite position of equilibrium, &c. The well-known laws of *direct* or *actual adaptation*, which we designate cumulative, correlative, divergent adaptation, &c., here explain a multitude of morphological phenomena. The connection is less distinct in the case of the laws of *indirect* or *potential adaptation*, although this must play as important a part in the formation of the Radiolaria as in that of other organisms (compare on this head my Generelle Morphologie, Bd. ii. pp. 202-222).

212. *Reproduction.*—The most common form of reproduction in the Radiolaria is the formation of spores in the central capsule, which in this respect is to be regarded as a sporangium (§ 215). In many Radiolaria (Polycyttaria and PHÆODARIA), however, there occurs in addition an increase of the unicellular organism by simple division (§ 213); upon this the formation of colonies in the social Radiolaria is dependent (§ 14). Reproduction by gemmation is much less common, and has hitherto been observed only in the Polycyttaria (§ 214). In this group alone there also occur at certain times two different forms of swarm-spores which copulate, and thus indicate the commencement of sexual reproduction (Alternation of Generations, § 216). The general organ of reproduction is in all cases the central capsule, whilst the extracapsulum never takes an active part in the process.

213. *Cell-Division.*—Increase by cell-division among the Radiolaria in the early stage, before the formation of the skeleton, is widely distributed (perhaps even general?); in the adults of this class it is rather rare and limited to certain groups. It is most readily observed in the Polycyttaria; the growth of the colonies in this social group depends mainly (and in many species exclusively) upon repeated spontaneous division of the central capsule; all the individuals of each colony (in so far as this has not arisen by the accidental fusion of two or more colonies) are descendants of a single central capsule, which has arisen from an asexual swarm-spore (§ 215) or from the copulation of two sexual swarm-spores (§ 216). Whilst the central capsules of the colonies continually increase by division, their calymma remains a common gelatinous sheath. Among the SPUMELLARIA reproduction by simple cell-division probably occurs also in many monozootic Collodaria. Among the ACANTHARIA the peculiar group Litholophida has perhaps arisen by the spontaneous division of Acanthonida (see p. 734). Among the PHÆODARIA increase by cell-division seems to occur commonly in many groups, as in the Phæocystina, which have no skeleton (Phæodinida, Pl. **101**,

fig. 2), or only an incomplete Beloid skeleton (Cannorrhaphida, Pl. 101, figs. 3, 6, and Aulacanthida, Pl. 104, figs. 1–3). The P h æ o s p h æ r i a also (Aulosphærida, Cœlacanthida) and the P h æ o g r o m i a (Tuscarorida, Challengerida) appear sometimes to divide; at all events, their central capsule often contains two nuclei. Of special interest is the spontaneous division of the P h æ o c o n c h i a, especially the Concharida (Pl. 124, fig. 6). In all monozootic Radiolaria, the nucleus first divides by a median constriction into two equal halves (usually by the mode of direct division); then the central capsule becomes constricted in the middle (in the PHÆODARIA in the vertical main axis), and each portion of the capsule retains its own nucleus. In the P h æ o-c o n c h i a each half or daughter-cell corresponds to one valve of the shell, dorsal or ventral, so that probably on subsequent separation each daughter-cell retains one valve of the mother-cell, and forms a new one for itself by regeneration (as in the Diatoms). In the polyzootic Radiolaria, which already contain many small nuclei, but usually only a single central oil-globule in each central capsule, the division of the latter is preceded by that of the oil-globule. In many Polycyttaria the colony as a whole multiplies by division.

The increase of the central capsule by division was first described in 1862 in my Monograph (L. N. 16, p. 146); since then R. Hertwig (L. N. 26, p. 24) and K. Brandt (L. N. 52, p. 144) have confirmed my statement. In the PHÆODARIA the division of the central capsule appears always to take place in the main axis; in the bilateral sometimes in the sagittal, sometimes in the frontal plane. In the Tripylea each daughter-cell seems to retain one parapyle and half the astropyle (compare the general description of the PHÆODARIA, Pl. 101, figs. 1–6, Pl. 104, figs. 1–3, and also Hertwig, L. N. 33, p. 100, Taf. x. figs. 2, 11). Regarding the spontaneous division of colonies of the Polycyttaria, see K. Brandt, L. N. 52, p. 142.

214. *Cell-Gemmation.*—Reproduction by gemmation has hitherto been observed only in the social Radiolaria, but in them it appears to be widely distributed, and in very young colonies is perhaps almost universally present. The gemmules or capsular buds (hitherto described as "extracapsular bodies") are developed on the surface of young central capsules before these had secreted a membrane. They grow usually in considerable numbers, from the surface of the central capsule, which is sometimes quite covered with them. Each bud usually contains a raspberry-like bunch of shining fatty globules, and by means of reagents a few larger or a considerable number of smaller nuclei may be recognised in them; the naked protoplasmic body of the bud is not enclosed by any membrane. As soon as the buds have reached a certain size they are constricted off from the central capsule and separated from it, being distributed in the meshes of the sarcoplegma by the currents in the exoplasm. Afterwards each bud becomes developed into a complete central capsule by surrounding itself with a membrane when it has attained a definite size. From the special relations of the processes of nuclear formation, which take place in the multiplication of the

social central capsules by gemmation and by cell-division, it would appear that the capsules produced by the former method afterwards produce anisospores, whilst those in the latter way yield isospores (§ 216).

The gemmules or capsular buds of the Polycyttaria were first accurately described by Richard Hertwig (L. N. 26, pp. 37–39), under the name "extracapsular bodies," and their significance rightly indicated; earlier observers had incidentally mentioned and figured them, but had not seen their origin from the central capsule. Quite recently Karl Brandt has given a very painstaking account of them in the different Polycyttarian genera (L. N. 52, pp. 179–198). In the Monocyttaria such a formation of buds has not yet been observed. The basal lobes of the central capsule, which occur in many NASSELLARIA, are not buds, but simple processes of the capsule, due to its protrusion through the collar pores of the cortinar septum (§ 55).

215. *Sporification.*—Asexual reproduction by the formation of movable flagellate spores has been hitherto observed only in a very small number of genera; but since these belong to very different groups, and since the comparative morphology of the capsule appears to be similar throughout as regards the structure and development of its contents, it may be safely assumed that this kind of reproduction occurs quite generally in the Radiolaria. In all cases it is the contents of the central capsule, from which the swarm-spores are formed, both nucleus and endoplasm taking an equal share in the process; in all cases the spores produced are very numerous, small, ovoid or reniform, and have one or two very long slender flagella at one extremity (see §§ 141, 142). Since the whole contents of the mature central capsule are used up in the formation of these flagellate zoospores, it discharges the function of a sporangium. The division of the simple primary nucleus into numerous small nuclei, which usually (serotinous Radiolaria) takes place only shortly before sporification, but sometimes (precocious Radiolaria, § 63) happens very early, is the commencement of the often repeated process of nuclear division, which terminates with the production of a very large number of small spore-nuclei. The nucleolus often divides very peculiarly (§ 69, C). Each spore nucleus becomes surrounded by a portion of endoplasm and usually receives in addition one or more fatty granules, and sometimes also a small crystal (hence the "crystal-spores"). The size of the flagellate zoospores which emerge from the ruptured central capsule and swim freely in the water by means of their flagellum, varies generally between 0·004 and 0·008 mm. The extracapsulum is not directly concerned in the sporification, but undergoes degeneration during the process and perishes at its conclusion.

The first complete and detailed observations on the formation of spores in the Radiolaria were published by Cienkowski in 1871 and related to two genera of Polycyttaria, the skeletonless *Collozoum* and the spherical-shelled *Collosphæra* (L. N. 22, p. 372, Taf. xxix.). These were subsequently continued and supplemented by R. Hertwig (1876, L. N. 26, pp. 26–42, and L. N. 33 p. 129), and a general summary of these results has been given by Bütschli (L. N. 41, pp. 449–455).

Recently Karl Brandt has given a very detailed and fully illustrated account of the sporification of the Polycyttaria (L. N. 52, pp. 145–178). I have also had the opportunity during my sojourn in the Canary Islands (1866), in the Mediterranean at Corfu (1877), and Portofino (1880), as well as in Ceylon (1881), of observing the development of flagellate zoospores from the central capsule of individuals of all four legions: among the SPUMELLARIA in certain Colloidea, Beloidea, Sphæroidea, and Discoidea, among the ACANTHARIA in several Acanthometra and Acanthophracta, among the NASSELLARIA in individuals belonging to the Stephoidea, Plectoidea, and Cyrtoidea, and among the PHÆODARIA in one Castanellid. In most zoospores I could distinctly observe only a single long flagellum; sometimes, however, two or even three appeared to be present, but the determination of their number is very difficult.

216. *Alternation of Generations.*—A peculiar form of reproduction, which may be designated "alternation of generations," appears to occur generally in the Polycyttaria, but has not yet been observed in the Monocyttaria. All Collozoida, Sphærozoida, and Collosphærida which have hitherto been carefully and completely examined with respect to their development, are distinguished by the production of two different kinds of swarm-spores, isospores and anisospores. The *Isospores* (or monogonous spores) correspond to the ordinary asexual zoospores of the Monocyttaria (§ 215); they possess a homogeneous, doubly refracting nucleus of uniform constitution and develop asexually, without copulation. The *Anisospores* (or amphigonous spores), on the other hand, are sexually differentiated and possess a heterogeneous, singly refracting nucleus of twofold constitution; they may therefore be distinguished as female macrospores and male microspores. The *Macrospores* (or gynospores, comparable with the female macrogonidia of many Cryptogams) are larger, less numerous, and possess larger nuclei, which are less easily stained, and have a fine filiform trabecular network. On the other hand the *Microspores* (or *androspores*, comparable with the male microgonidia) are much smaller and more numerous, and are distinguished by their smaller nuclei, which have thicker tuberculæ and become stained more intensely. The gynospores and androspores are developed in the Collozoida and Sphærozoida in the same individual, but not in the Collosphærida. It is very probable that these two forms of anisospores copulate with each other after their exit from the central capsule and thus produce a new cell by the simplest mode of sexual reproduction. But, since the same Polycyttaria, which produce these anisospores, at other times give rise to ordinary or asexual isospores, it is further possible that these two forms of reproduction alternate with each other, and that the Polycyttaria thus pass through a true alternation of generations. This has not yet been observed in the Monocyttaria, and hence these latter seem to bear to the Polycyttaria a relation similar to that in which the sexless solitary Flagellata (Astasiea) stand to the sexual social Flagellata (Volvocinea). In the two analogous cases the sexual differentiation may be regarded as a consequence of the social life in the gelatinous colonies.

The *sexual differentiation of the Polycyttaria* was first discovered in 1875 by R. Hertwig, and accurately described in the case of *Collozoum inerme* as occurring in addition to the formation of the ordinary crystal-spores (L. N. 26, p. 36); compare also the general discussion of Bütschli (L. N. 41, p. 52). Recently Karl Brandt has demonstrated the formation of both homogeneous isospores (crystal-spores) and heterogeneous anisospores (macro- and microspores) in seven different species of Polycyttaria, and has come to the conclusion that in all social Radiolaria there is a regular alternation between the former and latter generations. Compare his elaborate account of the colonial Radiolaria of the Gulf of Naples (L. N. 52, pp. 145–178).

217. *Inheritance.*—Inheritance is to be regarded as the most important accompaniment to the function of reproduction, and especially in the present case, because the comparative morphology of the Radiolaria furnishes abundant instances of the action of its different laws. The laws of *conservative inheritance* are illustrated by the comparative anatomy of the larger groups; thus, in the four legions the characteristic peculiarities of the central capsule are maintained unaltered in consequence of continuous inheritance, although great varieties appear in the skeleton in each legion. The individual parts of the skeleton furnish by their development on the one hand and their degeneration on the other, especially in the smaller groups, examples of *progressive inheritance*. Thus in the Spumellaria the constant formation of the primary lattice-shell (a central medullary shell) and its ontogenetic relation to the secondary one, which is developed concentrically round it, can only be explained phylogenetically by conservative inheritance, whilst on the other hand the characteristic differentiation of the axes in the various families of the Spumellaria is to be explained by progressive inheritance. In the Acantharia the arrangement of the twenty radial spines (in accordance with Müller's law, §§ 110, 172) was first acquired by a group of the most archaic Actinelida (Adelacantha) through hydrostatic adaptation, and has since been transmitted by inheritance to all the other families of the legion (Icosacantha). The morphology of the Nassellaria is not less interesting, because here several different heritable elements (the primary sagittal ring and the basal tripod) combine in the most manifold ways in the formation of the skeleton (compare §§ 123, 124, 182). The affinities of the genera in the different families yield an astonishing variety of interesting morphological phenomena, which can only be explained by progressive inheritance. The same is true also of the Phæodaria. In this legion the primary inheritance is especially manifested in the constant and firm structure of the central capsule with its characteristic double wall and astropyle, whilst the formation of the skeleton in this legion proceeds in different directions by means of divergent adaptation. The morphology of the Radiolaria thus proves itself a rich source of materials for the physiological study of adaptation and inheritance.

Chapter VIII.—ANIMAL FUNCTIONS.

(§§ 218–225.)

218. *Motion.*—In addition to the internal movements which appear generally in the unicellular Radiolaria and have already been mentioned as plasmatic currents in treating of the circulation (§§ 207–209), two different groups of external motor phenomena are to be observed in this class: first, the *contraction* of individual parts, which brings about modifications of form (§ 220), and secondly, voluntary or reflex *locomotion* of the whole body (§ 220). These movements are partly due to changes in form of undifferentiated plasmatic threads or sarcode filaments, partly to the actual contraction of differentiated filaments which are comparable to muscle fibrillæ, and must therefore be distinguished as myophanes. In addition to this, endosmose and exosmose probably play an important part in some of the locomotive phenomena, but nothing is yet certainly known regarding these osmotic processes. We are at present equally ignorant whether all the movements of the Radiolaria are simply reflex (direct consequences of irritation) or whether they are in part truly spontaneous.

219. *Suspension.*—From direct observation of living Radiolaria, as well as from deductive reasoning, based upon their morphology (and especially their promorphology, §§ 17–50), the conclusion appears justified that all Protista of this class in their normal condition float suspended in the sea-water, either at the surface or at a definite depth. A necessary condition of this hydrostatic suspension is that the specific gravity of the Radiolarian organism must be equal to, or but slightly greater than that of sea-water. The increase in specific gravity brought about by the production of the siliceous skeleton, is compensated by the lighter fatty globules, and partly perhaps by the calymma, especially when the latter contains vacuoles or alveoles. The fluid or jelly contained in the latter appears to be for the most part lighter than sea-water (containing no salt, or only a very small quantity?). But if the specific gravity of the whole body should be generally (or perhaps always) slightly greater than that of sea-water, then the organism would be prevented from sinking, partly by the increased friction, due to the radiating pseudopodia and the radial spines usually present, and partly perhaps by active (if only feeble) movements of the pseudopodia.

220. *Locomotion.*—Active locomotion of the whole body, which is very probably to be regarded as voluntary, occurs in the Radiolaria in three different modes; (1) the vibratile movement of the flagellate swarm-spores; (2) the swimming of the floating organisms; (3) the slow creeping of those which rest accidentally upon the bottom.

The *vibratile* movement of the swarm-spores is the result of active sinuous oscillation of the single or multiple flagellum, and is not essentially different from that of ordinary flagellate Infusoria (see note A). Of the active swimming of mature Radiolaria, only that form is known which is vertical in direction and causes the sinking and rising in the sea-water. This is probably, for the most part (perhaps exclusively), due to increase or diminution in the specific gravity, which is perhaps brought about by the retraction or protrusion of the pseudopodia; slow, oscillating, sinuous motions of these organs have been directly observed to take place (though very slowly) in suspended living Radiolaria. The most important hydrostatic organ is probably the calymma, by the contraction of which the specific gravity is increased, while it is diminished by its expansion; the contraction is probably brought about by active contraction of the sarcodictyum, and is connected with exosmosis, while the expansion is probably due to the elasticity of the calymma and the inception of water by endosmosis. In the Acanthometra (§ 96) the peculiar myophrises appear to be charged with the duty of distending the gelatinous envelope, and thus diminishing the specific gravity; the latter increases again when the myophrises are relaxed, and the calymma contracts by virtue of its own elasticity (see note B). The slow *creeping locomotion* exhibited by Radiolaria on a glass slide under the microscope, does not differ from that of the Thalamophora (Monothalamia and Polythalamia), but can only occur normally when the animal accidentally comes into contact with a solid surface or sinks to the bottom of the sea. Whether this actually occurs periodically is not known (see note C). The slow or gliding locomotion exhibited by creeping Monozoa on a glass slide is due to muscle-like contractions of bundles of pseudopodia, just as in the case of the social central capsules of Polyzoa, which live together in the same cœnobium and are able to move within their common calymma sometimes centrifugally to its surface, sometimes towards the centre where they aggregate into a roundish mass (see note D).

A. Regarding the movement of the flagella in mature swarm-spores compare L. N. 22, p. 375; L. N. 26, pp. 31, 35; L. N. 41, p. 452, and L. N. 52, p. 170.

B. On the active vertical swimming movements of mature Radiolaria, especially the cause of sinking and rising, see L. N. 16, p. 134; L. N. 41, p. 443, and L. N. 52, pp. 97–102.

C. On the active horizontal creeping movements of mature Radiolaria on a firm ground, compare L. N. 12, p. 10, and L. N. 16, pp. 132–134.

D. Regarding the motion of social central capsules within the same cœnobium and the changes thus brought about in the structure of the calymma, see L. N. 16, pp. 119–127 and L. N. 52, pp. 75–82.

221. *Contraction.*—Motions, which are due to the contraction of individual portions and cause changes in volume or form, have been partly already spoken of under the head of locomotion (§ 220) and are partly connected with other functions. Examples may be seen in the contraction of the central capsule and of the calymma. A certain

contraction of the central capsule is probably brought about by the myophanes, which arise by differentiation of the endoplasm and hence may assume different forms in the four legions. In the SPUMELLARIA, where numerous radial fibrillæ run from the central nucleus to the capsule membrane (§ 77), the endoplasm is probably driven out evenly through all the pores of the capsule membrane by their simultaneous contraction, and hence the volume of the capsule is diminished in all directions. The ACANTHARIA probably behave similarly, but are different, inasmuch as the number of their contractile radial fibrillæ is less, and special axial threads (§ 78) are already differentiated. In the NASSELLARIA it is probable that owing to the contraction of the divergent myophane fibrillæ in the podoconus the vertical axis of the latter is shortened, the opercular rods of the porochora are lifted, and the endoplasm driven out of its pores, so that the volume of the monaxon central capsule is diminished (§ 79). In the PHÆODARIA the same result is probably brought about by the contraction of the cortical myophane fibrillæ, which run meridionally along the inside of the capsule membrane from the apical to the basal pole of the vertical main axis, where they are inserted into the periphery of the astropyle; since the volume of the capsule is diminished by their contraction (their spheroidal figure becoming more nearly spherical) the endoplasm will be driven out through the proboscis of the astropyle. Whilst these contractions of the central capsule are largely due to differentiated muscle-like threads of endoplasm (myophanes), this appears to be but rarely the case with the contractions of the extracapsulum (*e.g.*, the myophrises of the Acanthometra, § 96). Most of the phenomena of contraction which can be observed in the calymma and pseudopodia depend upon exoplasmatic currents (§ 209).

222. *Protection.*—Of the utmost importance, both for the physiology and for the morphology of the Radiolaria are their manifold protective functions, which we now consider under the heading "protection." From the physiological point of view the consideration of the exposed situation in which the delicate, free-swimming Radiolarian organism lives, and the numerous dangers which beset it in the struggle for existence, would lead *a priori* to the expectation, that many special protective adaptations would be developed by natural selection. On the other hand, morphological experience shows us that this latter has been in action for immeasurable periods, and has gradually produced an abundance of the most remarkable protective modifications. Examples of these may be found in the formation of the voluminous calymma, as a gelatinous protective covering for the central capsule, and further, the formation of the capsule membrane itself, which separates the generative contents of the central capsule from the nutritive exoplasm. The phosphorescence of the central capsule, too (§ 223), may be regarded as a useful protective arrangement; as also the radiating of the numerous pseudopodia in all directions from the surface of the calymma; for they are of great significance to the

well-being of the organism, both as sensory organs and as prehensile organs. By far the most important and most varied means for the actual defence of the soft body is to be seen in the endless modifications of the skeleton; first, in the production of the enclosing lattice-shells and projecting radial spines, but especially also in the very varied structure of the individual parts of the skeleton, and in the special differentiation of the small appendicular organs which grow out from it (hairs, thorns, spines, scales, spathillæ, anchors, &c.). Finally "mimicry" possesses a considerable significance among the different forms of adaptation which are to be observed in this class.

223. *Phosphorescence.*—Many Radiolarians shine in the dark, and their phosphorescence presents the same phenomena as that of other luminous marine organisms; it is increased by mechanical and chemical irritation, or renewed if already extinguished. The light is sometimes greenish, sometimes yellowish, and appears generally (if not always) to radiate from the intracapsular fatty spheres (§ 73). Thus these latter unite several functions, inasmuch as they serve, firstly, as reserve stores of nutriment, secondly, as hydrostatic apparatus, and thirdly, as luminous organs for the protection of the Radiolaria; probably the light acts by frightening other animals, for the phosphorescent animals are provided with spines, nettle-cells, poison glands or other defensive weapons. The production of the light depends probably, as in other phosphorescent organisms, upon the slow oxidation of the fat-globules, which combine with active oxygen in the presence of alkalis. Phosphorescence is very likely widely distributed among the Radiolaria.

The shining of the Radiolaria in the dark has been noticed by the earliest observers of the class (see L. N. 1, p. 163, L. N. 16, p. 2, and L. N. 52, pp. 136–139). In the winter of 1859 I observed the production of light in the case of many monozootic and polyzootic Radiolaria, but inadvertently omitted to record the fact in my Monograph. I made more accurate observations in the winter of 1866 at Lanzerote in the Canary Islands, and convinced myself that the light emanates from the central capsule, and in particular from the fat-globules contained in it. In most Polycyttaria (both Collosphærida and Sphærozoida), when each central capsule contains a large central oil-globule the light radiates from it. In *Collozoum serpentinum* (Pl. 3, figs. 2, 3) each cylindrical central capsule contains a row of luminous spherules like a string of beads. In *Alacorys friderici* (Pl. 65, fig. 1) the four-lobed central capsule contains four shining points. Karl Brandt has recently made more detailed communication on this point (L. N. 52, p. 137).

224. *Sensation.*—The general irritability which we ascribe to all organisms, and as the basis of which we regard the protoplasm, remains at an inferior stage of development in the Radiolaria. For although they are subject to various stimuli, and certainly possess a power of discrimination, special sensory organs are not differentiated; the peripheral portions of the protoplasm, and especially the pseudopodia, rather act both as organs of the different kinds of sensation and various modes of motion. That different Radiolaria have attained different degrees of development in this respect may be seen

partly by direct observation of the reaction of the living organism towards various stimuli, and partly by the comparison of the different conditions of existence under which Radiolarians exist, both in the most various depths of the ocean and in all climatic zones (see note A). In general the Radiolaria seem to be sensitive to the following stimuli; (1) pressure (see note B); (2) temperature (see note C); (3) light (see note D); (4) chemical composition of the sea-water (see note E). The reaction towards these stimuli, corresponding to the sensation of pleasure or dislike which they call forth, is shown in various forms of motion of the protoplasm, changes in the currents in it, contraction of the central capsule, changes in the size, position, and form of the pseudopodia, changes in the volume of the calymma (by the evacuation of water), &c. Among the sensory functions of the Radiolaria must be especially mentioned their remarkably developed perception of hydrostatic equilibrium (see note F), as well as their perception of distances, so clearly shown in the production of equal lattice-meshes and other regularly formed skeletal structures (see note G).

A. I can add but little to the communication which I made twenty-four years ago regarding sensation in the Radiolaria (L. N. 16, pp. 128–131). The most important point would be the great difference in irritability which must obtain between the pelagic, zonarial and abyssal Radiolaria, which may be assumed from a consideration of their very different conditions of existence as regards pressure, light, warmth, nutrition, &c. It is natural to suppose that the numerous abyssal Radiolaria, discovered by the Challenger, which live at great depths (2000 to 4500 fathoms) in complete darkness, in icy cold and under an enormous pressure, must have quite different sensations of pleasure from their pelagic relatives which live at the surface of the sea under an equatorial sun. Karl Brandt has recently added much to our knowledge regarding the special action of different vital conditions upon the various Polycyttaria and the degrees of their irritability (L. N. 52, pp. 113–132).

B. Regarding the sensation of pressure or sensation of touch of the Radiolaria and the various degrees of their mechanical irritability, see L. N. 16, p. 129; L. N. 41, p. 464.

C. Regarding the sensation of warmth or temperature-sense and its dependence upon different climatic relations, see L. N. 16, p. 129; L. N. 52, pp. 114–129.

D. Regarding the sensation of light, compare L. N. 16, p. 128; L. N. 42, p. 304; L. N. 52, pp. 102–104, 114.

E. Regarding the sense of taste of the Radiolaria or their peculiar sensitiveness towards the different chemical composition of the water, change in its salinity, presence of organic impurities, &c., see L. N. 16, p. 130; L. N. 52, pp. 103, 113. This chemical irritability seems to be the most highly developed sense in the Radiolaria, even more so than their mechanical irritability.

F. The perception of hydrostatic equilibrium among the Radiolaria is immediately visible from the position which their bodies, floating freely in the water, assume spontaneously, and from the symmetrical development of the skeleton, which by its gravitation necessitates a definite position. It may be assumed that the development of the various geometrical ground forms which correspond to a definite position of equilibrium, is the result of this particular kind of perception (compare §§ 40–45).

G. The plastic perception of distance of the pseudopodia is shown by the symmetry with which the forms composing the regular skeletal structures (*e.g.*, the ordinary lattice-spheres with regular hexagonal meshes, the radial spines with equidistant branches) are excreted from the exoplasm. Both this form of sensation and the one first mentioned (note F) have hitherto received scarcely any attention, but are deserving of a thorough physiological investigation.

225. *The Cell-Soul (Zellseele).*—The common central vital principle, commonly called the "soul," which is considered to be the regulator of all vital functions, appears in the Radiolaria as in other Protista in its simplest form, as the cell-soul. By the continual activity of this central "psyche" all vital functions are maintained in unbroken action, and in uniform correlation. It is also probable that by it the stimulations which the peripheral portions of the cell receive from the outer world are first transmitted into true sensation, and that, on the other hand, the volition, which alone calls forth spontaneous movements, proceeds from it. The central capsule is most likely the sole organ of this cell-soul or central psychic organ, and the active portion may be either the endoplasm or the nucleus, or both. The central capsule may thus (apart from its function as a sporangium, § 215) be regarded as a simple ganglion cell, physiologically comparable to the nervous centre of the higher animals, whilst the exoplasm (sarcomatrix and pseudopodia) are to be compared to the peripheral nervous system and sense organs of the latter. The great simplicity of the functions of the cell-soul which appear in the Radiolaria, and the intimate connection of their different psychic activities, give to these unicellular Protista a special significance for the comprehension of the monistic elements of a natural psychology.

Regarding the theory of the cell-soul as the only psychological theory which is able to explain naturally the true nature of the life of the soul in all organisms as well as in man, see my address on cell-souls and soul-cells ("Zellseelen und Seelenzellen") in Gesammelte populäre Vorträge aus dem Gebiete der Entwickelungslehre, Heft 1, p. 143; Bonn, 1878.

CHOROLOGICAL SECTION.

CHAPTER IX.—GEOGRAPHICAL DISTRIBUTION.

(§§ 226-240.)

226. *Universal Marine Distribution.*—Radiolaria occur in all the seas of the world, in all climatic zones and at all depths. Probably under normal conditions they always float freely in the water, whether their usual position be at the surface (pelagic), or at a certain depth (zonarial), or near to the bottom of the sea (abyssal). This appears both from numerous direct observations, as well as from conclusions which may be drawn from their organisation (and especially their promorphology) regarding their floating life (compare §§ 40-50, 219, 220). Hitherto no observation has been recorded, which justifies the assumption that Radiolaria live anywhere upon the bottom of the sea (on stones, Algæ, or other firm substances), either sessile or creeping. They perform the latter action, however, when they fall accidentally upon a firm basis or are accidentally placed upon it, but they seem normally always to float freely in the water with pseudopodia radiating in all directions. Active free-swimming movements are only met with in the case of the flagellate zoospores (§ 142). The development of Radiolaria in large masses is very remarkable (see note A), and in many parts of the ocean is so great that they play an important part in the economy of marine life, especially as food for other pelagic and abyssal animals (see note B). Medium salinity of the water seems to be most favourable to their development in masses, although it is not unknown in seas of high and low salinity (see note C). There are no Radiolaria in fresh water (see note D).

A. The development of Radiolaria takes place in many parts of the ocean in astonishingly large masses on the surface, in different strata, and near the bottom. The C o l l o d a r i a (and especially the Sphærozoida) often cover the surface of the sea in millions, and form a shining layer, phosphorescent in the dark like the *Noctilucæ*, as I observed in 1859 in the Strait of Messina, in 1866 at the Canaries, and in 1881 in the Indian Ocean. Similar masses of *Sphærozoum* and *Acanthometron* were seen by Johannes Müller on the French and Ligurian coasts (L. N. 12), and John Murray found another in the Gulf Stream, off the Færöe Islands, from the surface to a depth of 600 fathoms; considerable masses of large PHÆODARIA live there also.

B. The alimentary canal of Medusæ, Salpæ, Crustacea, Pteropoda, and many other pelagic animals is a rich field for the discovery of Radiolaria, and many of the species hereinafter described are from such sources. Fossil coprolites too (*e.g.*, those from the Jura) often contain many Polycystina.

C. Some ACANTHARIA (A c a n t h o m e t r a) and PHÆODARIA (species of *Mesocena* and *Dictyocha*) live in the Baltic; I found their skeletons in the alimentary canal of *Aurelia*, Ascidians and Copepods.

D. The so-called "fresh-water Radiolaria," which have been described by Focke, Greeff, Grenacher and others, are all Heliozoa, without either central capsule or calymma.

227. *Local Distribution.*—As regards their local distribution and its boundaries the Radiolaria show in general the same relations as other pelagic animals. Since they are only to a very slight extent, if at all, capable of active horizontal locomotion, the dispersion of the different species from their point of development (or "centre of creation") is dependent upon oceanic currents, the play of winds and waves and all the accidental causes which influence the transport of pelagic animals in general. These passive migrations are here, however, as always, of the greatest significance, and bring about the wide distribution of individual species in a far higher degree than any active wanderings could do. Any one who has ever followed a stream of pelagic animals for hours and seen how millions of creatures closely packed together are in a short time carried along for miles by such a current, will be in no danger of underestimating the enormous importance of marine currents in the passive migration of the fauna of the sea. Such constant currents may, however, be recognised both near the bottom of the sea and at various depths, as well as at the surface, and are therefore of just as much significance for the abyssal and zonarial as for the pelagic Radiolaria. It is easy to explain by this means how it is that so many animals of this class (probably indeed the great majority) have a wide range of distribution. The number of *cosmopolitan* species which live in the Pacific, Atlantic and Indian Oceans is already relatively large. In each of these three great ocean basins, too, many species show a wide distribution. On the other hand, there are very many species which are hitherto known only from one locality, and probably many small local faunas exist, characterised by the special development of particular groups. The observations which we at present possess are too incomplete, and the rich material of the Challenger is too incompletely worked out, to enable any definite conclusions to be drawn regarding the local distribution of Radiolaria.

The statements made in the systematic portion of this Report regarding the distribution of the Challenger Radiolaria are very incomplete. In most cases only one locality is mentioned, and that is the station (§ 240) in the preparations or bottom deposit from which I first found the species in question. Afterwards I often found the same species again in one or more additional stations (not seldom in numerous preparations both from the Pacific and Atlantic), without the possibility of adding them to the habitat recorded under the description. The necessary accurate determination and identification of the species (measuring the different dimensions, counting the pores, &c.), would have occupied too much time, and the writing of this extensive Report would have lasted not ten but twenty or thirty years.

228. *Horizontal Distribution.*—From the extensive collections of the Challenger and from the other collections which have furnished a welcome supplement to them, it appears

that Radiolaria are distributed throughout all seas without distinction of zones and physical conditions, even though these latter may be the cause of differences in their qualitative and quantitative development. In the case of the Radiolaria as well as of many other classes of animals, the law holds good that the richest development of forms and the greatest number of species occurs between the tropics, whilst the frigid zones (both Arctic and Antarctic) exhibit great masses of individuals, but relatively few genera and species (see note A). In the Challenger collection the greatest abundance of species of Radiolaria is exhibited by those preparations which were collected at low latitudes in the immediate neighbourhood of the equator; this is true both of the Atlantic (Stations 346 to 349) and of the Pacific (Stations 266 to 274); in the former the richest of all is Station 347 (lat. 0° 15' S.), in the latter Station 271 (lat. 0° 33' S.) (see note B). From the tropics the abundance of species seems to diminish regularly towards the poles, and more rapidly in the northern than in the southern hemisphere; the latter also appears, considered as a whole, to possess more species than the former. A limit to the life of the Radiolaria towards the poles has not yet been found; the expeditions towards the North Pole (see note C), like those towards the South (see note D), have obtained bottom-deposits and ice enclosures which contained Radiolaria; in some of the most northerly and most southerly positions which were reached the number of Radiolaria enclosed in the ice was relatively great.

A. The greater abundance of Radiolaria in the tropical seas is probably to be explained by the more favourable conditions of existence, and in particular the larger quantity of nutritive material (especially of decayed animals) and not by the higher temperature of the surface, for at depths of from 2000 to 3000 fathoms where the abyssal Radiolaria live, the temperature is but little above the freezing point or even below it (compare the bottom temperatures in the list of Challenger Stations, § 240).

B. Station 271 of the Challenger Expedition, situated almost on the equator in the Mid Pacific (lat. 0° 33' S.), exceeds all other parts of the earth, hitherto known, in respect of its wealth in Radiolaria, and this is true of the pelagic as well as of the zonarial and abyssal forms. In the Station List the deposit at this point is stated to be "Globigerina ooze"; but after the calcareous matter has been removed by means of acid, the purest Radiolarian ooze remains, rich in varied and remarkable species. More than one hundred new species have been described from this Station alone.

C. Regarding the Arctic Radiolaria compare the contributions of Ehrenberg (L. N. 24, pp. 138, 139, 195) and Brady on the English North Polar Expedition, 1875–76 (Ann. and Mag. Nat. Hist., 1878, vol. i. pp. 425, 437).

D. Regarding the Antarctic Radiolaria, compare § 230, note A, and Ehrenberg, Mikrogeologie (L. N. 6, Taf. xxxv., A.), also L. N. 24, pp. 136–139.

229. *Fauna of the Pacific Ocean.*—From the splendid discoveries of the Challenger, and the supplementary observations obtained from other sources, the Pacific seems to be the ocean basin which is richest both quantitatively and qualitatively in Radiolarian life,

excelling both the Indian and Atlantic Oceans in this respect. It may be assumed with great probability that by far the largest portion of the Pacific has a depth of between 2000 and 3000 fathoms, and that its bottom is covered either with Radiolarian ooze (§ 237) or with a red clay (§ 239), which contains many SPUMELLARIA and NASSELLARIA, and has probably been derived for a great part from broken down and metamorphosed Radiolarian ooze (see note A). Pure Radiolarian ooze was found by the Challenger eastwards in the Central Pacific (over a wide area between lat. 12° N. and 12° S., Stations 265 to 274), and also westwards in the latitude of the Philippines, twenty degrees to the east of them (between lat. 5° N. and 15° N.). The great abundance of Radiolaria present in the neighbourhood of the Philippines and in the Sunda Sea was already known from other investigations (note B). The red clay also, which covers a great part of the bottom of the North Pacific, and which was obtained of very constant composition by the Challenger between lat. 35° N. and 38° N., from Japan to the meridian of Honolulu (from long. 144° E. to 156° W.), is so pre-eminently rich in Radiolaria that it often approaches in composition the Radiolarian ooze, and has probably been derived from it. The track of the Challenger through the tropical and northern parts of the Pacific describes nearly three sides of a rectangle, which includes about half of the enormous Pacific basin, and from this as well as from other supplementary observations it may with great probability be concluded that by far the largest part of the bed of the Pacific (at least three-fourths) is covered either with Radiolarian ooze or with red clay, which contains a larger or smaller amount of the remains of Radiolaria. With this agrees also the important fact that the numerous preparations of pelagic materials and collections of pelagic animals, which were collected by the Challenger in the Pacific, almost always indicate a corresponding amount of Radiolarian life on the surface. This is true in particular also of the South Pacific, between lat. 33° S. and 40° S. (from long. 133° W. to 73° W., Stations 287 to 301); the surface of this southern region and the different bathymetrical zones were rich in new and peculiar species of Radiolaria.

A. Many specimens of bottom-deposits from the Pacific, which are entered in the Challenger lists either as "red clay" or "Globigerina ooze," contain larger or smaller quantities of Radiolaria, and the number of different species of SPUMELLARIA and NASSELLARIA which they contain is often so great that the deposit might have been almost as appropriately termed "Radiolarian ooze," *e.g.*, Stations 241 to 245, and 270, 271 (compare §§ 236–239).

B. Pacific Radiolarian ooze was first obtained by Lieutenant Brooke (May 11, 1859) between the Philippines and Marianne Islands, from a depth of 3300 fathoms (lat. 18° 3′ N., long. 129° 11′ E.). Ehrenberg, who first described it, found seventy-nine different species of Polycystina in it, and reported "that their quantity and the number of different forms increased with the depth" (Monatsber. d. k. preuss. Akad. d. Wiss. Berlin, 1860, pp. 466, 588, 766).

230. *Fauna of the Indian Ocean.*—As regards its Radiolarian fauna the Indian Ocean is the least known of the three great basins. Still the few limited spots, regarding which

investigations are forthcoming, indicate a very rich development of Radiolarian life. Probably it approaches more nearly the fauna of the Pacific than that of the Atlantic, both as regards the abundance and the morphological characters of its species. The researches of the Challenger are very limited and incomplete as regards the Indian Ocean, for the expedition only just touched upon this great ocean basin (2000 to 3000 fathoms deep) at its two extremities (westwards at the Cape of Good Hope and eastwards at Tasmania), its course lying for the most part south of lat. 45° S. and extending beyond lat. 65° S. (from Station 149 to 158, south of lat. 50° S.). It is true that this portion of the South Indian Ocean was shown to contain Radiolaria everywhere, but these were more plentiful in individuals than in species. Only from Station 156 to Station 159 (between lat. 62° and 47° S., and long. 95° and 130° E.) was the bottom, which consisted partly of Diatom ooze and partly of Globigerina ooze, richer in species (see note A). The gaps left by the Challenger in the investigation of the Indian Ocean, have, however, been to some extent filled from other sources. As early as 1859 the English "Cyclops" expedition had shown that the bottom of the Indian Ocean to the east of Zanzibar (lat. 9° 37' S., long. 61° 33' W.) is covered with pure Radiolarian ooze (see note B). Also since the Tertiary rocks of the Nicobar Islands are for the most part of the same composition, and since a great abundance of Radiolaria has been shown to be present both in the east part of the ocean, between the Cocos Islands and the Sunda Archipelago (see note C), and in the northern part or Arabian Sea between Socotra and Ceylon (see note D); it may be assumed with great probability that the greater part of the basin of the Indian Ocean, like that of the Pacific, is covered either with Radiolarian ooze or with the characteristic red clay. With this agrees the richness of the surface of the Indian Ocean in Radiolaria of the most various groups, which has been more extensively demonstrated.

A. The Radiolarian fauna collected by the Challenger on the voyage from the Cape to Melbourne, shows in part, namely, from Station 156 to Station 158, very peculiar and characteristic composition; in particular, the Diatom ooze of Station 157 passes over in great part into a Radiolarian ooze, mainly composed of S p h æ r e l l a r i a. This is worthy of a more thorough investigation than I was able, owing to lack of material and time, to give it.

B. The remarkably pure Radiolarian ooze of Zanzibar, discovered by Ehrenberg in 1859, was the earliest known recent example of that deposit. It was brought up by Captain Pullen of the English man-of-war "Cyclops," from a depth of 2200 fathoms, between Zanzibar and the Seychelles, and "under a magnifying power of 300 diameters, showed at the first glance a mass of almost pure Polycystina, such as no sample of a deep-sea deposit has hitherto shown. It is very noticeable that in the whole of this mass of living forms, no calcareous shells are to be seen" (Ehrenberg, L. N. 24, pp. 148, 149).

C. For the most important material from the Indian Ocean, I am indebted to Captain Heinrich Rabbe of Bremen, who during many voyages in the Indian Ocean, in his ship "Joseph Haydn," made numerous collections in different localities with the tow-net and the trawl, and admirably preserved the rich collections thus made. The greatest abundance of Radiolaria was found in those obtained to the east of Madagascar, and next in those from the neighbourhood of the Cocos Islands. I take this opportunity of expressing my thanks to Captain Rabbe for the liberality with which he placed all this valuable material at my disposal.

D. On my voyage from Aden to Bombay, and thence to Ceylon (1881), and especially on my return journey from Ceylon, between the Maldive Islands and Socotra (1882), I carried on a number of experiments with a surface net, which yielded a rich fauna of pelagic animals, and among them many new species of Radiolaria, for observation. On several nights when the smooth surface of the Indian Ocean, unrippled by any wind, shone with the most lovely phosphorescent light, I drew up water from the surface with a bucket, and obtained a rich booty. A number of other new species of Radiolaria from very various parts of the Indian Ocean I obtained from the alimentary canal of pelagic animals, such as Medusæ, Salpæ, Crustacea, &c. Although the total number of Radiolaria known to me from the Indian Ocean is much less than from the Atlantic and Pacific, there are several new genera and numerous species among them, which show that a careful study of this fauna will be of wide interest.

231. *Fauna of the Atlantic Ocean.*—The Atlantic Ocean in all parts, of which the pelagic fauna has been examined, has shown the same constant presence of Radiolaria, and in certain parts of its abyssal deposits a larger or smaller quantity of different types belonging to this class; on the whole, however, its Radiolarian fauna is inferior to that of the Pacific, and probably also to that of the Indian Ocean, both in quantity and quality. Pure Radiolarian ooze, such as is so extensively found on the floor of the Pacific, and in certain places in that of the Indian Ocean, has not yet been found in the Atlantic (see § 237). The red clay, too, of the deep Atlantic does not seem to be so rich in Radiolaria as that of the Pacific; nevertheless, the number of species peculiar to the Atlantic is very large, and at certain points the abundance of species as well as of individuals seems to be scarcely less than in the Pacific. This is especially true of the eastern equatorial zone not far from Sierra Leone, Stations 347 to 352 (see note A); also of the South Atlantic between Buenos Ayres and Tristan da Cunha, Stations 324, 325, 331 to 333 (see note B); and, lastly, in the North Atlantic in the Gulf Stream and near the Canary Islands (see note C). The fauna of the latter agrees for the most part with that of the Mediterranean (see note D). In addition to the material collected by the Challenger, other deep-sea investigations have furnished bottom-deposits from different parts of the ocean, which have proved very rich in Radiolaria (see note E). Furthermore, since the island of Barbados consists for the most part of fossil Radiolarian ooze, it is very probable that at certain parts of the tropical Atlantic true Radiolarian ooze, like that of the Pacific and Indian Oceans, will eventually be found in depths between 2000 and 3000 fathoms, perhaps over a considerable area.

A. The tropical zone of the eastern Atlantic seems to be especially rich in peculiar Radiolaria of different species. This is shown by numerous preparations from the surface, and from various depths (between lat. 3° S. and 11° N., and long. 14° W. to 18° W.), which were made towards the end of the cruise. Unfortunately no bottom-deposits were obtained from the most important stations (except Nos. 346 and 347, depths 2350 and 2250 fathoms) in this region; at these the deposit was a Globigerina ooze containing numerous different species of Radiolaria.

B. In the South Atlantic, between Buenos Ayres and Tristan da Cunha (between lat. 35° S. and 43° S., long. 8° W. and 57° W.) there appears to be a long stretch covered partly with Globigerina ooze (Stations 331 to 334), or red clay (Stations 329, 330), partly with blue mud (Stations 318 to 328), which contains not only large masses of individuals but numerous peculiar species of SPUMELLARIA and NASSELLARIA. The preparations from the surface-takings of this region are also rich in these, as well as in peculiar PHÆODARIA.

C. The northern part of the Atlantic appears on the whole to be inferior to the tropical and southern portions as regards its richness in Radiolaria, and from the western half more especially, only few species are known. From my researches at Lanzerote in 1866–67, it appears that the pelagic fauna of the Canary Islands is very rich in them, as is also the Gulf Stream in the neighbourhood of the Færöe Channel, according to the investigations of John Murray (see his Report on the "Knight-Errant" Expedition, Proc. Roy. Soc. Edin., vol. xi., 1882).

D. The Radiolaria of the Mediterranean are of special interest, because almost all our knowledge of these organisms in the living conditions and of their vital functions has been derived from investigations conducted on its shores. Johannes Müller laid the foundation of this knowledge by his investigations at Messina, and on the Ligurian and French coasts at Nice, Cette, and St. Tropez (L. N. 10). The many new Radiolaria which I described in my Monograph (L. N. 16, 1862), were for the most part taken at Messina, the place which possesses a richer pelagic fauna than any other, so far as is yet known, in the Mediterranean. Other new species I found afterwards at Villafranca near Nice, in 1864 (L. N. 19), at Portofino near Genoa (1880), at Corfu (1877), and at other points on the coast. In Messina also, Richard Hertwig collected the material for his valuable treatise on the Organisation of the Radiolaria (L. N. 33), after he had previously made investigations into their histology at Ajaccio in Corsica (L. N. 26). Lastly, at Naples, Cienkowski (L. N. 22) and Karl Brandt (L. N. 38, 39, 52) carried out their important investigations into the reproduction and symbiosis of the Radiolaria. With respect to the character of its Radiolaria, the Mediterranean fauna is to be regarded as a special province of the North Atlantic.

E. Among the smaller contributions which have been made towards our knowledge of the Atlantic Radiolarian fauna, the communications of Ehrenberg on the deposits obtained in sounding for the Atlantic cable, and on the Mexican Gulf Stream near Florida, deserve special mention (L. N. 24, pp. 138, 139–145).

232. *Vertical Distribution.*—The most important general result of the discoveries of the Challenger, as regards the vertical or bathymetrical distribution of the Radiolaria, is the interesting fact that numerous species of this class are found living at the most various depths of the sea, and that certain species are limited to particular bathymetrical zones, *i.e.*, are adapted to the conditions which obtain there. In this respect three different Radiolarian faunas may be distinguished, which may be shortly termed "pelagic," "zonarial," and "abyssal." The *pelagic* Radiolaria swim at the surface, and when they sink (*e.g.*, in a stormy sea), only descend to a small depth, probably not more than from

20 to 30 fathoms (§ 233). The complicated conditions of existence created by the keen struggle for existence at the surface of the sea, give rise to the formation of very numerous pelagic species, especially of Porulosa (SPUMELLARIA and ACANTHARIA). The *abyssal* Radiolaria are very different from those just mentioned; they live at the bottom of the deep-sea, not resting upon nor attached to it, but probably floating at a little distance above it, and are adapted to the conditions of existence which obtain there (§ 235). Here the Osculosa (NASSELLARIA and PHÆODARIA) seem to predominate. The *zonarial* Radiolaria live floating at various depths between the pelagic and abyssal species (§ 234). In their morphological characters they gradually approach the pelagic forms upwards and the abyssal downwards.

The views which have hitherto been held regarding the bathymetrical or vertical distribution of the Radiolaria have been entirely altered by the magnificent discoveries of the Challenger, and especially by the important observations of Sir Wyville Thomson (L. N. 31) and John Murray (L. N. 27). These two distinguished deep-sea explorers have, as the result of their wide experience, been convinced that Radiolaria exist at all depths of the ocean, and that there are large numbers of true deep-sea species which are never found at the surface of the sea nor at slight depths (L. N. 31, vol. i. pp. 236–238; L. N. 27, pp. 523, 525). The result of my ten years' work upon the Challenger Radiolaria, and the comparative study of more than a thousand mountings from all depths, has only been to confirm this opinion, and I am further persuaded that it will some day be possible by the aid of suitable nets (not yet invented) to distinguish different faunistic zones in the various depths of the sea. In this connection may be mentioned the specially interesting fact that the species of Radiolaria of one and the same family present in the different depths characteristic morphological distinctions, which obviously correspond to their different physiological relations in the struggle for existence. Owing to those extensive discoveries, the representation which I gave in my Monograph (1862, L. N. 16, pp. 172–196) of the vertical distribution of the Radiolaria, and of their life in the greatest depths of the sea, has been entirely changed. Compare also Bütschli (L. N. 41, p. 466).

233. *The Pelagic Fauna.*—The surface of the open ocean seems everywhere, at a certain distance from the coast at least, to be peopled by crowds of living Radiolaria. In the tropical zone these pelagic crowds consist of many different species, whilst in the frigid zones, on the other hand, they are made up of many individuals belonging to but few species. Most of these inhabitants of the surface may be regarded as truly pelagic species, which either remain always at the surface or descend only very slightly below it. Probably most Porulosa (both SPUMELLARIA and ACANTHARIA) belong to this group; whilst but few Osculosa occur in it, and fewer PHÆODARIA than NASSELLARIA. In general the pelagic Radiolaria are distinguished from the abyssal by the more delicate and slender structure of their skeletons; the pores of the lattice-shells are larger, the intervening trabeculæ thinner; the armature of spines, spathillæ, anchors, &c., is more various and more highly developed. Numerous forms are to be found among the pelagic

Radiolaria which have either an incomplete skeleton or none at all. When the pelagic forms leave the surface on account of unfavourable weather, they appear only to sink to slight depths (probably not below 20 or 30 fathoms). Within the limits of the same family the size of the pelagic species seems to be on an average greater than that of the related abyssal forms.

234. *The Zonarial Fauna.*—Between the pelagic fauna living at the surface of the open sea and the abyssal, which floats immediately over the bottom, there appears to be usually a middle fauna, which inhabits the different bathymetrical zones of the intermediate water, and hence may be shortly called the "zonarial" fauna. The different species of Radiolaria which inhabit these different strata in the same vertical column of water present differences corresponding to those of the plants composing the several zones of vegetation, which succeed each other at different heights on a mountain; they correspond to the different conditions of existence which are presented by the different strata of water, and to which they have become adapted in the struggle for existence. The existence of such bathymetrical zones has been shown by those important, if not numerous, observations of the Challenger, in which the tow-net was used at different depths at one and the same Station. In several cases the character of the Radiolarian fauna at different depths presented characteristic differences.

For the present, and until we are better acquainted with the characters of the Radiolarian fauna at different depths, we may distinguish provisionally the following *five bathymetrical zones*:—(1) The *pelagic* zone, extending from the surface to a depth of about 25 fathoms; (2) the *pellucid* zone, extending from 25 to 150 fathoms, or as far as the influence of the sunlight makes itself felt; (3) the *obscure* zone, extending from 150 to 2000 fathoms, or from the depth at which sunlight disappears to that at which the influence of the water containing carbonic acid begins and the calcareous organisms vanish; (4) the *siliceous* zone, extending from 2000 or 2500 to about 3000 fathoms, in which only siliceous not calcareous Rhizopoda are found, and in which the peculiar conditions of the lowest regions have not yet appeared; (5) the *abyssal* zone, in which the accumulation of the oceanic deposits, and the influence of the bottom currents, create new conditions of existence. So far as our isolated and incomplete observations of the zonarial Radiolarian fauna extend, it appears that the subclass Porulosa (SPUMELLARIA and ACANTHARIA) predominates in the two upper zones, and as the depth increases is gradually replaced by the subclass Osculosa (NASSELLARIA and PHÆODARIA), so that the latter predominates in the two lowest zones. The obscure zone which lies in the middle is probably the poorest in species. In general, the morphological characters of the zonarial fauna appear to change gradually upwards into the delicate form of the pelagic and downwards into the robust constitution of the abyssal; so also the average size of the individuals (within the limits of the same family) appears to increase upwards and decrease downwards.

235. *The Abyssal Fauna.*—The great majority of Radiolaria which have hitherto been observed, and which are described in the systematic portion of this Report, have been obtained from the bottom of the deep-sea, and more than half of all the species have been

derived from the pure Radiolarian ooze, which forms the bed of the Central Pacific at depths of from 2000 to 4000 fathoms (§ 237). Many of these abyssal forms were brought up with the malacoma uninjured, and they show, both when mounted immediately in balsam, and when preserved in alcohol, all the soft parts almost as clearly as fresh preparations of pelagic Radiolaria. These species are to be regarded as truly abyssal, *i.e.*, as forms which live floating only a little distance above the bottom of the deep-sea, having become adapted to the peculiar conditions of life which obtain in the lowest regions of the ocean. Probably the majority of the PHÆODARIA belong to these abyssal Radiolaria, as well as a large number of NASSELLARIA, but on the other hand, only a small number of ACANTHARIA and SPUMELLARIA are found there. A character common to these abyssal forms, and rarely found in those from the surface or from slight depths, is found in their small size and their heavy massive skeletons, in which they strikingly resemble the fossil Radiolaria of Barbados and the Nicobar Islands. The lattice-work of the shell is coarser, its trabeculæ thicker and its pores smaller than in pelagic species of the same group; also the apophyses (spines, spathillæ, coronets, &c.), are much less developed than in the latter. From these true abyssal Radiolaria must be carefully distinguished those species whose empty skeletons, devoid of all soft parts, occur also in the Radiolarian ooze of the deep-sea, but are clearly only the sunken remains of dead forms, which have lived at the surface or in some of the upper zones.

236. *Deposits containing Radiolaria.*—The richest collection of Radiolaria is found in the deposits of ooze which form the bed of the ocean. Although the pelagic material skimmed from the surface of the sea, and the zonarial material taken by sinking the tow-net to various depths, are always more or less rich in Radiolaria, still the number of species thus obtained is, on the whole, much less than has hitherto been got merely from deep-sea deposits. Of course the skeletons found in the mud of the ocean-bed, may belong either to the abyssal species which live there (§ 235), or to the zonarial (§ 234), or to the pelagic species (§ 233), for the siliceous skeletons of these latter sink to the bottom after their death. Almost all these remains found in the deposits belong to the siliceous "Polycystina" (SPUMELLARIA and NASSELLARIA); PHÆODARIA occur but sparingly, and ACANTHARIA are entirely wanting, for their acanthin skeleton readily dissolves. The abundance of Radiolaria varies greatly according to the composition and origin of the deposits. In general marine deposits may be divided into two main divisions, terrigenous and abyssal, or, more shortly, muds and oozes. The *terrigenous* deposits (or muds) include all those sediments which are made up for the most part of materials worn away from the coasts of continents and islands, or brought down into the sea by rivers. Their greatest extent from the coast is about 200 nautical miles. They contain varying quantities of Radiolaria, but much fewer than those of the next group. The *abyssal* deposits (or oozes) usually commence at a distance of from 100 to 200 nautical miles

from the coast. In general they are characterised by great uniformity, corresponding to the constancy of the conditions under which they are laid down; they may be divided into three categories, the true Radiolarian ooze (§ 237), Globigerina ooze (§ 238), and red clay (§ 239). Of these three most important deep-sea formations the first is by far the richest in Radiolaria, although the other two contain often very many siliceous shells.

The marvellous discoveries of the Challenger have thrown upon the nature of marine deposits an entirely new light, which justifies most important conclusions regarding the geographical distribution and geological significance of the Radiolaria. Since Dr. John Murray and the Abbé Renard will treat fully of these interesting relations in a forthcoming volume of the Challenger series (Report on the Deep-Sea Deposits), it will be sufficient here to refer to their preliminary publication already published (Narrative of the Cruise of H.M.S. Challenger, 1885, vol. ii. part ii. pp. 915–926); see also the earlier communications by John Murray (1876, L. N. 27, pp. 518-537), and by Sir Wyville Thomson (The Atlantic, L. N. 31, vol. i. pp. 206–246). In the Narrative (*loc. cit.*, p. 916) the following table of marine deposits is given:—

Terrigenous deposits.	Shore formations, Blue mud, Green mud and sand, Red mud,	Found in inland seas and along the shores of continents.
	Volcanic mud and sand, Coral mud and sand, Coralline mud and sand,	Found around oceanic islands and along the shores of continents.
Abysmal deposits.	Globigerina ooze, Pteropod ooze, Diatom ooze, Radiolarian ooze, Red clay,	Found in the abysmal regions of the ocean basins.

237. *Radiolarian Ooze.*—By Radiolarian ooze, in the strict sense of the term, are understood those oceanic deposits, the greater part of which (often more than three-quarters) is composed of the siliceous skeletons of this class. Such *pure* Radiolarian ooze has only been found in limited areas of the Pacific and Indian Oceans. It is most conspicuous in the Central Pacific, between lat. 12° N. and 8° S., long. 148° W. to 152° W., the depth being everywhere between 2000 and 3000 fathoms (Stations 266 to 268 and 272 to 274). In the deepest of the Challenger soundings (Station 225, 4475 fathoms) the bottom is composed of pure Radiolarian ooze, as well as at the next Station in the Western Tropical Pacific (Station 226, 2300 fathoms), the latitude varying from 12° N. to 15° N., and the longitude from 142° E. to 144° E. In the Indian Ocean also, pure Radiolarian ooze was found in the year 1859 between Zanzibar and the Seychelles, this being the first known example of it (§ 230). On the other hand, it has not yet been found in the bed of the Atlantic; but the Tertiary formations of Barbados (Antilles, § 231) like those of the Nicobar Islands (Further India), are to be regarded as pure Radiolarian

ooze in the fossil condition. *Mixed* Radiolarian ooze is the name given to those deposits in which the Radiolaria exceed any of the other organic constituents, although they do not make up half the total mass. To this category belong a large number of the Challenger soundings which are entered in the Station list either as red clay or Globigerina ooze. Such mixed Radiolarian ooze has been discovered (A) in the North Pacific in an elongated area of red clay extending from Station 241 to Station 245 (perhaps even from Station 238 to Station 253), that is, at least, from long. 157° E. to 175° E., between lat. 35° N. and 37° N.; (B) in the tropical Central Pacific in the Globigerina ooze of Stations 270 and 271. The ooze from the latter station, situated almost on the equator (lat. 0° 33' S., long. 151° 34' W.), is specially remarkable, for it has yielded more new species of SPUMELLARIA and NASSELLARIA than any other Station, not excluding even the neighbouring Stations 268, 269, and 272. Probably such mixed Radiolarian ooze is very widely distributed in the depths of the ocean, as, for example, in the South Pacific (Stations 288, 289, 300, and 302), and in the Southern Ocean (Stations 156 to 159); also in the South Atlantic (Stations 324, 325, 331, 332) and in the tropical Atlantic (Stations 348 to 352). When carefully purified and decalcified by acids, Radiolarian ooze appears as a fine shining white powder; in the raw state it is yellowish or reddish, sometimes reddish-brown or dark brown in colour, according to the quantity of oxides of iron, manganese, &c., which it contains. Calcareous skeletons (especially the tests of pelagic Foraminifera) do not occur at all or only in very minute quantities in *pure* Radiolarian ooze from more than 2000 fathoms, whilst specimens of *mixed* ooze often contain considerable quantities of them.

Pure Radiolarian ooze was first described by Dr. John Murray as regards its peculiar nature and composition under the name "Radiolarian ooze" (1876, L. N. 27, pp. 525, 526); compare also Sir Wyville Thomson (The Atlantic, L. N. 31, vol. i. pp. 231–238), and John Murray (Narr. Chall. Exp., L. N. 53, vol. i. pt. ii. pp. 920–926, pl. N. fig. 2). The different specimens of pure Radiolarian ooze obtained by the Challenger from the Pacific, and handed to me for investigation, are from depths of from 2250 fathoms to 4475 fathoms, and may be divided according to their composition into three different groups :—I. The Radiolarian ooze of the Western Tropical Pacific, Stations 225 and 226, from depths of 4475 and 2300 fathoms (lat. 11° N. to 15° N., and long. 142° E. to 144° E). II. The Radiolarian ooze of the northern half of the Central Pacific, Stations 265 to 269, from depths of 2550 to 2900 fathoms. III. The Radiolarian ooze of the southern half of the Central Pacific, Stations 270 to 274, from depths of 2350 to 2925 fathoms. A fourth group would be constituted by the Radiolarian ooze from the Philippines, which was brought up by Brooke in 1860 near the Marianne Islands from 3300 fathoms, and described by Ehrenberg (Monatsber. d. k. preuss. Akad. d. Wiss. Berlin, 1860, p. 765). The Diatom ooze, too, found by the Challenger in the Antarctic regions (Stations 152 to 157) is in some parts so rich in Radiolaria that it passes over into true Radiolarian ooze. Regarding the Radiolarian ooze from Zanzibar, obtained by Captain Pullen in 1859 from 2200 fathoms (§ 230), we have only the incomplete communications of Ehrenberg (L. N. 24, p. 147). A more accurate knowledge of these deposits from the Indian Ocean, and of those which we may with probability expect from the tropical eastern Atlantic, will be sure to increase very widely our knowledge of the class.

238. *Globigerina Ooze.*—Next to the Radiolarian ooze proper the Globigerina ooze is the deposit which is richest in the remains of Radiolaria. Often these are so abundant that it is doubtful to which category the specimen should be referred (*e.g.*, Stations 270 and 271, see § 237). In fact, the two pass without any sharp boundary into each other, and both present transitions to the Diatom ooze. Next to red clay (§ 239), Globigerina ooze is the most widely distributed of all sediments, and forms a large part of the bed of the ocean at depths of 250 to 2900 fathoms (especially between 1000 and 2000 fathoms). It covers extensive areas at depths below 1800 fathoms, and in still deeper water is replaced by red clay. It is a fine-grained white, grey, or yellowish powder, which sometimes becomes coloured rose, red, or brown owing to the admixture of oxides of iron and manganese. True Globigerina ooze consists for the most part of the accumulated calcareous shells of pelagic Foraminifera, principally *Globigerina* and *Orbulina*, but also *Hastigerina, Pulvinulina*, &c. It contains usually from 50 to 80 per cent. of calcium carbonate, the extreme values being 40 and 95 per cent. After this has been removed by acids, there remains a residue, which consists partly of the siliceous shells of Radiolaria and Diatoms, and partly of mineral particles identical with the volcanic elements of the red clay.

Regarding the composition and significance of the Globigerina ooze, see John Murray (L. N. 27, pp. 523–525, and L. N. 53, vol. i. p. 919). Recently this author has separated from the Globigerina ooze (*sensu stricto*), the *Pteropod ooze*, distinguished from the former by the greater abundance of Pteropod shells and calcareous shells of larger pelagic organisms which it contains. It is found in moderate depths (at most 1500 fathoms), and contains fewer Radiolaria.

239. *Red Clay.*—This is quantitatively the most important of all deep-sea deposits, covering by far the greatest extent of the three great ocean basins at depths greater than 2200 fathoms. It thus far surpasses in area the other deposits, both Radiolaria and Globigerina oozes, and commonly forms a still deeper layer beneath them. Probably these three deep-sea deposits together cover about three-eighths of the whole surface of the earth, that is, about as much as all the continents together, whilst only two-eighths are covered by the terrigenous deposits. Red clay is principally composed of silicate of alumina, mixed in various proportions with other finely granular substances; its usual red colour, which sometimes passes over into grey or brown, is more especially due to admixture of oxides of iron and manganese. Calcareous matter is usually entirely wanting, or present only in traces, whilst free silica is found in very variable, often considerable quantities. The chief mass of the red clay consists of volcanic ashes, pumice, fragments of lava, &c., whilst a large part of it is generally composed of shells of Radiolaria or fragments of

them; in many places the number of well-preserved skeletons contained in the red clay is very considerable, so that it passes over gradually into the Radiolarian ooze (*e.g.*, in the North Pacific, Stations 238 to 253, see § 237). Hence it may be supposed that a large part of the red clay consists of decomposed Radiolarian ooze.

The characteristic composition and fundamental significance of the red clay in the formation of the deep-sea bed were first made known by the discoveries of the Challenger (compare John Murray, 1876, L. N. 27, p. 527, and Narr. Chall. Exp., L. N. 53, vol. i. pt. ii. pp. 920–926, pl. N; also Wyville Thomson, The Atlantic, L. N. 31, vol. i. pp. 226–229). The mineral components of the red clay are for the most part of volcanic origin, due to the decomposition of pumice, lava, &c. Among the organic remains found in it, the siliceous skeletons of Radiolaria are by far the most important, and their number is often considerable. A large portion of the red clay appears to me to consist of broken down Radiolarian shells, in which a peculiar metamorphism probably has taken place. Sir Wyville Thomson was of opinion that a considerable proportion of it consisted of the remains of Globigerina ooze, the calcareous constituents of which had been removed by the carbon dioxide in the deep-sea water (L. N. 31, *loc. cit.*). Among these remains, however, the siliceous skeletons of the Radiolaria play a significant and often the most important part. Furthermore, John Murray has called attention to the fact that in many deep-sea deposits yellow and red insoluble particles remain, which unmistakably present the form of Radiolarian shells (L. N. 27, p. 513). At Station 303 he found "amorphous clayey matter, rounded yellow minerals, many Radiolaria-shaped;" at Station 302 there was sediment "consisting almost entirely of small rounded red mineral particles; many of these had the form of both Foraminifera and Radiolaria; and it seemed as if some substance had been deposited in and on these organisms." Similar transitions from well-preserved Radiolarian shells into amorphous mineral particles I have found in several other specimens of Challenger soundings, and consider them a further argument for the supposition that the Radiolaria often take an important share in the formation of the red clay.

240. *List of Stations at which Radiolaria were observed on the Challenger Expedition.*—The 168 Stations recorded below, in soundings or surface preparations from which I found Radiolaria, belong to the most various parts of the sea which the Challenger traversed during her voyage round the world; they constitute about half of the (364) observing Stations contained in the official list published in the Narrative of the Cruise (Narr. Chall. Exp., vol. i. part ii. Appendix ii.).

In addition to the particulars given in the list regarding the geographical position of the Station, depth, temperature, and composition of the bottom deposit, I have added the result of my investigations as regards the relative abundance of the Radiolaria in each. The five letters (A to E) denote the following degrees of frequency:—A, abundant Radiolaria (AI, pure Radiolarian ooze; AII, mixed Radiolarian ooze); B, very numerous Radiolaria (but not a predominating quantity); C, many Radiolaria (medium quantity); D, few Radiolaria; E, very few Radiolaria (as they occur almost always). In using these symbols regard has been had to abundance of the abyssal as well as of the zonarial and pelagic forms (§ 232); sometimes also the estimated number of Radiolaria has been inserted, based upon information given by John Murray in his Preliminary Report (L. N. 27), and in the Narrative of the Cruise (L. N. 53), as well as by Henry B. Brady in his Report on the

Foraminifera (Zool. Chall. Exp., part xxii., 1884). From Stations 348 to 352 in the Eastern Tropical Atlantic no specimens of the bottom were obtained, but a rich pelagic Radiolarian fauna was demonstrated by numerous preparations from the surface. The depths are given in fathoms and the temperature in degrees Fahrenheit. In the column describing the nature of the bottom the following abbreviations are used:—

rad. oz. = Radiolarian ooze (§ 237).
gl. oz. = Globigerina ooze (§ 238).
r. cl. = red clay (§ 239).
pt. oz. = Pteropod ooze (see p. clviii).
di. oz. = Diatom ooze (see p. clvii).

bl. m. = blue mud,
gr. m. = green mud,
volc. m. = volcanic mud,
r. m. = red mud.
} terrigenous deposits (see p. clvi).

Challenger Station.	Locality.	Depth in Fathoms.	Bottom Temperature, F.	Nature of Bottom.	Relative Abundance of Radiolaria.	Date.	Latitude and Longitude.	Nearest Land.
						1873.		
1.	N. Atl.	1890	36·8	gl. oz.	D few	Feb. 15	27° 24′ N., 16° 55′ W.	S. of Tenerife.
2.	,,	1945	36·8	gl. oz.	E very few	,, 17	25° 52′ N., 19° 22′ W.	S.W. of the Canary Islands.
5.	,,	2740	37·0	r. cl.	D few	,, 21	24° 20′ N., 24° 28′ W.	S.W. of the Canary Islands.
9.	,,	3150	36·8	r. cl.	E very few	,, 26	23° 23′ N., 35° 11′ W.	(Ocean).
24.	Tr. Atl.	390	...	pt. oz.	D few	Mar. 25	18° 38′ N., 65° 5′ W.	Culebra (Antilles).
32.	N. Atl.	2250	36·7	gl. oz.	E very few	April 3	31° 49′ N., 64° 55′ W.	Bermuda.
45.	,,	1240	37·2	bl. m.	E ,,	May 3	38° 34′ N., 72° 10′ W.	S. of New York.
50.	,,	1250	38·0	bl. m.	E ,,	,, 21	42° 8′ N., 63° 39′ W.	S. of Halifax.
64.	,,	2700	...	r. cl.	D few	June 20	35° 35′ N., 50° 27′ W.	(Ocean).
76.	,,	900	40·0	pt. oz.	D ,,	July 3	38° 11′ N., 27° 9′ W.	Azores.
98.	Tr. Atl.	1750	36·7	gl. oz.	C many	Aug. 14	9° 21′ N., 18° 28′ W.	W. of Sierra Leone.
106.	,,	1850	36·6	gl. oz.	C ,,	,, 25	1° 47′ N., 24° 26′ W.	(Ocean).
108.	,,	1900	36·8	gl. oz.	C ,,	,, 27	1° 10′ N., 28° 23′ W.	(Ocean).
111.	,,	2475	33·7	gl. oz.	C ,,	,, 31	1° 45′ S., 30° 58′ W.	(Ocean).
120.	,,	675	...	r. m.	D few	Sept. 9	8° 37′ S., 34° 28′ W.	Pernambuco.
132.	S. Atl.	2050	35·0	gl. oz.	C many	Oct. 10	35° 25′ S., 23° 40′ W.	Tristan da Cunha.
134.	,,	2025	36·0	gl. oz.	C ,,	,, 14	36° 12′ S., 12° 16′ W.	Tristan da Cunha.
137.	,,	2550	34·5	r. cl.	D few	,, 23	35° 59′ S., 1° 34′ E.	(Ocean).
138.	,,	2650	35·1	r. cl.	D ,,	,, 25	36° 22′ S., 8° 12′ E.	(Ocean).
143.	S. Ind.	1900	35·6	gl. oz.	E very few	Dec. 19	36° 48′ S., 19° 24′ E.	Cape of Good Hope.
144.	,,	1570	35·8	gl. oz.	E ,,	,, 24	45° 57′ S., 34° 39′ E.	(Ocean).
145.	,,	140	...	volc. s.	D few	,, 27	46° 43′ S., 38° 4′ E.	Prince Edward Island.
146.	,,	1375	35·6	gl. oz.	C many	,, 29	46° 46′ S., 45° 31′ E.	(Ocean).
147.	,,	1600	34·2	di. oz.	C ,,	,, 30	46° 16′ S., 48° 27′ E.	W. of the Crozet Islands.
						1874.		
148.	,,	210	...	{gravel, shells}	D few	Jan. 3	46° 47′ S., 51° 37′ E.	E. of the Crozet Islands.
149H.	,,	127	...	volc. m.	D ,,	,, 29	48° 45′ S., 69° 14′ E.	Kerguelen Island.
150.	,,	150	35·2	gravel	D ,,	Feb. 2	52° 4′ S., 71° 22′ E.	N. of Heard Island.
151.	,,	75	...	volc. m.	D ,,	,, 7	52° 59′ S., 73° 33′ E.	Heard Island.
152.	,,	1260	...	di. oz.	C many	,, 11	60° 52′ S., 80° 20′ E.	(Ocean).
153.	,,	1675	...	bl. m.	C ,,	,, 14	65° 42′ S., 79° 49′ E.	Antarctic Ice.
154.	,,	1800	...	bl. m.	C ,,	,, 19	64° 37′ S., 85° 49′ E.	Antarctic Ice.
155.	,,	1300	...	bl. m.	C ,,	,, 23	64° 18′ S., 94° 47′ E.	Antarctic Ice.
156.	,,	1975	...	di. oz.	B numerous	,, 26	62° 26′ S., 95° 44′ E.	(Ocean).
157.	,,	1950	32·1	di. oz.	B ,,	Mar. 3	53° 55′ S., 108° 35′ E.	(Ocean).
158.	,,	1800	33·5	gl. oz.	B ,,	,, 7	50° 1′ S., 123° 4′ E.	(Ocean).
159.	,,	2150	34·5	gl. oz.	B ,,	,, 10	47° 25′ S., 130° 22′ E.	(Ocean).
160.	,,	2600	33·9	r. cl.	C many	,, 13	42° 42′ S., 134° 10′ E.	(Ocean).
162.	,,	38	...	sand	E very few	April 2	39° 10′ S., 146° 37′ E.	Bass Strait.
163.	S. Pac.	2200	34·5	gr. m.	E ,,	,, 4	36° 57′ S., 150° 34′ E.	Port Jackson.
164A.	,,	1200	...	gr. m.	E ,,	June 13	34° 9′ S., 151° 55′ E.	W. of Sydney.

REPORT ON THE RADIOLARIA.

Challenger Station.	Locality.	Depth in Fathoms.	Bottom Temperature °F.	Nature of Bottom.	Relative Abundance of Radiolaria.	Date.	Latitude and Longitude.	Nearest Land.
						1874.		
165.	S. Pac.	2600	34·5	r. cl.	D few	June 17	34° 50′ S., 155° 28′ E.	(Ocean).
166.	,,	275	50·8	gl. oz.	D ,,	,, 23	38° 50′ S., 169° 20′ E.	W. of New Zealand.
169.	,,	700	40·0	bl. m.	D ,,	July 10	37° 34′ S., 179° 22′ E.	E. of New Zealand.
175.	Tr. Pac.	1350	36·0	gl. oz.	E very few	Aug. 12	19° 2′ S., 177° 10′ E.	Fiji Islands.
181.	,,	2440	35·8	r. cl.	E ,,	,, 25	13° 50′ S., 151° 49′ E.	Louisiades.
193.	,,	2800	38·0	bl. m.	D few	Sept. 28	5° 24′ S., 130° 37′ E.	Banda Sea.
195.	,,	1425	38·0	bl. m.	C many	Oct. 3	4° 21′ S., 129° 7′ E.	Banda Sea.
197.	,,	1200	35·9	bl. m.	D few	,, 14	0° 41′ N., 126° 37′ E.	E. of Celebes.
198.	,,	2150	38·9	bl. m.	C many	,, 20	2° 55′ N., 124° 53′ E.	N. of Celebes.
200.	,,	250	...	gr. m.	B numerous	,, 23	6° 47′ N., 122° 28′ E.	W. of Mindanao.
201.	,,	82	...	st. & gra.	C many	,, 26	7° 3′ N., 121° 48′ E.	W. of Mindanao.
202.	,,	2550	50·5	bl. m.	B numerous	,, 27	8° 32′ N., 121° 55′ E.	W. of Mindanao.
205.	,,	1050	37·0	bl. m.	C many	Nov. 13	16° 42′ N., 119° 22′ E.	W. of Luzon.
						1875.		
206.	,,	2100	36·5	bl. m.	B numerous	Jan. 8	17° 54′ N., 117° 14′ E.	W. of Luzon.
211.	,,	2225	50·5	bl. m.	B ,,	,, 28	8° 0′ N., 121° 42′ E.	W. of Mindanao.
213.	,,	2050	38·8	bl. m.	C many	Feb. 8	5° 47′ N., 124° 1′ E.	S. of Mindanao.
214.	,,	500	41·8	bl. m.	C ,,	,, 10	4° 33′ N., 127° 6′ E.	N. of Gilolo.
215.	Tr. Pac.	2550	35·4	r. cl.	C many	Feb. 12	4° 19′ N., 130° 15′ E.	N. of Gilolo.
216A.	,,	2000	35·4	gl. oz.	B numerous	,, 16	2° 56′ N., 134° 11′ E.	S. of Pelew Islands.
217.	,,	2000	35·2	bl. m.	C many	,, 22	0° 39′ S., 138° 55′ E.	N. of New Guinea.
218.	,,	1070	36·4	bl. m.	C ,,	Mar. 1	2° 33′ S., 144° 4′ E.	N. of New Guinea.
220.	,,	1100	36·2	gl. oz.	C ,,	,, 11	0° 42′ S., 147° 0′ E.	N. of New Guinea.
221.	,,	2650	35·4	r. cl.	B numerous	,, 13	0° 40′ N., 148° 41′ E.	(Ocean).
222.	,,	2450	35·2	r. cl.	B ,,	,, 16	2° 15′ N., 146° 16′ E.	(Ocean).
223.	,,	2325	35·5	gl. oz.	B ,,	,, 19	5° 31′ N., 145° 13′ E.	Carolines.
224.	,,	1850	35·4	gl. oz.	B ,,	,, 21	7° 45′ N., 144° 20′ E.	Carolines.
225.	,,	4475	35·2	rad. oz.	A very many	,, 23	11° 24′ N., 143° 16′ E.	Ocean ⎫
226.	,,	2300	35·5	rad. oz.	A ,,	,, 25	14° 44′ N., 142° 13′ E.	Ocean ⎬ North-West Pacific,
230.	N. Pac.	2425	35·5	r. cl.	C many	April 5	26° 29′ N., 137° 57′ E.	Ocean ⎪ between Carolines
231.	,,	2250	35·2	bl. m.	C ,,	,, 9	31° 8′ N., 137° 8′ E.	Ocean ⎭ and Japan.
232.	,,	345	41·1	gr. m.	C ,,	May 12	35° 11′ N., 139° 28′ E.	Ocean.
234.	,,	2675	35·8	bl. m.	B numerous	June 3	32° 31′ N., 135° 39′ E.	S. of Japan.
235.	,,	565	38·1	bl. m.	D few	,, 4	34° 7′ N., 138° 0′ E.	S. of Japan.
236.	,,	775	37·6	gr. m.	C many	,, 5	34° 58′ N., 139° 29′ E.	S. of Japan.
237.	,,	1875	35·3	bl. m.	C ,,	,, 17	34° 37′ N., 140° 32′ E.	S. of Japan.
238.	,,	3950	35·0	r. cl.	B numerous	,, 18	35° 18′ N., 144° 8′ E.	Ocean ⎫
239.	,,	3625	35·1	r. cl.	B ,,	,, 19	35° 18′ N., 147° 9′ E.	Ocean ⎪
240.	,,	2900	34·9	r. cl.	B ,,	,, 21	35° 20′ N., 153° 39′ E.	Ocean ⎬
241.	,,	2300	35·1	r. cl.	A very many	,, 23	35° 41′ N., 157° 42′ E.	Ocean ⎪
242.	,,	2575	35·1	r. cl.	A11 ,,	,, 24	35° 29′ N., 161° 52′ E.	Ocean ⎭
243.	,,	2800	35·0	r. cl.	A11 ,,	,, 26	35° 24′ N., 166° 35′ E.	Ocean ⎫
244.	,,	2900	35·3	r. cl.	A11 ,,	,, 28	35° 22′ N., 169° 53′ E.	Ocean ⎪ North Pacific, between
245.	,,	2775	34·9	r. cl.	A11 ,,	,, 30	36° 23′ N., 174° 31′ E.	Ocean ⎬ Japan and San Fran-
246.	,,	2050	35·1	gl. oz.	B numerous	July 2	36° 10′ N., 178° 0′ E.	Ocean ⎪ cisco (35°–38° N. lat.,
247.	,,	2530	35·2	r. cl.	C many	,, 3	35° 49′ N., 179° 57′ W.	Ocean ⎭ 144°–156° W. long.).
248.	,,	2900	35·1	r. cl.	C ,,	,, 5	37° 41′ N., 177° 4′ W.	Ocean
249.	,,	3000	35·2	r. cl.	B numerous	,, 7	37° 59′ N., 171° 48′ W.	Ocean
250.	,,	3050	35·0	r. cl.	B ,,	,, 9	37° 49′ N., 166° 47′ W.	Ocean
251.	,,	2950	35·1	r. cl.	B ,,	,, 10	37° 37′ N., 163° 26′ W.	Ocean
252.	,,	2740	35·3	r. cl.	B ,,	,, 12	37° 52′ N., 160° 17′ W.	Ocean
253.	,,	3125	35·1	r. cl.	B ,,	,, 14	38° 9′ N., 156° 25′ W.	Ocean ⎫
254.	,,	3025	35·0	r. cl.	C many	,, 17	35° 13′ N., 154° 43′ W.	Ocean ⎪
255.	,,	2850	35·0	r. cl.	C ,,	,, 19	32° 28′ N., 154° 33′ W.	Ocean ⎬ North Pacific (35°–23°
256.	,,	2950	35·2	r. cl.	B numerous	,, 21	30° 22′ N., 154° 56′ W.	Ocean ⎪ N. lat., 154°–156°
257.	,,	2875	34·9	r. cl.	C many	,, 23	27° 33′ N., 154° 55′ W.	Ocean ⎭ W. long.).
258.	,,	2775	35·2	r. cl.	C ,,	,, 24	26° 11′ N., 155° 12′ W.	Ocean
259.	Tr. Pac.	2225	34·9	r. cl.	C ,,	,, 26	23° 3′ N., 156° 6′ W.	Ocean

Challenger Station.	Locality.	Depth in Fathoms.	Bottom Temperature °F.	Nature of Bottom.	Relative Abundance of Radiolaria.	Date.	Latitude and Longitude.	Nearest Land.
						1875.		
261.	Tr. Pac.	2050	35·2	volc. m.	C many	Aug. 12	20° 18′ N., 157° 14′ W.	Sandwich Islands.
262.	,,	2875	35·2	r. cl.	C ,,	,, 20	19° 12′ N., 154° 14′ W.	Sandwich Islands.
263.	,,	2650	35·1	r. cl.	B numerous	,, 21	17° 33′ N., 153° 36′ W.	Ocean
264.	,,	3000	35·2	r. cl.	C many	,, 23	14° 19′ N., 152° 37′ W.	Ocean
265.	,,	2900	35·0	r. cl.	A very many	,, 25	12° 42′ N., 152° 1′ W.	Ocean
266.	,,	2750	35·1	rad. oz.	A ,,	,, 26	11° 7′ N., 152° 3′ W.	Ocean
267.	,,	2700	35·0	rad. oz.	A ,,	,, 28	9° 28′ N., 150° 49′ W.	Ocean
268.	,,	2900	34·8	rad. oz.	A ,,	,, 30	7° 35′ N., 149° 49′ W.	Ocean — Tropical Central Pacific,
269.	,,	2550	35·2	rad. oz.	A ,,	Sept. 2	5° 54′ N., 147° 2′ W.	Ocean — between Sandwich and
270.	,,	2925	34·6	gl. oz.	A ,,	,, 4	2° 34′ N., 149° 9′ W.	Ocean — Paumotu (17° N. lat. to 11° S. lat.).
271.	,,	2425	35·0	gl. oz.	A ,,	,, 6	0° 33′ S., 151° 34′ W.	Ocean
272.	,,	2600	35·1	rad. oz.	A ,,	,, 8	3° 48′ S., 152° 56′ W.	Ocean
273.	,,	2350	34·5	rad. oz.	A ,,	,, 9	5° 11′ S., 152° 56′ W.	Ocean
274.	,,	2750	35·1	rad. oz.	A ,,	,, 11	7° 25′ S., 152° 15′ W.	Ocean
275.	,,	2610	35·0	r. cl.	B numerous	,, 14	11° 20′ S., 150° 30′ W.	Ocean
276.	,,	2350	35·1	r. cl.	C many	,, 16	13° 28′ S., 149° 30′ W.	Paumotu.
280.	,,	1940	35·3	gl. oz.	D few	Oct. 4	18° 40′ S., 149° 52′ W.	S. of Tahiti.
281.	,,	2385	34·9	r. cl.	C many	,, 6	22° 21′ S., 150° 17′ W.	Tubuai Islands.
282.	S. Pac.	2450	35·1	r. cl.	C ,,	,, 7	23° 46′ S., 149° 59′ W.	Tubuai Islands.
283.	,,	2075	35·4	gl. oz.	D few	,, 9	26° 9′ S., 145° 17′ W.	N. of Oparo Island.
284.	,,	1985	35·1	gl. oz.	C many	,, 11	28° 22′ S., 141° 22′ W.	S. of Oparo Island.
285.	,,	2375	35·0	r. cl.	D few	,, 14	32° 36′ S., 137° 43′ W.	Ocean
286.	,,	2335	34·8	r. cl.	D ,,	,, 16	33° 29′ S., 133° 22′ W.	Ocean
287.	,,	2400	34·7	r. cl.	D ,,	,, 19	36° 32′ S., 132° 52′ W.	Ocean
288.	,,	2600	34·8	r. cl.	B numerous	,, 21	40° 3′ S., 132° 58′ W.	Ocean
289.	,,	2550	34·8	r. cl.	B ,,	,, 23	39° 41′ S., 131° 23′ W.	Ocean
290.	,,	2300	34·9	r. cl.	C many	,, 25	39° 16′ S., 124° 7′ W.	Ocean — Open South Pacific
291.	,,	2250	34·6	r. cl.	C ,,	,, 27	39° 13′ S., 118° 49′ W.	Ocean — Ocean, between New
292.	,,	1600	35·2	gl. oz.	C ,,	,, 29	38° 43′ S., 112° 31′ W.	Ocean — Zealand and Valparaiso.
293.	,,	2025	34·4	gl. oz.	C ,,	Nov. 1	39° 4′ S., 105° 5′ W.	Ocean
294.	,,	2270	34·6	r. cl.	D few	,, 3	39° 22′ S., 98° 46′ W.	Ocean
295.	,,	1500	35·3	gl. oz.	C many	,, 5	38° 7′ S., 94° 4′ W.	Ocean
296.	,,	1825	35·3	gl. oz.	D few	,, 9	38° 6′ S., 88° 2′ W.	Ocean
297.	,,	1775	35·5	gl. oz.	D ,,	,, 11	37° 29′ S., 83° 7′ W.	Ocean
298.	,,	2225	35·6	bl. m.	C many	,, 17	34° 7′ S., 73° 56′ W.	W. of Valparaiso.
299.	,,	2160	35·2	bl. m.	C ,,	Dec. 14	33° 31′ S., 74° 43′ W.	W. of Valparaiso.
300.	,,	1375	35·5	bl. m.	B numerous	,, 17	33° 42′ S., 78° 18′ W.	N. of Juan Fernandez.
302.	,,	1450	35·6	gl. oz.	B ,,	,, 28	42° 43′ S., 82° 11′ W.	(Ocean).
303.	,,	1325	36·0	bl. m.	D few	,, 30	45° 31′ S., 78° 9′ W.	W. of Patagonia.
304.	,,	45	...	gr. s.	E very few	,, 31	46° 53′ S., 75° 12′ W.	W. of Patagonia.
						1876.		
318.	S. Atl.	2040	33·7	bl. m.	C few	Feb. 11	42° 32′ S., 56° 29′ W.	(Ocean).
319.	,,	2425	32·7	bl. m.	C ,,	,, 12	41° 54′ S., 54° 48′ W.	(Ocean).
323.	,,	1900	33·1	bl. m.	C ,,	,, 28	35° 39′ S., 50° 47′ W.	W. of Buenos Ayres.
324.	,,	2800	32·6	bl. m.	B numerous	,, 29	36° 9′ S., 48° 22′ W.	Ocean
325.	,,	2650	32·7	bl. m.	B ,,	Mar. 2	36° 44′ S., 46° 16′ W.	Ocean
326.	,,	2775	32·7	bl. m.	C many	,, 3	37° 3′ S., 44° 17′ W.	Ocean — Open South Atlantic
327.	,,	2900	32·8	bl. m.	C ,,	,, 4	36° 48′ S., 42° 45′ W.	Ocean — Ocean, between Buenos
328.	,,	2900	32·9	bl. m.	B numerous	,, 6	37° 38′ S., 39° 36′ W.	Ocean — Ayres and Tristan
329.	,,	2675	32·3	r. cl.	C many	,, 7	37° 31′ S., 36° 7′ W.	Ocean — da Cunha (35°–37° S.
330.	,,	2440	32·7	r. cl.	C ,,	,, 8	37° 45′ S., 33° 0′ W.	Ocean — lat., 21°–48° W. long.).
331.	,,	1715	35·4	gl. oz.	B numerous	,, 9	37° 47′ S., 30° 20′ W.	Ocean
332.	,,	2200	34·0	gl. oz.	B ,,	,, 10	37° 29′ S., 27° 31′ W.	Ocean
333.	,,	2025	35·3	gl. oz.	B ,,	,, 13	35° 36′ S., 21° 12′ W.	Ocean
334.	,,	1915	35·8	gl. oz.	C many	,, 14	35° 45′ S., 18° 31′ W.	W. of Tristan da Cunha.
335.	,,	1425	37·0	pt. oz.	D few	,, 16	32° 24′ S., 13° 5′ W.	N. of Tristan da Cunha.

REPORT ON THE RADIOLARIA.

Challenger Station.	Locality.	Depth in Fathoms.	Bottom Temperature F.	Nature of Bottom.	Relative Abundance of Radiolaria.	Date.	Latitude and Longitude.	Nearest Land.
						1876.		
338.	Tr. Atl.	1990	36·3	gl. oz.	D few	Mar. 21	21° 15′ S., 14° 2′ W.	(Ocean).
340.	,,	1500	37·6	pt. oz.	E very few	,, 24	14° 33′ S., 13° 42′ W.	Ocean ⎫
341.	,,	1475	38·2	pt. oz.	E ,,	,, 25	12° 16′ S., 13° 44′ W.	Ocean ⎬ W. of St. Helena.
342.	,,	1445	37·5	pt. oz.	D few	,, 26	9° 43′ S., 13° 51′ W.	Ocean ⎭
343.	,,	425	40·3	volc. s.	E very few	,, 27	8° 3′ S., 14° 27′ W.	Ascension Island.
344.	,,	420	...	volc. s.	E ,,	April 3	7° 54′ S., 14° 28′ W.	Ascension Island.
345.	,,	2010	36·8	gl. oz.	D few	,, 4	5° 45′ S., 14° 25′ W.	Ocean ⎫
346.	,,	2350	34·0	gl. oz.	C many	,, 6	2° 42′ S., 14° 41′ W.	Ocean ⎬ Tropical Atlantic,
347.	,,	2250	36·2	gl. oz.	B numerous	,, 7	0° 15′ S., 14° 25′ W.	Ocean ⎬ between Ascension and
348.	,,	(2450)	...	(Pelag.)	B ,,	,, 9	3° 10′ N., 14° 51′ W.	Ocean ⎬ Sierra Leone.
349.	,,	...	...	(Pelag.)	B ,,	,, 10	5° 28′ N., 14° 38′ W.	Ocean ⎭
350.	,,	...	...	(Pelag.)	B ,,	,, 11	7° 33′ N., 15° 16′ W.	W. of Sierra Leone.
351.	,,	...	...	(Pelag.)	B ,,	,, 12	9° 9′ N., 16° 41′ W.	W. of Sierra Leone.
352.	,,	...	...	(Pelag.)	B ,,	,, 13	10° 55′ N., 17° 46′ W.	W. of Sierra Leone.
353.	N. Atl.	2965	37·6	r. cl.	C many	May 3	26° 21′ N., 33° 37′ W.	W. of Canary Islands.
354.	,,	1675	37·8	gl. oz.	D few	,, 6	32° 41′ N., 36° 6′ W.	S. of Azores.

CHAPTER X.—GEOLOGICAL DISTRIBUTION.

(§§ 241-250.)

241. *Historical Distribution.*—Radiolaria are found fossil in all the more important groups of the sedimentary rocks of the earth's crust. Whilst a few years ago their well-preserved siliceous skeletons were only known in considerable quantity from Cainozoic marls (§ 242), very many SPUMELLARIA and NASSELLARIA have recently been found in Mesozoic and a few in Palæozoic strata. By the aid of improved modern methods of investigation (especially by the preparation of thin sections of very hard rocks) it has been shown that many hard siliceous minerals, especially cryptocrystalline quartz, contain numerous well-preserved Radiolaria, and sometimes are mainly composed of closely compacted masses of such siliceous shells; of this kind are many quartzites of the Jura (§ 243). These Jurassic quartzes (Switzerland), as well as the Tertiary marls (Barbados) and clays (Nicobar Islands), are to be regarded as "fossil Radiolarian ooze" (§ 237). Dense masses of compressed SPUMELLARIA and NASSELLARIA form the principal part of these rocks. Isolated or in smaller quantities, fossil Polycystina, belonging to different families of SPUMELLARIA and NASSELLARIA, also occur in other rocks, and even in some of Palæozoic origin. Since specimens have also been recently found both in Silurian and Cambrian strata, it may be stated that as regards their historical distribution, Radiolaria occur in all fossiliferous sedimentary deposits, from the oldest to those of the present time.

242. *Cainozoic Radiolaria.*—The great majority of fossil Radiolaria which, have hitherto been described, belong to the Cainozoic or Tertiary period, and in fact, to its middle portion, the Miocene period. At this period the richest and most important of all the Radiolarian formations were deposited, such as the pure "Polycystine marl" of Barbados (see note A), also that of Grotte in Sicily (see note B), and the clay of the Nicobar Islands (see note C). Besides the above-mentioned deposits, which may be designated "pure" fossil Radiolarian ooze, many deposits containing these organisms have recently been discovered in widely separated parts of the earth, partly of the nature of tripoli or marl, partly resembling clay. Among these may be mentioned in the first place many coasts and islands of the Mediterranean, both on the south coast of Europe (Sicily, Calabria, Greece), and the north coast of Africa (from Oran to Tripoli). The extensive layers of tripoli which are found in these Mediterranean Tertiary mountains belong to the upper Miocene (Tortona stage), and consist partly of marl rich in calcareous matter, and resembling chalk, partly passing over into plastic clay or "Kieselguhr" (§ 246). The quantity of Radiolaria contained varies, and is more conspicuous the fewer the calcareous shells of Foraminifera present. Similar Tertiary Polycystine formations occur in some parts of America (see note D); probably they have a very wide distribution. In their general morphological characters, the Tertiary SPUMMELLARIA and NASSELLARIA

are related to those forms which are found in the recent Radiolarian ooze of the depths of the Pacific, especially to the species which are characteristic of the Challenger Stations 225, 226, 265 and 268. Many living genera and families (e.g., most Larcoidea and Stephoidea) have not yet been found in the Tertiary formations.

A. The famous Polycystine marl of Barbados in the Antilles, which Robert Schomburgk discovered forty years ago, belongs to the Miocene formation, and is the richest and best known of all the important Radiolarian deposits (see L. N. 16, pp. 5–8). After Ehrenberg had published in December 1846 the first preliminary communication regarding its composition out of masses of well-preserved Polycystina, he was able in the following year to describe no less than 282 species from it; he distributed these in 44 genera and 7 families (L. N. 4, 1847, p. 54). In the year 1854 Ehrenberg published figures of 33 species in his Mikrogeologie (L. N. 6, Taf. xxxvi.); but it was only in 1873 that he published descriptions of 265 species (Monatsber. d. k. preuss. Akad. d. Wiss. Berlin, Jan. 30, pp. 213–263). Finally there followed in 1875 his Fortsetzung der Mikrogeologischen Studien, mit specieller Rücksicht auf den Polycystinen-Mergel von Barbados (L. N. 25). On the thirty plates which accompany this the last work of Ehrenberg, 282 species are figured and named, of which 54 are SPUMELLARIA (13 Sphæroidea, 8 Prunoidea, 33 Discoidea), and 228 NASSELLARIA (2 Stephoidea, 38 Spyroidea, and 188 Cyrtoidea). The fourth section of this memoir contains a survey of the Polycystine formation of Barbados (pp. 106–115), and the fifth section the special description of a large specimen of rock from Mount Hillaby in Barbados (see also L. N. 28, p. 117, and L. N. 41, pp. 476–478). The account given by Ehrenberg of the Polycystina of Barbados is in many respects very incomplete, and very far from exhausting this rich mine of remarkable forms. This may be readily seen from the twenty-five plates of figures of Polycystins in the Barbados Chalk Deposit published by Bury in 1862 (L. N. 17). The number of species here figured (140 to 142) is about half of those given by Ehrenberg; and there are among them numerous generic types, some of great interest, which were entirely overlooked by the latter; e.g. Saturnalis (Sphæroidea), Cannartidium (Prunoidea), Tympanidium (Stephoidea), Cinclopyramis (Cyrtoidea), &c. Finally, Ehrenberg always (until 1875) ignored Bury's atlas, which had been published thirteen years ago and was quite accessible to him. How different were the contents of the two works may easily be seen from the following abstract.

Comparative View of the Species of Fossil Radiolaria from Barbados made known by the figures of Bury in 1862 and of Ehrenberg in 1875.

Legion.	Order.	Bury.	Ehrenberg.	Total.
I. Legion SPUMELLARIA (PERIPYLEA).	1. Sphæroidea	16	13	29
	2. Prunoidea	10	8	18
	3. Discoidea	37	33	70
II. Legion NASSELLARIA (MONOPYLEA).	4. Stephoidea	5	2	7
	5. Spyroidea	13	38	51
	6. Cyrtoidea	60	188	248
	Total,	141	282	423

In 1882 Bütschli still further increased the number of known Radiolaria from Barbados both by figures and descriptions (L. N. 40), and gave in particular a very accurate morphological analysis of 12 new NASSELLARIA (3 Stephoidea, 3 Spyroidea, and 6 Cyrtoidea; L. N. 40 Taf. xxxii., xxxiii.). The number of the fossil species collected in the Barbados marl is, however, greater than would appear from the above-quoted communications. My respected friend, Dr. R. Teuscher, of Jena, has, at my request, made a large number (about a thousand) of very accurate drawings with the camera lucida of Polycystina from Barbados (see p. 1760). From these it appears that the variations in the structure of the shells, with respect to number, size, and form of the lattice-pores, of the spines, &c., is much greater than would be supposed from the figures of Ehrenberg and Bury. I have thus come to the conviction that the number of species from Barbados (using the word "species" in the sense understood by those authors) is not less than 400 and probably more than 500. Descriptions of some particularly interesting new species from this series have been included in the systematic account of the Challenger Radiolaria. A complete critical investigation of the Radiolaria of Barbados, and especially an accurate comparison of these Cainozoic species with the Mesozoic forms from the Jura, on the one hand, and with recent types on the other, must be left to the future for its accomplishment (see § 246).

B. The Cainozoic Polycystine tripoli or marl of the Mediterranean coast, which is probably always of Miocene origin, forms very extensive mountain ranges both in the south of Europe (Sicily, Calabria, Greece) and in the north of Africa (from Oran to Tripoli) (§ 246). Hitherto, however, only one locality has been thoroughly investigated, namely, Grotte in the province of Girgenti in Sicily (L. N. 35). In the accurate account which was given of it by Stöhr in 1880, 118 species were described, distributed in 40 genera (L. N. 35; pp. 72–84); of these 118 species 78 are quite new, 25 are identical with previously known fossils, and 29 identical with living forms. Among them are 73 SPUMELLARIA (28 Sphæroidea, 8 Prunoidea, and 37 Discoidea), but only 40 NASSELLARIA (1 Stephoidea, 6 Spyroidea, and 33 Cyrtoidea), and 5 PHÆODARIA (Dictyochida). The other parts of Sicily from which the same upper Miocene tripoli has been investigated (belonging to the Tortona stage) have proved less rich than Grotte. The best known of these places is Caltanisetta, since upon three genera discovered here (*Haliomma, Cornutella, Lithocampe*) the group Polycystina was founded by Ehrenberg in 1838 (see L. N. 16, p. 3). Afterwards 31 species were described from this locality, of which 23 were again found in Grotte. The richest deposit on the Mediterranean coast, however, appears to be at Oran. A small specimen of the Kieselguhr found there, which was recently sent to me by Professor Steinman, proved to be pure Radiolarian ooze, very similar to that now found in the Central Pacific, and contained many hitherto undescribed species; it is deserving of careful investigation and comparison.

C. Regarding the Tertiary Radiolarian clay of the Nicobar Islands, see § 247 and L. N. 25, pp. 116–120. Its fauna is incompletely known; probably it is of Miocene or Oligocene origin.

D. Cainozoic tripoli, containing larger or smaller quantities of Radiolaria, appears to be rather widely distributed in America. Ehrenberg has described such from South America (polishing-slate from Morro di Mijellones, on the coast between Chili and Bolivia), and from North America (Richmond and Petersburg in Virginia, Piscataway in Maryland). Similar deposits are also found in the Bermuda Islands (L. N. 4, 1855–56; L. N. 6, Taf. 18; L. N. 16, pp. 3–9; L. N. 41, pp. 475–478, and L. N. 25, pp. 2–6).

243. *Mesozoic Radiolaria.*—From the Mesozoic or Secondary period numerous well-preserved Radiolaria have recently been described. They belong for the most part to the Jurassic formation (see notes A, B, C), whilst the more recent Chalk (see note D) and the older Trias (see note E) have hitherto yielded but few species. All the main divisions of the Jura, both the upper (Malm) and the middle (Dogger), and especially the lower (Lias) appear in certain localities to be very rich in well-preserved shells of fossil Polycystina. Most of these are aggregated together in coprolites and quartzites (jasper, chert, flint, &c., § 248). The majority are Cyrtoidea, the minority Sphæroidea and Discoidea in almost equal proportions; a few Beloidea (*Sphærozoum*) and Phæocystina (Dictyocha) are also found among them. The general morphological character of these Jurassic Radiolaria is very different from that of the nearly related Tertiary and living forms. In general, their siliceous shells are firmer and more massive, usually also somewhat larger, but of simpler structure. The manifold delicate appendages (spines, bristles, feet, wings, &c.) which are so richly developed in the living Spumellaria and Nassellaria, and are also well shown in the Tertiary species, are entirely wanting in the majority of the Jurassic Polycystina. The Sphæroidea and Prunoidea are all simple spherical or ellipsoidal lattice-shells (Monosphærida); concentric lattice-shells (Polysphærida) are entirely wanting. The Cyrtoidea are, for the most part, devoid of radial processes or basal feet (Eradiata); triradiate and multiradiate forms, such as are found abundantly in the recent and Tertiary formations, are very rare. The large number of many-jointed forms (Stichocyrtida) and of Cyrtoidea with latticed basal opening is very striking.

A. The most important work on the Jurassic Radiolaria, regarding which but little was known prior to the year 1885, is the valuable and in some respects very interesting Beiträge zur Kenntniss der fossilen Radiolarien aus Gesteinen des Jura, by Dr. Rüst of Freiburg i. B. (1885, Palæontographica, Bd. xxxi. 51 pp. with 12 plates). Unfortunately this important work was issued only when about half of the present Report was printed off, so that it was no longer possible to include the 234 species there described in its systematic part. I have therefore elsewhere given a list of the Jurassic Radiolaria, and at present only make the following remarks:—Of the 234 species described, the larger half (130) belong to the Nassellaria (Cyrtoidea), the smaller half (102) to the Spumellaria (38 Sphæroidea, 14 Prunoidea, and 50 Discoidea). In addition, there are 2 Phæodaria depicted, and several spicules which are probably to be referred to the Beloidea. Among the 130 Cyrtoidea (of which 2 are described as Botryodea), there are 24 Monocyrtida, 14 Dicyrtida, 22 Tricyrtida, and 70 Stichocyrtida. Just as striking as the predominant number of the last is the fact that there are only very few triradiate (9) and multiradiate (4) species found among these 130 Cyrtoidea, as also the large number of species with latticed basal opening; Stephoidea appear to be entirely wanting. The rich material of jasper, chert, flint, and coprolites in which Dr. Rüst found these Radiolaria, is derived for the most part from the Jurassic rocks of Germany (Hanover, South Bavaria), Tyrol, and Switzerland (compare § 248).

B. Jurassic Radiolaria from Italy, also found in jasper, which are closely related to the forms from Germany and Switzerland described by Dr. Rüst, were made known so long ago as 1880 by Dante Pantanelli in his treatise I Diaspri della Toscana e i loro Fossili (Rome, 1880, 33 pp. 60 figs.). Pantanelli believes, however, that this jasper is for the most part of Eocene origin; but from his description, and especially from the morphological character of the forms which he figures, it appears very probable "that these Tuscan jaspers from Galestro, like those of the Swiss conglomerates, are found in a secondary locality and belong to the Jurassic period" (Rüst, L. N. 51, p. 3). Unfortunately the figures of Pantanelli are so small and incomplete that a reliable determination of the species is hardly possible; for example, the lattice-work is only given in ten of the sixty figures. Among the 32 recorded species 15 are SPUMELLARIA (6 Sphæroidea and 9 Discoidea) and 17 NASSELLARIA (4 Stephoidea and 13 Cyrtoidea); many of which seem to be identical with the forms more accurately described by Dr. Rüst (compare p. 1762).

C. From the Lias of the Alps and more particularly "from the lower Liassic beds of the Schafberg near Salzburg," Dr. Emil von Dunikowski in 1882 described 18 species of fossil Radiolaria (L. N. 44, pp. 22–34, Taf. iv.–vi.); most of these are Sphæroidea and Discoidea and appear to have been more or less altered by petrological changes; their spongy structure is probably secondary.

D. Cretaceous Radiolaria have been hitherto described only in very small numbers; quite recently Dr. Rüst has found a larger number chiefly in flints from the English chalk, but they have not yet been published. In 1876 Zittel described 6 very well-preserved species from the upper chalk of North Germany (L. N. 29, pp. 76–96, Taf. ii.); among them were 1 Sphæroidea, 1 Discoidea, 1 Dictyocha, and 3 Cyrtoidea.

E. Triassic Radiolaria have recently been discovered by Dr. Rüst in chert, but have not yet been described.

244. *Palæozoic Radiolaria.*—The number of Radiolaria which are known from the Palæozoic or Primary formations is much less than from either the Mesozoic or Cainozoic periods. Here, however, the investigations of recent times have yielded important information; a few species, at all events, of Polycystina (mostly Sphæroidea) are now known from various Palæozoic formations, and not only from the Permian ("Zechstein") and the Coal-measures, but also from the older Devonian and Silurian systems. Even in the still older Cambrian rocks a few fossil Radiolaria have been found. All these Palæozoic Radiolaria are Polycystina of very simple form and primitive structure, mostly simple SPUMELLARIA (latticed spheres, ellipsoids, lenses, &c.), but partly also simple NASSELLARIA.

The important discoveries which have recently been made by Dr. Rüst regarding the occurrence of Radiolaria in all the Palæozoic formations have not yet been published. From conversations with this estimable palæontologist I have learned, however, that he has pursued his fruitful investigation of the Mesozoic quartzites (§ 243), and has met with no less success in the case of similar Palæozoic structures. Although the number of species hitherto discovered is relatively small, the important conclusion appears to be warranted that they extend as far as the Silurian and Cambrian systems. All these very ancient SPUMELLARIA (Sphæroidea) and NASSELLARIA (Cyrtoidea) exhibit very primitive structural relations. The occurrence of fossil Polycystina in the Carboniferous formation of England has been incidentally mentioned by W. J. Sollas:—" In the carboniferous beds of North Wales pseudomorphs of Radiolaria in calcite occur, along with minute quartz crystals" (Ann. and Mag. Nat. Hist., 1880, ser. 5, vol. vi. p. 439); and in the siliceous slate-beds of Saxony Rothpletz has shown the existence of a few Sphæroidea (Zeitschr. d. Deutsch. Geol. Gesellsch., 1800, p. 447).

245. *Abundance of Radiolaria in the Various Rocks.*—The relative quantity of well-preserved or at all events recognisable Radiolaria in the different rocks is very variable. In this respect three different degrees may be distinguished, which may be called shortly "pure, mixed, and poor" Radiolarian formations. The *pure* Radiolarian rocks consist for the greater part (usually much more than half, sometimes even more than three-quarters) of closely compacted often calcined masses of siliceous Polycystine shells. To this category belong the pure Miocene Polycystine marls of Barbados (§ 246), the Tertiary Polycystine clay of the Nicobar Islands (§ 247), and the Polycystine quartz of the Jura (§ 248). All these pure Radiolarian rocks may be regarded as fossil Radiolarian ooze (§ 237), and are certainly of deep-sea origin, having probably been deposited at depths greater than 2000 fathoms. Their palæontological character also is in favour of this view, for the abyssal Osculosa (§ 235) are more abundant and richer in species than the pelagic Porulosa (§ 233). The elevation of this deep-sea layer above the surface of the sea appears to have taken place but seldom; it has only been observed on a large scale at Barbados and in the Nicobar Islands. The *mixed* Radiolarian rocks are much more common; they were probably deposited at much less depths, or perhaps are not true deep-sea formations at all. The siliceous shells of Polycystina always constitute less than half (sometimes less than one-tenth) of their mass, and are less prominent than other siliceous remains (Diatoms), or calcareous remains (Foraminifera), or in some cases than the mineral constituents (pumice, &c.). To this group belong many of the above-mentioned Tertiary marls and clays (especially the Mediterranean Tripoli), also many flints, cherts, and other quartzites from Mesozoic strata (especially from the Jura), and probably also some palæozoic quartzites. The marine ooze from which they have originated may have been deposited at very various, even at slight, depths of the ocean. Formations *poor* in Radiolaria, which contain only a few species of SPUMELLARIA and NASSELLARIA mingled with other fossil remains and mineral particles, occur in all formations and are probably very widely distributed. Further careful examination of thin sections (especially of coprolites) will yield here a rich harvest of new forms. Both the mixed and the pure Radiolarian formations may be divided according to their petrographic characters into three groups, which, however, are connected by intermediate varieties— (1) soft, chalky marl (§ 246), (2) plastic clay (§ 247), and (3) hard, flinty quartz (§ 248).

246. *Radiolarian Marl.*—Those soft, friable rocks, which contain a large quantity of calcareous matter, but consist for the most part of the shells of SPUMELLARIA and NASSELLARIA, are called Radiolarian or Polycystine marl, often more correctly Polycystine tripoli; the best known example of them is the chalky marl of Barbados in the Antilles (§ 242). The Tertiary mountain system of this island, which in Mount Hillaby rises to a height of 1147 feet and includes about 15,800 acres, consists almost exclusively of these remarkable masses of rock. Most of it appears as a soft, earthy, often chalky marl, with a considerable but variable amount of calcareous matter. Those specimens, the greater half of which is composed of well-preserved siliceous shells of Polycystina, and which contain little lime, approach the tripoli and "Kieselguhr." Those specimens, however, which contain the largest amount of calcareous matter resemble common writing chalk in consistency, and consist for the most part of shells of Foraminifera and their fragments; of these there are only few species but large numbers of individuals, generally in small fragments with a fine calcareous powder between them. They may be regarded as fossil Globigerina ooze (§ 238). In a third group of specimens from Barbados the quantity of fragments of pumice and other volcanic matters predominates; the amount of clay is also very considerable; these deposits pass over partly into actual clay partly into volcanic tuff. A fourth group exhibits relations to a coarser, often ferruginous material, and although the shells of Polycystina are less abundant in it, still it may be shown to be composed largely of fragments and metamorphosed remains of them. The colour of this deposit, which in some places passes over into sandstone, in others into clay, is usually rather dark, grey, brown, sometimes red and occasionally black (bituminous). The Radiolarian marls of the first two groups, which sometimes approach the white chalk, sometimes the Kieselguhr, are grey, or even pure white (see note A). The same constitution is exhibited by the yellowish or white, very light and friable Polycystine marls of Sicily, which in Caltanisetta approach the chalk, and in Grotte the Kieselguhr. In Greece (Ægina, Zante, &c), on the other hand, they pass over into plastic clay, and the same occurs in the Baden marl of the Vienna basin. In North Africa, however, on the Mediterranean shores of which the Radiolarian marl seems to be very widely distributed (from Tripoli to Oran), it sometimes becomes changed into actual firm polishing slate, sometimes into pulverulent Kieselguhr or tripoli (Terra tripolitana, see note B). Most of these Radiolarian marls appear to date from the middle Tertiary (Miocene) period, and to be deep-sea formations.

A. The Polycystine marl of Barbados appears at different parts of the island to present greater variations in its petrographical and zoographical composition than would appear from Ehrenberg's description (1875, L. N. 25, pp. 106–116). Through the kindness of one of my former students, Dr. Dorner, to whom I take this opportunity of expressing my thanks for the favour, I received a large number of specimens of Barbados rock, taken from various parts of the island, and they exhibit very great variations in their external appearance, their chemical composition, and the

Radiolaria which they contain. The white specimens resembling Kieselguhr contained approximately 60 to 70 per cent. by volume of Radiolarian shells, the yellowish marl 40 to 50 per cent., and the brown and black (bituminous) marl 10 to 20 per cent. or less. Two analyses of the first, which my friend Dr. W. Weber was good enough to carry out, yielded different results from those which are given by Ehrenberg on the basis of Rammelsberg's analyses (L. N. 25, p. 116). The results of both are here given for comparison.

Ehrenberg-Rammelsberg (Fragment from Hillaby).		Weber I. (Chalk-like Fragment).		Weber II. (Tripoli-like Fragment).
Silicate of alumina,	59·47	Silica,	52·2	71·3
Alumina and oxide of iron,	1·95	Alumina (with traces of oxide of iron),	12·3	11·2
Calcium carbonate,	34·31	Lime and magnesia,	31·9	14·8
Water,	3·67	Carbon dioxide	3·2	2·7
Total,	99·40	Total,	99·6	100·0

For further comparison I here add the three different analyses of Miocene Tripoli-marls from Sicily, given by Stöhr on the authority of Fremy, Schwager, and Mottura (Tagebl. d. fünfzigsten Versamml. Deutsch. Naturf. u. Aertzte in München, 1877, p. 163).

Composition.	Tripoli from Licata (Fremy).	Tripoli from Grotte (Schwager).	Tripoli from Caltanisetta (Mottura).
Silica,	30·98	58·58	68·6
Alumina,	17·54	11·51	3·6
Oxide of iron,	0·33	1·84	
Lime,	38·09	8·49	12·1
Magnesia,		0·41	
Water and Organic matter,	13·06	11·26	15·2
Carbonic acid,		7·12	
	100·00	99·21	99·5

B. The Radiolarian marl of the Mediterranean appears, judging by the accounts already published, to stretch along a considerable part of the coast in the earlier and middle Tertiary formations; thus it occurs of similar composition in widely separated localities, in Sicily, Calabria, Zante, and Greece; in North Africa from Tripoli to Oran and probably much farther. So long ago as 1854 Ehrenberg, in his Mikrogeologie (L. N. 6) gave a series of important, even if incomplete, communications regarding the "chalky white calcareous marl of Caltanisetta" (Taf. xxii.), the "Platten marl of Zante" (Taf. xx.), the "plastic clay of Ægina" (Taf. xix.), and the "polishing slate of Oran" (Taf. xxi.). In 1880 Stöhr showed in his fundamental description of the Tripoli from

Grotte in Sicily (L. N. 35) that its Radiolarian fauna is much richer than Ehrenberg supposed. The same is the case in the Tripoli of Caltanisetta, and also in the Baden marl of the Vienna basin. The richest deposit appears to be the pure Kieselguhr-like Tripoli from Oran; a small specimen, which was recently sent to me by Professor Steinmann of Freiburg, i. B., contained many hitherto undescribed species, and was at least as rich as the purest Barbados marl.

247. *Radiolarian Clays.*—Among the Radiolarian or Polycystine clays we include the firm, often plastic, formations, which contain a larger proportion of Radiolaria than of other organic remains. The first of these to be mentioned is the Cainozoic formation of the Nicobar Islands in Further India, which rises to a height of 2000 feet above the level of the sea, and consists for the most part of coloured masses of clay of varying constitution; on Car Nicobar these are mostly grey or reddish, on the Island of Camorta they are partly strongly ferruginous and red and yellow (*e.g.* at Frederickshaven), partly white and light, like meerschaum (*e.g.* at Mongkata). The latter varieties appear to pass over into pure loose Polycystine marl like that of Barbados, the former into calcareous sandstone. Although the Polycystine clays of the Nicobar Islands are as yet only very incompletely known, it may be concluded with great probability that they are true deep-sea formations and nearly allied to those recent forms of red clay, which by their abundance in Radiolaria most nearly approach the Radiolarian ooze, such for example as the red clay of the North Pacific between Japan and the Sandwich Islands (Stations 241 to 245, compare §§ 229 and 239). With this view agrees also the greater or less quantity of pumice dust and other volcanic products. Probably Radiolarian clays like those of the Nicobar Islands occur also in other Tertiary rocks; part of the Barbados marl passes by gradually increasing content of clay into such; and in this case also the amount of included pumice is often considerable. Many mixed Radiolarian marls of the Mediterranean (*e.g.*, of Greece and Oran) also appear to pass over at certain points into Radiolarian clay.

The Radiolarian clays of the Nicobar Islands are unfortunately very incompletely known both as regards their geological nature and their palæontological composition. The communications of Rink (Die Nikobaren-Inseln, eine geographische Skizze, Kopenhagen, 1847) and of Ehrenberg (L. N. 6, p. 160 and L. N. 25, pp. 116 to 120) leave many important questions unanswered. The latter has only figured twenty-three species in his Mikrogeologie (L. N. 6, Taf. xxxvi.). In his tabular list of names (L. N. 25, p. 120) he only incompletely records thirty-nine species, although in 1850, immediately after the first examination of the Nicobar clay, he had distinguished "more than a hundred species, partly new, partly identical with those of Barbados" (L. N. 16, p. 8). I have unfortunately been unable in spite of many efforts, to obtain for investigation a specimen of Nicobar clay. The only microscopical preparation (from Ehrenberg's collection), which I was able to examine, contained several hitherto undescribed species. A thorough systematic examination of these important Radiolarian clays is a pressing necessity, especially as they seem to be markedly different from those of the Mediterranean (from Ægina, Zante, &c.).

248. *Radiolarian Quartzes.*—Under the name Radiolarian or Polycystine quartzes are included those hard, siliceous rocks, which consist for the most part of the closely compacted shells of SPUMELLARIA and NASSELLARIA. To these "cryptocrystalline quartzes," or better, quartzites, belong more especially the pure Radiolarian formations of the Jura, which have been described as flint, chert, jasper, as well as other cryptocrystalline quartzites. Most of the rocks of this nature hitherto examined are from Germany (Hanover, South Bavaria), Hungary, Tyrol, and Switzerland; others are known from Italy (Tuscany). They occur both in the upper and middle, but especially in the lower Jurassic formation (also in the lower layers of the Alpine Lias). A small part of them has been examined in their primary situation (the red jaspers of Allgäu and Tyrol), the greater part, however, only as loose rolled stones in secondary situations (thus in Switzerland in the breccia of the Rigi, in the conglomerate of the Uetli-Berg, and in many boulders of the Rhine, the Limmat, the Reuss, and the Aar). The greatest abundance, however, of Jurassic Radiolaria has been yielded by the silicified coprolites from the Lias of Hanover. These "Radiolarian coprolites" are roundish or cylindrical bodies, which may attain the size of a goose-egg; they probably originated from Fish or Cephalopods, which had fed upon Crustacea, Pteropoda, and similar pelagic organisms, whose stomachs were already full of Radiolarian skeletons. Next to the coprolites the richest is the red jasper, whose colour varies from bright to dark red; it constitutes a true "silicified deep-sea Radiolarian ooze." The "*Aptychus* beds" also of South Bavaria and Tyrol are very rich, and have furnished about one-third of all the Radiolaria known from the Jura; most of the species too are very well preserved (compare § 243).

Regarding the remarkable composition and manifold varieties of the Jurassic Radiolarian quartz, the very full treatise of Dr. Rüst may be consulted (L. N. 51). The very interesting Radiolarian coprolites, which that author has discovered in the lower and middle Jura of Hanover, occur in astonishing numbers in the iron mines at the village of Gross-Ilsede, four and a half miles south of the town of Peine. They constitute from 2 to 5 per cent. by weight of the Liassic iron ore; of this latter, in the year 1883 alone, not less than two hundred and eighty million kilograms were excavated. It is very probable that the careful microscopic examination of thin sections of coprolites, as well as of flints, chert, jasper, and other quartzites, would yield a rich harvest of fossil Radiolaria in other formations also. In Italy Dante Pantanelli has discovered interesting Polycystine jaspers in Tuscany (L. N. 36, 45); these also appear to occur in the Jura (compare § 243, and L. N. 51, pp. 3–10).

249. *Fossil Groups.*—The preservation of Radiolaria in the fossil state is, of course, primarily dependent on the composition of their skeleton. Hence the ACANTHARIA, whose acanthin skeleton although firm is readily soluble, are never found fossil. The same is true of the skeletons of the PHÆODARIA, which consist of a silicate of carbon; here, however, a single exception is found in the Dictyochida, a subfamily of the Cannorrhapida, the isolated parts of whose skeletons appear to consist of pure silica, and

are often found fossil. Of the two other legions those families which possess no skeleton are of course excluded; the Nassellida among the NASSELLARIA, and the Thalassicollida and Collozoida among the SPUMELLARIA. Thus of the 85 known families there remain scarcely 55 of which the skeletons may be expected in the fossil state; and of these scarcely half have been actually observed in this condition. Of the 20 orders of this class enumerated in § 155, the following 9 may be, for palæontological and geological purposes, completely excluded:—(A) The 4 orders of ACANTHARIA (1, Actinelida 2, Acanthonida; 3, Sphærophracta; 4, Prunophracta); (B) 3 orders of PHÆODARIA (5, Phæosphæria; 6, Phæogromia; 7, Phæoconchia); (C) order of NASSELLARIA (8, Nassoidea); (D) 1 order of SPUMELLARIA (9, Colloidea). From a geological point of view the following 6 orders, although occasionally found fossil are of quite subordinate importance:—(A) Among the SPUMELLARIA (10, Beloidea and 11, Larcoidea); (B) among the NASSELLARIA (12, Plectoidea; 13, Stephoidea; 14, Botryodea); (C) among the PHÆODARIA (15, the Phæocystina). On the other hand the following 5 orders, which are the main constituents of Radiolarian rocks, are of pre-eminent geological importance:—(A) Among the SPUMELLARIA (16, Sphæroidea; 17, Prunoidea; 18, Discoidea); (B) among the NASSELLARIA (19, Spyroidea, and 20, Cyrtoidea). The numerical relation in which the different families of these orders appear in the Radiolarian formations may be seen on consulting § 157.

250. *Fossil and Recent Species.*—The fact that there are many Radiolaria living at the present day, whose shells are found fossil in Tertiary rocks, is of great phylogenetic and geological significance. This appeared to be the case even from the older observations upon the Polycystina of the Barbados marl (see note A), but more recent and extensive observations both upon these and upon the Miocene Radiolaria of Sicily, have shown that the number of these "living fossil" forms is much greater than was previously supposed (see note B). Among the Miocene Radiolaria numerous species, both SPUMELLARIA (especially Sphæroidea and Discoidea) and of NASSELLARIA (especially Spyroidea and Cyrtoidea) are not to be distinguished from the corresponding still living forms (see notes C, D). On the other hand, those genera, which are rich both in species and individuals (recent as well as fossil), present continuous series of forms which lead gradually and uninterruptedly from old Tertiary species to others still living, which are specifically indistinguishable from them. These interesting morphological facts are capable of direct phylogenetic application, and furnish valuable proof of the truth of the theory of descent.

A. Ehrenberg, in his list of fossil Polycystina (L. N. 25, pp. 64–85, 1875), records 325 species of which 26 are still living.

B. Stöhr, in his list of Miocene Radiolaria from Grotte (L. N. 35, p. 84, 1880), records 118 species, of which 29 are still living.

C. Teuscher, who at my request has made a large number of comparative measurements and drawings, both of fossil and living Radiolaria, comes to the conclusion that numerous SPUMELLARIA and NASSELLARIA from Barbados are to-day extant and unchanged in the Radiolarian ooze of the deep Pacific Ocean (compare § 242A, and p. 1760, Note).

D. From the comparative investigations, which I have made during the last ten years into the recent deep-sea Radiolaria of the Challenger collection and the Miocene Polycystina of Barbados, it appears that about a quarter of the latter are identical with living species of the former.

BIBLIOGRAPHICAL SECTION.

CHAPTER XI.—LITERATURE AND HISTORY.

251. *List of Publications from 1834 to 1884* :—

Note.—In the text the references to the following publications are indicated by the letters L. N.

1. 1834. MEYEN, F., Palmellaria (Physematium, Sphærozoum), in Beiträge zur Zoologie, gesammelt auf einer Reise um die Erde. *Nova Acta Acad. Cæs. Leop.-Carol.*, vol. xvi., Suppl., p. 160, Taf. xxviii. figs. 1–7.
2. 1838. EHRENBERG, G., Polycystina (Lithocampe, Cornutella, Haliomma) in Ueber die Bildung der Kreidefelsen und des Kreidemergels durch unsichtbare Organismen. *Abhandl. d. k. Akad. d. Wiss. Berlin*, p. 117.
3. 1839. EHRENBERG, G., Ueber noch jetzt lebende Thierarten der Kreidebildung (Haliomma radians). *Abhandl. d. k. Akad. d. Wiss. Berlin*, p. 154.
4. 1844–1873. EHRENBERG, G. Vorläufige Mittheilungen über Beobachtungen von Polycystinen. *Monatsber. d. k. preuss. Akad. d. Wiss. Berlin.* Republished with illustrations in the Mikrogeologie (L. N. 6) and in the two treatises of 1872 (L. N. 24) and 1875 (L. N. 25). Compare the *Monatsberichte* of 1844 (pp. 57, 182, 257), of 1846 (p. 382), of 1847 (p. 40), of 1850 (p. 476), of 1854 (pp. 54, 205, 236), of 1855 (pp. 292, 305), of 1856 (pp. 197, 425), of 1857 (p. 142, 538), of 1858 (pp. 12, 30), of 1859 (p. 569), of 1860 (pp. 765, 819), of 1861 (p. 222), of 1869 (p. 253), of 1872 (pp. 300–321), of 1873 (pp. 214–263). Only one of these small papers is of permanent value, The First Systematic Arrangement of the Polycystina in 7 families, 44 genera, and 28 species (*Monatsber. d. k. preuss. Akad. d. Wiss. Berlin*, 1847, p. 54). Compare my Monograph (1862, L. N. 16), pp. 3–12, 214–219.
5. 1851. HUXLEY, TH., Upon Thalassicolla, a new Zoophyte. *Ann. and Mag. Nat. Hist.*, ser. 2, vol. vii. pp. 433–442, pl. xvi.
6. 1854. EHRENBERG, G., Mikrogeologie. Figures of numerous Polycystina on 8 plates ; Taf. xviii. figs. 1–111 ; Taf. xix. figs. 48–56, 60–62 ; Taf. xx. Nr. i., figs. 20–25, 42 ; Taf. xxi. figs. 51–56 ; Taf. xxii. figs. 20–40 ; Taf. xxxv. A., Nr. xix. A. fig. 5 ; Taf. xxxv. B. figs. 16–23 : Taf. xxxv. figs. 1–33.
7. 1855. BAILEY, J. W., Notice of Microscopic Forms of the Sea of Kamtschatka. *Amer. Journ. Sci. and Arts*, vol. xxii. p. 1, pl. i.
8. 1855. MÜLLER, JOHANNES, Ueber Sphærozoum und Thalassicolla. *Monatsber. d. k. preuss. Akad. d. Wiss. Berlin*, p. 229.
9. 1855. MÜLLER, JOHANNES, Ueber die im Hafen von Messina beobachteten Polycystinen (Haliomma, Eucyrtidium, Dictyospyris, Podocyrtis). *Monatsber. d. k. preuss. Akad. d. Wiss. Berlin*, p. 671.
10. 1856. MÜLLER, JOHANNES, Ueber die Thalassicollen, Polycystinen und Acanthometren des Mittelmeeres. *Monatsber. d. k. preuss. Akad. d. Wiss. Berlin*, p. 474.
11. 1858. MÜLLER, JOHANNES, Erläuterung einiger bei St. Tropez am Mittelmeer beobachteter Polycystinen und Acanthometren. *Monatsber. d. k. preuss. Akad. d. Wiss. Berlin*, p. 154.
12. 1858. MÜLLER, JOHANNES, Ueber die Thalassicollen, Polycystinen und Acanthometren des Mittelmeeres. *Abhandl. d. k. Akad. d. Wiss. Berlin*, pp. 1–62, Taf. i.–xi. (The fundamental treatise on the Radiolaria.)

13. 1858. SCHNEIDER, ANTON, Ueber zwei neue Thalassicollen von Messina. *Archiv f. Anat. u. Physiol.*, p. 38, Taf. iii. B, figs. 1–4.
14. 1858. CLAPARÈDE et LACHMANN, Echinocystida (Plagiacantha et Acanthometra). Études sur les Infusoires et les Rhizopodes, p. 458, pl. xxii. figs. 8, 9 ; pl. xxiii. figs. 1–6.
15. 1860. HAECKEL, ERNST, Ueber neue lebende Radiolarien des Mittelmeeres. *Monatsber. d. k. preuss. Akad. d. Wiss. Berlin*, pp. 794, 835.
16. 1862. HAECKEL, ERNST, Die Radiolarien (Rhizopoda radiaria). Eine Monographie. 572 pp. fol. with an Atlas of 35 Copperplates.
17. 1862. BURY, Mrs., Polycystins, figures of remarkable forms in the Barbados Chalk Deposit. Ed. ii. By M. C. Cooke, 1868. 25 quarto plates, photographed from drawings by hand, containing many forms overlooked by Ehrenberg from Barbados.
18. 1863. HARTING, PAUL, Bijdrage tot de Kennis der mikroskopische Fauna en Flora van de Banda-Zee (Diep-Zee-Polycystinen). *Verhandl. d. Kon. Akad. van. Wetensch. Amsterdam*, vol. ix. p. 30, pls. i.–iii.
19. 1865. HAECKEL, ERNST, Ueber den Sarcode-Körper der Rhizopoden (Actinelius, Acanthodesmia, Cyrtidosphæra, &c.). *Zeitschr. f. wiss. Zool.*, Bd. xv. p. 342, Taf. xxvi.
20. 1867. SCHNEIDER, ANTON, Zur Kenntniss des Baues der Radiolarien (Thalassicolla). *Archiv f. Anat. u. Physiol.*, 1867, p. 509.
21. 1870. HAECKEL, ERNST, Beiträge zur Plastiden Theorie (Myxobrachia; Amylum in den gelben Zellen). *Jenaische Zeitschr. für Naturw.*, Bd. v. p. 519–540, Taf. xviii.
22. 1871. CIENKOWSKI, L., Ueber Schwärmer-Bildung bei Radiolarien. *Archiv f. mikrosk. Anat.*, Bd. vii. p. 372–381, Taf. xxix.
23. 1872. WAGNER, N., Myxobrachia Cienkowskii. *Bull. d. Acad. St. Petersburg*, vol. xvii. p. 140.
24. 1872. EHRENBERG, GOTTFRIED, Mikrogeologische Studien über das kleinste Leben der Meeres-Tiefgründe aller Zonen und dessen geologischen Einfluss. *Abhandl. d. k. Akad. d. Wiss. Berlin*, 1872. Mit 12 Tafeln. (The Latin diagnoses of 113 new species here mentioned are given in the *Monatsberichte* of April 25, 1872, pp. 300–321.)
25. 1875. EHRENBERG, GOTTFRIED, Polycystinen-Mergel von Barbados (Fortsetzung der Mikrogeologischen Studien). *Abhandl. d. k. Akad. d. Wiss. Berlin*, 1875, 168 pag. mit 30 Tafeln. (The Latin diagnoses of 265 species here recorded are given in Namensverzeichniss der fossilen Polycystinen von Barbados. *Monatsber. d. k. preuss. Akad. d. Wiss. Berlin*, Jan. 30, 1873, pp. 213–263.)
26. 1876. HERTWIG, RICHARD, Zur Histologie der Radiolarien. Untersuchungen über den Bau und die Entwickelung der Sphærozoiden und Thalassicolliden. 91 pp. with 5 plates.
27. 1876. MURRAY, JOHN, Challengerida. Preliminary Reports on Work done on board the Challenger. *Proc. Roy. Soc. Lond.*, vol. xxiv. pp. 471–536, pl. xxiv.
28. 1876. ZITTEL, KARL, Palæozoologie, Bd. i. pp. 114–126, figs. 46–56.
29. 1876. ZITTEL, KARL, Ueber fossile Radiolarien der oberen Kreide. *Zeitschr. d. deutsch. geol. Gesellsch*, Bd. xxviii. pp. 75–96, Taf. ii. (with figures of six Cretaceous species).
30. 1877. MIVART, ST. GEORGE, Notes touching recent researches on the Radiolaria. *Journ. Linn. Soc. Lond.* (Zool.), vol. xiv. pp. 136–186. (Historical sketch of previous literature.)
31. 1877. WYVILLE THOMSON, The Voyage of the Challenger—The Atlantic, vol. i. pp. 231–237, figs. 51–54 ; vol. ii. pp. 340–343, figs. 58, 59, &c.
32. 1878. HAECKEL, ERNST, Das Protistenreich, eine populäre Uebersicht über das Formengebiet der niedersten Lebewesen, pp. 101–104.
33. 1879. HERTWIG, RICHARD, Der Organismus der Radiolarien. *Jenaische Denkschriften*, Bd. ii. Taf. vi.–xvi. pp. 129–277.
34. 1879. HAECKEL, ERNST, Ueber die Phæodarien, eine neue Gruppe kieselschaliger mariner Rhizopoden. *Sitzungsb. med.-nat. Gesellsch. Jena*, December 12, 1879.
35. 1880. STÖHR, EMIL, Die Radiolarien-Fauna der Tripoli von Grotte (Provinz Girgenti in Sicilien). *Palæontographica*, Bd. xxvi. pp. 71–124, Taf. xvii.–xxiii. A preliminary communication regarding this fauna from the tripoli is given in *Tagebl. d. Naturf. Versamml. München*, 1877.

36. 1880. PANTANELLI, DANTE, I Diaspri della Toscana e i loro fossili. *Real. Accad. dei Lincei*, ser. 3, voL vii. pp. 13–34, Tab. i. Radiolaria di Calabria. *Atti. Soc. Tosc.*, p. 59.
37. 1881. HAECKEL, ERNST, Prodromus Systematis Radiolarium, Entwurf eines Radiolarien-Systems auf Grund von Studien der Challenger-Radiolarien. *Jenaische Zeitschr. für Naturw.*, Bd. xv. pp. 418–472.
38. 1881. BRANDT, KARL, Untersuchungen an Radiolarien. *Monatsber. d. k. preuss. Akad. d. Wiss. Berlin*, (April 21), pp. 388–404, Taf. i.
39. 1882. BRANDT, KARL, Ueber die morphologische und physiologische Bedeutung des Chlorophylls bei Thieren. I. Artikel. *Archiv f. Anat. u. Physiol.*, pp. 125–151, Taf. i. II. Artikel. *Mittheil. a. d. Zool. Station zu Neapel*, Bd. iv. pp. 193–302, Taf. xix., xx.
40. 1882. BÜTSCHLI, OTTO, Beiträge zur Kenntniss der Radiolarien-Skelette, insbesondere der der Cyrtida. *Zeitschr. f. wiss. Zool.*, Bd. xxxvi. pp. 485–540, Taf. xxxi.–xxxiii.
41. 1882. BÜTSCHLI, OTTO, Radiolaria. In Bronn's Klassen und Ordnungen des Thierreichs. Bd. i., Protozoa, pp. 332–478, Taf. xvii.–xxxii.
42. 1882. GEDDES, PATRICK, Further Researches on Animals containing Chlorophyll. *Nature*, pp. 303–305.
43. 1882. GEDDES, PATRICK, On the Nature and Functions of the "Yellow Cells" of Radiolarians and Coelenterates. *Proc. Roy. Soc. Edin.*, p. 377.
44. 1882. DUNIKOWSKI, EMIL, Die Spongien, Radiolarien und Foraminiferen der Unter-Liassischen Schichten vom Schafberg bei Salzburg. *Denkschr. d. k. Akad. d. Wiss. Wien*, Bd. xlv. pp. 22–34. Taf. iv.–vi.
45. 1882. PANTANELLI, DANTE, Fauna miocenica di Radiolari del Appennino settentrional. *Boll. Soc. Geol. Ital.*
46. 1883. HAECKEL, ERNST, Die Ordnungen der Radiolarien (Acantharia, Spumellaria, Nassellaria, Phæodaria). *Sitzungsb. med.-nat. Gesellsch. Jena*, February 16, 1883.
47. 1883. HERTWIG, OSCAR, Die Symbiose oder das Genossenschaftsleben im Thierreich. 56. *Versamml. Deutscher Naturf. u. Aerzte*, Freiburg i/B.
48. 1883. RÜST, WILHELM, Ueber das Vorkommen von Radiolarien-Resten in kryptokrystallinischen Quarzen aus dem Jura und in Koprolithen aus dem Lias. 56. *Versamml. Deutscher Naturf. u. Aerzte*, Freiburg i/B.
49. 1884. CAR, LAZAR, Acanthometra hemicompressa (= Zygacantha semicompressa). *Zool. Anzeiger*, p. 94.
50. 1884. HAECKEL, ERNST, Ueber die Geometrie der Radiolarien (Promorphologie). *Sitzungsb. med.-nat. Gesellsch. Jena*, November 22, 1883.

251 A. *Supplementary List of Works Published in* 1885 :—

51. 1885. D. RÜST, Beiträge zur Kenntniss der fossilen Radiolarien aus Gesteinen des Jura. 45 pp. 4to, and 20 plates. *Palæontographica*, Bd. xxxi. (oder iii. Folge, vii. Band).
52. 1885. KARL BRANDT, Die koloniebildenden Radiolarien (Sphærozoeen) des Golfes von Neapel und der angrenzenden Meeres-Abschnitte. 276 pp. 4to, and 8 plates.
53. 1885. JOHN MURRAY, Narrative of the Cruise of H.M.S. Challenger, with a general account of the scientific results of the Expedition. Vol. i. First part, pp. 219–227, pl. A. Second part, pp. 915–926, pl. N. fig. 2.
54. 1885. ERNST HAECKEL, System der Acantharien. *Sitzungsb. med.-nat. Gesellsch. Jena.*, November 13.

Since the printing of this Report began in 1884 and was far advanced in 1885, it was impossible to include the important works of Rüst and Brandt (L. N. 51, 52) in the descriptive portion, so that they are only referred to in the Introduction.

251 B. *Phaulographic Appendix* :—

A list of absolutely worthless literature, which contains either only long known facts or false statements, and may hence be entirely neglected with advantage. Compare § 252, and also L. N. 26, p. 9.

55. 1865. WALLICH, G. C., On the structure and affinities of Polycystina. *Trans. Micr. Soc. Lond.*, voL xiii. pp. 57–84. (Compare L. N. 26, p. 9.)

56. 1879. WALLICH, G. C., Observations on the Thalassicollidæ. *Ann. and Mag. Nat. Hist.*, ser. 4, vol. iii. p. 97.
57. 1866. STUART, ALEXANDER, Ueber Coscinosphæra ciliosa, eine neue Radiolarie (= Globigerina echinoides !!). *Zeitschr. f. wiss. Zool.*, Bd. xvi. p. 328, Taf. xviii. (Compare L. N. 26, p. 9.)
58. 1870. STUART, ALEXANDER, Neapolitanische Studien. *Göttinger Nachr.*, p. 99, and *Zeitschr. f. wiss. Zool.*, Bd. xxii. p. 290 ("Blue Siliceous Crystals" in Collozoum inerme !).
59. 1871. MACDONALD, JOHN DENIS, Remarks on the Structure of Polycystina (Astromma Yelvertoni = Euchitonia Mülleri). *Ann. and Mag. Nat. Hist.*, ser. 4, vol. viii. p. 226.
60. 1871. DOENITZ, W., Beobachtungen über Radiolarien. *Archiv f. Anat. u. Physiol.*, 1871, p. 71, Taf. ii. (Compare L. N. 26, p. 7.)

252. *Progress of our Knowledge of the Radiolaria from* 1862 *to* 1885.—The history of our scientific knowledge of the Radiolaria extends over about half a century (from 1834 to 1885). A historical and critical discussion of the works which appeared within the first twenty-eight years of this period (from 1834 to 1862) is contained in the historical introduction to my Monograph (L. N. 16, pp. 1–24); I shall therefore give here only a brief survey of the investigations published during the last twenty-three years (from 1862 to 1885). The most important steps in our progress during this period we owe to the following naturalists :—Cienkowski (1871), Ehrenberg (1872 and 1875), Richard Hertwig (1876 and 1879), Karlt Brandt (1881 and 1885), Bütschli (1882), and Rüst (1885). To the valuable works of these authors must be added a number of smaller contributions, which are recorded in the foregoing Bibliography. Some communications from dilettanti, written with insufficient knowledge of the subject, and hence of no value, are mentioned for the sake of completeness in the "Phaulographic Appendix" (compare L. N. 55–60, also L. N. 26, p. 9).

The first important advance in our knowledge of the organisation of the Radiolaria, made after the publication of my Monograph (1862), was the demonstration of the nature of the extracapsular "yellow cells." In the year 1870 I showed that these yellow cells contain starch (L. N. 21, p. 519). I regarded them, as did all authors up to that time, as integral parts of the Radiolarian organism, and hence considered this to be multicellular; for no doubt was possible regarding the true cellular nature of these remarkable, nucleated, yellow globules, which I had thoroughly studied in 1862. It was first shown by Cienkowski in 1871 that the yellow cells of the **Collodaria** remain unchanged even after the death of these organisms, "that they continue to grow uninterruptedly, and eventually multiply by division" (L. N. 22, pp. 378–380, Taf. xix. figs. 30–36). Cienkowski concluded from these important observations that the yellow cells are not integral parts of the Radiolarian body, but "parasitic structures," independent, unicellular organisms, which live only as parasites in the body of the Radiolaria (compare § 90).

This important recognition underwent ten years later a further development and complete establishment by the extensive investigations of Karl Brandt (L. N. 38, 39 and Patrick Geddes (L. N. 42, 43). This arrangement was compared by Brandt to the remarkable symbiosis of the Algoid gonidia and Fungoid hyphæ in the organisation of the Lichens, which had been recently discovered, and since he recognised the independent nature of the yellow cells, as unicellular Algæ, in all divisions of the Radiolaria, he founded for them the genus *Zooxanthella.* Geddes named them *Philozoon,* and showed experimentally that they give out oxygen under the influence of sunlight (compare § 90). The great physiological importance of the yellow cells in the metastasis of the Radiolaria, and, when they are developed in large quantities, in the economy of marine organisms in general, has recently been insisted upon by Brandt (see § 205 and L. N. 52, pp. 65–71, 86–94).

The proof that the yellow cells do not belong to the Radiolarian organism itself, but only live parasitically in it, was a necessary preliminary to the very important step which next took place in our knowledge of the organisation of the Radiolaria. This step consisted in the demonstration that the whole body of the Radiolaria, like that of all other Protista, is only a single cell. It was Richard Hertwig who in two remarkable works (L. N. 26, 33) firmly established this fundamental theorem of the unicellular nature of the Radiolaria. In his treatise on the histology of the Radiolaria (L. N. 26, 1876) he published complete investigations into the structure and development of the Sphærozoida and Thalassicollida. Since he made use of the modern methods of histological examination, and especially of staining fluids, which he was the first to apply to the study of the Radiolaria, he was able to show that no true cells (apart from the parasitic yellow cells) are to be found in their bodies, but rather that all their morphological components are to be regarded as differentiated parts of a single true cell, and in particular that the central capsule includes a genuine nucleus.

A wider foundation for this important discovery and its applicability to all divisions of this extensive class, was given by Hertwig in a second work on the organisation of the Radiolaria (L. N. 33, 1879). Among the numerous discoveries by which this work enriched the natural history of the Radiolaria must be specially mentioned the recognition of the fundamental differences exhibited by the main divisions of the class in the structure of their central capsule. Hertwig first observed that the capsular membrane is double in the PHÆODARIA but single in the other Radiolaria (§ 56); the former he named "TRIPYLEA" because he discovered in their capsular membrane a large, peculiarly constructed main opening and two small accessory openings. The NASSELLARIA, in which he found a single porous area at the basal pole of the main axis, with a cone of pseudopodia rising from it, he called on this account "MONOPYLEA"; whilst the other Radiolaria, whose capsular membrane is perforated on all sides with fine pores, were termed "PERIPYLEA." Besides the central capsule, Hertwig laid stress upon the significance of the gelatinous envelope as a constant and important constituent of the body. He also devoted attentive consideration to the morphology of the skeleton, and on the basis of certain phylogenetic conclusions which he drew from it, he arrived at an improved systematic arrangement in which he distinguished six orders:—(1) Thalassicollea, (2) Sphærozoea, (3) Peripylea, (4) Acanthometrea, (5) Monopylea, (6) Tripylea. The numerous isolated discoveries with which Hertwig enriched the morphology of the Radiolaria, have been already alluded to in the appropriate paragraphs in the anatomical portion of this Introduction (see L. N. 42, pp. 340, 341).

The new and interesting group, which was thus erected into an order under the name TRIPYLEA, I had already a year previously separated from the other Radiolaria as "*Pansolenia*" in my Protistenreich (L. N. 32, p. 102). Since, however, neither the three capsular openings of the TRIPYLEA nor the skeletal tubes of the Pansolenia are present in all the families of this extensive order, I substituted in 1879 the more suitable name PHÆODARIA, which is applicable to all members of the group (L. N. 34). In the preliminary memoir then published regarding the Phæodaria, a New Group of Siliceous Marine Rhizopods, I distinguished four orders, ten families, and thirty-eight genera. The great majority of these new forms (among which were no less than 465 different species) were first discovered by the deep-sea investigations of the Challenger. John Murray was the first who called attention to the great abundance in the deep sea of these remarkable Rhizopods, and to the constant presence of their peculiar, dark, extracapsular pigment body (phæodium); even in 1876 he described a portion of them as Challengerida (L. N. 27, p. 536; L. N. 53, p. 226). The earliest observations on the PHÆODARIA were made at Messina in 1859, where I examined five genera of this remarkable group alive (compare p. 1522 and L. N. 16).

By the discovery that the PHÆODARIA, although differing in important respects from the other Radiolaria, still conform to the definition of the class, a new and extensive series of forms was added to this latter, and by their closer investigation a fresh source of interesting morphological problems was disclosed. In other groups, however, morphology was advanced by comparative anatomical studies. In addition to the smaller contributions of various authors, mentioned in the foregoing bibliography, I may specially refer to the valuable Beiträge zur Kenntniss der Radiolarien-Skelete, insbesondere der der Cyrtida by O. Bütschli (L. N. 40, 1882). On the basis of careful comparative anatomical studies, investigations into the skeletal structure of a number of fossil Cyrtoidea and critical application of the recently published researches of Ehrenberg into the Polycystina of Barbados (L. N. 25), Bütschli attempted to derive the complicated relations of the Monopylean skeletons phylogenetically from a simple primitive form,—the primary sagittal ring. Even if this attempt did not actually solve the very difficult morphological problem in question, still the critical and synthetic mode in which it was carried out deserves full recognition, and furnishes the proof that the comparative anatomy of the skeleton in the Radiolaria not less than in the Vertebrata, is a most interesting and fruitful field of phylogenetic investigation. A further demonstration of this was furnished by Bütschli in the general account of the organisation of the Radiolaria which he published in 1882 in Bronn's Klassen und Ordnungen des Thierreichs (L. N. 41).

In our knowledge of the developmental history of these Protista the last two decades have witnessed less progress than in their comparative anatomy. The most important advance in this direction has been the proof that in all the main groups of the class the contents of the central capsule are used in the formation of swarm-spores. The movements of these zoospores in the central capsule had indeed been observed by several previous authors in the case of the SPUMELLARIA and ACANTHARIA (L. N. 10, 13, 16 ; compare also § 142, Note A). The origin of the flagellate spores from the contents of the central capsule and their peculiar constitution were, however, first described fully by Cienkowski in 1871 (L. N. 22, p. 372). Soon after this, R. Hertwig discovered that in the social Radiolaria (Polycyttaria or Sphærozoea) two different forms of zoospores are formed, one with, the other without crystals, and that the latter are also divided into macrospores and microspores (compare L. N. 26, and § 142). Recently this sexual differentiation has been shown by Karl Brandt to exist in all the groups of Sphærozoea, and its regular interchange with the formation of crystal-spores has been interpreted as a true "alternation of generations" (compare L. N. 52 and also § 216). The other forms of development also, especially reproduction by cell-division (§ 213) and gemmation (§ 214), have been elucidated by the recent investigations of the same author.

The palæontology of the Radiolaria has of late made important and interesting advances. Until ten years ago fossil remains of this class were known exclusively from the Tertiary period; almost the only source of our information was to be found in the researches of Ehrenberg, commenced in 1838, continued in his Mikrogeologie in 1854, and concluded in his last work (L. N. 25) published in 1875 (compare L. N. 16, pp. 3–9, 191–193). In the year 1876 a number of Mesozoic Radiolaria from the chalk were described by Zittel (L. N. 28), and afterwards others from the Jura by Dunikowski (L. N. 44). That fossil Radiolaria occur in Mesozoic formations, especially in the Jura, as well preserved and as abundantly as in the Tertiary rocks of Barbados, was shown in 1883 by Rüst (L. N. 48). By the examination of numerous thin sections he discovered that in all the main divisions of the Jurassic formation (Lias, Dogger, Malm) there are distributed jaspers, flints, cherts, and other quartzites, which consist largely of the siliceous shells of Polycystina ; the same is true also of many Coprolites found in the Jura. The full account of these and the descriptions and figures of 234 Jurassic species, distributed in 76 genera, are contained in the Beiträge zur Kentniss der fossilen Radiolarien aus Gesteinen des Jura (L. N. 51, 1885). But even in the older rocks, the Trias, the Permian, and Carboniferous systems, and even as far downwards as the Silurian and Cambrian formations, Rüst has recently shown the existence of fossil Radiolaria, and thus increased the known period of the developmental history of the class by many millions of years (§ 244).

The great significance of the Radiolaria in geology and palæontology has been brought into new light not only by these extensive discoveries, but also by the important relations which have been shown to exist between the Radiolarian rocks and the deep-sea deposits of the present day. In this direction the wonderful discoveries of the Challenger, and especially the investigations of the deep-sea deposits by Wyville Thomson (L. N. 31) and John Murray (L. N. 27), have furnished us with new and valuable information (compare §§ 236–239, and §§ 245–250). The Tertiary Polycystine formations of Barbados and the Nicobar Islands, with which we have been acquainted for the last forty years, as also the Mesozoic Radiolarian quartzes, which have only recently been made known to us from the Jura, are ascertained to be fossil representatives of the same deep-sea deposits which now occur in the form of Radiolarian ooze (§ 237), and to some extent also of Globigerina ooze and red clay (§§ 238, 239), on the bottom of the ocean, at depths of from 2000 to 4500 fathoms.

These investigations into fossil Radiolaria and their comparison with recent deep-sea forms have a further general significance, inasmuch as the identity of many living and fossil species from the Tertiary formation has been shown beyond all doubt. In this direction the numerous measurements and accurate comparisons which I have made during the last ten years of the abyssal forms in the Challenger collection, and of fossil species from Barbados and Caltanisetta, have brought to light many important facts. In this I had the able assistance of my friend, Dr. Reinhold Teuscher (compare § 250, and p. 1760). Further valuable contributions in this direction are found in the careful observations and comparative measurements recently published by Emil Stöhr (L. N. 35, 1880), regarding the Radiolarian fauna of the Tripoli of Grotte in the province of Girgenti, Sicily. From these it appears that the number of Miocene species which are still extant, is much greater than would appear from the results of Ehrenberg.

Ehrenberg himself, towards the end of his long and laborious life, collected the results of the systematic and palæontological researches, which he had begun thirty-seven years previously (L. N. 16, pp. 3–12) into the Polycystina, in two large works (L. N. 24, 25). The first treatise (L. N. 24, 1872) contains the Mikrogeologische Studien über das Kleinste Leben der Meeres-Tiefgründe aller Zonen und dessen geologischen Einfluss, with a list of 279 Polycystina observed by him from the deep-sea, as well as figures of 127 species. The second work (L. N. 25, 1875) contains the Fortsetzung der Mikrogeologischen Studien, mit specieller Rücksicht auf den Polycystinen-Mergel von Barbados; the list of fossil Polycystina observed by him includes 325 species, of which 26 are still extant; 282 of them are figured on the thirty plates accompanying the memoir. By means of these numerous figures, as well as by the appended systematic and chorological tables, Ehrenberg furnished a welcome supplement to the numerous communications regarding the Polycystina, which he had made to the Berlin Academy since 1838, and which he had published in his Mikrogeologie in 1854. It will always be the merit of this zealous and indefatigable microscopist that he first called attention to the great wealth of forms existing in this class; he separated systematically about 500 species, and published drawings of about 400; in addition to which he was the first to lay stress upon the great chorological and geological importance of the Radiolaria.

With these systematic and descriptive, chorological and palæontological works, however, which relate exclusively to the Polycystina, the merits of the famous naturalist of Berlin are exhausted as regards this class of animals. Of the organisation of the Radiolaria, Gottfried Ehrenberg remained entirely ignorant up till his death in 1876. All that a number of famous naturalists had observed during a quarter of a century as to the structure and life-history of the Radiolaria, all the important discoveries of Huxley (1851), Johannes Müller (1858), Claparède (1858), Cienkowski (1871), and many others (L. N. 1–22), and all that I had published in my Monograph (1862) on the basis of three years' study of their anatomy and physiology—all this Ehrenberg ignored, or rather, he regarded it all as worthless rubbish of science, as a chaos of devious errors, resting upon incomplete observations and false conclusions. His strange "special considerations regarding the Polycystina" (L. N. 24, pp. 339–346) and the general "concluding remarks" (L. N. 25, pp. 146–147) leave no room for doubt on this point. Ehrenberg indeed doubted to the last whether any observer had seen living Radiolaria at all (L. N. 25, p. 108).

The invincible obstinacy with which Ehrenberg maintained his preconceived opinion of the high organisation of the Radiolaria, and entirely ignored the contrary observations of other naturalists, is explained by the consistency with which he held to the end the "principle peculiar to himself of the universally equal development of the animal kingdom" (L. N. 16, p. 7). From the complicated arrangement of their siliceous shells he concluded that the animals inhabiting them must possess a structure correspondingly complex, and nearly related to that of the Echinodermata (Holothuria). Like all other animals the Radiolaria must possess systems of organs for locomotion, sensation, nutrition, circulation, and reproduction. Whilst Ehrenberg originally interpreted the Polycystina as siliceous Infusoria polygastrica, and regarded them as compound Arcellina, he afterwards classed them sometimes with the Echinodermata (Holothuria), sometimes with the Bryozoa, sometimes with the Oscillaria (see L. N. 41, p. 336). Although a decided opponent of the cell-theory he called them "multicellular animalcules" (Polycystina), interpreting the pores of the siliceous shell as cells. To-day the opposite term (Monocystina) might be adopted to express their unicellular organisation. It was a remarkable irony of fate that in the self-same year (1838) in which Schwann of Berlin made by his foundation of the cell theory the greatest advance in the whole of Biological Science, that Ehrenberg, all his life the most zealous opponent of that theory, published his great work on the Infusoria, and at the same time established the "family of multicellular animalcules or Polycystina" (L. N. 16, p. 4).

The "short systematic survey of the genera of cellular animalcules" given by Ehrenberg in 1875 (L. N. 25, p. 157), is only a new edition, increased by sixteen genera, of his first systematic arrangement of the Polycystina of 1847 (L. N. 4, p. 53). Since I have already given a full discussion of this in my Monograph (L. N. 16, pp. 214–219), I need only here remark that a correct understanding of his very inadequate generic diagnoses is only possible by the aid of his figures. Relying upon these I have retained almost all Ehrenberg's genera, although entirely new definitions of most of them have been necessary.

The same is true also of the two orders which Ehrenberg distinguished in his class of "Zellenthierchen." The first order is constituted by his "Netzkörbchen" (Monodictya or NASSELLARIA) formerly known as "Polycystina solitaria"; they include our Cyrtoidea, the greater part of Hertwig's Monopylea. Ehrenberg's second order is the "Schaumsternchen" (Polydictya or SPUMELLARIA), previously called "Polycystina composita"; they include the Peripylea of Hertwig, as well as the Spyridina (our Spyroidea), which belong properly to the NASSELLARIA. Although Ehrenberg's statements regarding the organisation of both these orders were quite erroneous, and his knowledge even of the structure of their shells very defective, I still thought it advisable to retain his names for the groups, since they constituted his one successful effort in the systematic treatment of the Radiolaria (compare L. N. 41, p. 336).

The sketch of a systematic arrangement of the Radiolaria (L. N. 37), which I published in 1881 on the basis of the study of the Challenger Radiolaria, resembles, in respect of seven orders being distinguished, the new system which R. Hertwig founded in 1879, in consequence of the variations which he discovered in the structural relations of the central capsule (L. N. 33, p. 133). It differs, however, inasmuch as his Sphærozoea (my Polycyttaria) are here divided into two orders, Symbelaria (Collosphærida) and Syncollaria (Sphærozoida). In that sketch too I separated for the first time the two subclasses Holotrypasta (Porulosa) and Merotrypasta (Osculosa). The fifteen families established by Hertwig were then raised to twenty-four. The six hundred and thirty genera, which I then distinguished, are still for the most part retained, some, however, in a restricted sense, or with amended definitions.

The differential characters of the orders and families of the Radiolaria, given in the Prodromus in 1881, were amended in a further communication which I gave in 1883 regarding the orders of the Radiolaria (L. N. 46, p. 17). There I reduced the seven orders to four, the structural relations of the central capsule being precisely the same in the Polycyttaria and Collodaria as in the Peripylea. The survey of the affinities of the class was thus rendered much simpler and clearer, and the hypothetical genealogical tree, which I then published, has been still further carried out in Chapter VI. of the present Introduction (see §§ 153–200).

253. *General Survey of the Growth of our Systematic Acquaintance with the Radiolaria from 1834 to 1885.*

1834. MEYEN (L. N. 1) describes 2 genera and species of Collodaria:—*Sphærozoum fuscum* and *Physematium atlanticum.*

1838. EHRENBERG (L. N. 2) founds the family Polycystina upon 3 fossil genera (with 6 species):—*Lithocampe, Cornutella, Haliomma.*

1847. EHRENBERG (L. N. 4) publishes his preliminary communications regarding the fossil Polycystina of Barbados and distinguishes 282 species, distributed in 44 genera and 7 families. In the tabular view of the genera he distinguishes two orders:—I. Solitaria— (1) Halicalyptrina, (2) Lithochytrina, (3) Eucyrtidina; and II. Composita— (4) Spyridina, (5) Calodictya, (6) Haliommatina, (7) Lithocyclidina (compare L. N. 16 pp. 214–219).

1851. HUXLEY (L. N. 5) gives the first accurate account of living Radiolaria, and describes 2 species of the genus *Thalassicolla* (*nucleata* and *punctata*); under the latter are included 4 genera of Sphærozoea:—*Collozoum, Sphærozoum, Collosphæra, Siphonosphæra* (compare L. N. 16, pp. 12–14).

1854. EHRENBERG (L. N. 6) publishes in his Mikrogeologie, figures of seventy-two species of fossil Polycystina (without descriptions).

1855. JOHANNES MÜLLER (L. N. 8, p. 248) describes the first *Acanthometra,* and elucidates its affinity to Huxley's *Thalassicolla* and Ehrenberg's Polycystina.

1858. JOHANNES MÜLLER (L. N. 12) establishes the new group Radiolaria as a special order of the Rhizopoda, and includes in it the Thalassicolla, Polycystina, and Acanthometra as closely related families. He opposes these radiate Rhizopoda to the Polythalamia, and describes 50 species observed by him living in the Mediterranean, these he arranges in 20 genera, of which 10 are new. The figures are contained in eleven plates (see L. N. 16, pp. 22–24).

1858. CLAPARÈDE (L. N. 14) describes the first Plectoidean (*Plagiacantha arachnoides*) and two species of *Acanthometra,* which he had observed living in Norway (see L. N. 16, p. 18).

1860. EHRENBERG (L. N. 4) gives a short diagnosis of 22 new genera of Polycystina, based on the investigation of numerous deep-sea species brought up by Brooke from the depths of the Pacific Ocean. The number of his genera is thus increased to 66 (compare L. N, 16, pp. 10, 11).

1862. ERNST HAECKEL (L. N. 16) embraces in his Monograph of the Radiolaria all the species hitherto known either by figures or descriptions, and arranges them in 15 families and 113 genera; of which latter 46 are new. The number of new species observed living amounts to 144. In a "survey of the Radiolarian fauna of Messina" (p. 565) he records 72 genera and 169 species. Most of these are figured in the accompanying atlas of thirty-five plates.

1862. BURY (L. N. 17) gives in an atlas of twenty-five plates, photographed from drawings, the figures of numerous fossil Polycystina of Barbados (without descriptions), of which many are new species overlooked by Ehrenberg (compare § 242, above).

1872. EHRENBERG (L. N. 24) gives a list of names (without description) of all the Polycystina observed by him from the bottom of the sea, 279 species, of which 127 are figured on twelve plates.

1875. EHRENBERG (L. N. 25) gives a list of names of all the fossil Polycystina observed by him (from Barbados, the Nicobar Islands and Sicily), 326 species, of which 282 are figured (compare § 242 above). In a new "Systematic Survey of the Genera" the number of these is given as 63. The 7 families are the same as given in 1847 (see above), as also the two orders (NASSELLARIA = Solitaria, SPUMELLARIA = Composita).

1876. ZITTEL (L. N. 29) describes the first fossil Radiolaria from the chalk (6 species) and establishes the new Cyrtoid genus *Dictyomitra*.

1876. JOHN MURRAY (L. N. 27) establishes the new family Challengerida, and figures 6 new generic types of PHÆODARIA.

1879. RICHARD HERTWIG (L. N. 33) first describes the fundamental differences in the structure of the central capsule, and in accordance with them divides the Radiolaria into six orders:— (1) Thalassicollea, (2) Sphærozoea, (3) Peripylea, (4) Acanthometrea, (5) Monopylea, (6) Tripylea (p. 133). These are subdivided into 18 families, and their phylogenetic affinities discussed (p. 137). On the ten plates, several new species from Messina are figured, among them the types of several new genera (*Cystidium, Cœlacantha, Echinosphæra*) (compare § 252).

1879. ERNST HAECKEL (L. N. 34) founds the order PHÆODARIA as a "new group of marine siliceous Rhizopods," and distinguishes in it 4 suborders, 10 families and 38 genera.

1880. EMIL STÖHR (L.N. 35) describes the Miocene "Radiolarian fauna of the tripoli from Grotte in Sicily," 118 species, of which 78 are new; among them is the new genus *Ommatodiscus*, the type of a new family, Ommatodiscida. The new species are figured on seven plates.

1880. DANTE PANTANELLI (L. N. 36) describes 30 species of fossil Polycystina from the jaspers of Tuscany, which he regarded as Eocene, but which were probably of Jurassic origin (compare § 243, note B, above).

1881. ERNST HAECKEL (L. N. 37) publishes a "Sketch of a classification of the Radiolaria on the basis of the study of the Challenger Collection," and distinguishes in his "conspectus ordinum" (p. 421) 2 subclasses and 7 orders, and in the "prodromus systematis Radiolarium" (pp. 423-472) 24 families with 630 genera, among which are more than 2000 new species.

1882. BÜTSCHLI (L. N. 40) on the basis of studies of the fossil Monopylea of Barbados, investigates the "mutual relations of the Acanthodesmida, Zygocyrtida and Cyrtida," and gives a critical revision of the genera of these "Cricoidea;" a number of new species are described and figured (Tafs. xxxii., xxxiii.), and some new genera of Stichocyrtida established (*Lithostrobus, Lithomitra*, &c.).

1882. DUNIKOWSKI (L. N. 44) describes 18 new fossil Polycystina from the lower lias of the Salzburg Alps, among them the types of 3 new genera (*Ellipsoxiphus, Triactinosphæra,* and *Spongocyrtis*).

clxxxviii   THE VOYAGE OF H.M.S. CHALLENGER.

1883. ERNST HAECKEL (L. N. 46) revises the 4 orders and 32 families of Radiolaria, and gives more accurate definitions of them, as well as of the 2 subclasses (I. *Holotrypasta* = ACANTHARIA and SPUMELLARIA; II. *Merotrypasta* = NASSELLARIA and PHÆODARIA).

1885. D. RÜST (L. N. 51) describes 234 new species of fossil Radiolaria from the Jura, and illustrates them by twenty plates. Among them are 103 SPUMELLARIA, 130 NASSELLARIA, and 1 PHÆODARIA; these are contained in 35 genera, of which 20 belong to the Porulosa, and 15 to the Osculosa.

254. Statistical Synopsis of the Twenty Orders:—

Legion.	Sublegion.	Order.	Number of Families.	Number of Genera.	Number of Species.	Previously known Species.	New Species.	Fossil Species.	Pelagic Abundance.	Abyssal Abundance.	Figured Plate
I. Legion Spumellaria (Porulosa peripylea)	I. Collodaria (Spumellaria palliata)	1. Colloidea	2	6	36	9	27	0	A	E	1,
		2. Beloidea	2	8	56	9	47	0	A	D	2,
	II. Sphærel'aria (Spumellaria loricata)	3. Sphæroidea	6	107	660	105	555	66	A	B	5
		4. Prunoidea	7	53	280	35	245	36	B	B	11-16,
		5. Discoidea	6	91	503	126	376	102	B	A	39, 31-41
		6. Larcoidea	9	51	260	8	252	0	E	B	9, 49,
II. Legion Acantharia (Porulosa actipylea)	III. Acanthometra (Acantharia palliata)	7. Actinelida	3	6	22	6	16	0	E	E	129 (fig
		8. Acanthonida	3	21	138	50	88	0	A	C	130-
	IV. Acanthophracta (Acantharia loricata)	9. Sphærophracta	3	27	149	9	140	0	B	B	133-
		10. Prunophracta	3	11	63	5	58	0	D	B	139,
III. Legion Nassellaria (Osculosa monopylea)	V. Plectellaria (Nassellaria palliata)	11. Nassoidea	1	2	5	1	4	0	E	E	91 (fi
		12. Plectoidea	2	17	61	5	56	0	D	C	91 (figs.
		13. Stephoidea	4	40	205	14	191	17	C	B	81, 92
	VI. Cyrtellaria (Nassellaria loricata)	14. Spyroidea	4	45	239	51	188	53	C	A	83-
		15. Botryodea	3	10	55	15	40	10	E	C	9
		16. Cyrtoidea	12	160	1122	328	794	250	C	A	51-
IV. Legion Phæodaria (Osculosa cannopylea)	VII. Phæocystina (Phæodaria palliata)	17. Phæocystina	3	15	112	30	82	24	C	B	101-
	VIII. Phæocoscina (Phæodaria loricata)	18. Phæosphæria	4	22	121	5	116	0	C	A	106-
		19. Phæogromia	5	27	159	5	154	0	C	A	99, 113-
		20. Phæoconchia	3	20	73	4	69	0	D	B	121-
		Total,	85	739	4318	810	3508	558	...	...	14

Note.—In the tenth and eleventh columns the relative abundance of each order at or near the surface and near the bottom is mately indicated by the letters A–E, which have the following significance:—**A**, abundant; **B**, very numerous; **C**, many (medium qu D, few; E, very few.

PLANKTONIC STUDIES

Ernst Haeckel

6.—PLANKTONIC STUDIES: A COMPARATIVE INVESTIGATION OF THE IMPORTANCE AND CONSTITUTION OF THE PELAGIC FAUNA AND FLORA.

By Ernst Hæckel.

[Translated by George Wilton Field.]

TRANSLATOR'S PREFACE.

Prof. Hæckel's "Plankton-Studien" first appeared in the *Jenaische Zeitschrift*, vol. XXV, first and second parts, 1890. It was immediately published in separate form by Gustav Fischer, of Jena, and attracted much attention on the Continent and in England. The subject, "a comparative study of the importance and constitution of the marine fauna and flora," is presented in Prof. Hæckel's usual pleasing style, and the work can not fail to be of value to all interested in the biological sciences, to the general reader as well as to the specialist. It derives especial interest in connection with the work of the Fish Commission, from its broad discussion of those many important elements which enter into the food supply of all pelagic fishes, such as the mackerel and menhaden, and, considering the extensive physical investigations now being conducted in our coast waters by the schooner *Grampus*, its publication at the present time will prove exceedingly advantageous.

The terminology used by Prof. Hæckel may at first seem formidable, but this difficulty is more fancied than real. The terms are formed upon correct analogies, and most of them will probably find a permanent place. The definite restriction of the meaning of terms is a fundamental necessity in every science, and for the lack of this the branch of biology here considered is in a very unsatisfactory condition. The author, first of all, proposes certain terms with a definite meaning. The word "plankton," from the Greek πλαγκτός, *wandering, roaming*, was, I believe, first employed by Hensen in place of the German "Auftrieb," to designate all plants and animals found at the surface of the ocean which are carried about involuntarily in the water. Hæckel adopts this term, but objects somewhat to the meaning at present attached to it.

Particularly valuable for us is the general review which the author gives of the discovery and growth of our knowledge of this branch, which he names "planktology"; the distinctions which he points out between the varied constituents and distribution of the plankton; and finally his extremely valuable suggestions for further work in the field which he so justly terms "a wonder-land."

In the translation the liberty of omitting a few personal references was taken, for the reason that we in this country know very little of the facts which have called them forth.

In the case of several German words it has been found necessary for the sake of clearness to use a circumlocution. For instance, I can recall no English equivalent for "*Stoffwechsel des Meeres*," which would convey its meaning in a single word. The "cycle of matter in the sea," *i. e.*, the change of inorganic matter into vegetable and animal organic matter, and this finally again into inorganic matter, seemed the best rendering, though even this does not include all which the German term implies.

I.—HISTORICAL EXPLANATIONS.

For the great progress made in the last half century in our knowledge of organic life, we are indebted—next to the theory of development—in a great measure to the investigation of the so-called "pelagic animal world." These wonderful organisms, which live and swim at the surface of the sea and at various depths, have long aroused the interest of seafarer and naturalist, by the wealth of the manifold and strange forms, as well as by the astonishing number of individuals—these have been referred to in many old as well as in recent narratives. A considerable number of these, especially of the larger and more remarkable forms, were described and figured in the last, or in the first half of the present, century. The new and comprehensive investigation of the "pelagic world" began in the fifth decade of our century, and is therefore not yet 50 years old.

Into this, as into so many other regions of biology, the great Johannes Müller, of Berlin, equally distinguished in the realms of morphology and physiology, entered as a pioneer. He was the first who systematically and with great results carried on the "pelagic fishery by means of a fine net." In the autumn of 1845, at Helgoland, he began his celebrated investigations upon the development of echinoderms, and obtained the small pelagic larvæ of the echinoderms, and other small pelagic animals living with them, as sagitta, worm larvæ, etc., at first by "microscopical examination of the sea water, which was brought in" (1). This wearisome and thankless method was soon displaced by the successful use of the "fine pelagic net." In the treatise "on the general plan in the development of the echinoderms,"

NOTE.—Citations inclosed in parentheses which occur in the text refer to the list of publications at the end of this paper (pp. 640, 641).

Müller compares the different methods of obtaining them, and chooses, above all, "fishing with a fine net at the surface of the sea." He says:

> I have used this method for many years with the best results; for the advanced stages of the swimming larvæ and for the time of maturity and metamorphosis it is quite indispensable, and in no way to be replaced.

The students who, in 1845–46, as well as in the following years, accompanied Johannes Müller to Helgoland and Trieste (Max Müller, Busch, Wilms, Wagener, and others) were introduced into this method of "pelagic fishery" and into the investigation of "pelagic tow-stuff" (*pelagische Auftrieb*) obtained thereby. It was soon employed at sea with excellent results by other zoölogists—by T. H. Huxley, by Krohn, Leuckart, Carl Vogt, and others, and especially by the three Würtsburg naturalists, A. Kolliker, Heinrich Müller, and C. Gegenbaur, who in 1852 examined with such brilliant success the treasures of the Straits of Messina. At this time, in the beginning of the second half of our century, the astonishing wealth of interesting and instructive forms of life which the surface of the sea offers to the naturalist first became known, and that long series of important discoveries began which in the last forty years have filled so many volumes of our rapidly increasing zoölogical literature. A new and inexhaustibly rich field was thus opened to zoötomical and microscopical investigation, and anatomy and physiology, organology and histology, ontogeny and systematic zoölogy have been advanced to a surprising degree. The investigation of the lower animals has since then been recognized as a wide field of work, whose exploration is of great significance for all branches of science and to which we owe numberless special and the most important general conclusions.

The general belief of zoölogists regarding the extent of this rich pelagic animal world arose as the result of the discovery that a special "pelagic fauna" exists, composed of many characteristic forms, fundamentally different from the littoral fauna. This pelagic fauna is made up of animals (some floating passively, others actively swimming) which remain at the surface of the sea and never leave it, or only for a short time descend to a slight depth. Among such true "pelagic animals" are the radiolaria, peridinia, noctiluca, medusæ, siphonophores, ctenophores, sagitta, pteropods, heteropods, a greater part of the crustacea, the larvæ of echinoderms, of many worms, etc.

Important changes were first made in the prevailing idea of the "pelagic fauna" by the remarkable discoveries of the epoch-making *Challenger* expedition (1873–1876). The two leaders of this, Sir Wyville Thompson and Dr. John Murray, did not limit themselves to their chief object, the general physical and biological investigation of the deep sea, but studied with equal care and perseverance the conditions of organic life at the surface of the ocean and in zones of various depths. As the most significant general result Murray, in his "Preliminary Report" (1876), says:

Everywhere we have found a rich organic life at and below the surface of the ocean. If living individuals are scarce at the surface, below it the tow net commonly discloses numerous forms, even to a depth of 1,000 fathoms and more (5, p. 536).

In 1875, on the journey through the North Pacific Ocean (from Japan to the Sandwich Islands), the extremely important fact was established that the pelagic organisms in oceanic zones of different depths belong to different species; fine pelagic nets (or tow nets) "on many occasions were let down even to depths of 500, 1,000, and 2,000 fathoms, and thereby were discovered many swimming organisms which had never been captured hitherto, either at the surface of the ocean or at slight depths (up to 100 fathoms below the surface)" (6, p. 758). The most characteristic forms of these zones of different depths belong chiefly to the class of the *Radiolaria*, especially to the order of the *Phæodaria*.

Through the investigation of the *Challenger* radiolaria, which occupied for ten years the greater part of my time and attention, I was led to study anew these conditions of distribution; and I reached the conviction that the differences discovered by Murray in the pelagic fauna, at different depths of the ocean, were still more significant than he assumed, and that they had the greatest significance, not merely for the radiolaria, but also for other groups of swimming oceanic organisms. In 1881, in my "*Entwurf eines Systems der Challenger Radiolarien*," p. 422, I distinguished three groups: (*a*) pelagic, living at the surface of the calm sea; (*b*) zonary, living in distinct zones of depth (to below 20,000 feet); and (*c*) profound (or abyssal) animals living immediately above the bottom of the deep sea. In general, the different characteristic forms correspond (to below 27,000 feet) to the different zones.

In my "General Natural History of the *Radiolaria*" (4, p. 129) I have established this distinction, and have expressed my conviction that it is possible, by the aid of a suitable bathygraphic net, to demonstrate many different faunal belts overlying one another in the great deep-sea zones.

The existence of this "intermediate pelagic fauna," discovered by Murray, inhabiting the zones of different depths of the ocean between the surface and the deep-sea bottom, which I have briefly called "zonary fauna," has been decidedly contradicted by Alexander Agassiz. He claimed, on the ground of "exact experiments" carried on during the *Blake* expedition, in 1878, that the greater part of the ocean contains absolutely no organic life, and that the pelagic animals go down no deeper than 100 fathoms. "The experiments finally show that the surface fauna of the sea is actually limited to a relatively thin layer, and that no intermediate zone of animal life, so to speak, exists between the fauna of the sea bottom and of the surface" (15, pp. 46, 48).

Although these negative conclusions from the so-called "exact experiments" of Agassiz are contradicted by the foregoing results of the *Challenger* investigator, yet against the latter, with some show of right, Agassiz might have raised the objection that the "tow net" used could establish no safe conclusion.* This objection could only be finally removed by the construction of a new tow net, which could be let down closed to a certain depth, and then opened and closed again. The merit of inventing such a closible net, and of the immediate successful use of it, belongs to two distinguished Italian naval officers: G. Palumbo, commander of the Italian war corvette *Vettor Pisani*, first constructed such a closible pelagic net or "bathygraphical zone net;" and Naval Lieutenant Gaetano Chierchia, who during the three years' voyage of the *Vettor Pisani* around the world made a very valuable collection of pelagic animals, used the new closible net with fine results, even at a depth of upwards of 4,000 meters (8, p. 83).

Chierchia's first trial with this "deep-sea closible net" was June 5, 1884, in the East Pacific Ocean, directly under the equator, 15° west of the Galapagos Islands. Fourteen days later, June 19, midway between the Galapagos and the Sandwich Islands, this closible net was sunk to 4,000 meters. In this and in many other trials these Italian naval officers captured an astonishing wealth of new and interesting zonary animals, whose description has for a long time busied zoölogists. The collections brought back to Naples by the *Vettor Pisani* are, next to those of the *Challenger*, the most important materials from the region under consideration.

A few faults which pertained to Palumbo's net were soon done away with by improvements, for which we are indebted to the engineer Petersen and to Prof. Carl Chun, of Breslau. The latter, in 1886, made trials in the Gulf of Naples with the improved closible net which showed "a still more astonishing richness of pelagic animals in greater

---

* The "tow nets" used by the *Challenger* were the ordinary Müller's net (or the "fine pelagic net" of Joh. Müller), a round bag of Müller gauze or silk mull, the mouth being kept open by a circular metallic ring. This ring is in ordinary pelagic fishing fastened to a handle 2 or 3 meters long (like the ordinary butterfly net). While the boat moves along, the opening of this net is held at the surface in such a way that the swimming animals are taken into the bag. They remain hanging in the bottom of this, while the water passes through the narrow meshes of the net. After a time the net is carefully inverted and the tow stuff (*Auftrieb*) is emptied into a glass vessel filled with sea water. If one wishes to fish below the surface, the ring of the net is fastened by means of three strings, equally distant from one another, which at a point (about 1 meter distant from the opening of the net) are joined to a longer line which is sunk by weights to a definite distance, corresponding to the desired depth. When Murray fastened such a tow net to the deep-sea sounding line or to the long line of the deep-sea dredge, he first obtained the inhabitants of the "intermediate ocean zones," but he could not thereby avoid the objection that, since this tow net always remained open, the contents might come from very different depths or even only from the surface. For in drawing up the open tow net animals from the most different zones of depth might occasionally be taken in.

depths, and completely overthrew the assumption that an azoic layer of water exists between the surface and the sea bottom" (15, p. 2). Chun embraced the general results of his important bathypelagic investigations under the four following heads:

(1) The portion of the Mediterranean investigated showed a rich pelagic fauna at the surface as well as at all depths up to 1,400 meters.

(2) Pelagic animals which during the winter and spring appear at the surface seek deep water at the beginning of summer.

(3) At greater depths occur pelagic animals which have hitherto been seldom or never observed at the surface.

(4) A number of pelagic animals also remain at the surface during the summer, and never sink into deep water (15, p. 44).

Among the remarks which Chun made on the vertical distribution of the pelagic fauna and the astonishing planktonic wealth of the depths of the sea (at 1,000 to 2,000 meters), he justly throws out the question, "Who knows, whether in the course of time our views will not undergo a complete reversal, and whether the depths will not show themselves as the peculiar mother earth of pelagic life, from which, for the time being, swarms are sent out to the surface as well as to the sea bottom! There are only a few forms which can so completely adapt themselves to the changing conditions of existence at the surface that they no more seek the deeper levels" (15, p. 49). In consequence of his observations on the periodic rising and sinking of pelagic animals, Chun "can not resist the impression that from the abundance of animal life in the depths the surface fauna represents relatively only an advance guard of the whole, which sometimes to a greater, sometimes to a less extent, and occasionally completely, withdraws itself into more protected regions. Facts plainly speak for this, that the periodical wandering of pelagic animals in the vertical direction is especially conditioned by the changes in temperature. Only a few pelagic animal groups can endure the high temperature of the surface water during the summer; the majority withdraw from the influence of this by sinking, and, finally, whole groups pass their life in the cool deep regions without ever rising to the surface" (15, p. 54).

The general ideas which Chun had obtained by this deep-sea investigation of the Mediterranean he was able to confirm for the Atlantic Ocean on a trip made in the winter of 1887-88 to the Canary Islands (16, p. 31). At this time he made the observation that the periodical wandering of pelagic animals in a vertical direction was influenced in great part by ocean currents (at the surface as well as in deep water), and that among other things the occurrence of the full moon exerted a significant action (16, p. 32). Chun's special observation in the sea of Orotava, upon the poverty of the Canary plankton in November and December and the sudden appearance of great numbers and many species of pelagic animals in January and February, agrees completely with the observations which I myself made twenty years before at the Canary island Lanzarote. I also entirely agree with Chun in regard to his general views upon the chorology of the plankton, and consider his investigations upon the pelagic animal world and its relation to the surface fauna as the most important contribution which planktology has received since the pioneer discoveries of the *Challenger* and of the *Vettor Pisani.*

Entirely new aspects and methods have been introduced into pelagic biology in the last three years by Dr. Victor Hensen, professor of physiology at Kiel (9 and 22). He has for a number of years thoroughly studied the conditions of life of the fauna and flora of the bay of Kiel, and as a member of the commission for the scientific investigation of the German Ocean (at Kiel) has endeavored to improve and extend the fisheries there, and by counting the fish eggs collected to get an approximate idea of the number of fish in corresponding districts (9, p. 2). This investigation led him to the conclusion that it was necessary and possible to come nearer to the fundamental food supply of marine animals and to determine this quantitatively. For solving this problem Hensen invented a new mathematical method (2, p. 33). He constructed a new pelagic net (p. 3), and in July, 1884, in company with three other naturalists of Kiel, undertook a nine-day excursion in the North Sea and Atlantic Ocean, which was extended to the Hebrides and to the Gulf Stream (57° 42′ N. Lat.) (p. 30). In 1887 he published the results of this investigation in a comprehensive work containing many long numerical tables, "On the Determination of the Plankton, or the Animal and Vegetable Material found in the sea" (9). He used the term "plankton" in place of "*Auftrieb,*" the word hitherto commonly used, because this name is not sufficiently comprehensive and suitable (9, p. 1). To be sure, the German term "*Auftrieb*" or "*pelagischer Mulder,*" introduced by Johannes Müller forty years ago, was in general use and has many times been used in English, French, and Italian works. But I agree with Hensen that in this, as in other scientific terms, a Greek *terminus technicus,* capable of easier flexion, is preferable. I adopt the term Plankton in place of "*Auftrieb,*" and form from it the adjective planktonic (*planktonisch*). The whole science which treats of this important division of biology is briefly called planktology.

Hensen regards the *mathematical determination of the plankton* as the chief aim of planktology from a physiological standpoint. By it he hopes to solve the somewhat neglected question of the cycle of matter in the sea. For the purpose of solving this, and to make a trial of his new method on a larger scale, Hensen, in the summer of 1889, arranged a more extensive expedition in the Atlantic, which was most liberally supported by the German government and by the Berlin Academy of Sciences. The German Emperor furnished 70,000 marks; the Berlin

Academy gave, from the income of the Humboldt fund, 24,600 marks, and by further contributions the entire sum at the disposal of the expedition was raised to 105,600 marks—a sum never before made available in Germany for a biological expedition. The new steamer *National*, of Kiel, was chartered for three months, and was fitted out " with all the admirable contrivances for obtaining plankton, for deep-sea fishing, and for sounding." Besides the leader of the expedition, Prof. Hensen, five other naturalists participated: the zoölogists Brandt and Dahl; the botanist Schütt; the bacteriologist Fischer; the geographer Krümmel; and the marine artist Richard Eschke. The voyage of the *National* lasted 93 days (July 7 to November 15). The course was westward through the north Atlantic (Gulf Stream, Sargasso Sea), then southward (Bermudas, Cape Verde, Ascension) to Brazil, and eastward back by the Azores. During this voyage 400 casts were made, 140 with the plankton nets, 260 with other nets.

Our German navy has been but little used for scientific, still less for biological, investigations; much less than the navies of England, France, Italy, Austria, and the United States. The remarkable services which many distinguished German zoölogists have rendered in the last half century for the advancement of marine biology have been carried on almost entirely without government aid. The German government has hitherto had very little means available for this branch of science. Therefore, great was the satisfaction when, by the liberal endowment of the plankton expedition of Kiel, the first step was taken for the more extensive investigation, with better apparatus, of the biology of the ocean, and for emulation of the results which the English *Challenger* and the Italian *Vettor Pisani* had lately obtained in this region.

Accounts have been published of the results of the plankton expedition of Kiel, by Victor Hensen (22), Karl Brandt (23), E. du Bois Reymond (21), and Krümmel. The essential details of these accounts have been repeatedly published in the German newspapers, to the general effect that the proposed goal was reached and the most important question of the plankton was happily solved. I very greatly regret that I can not agree with this favorable verdict. (1) The most important generalizations which the plankton expedition of Kiel obtained on the composition and distribution of the plankton in the ocean stand in sharp contradiction to all previous experience; one or the other is wrong. (2) It seems to me that Hensen has incautiously founded a number of far-reaching erroneous conclusions on very insufficient premises. Finally, I am convinced that the whole method employed by Hensen for determining the plankton is utterly worthless, and that the general results obtained thereby are not only false, but also throw a very incorrect light on the most important problems of pelagic biology. Before I establish this dissenting opinion let me give an account of my own planktonic studies and their results.

II.—PLANKTONIC STUDIES.

My own investigations on the organisms of the plankton were begun thirty-six years ago, when I got my first conception of the wonderful richness of the marine fauna and flora in the North Sea. Accepting the kind invitation of my ever-remembered teacher, Johannes Müller, I accompanied him in the autumn of 1854 on a vacation trip to Helgoland, and was introduced by him personally into the methods of plankton fishery and the investigation of the pelagic fauna. There, during August and September, I accompanied him daily on his boating trips, and under all conditions of the rich planktonic captures I received from him the most competent instruction, and pressed with corresponding eagerness into the mysteries of this wonderful world. Never will I forget the astonishment with which I first beheld the swarms of pelagic animals which Müller emptied by inversion of his "fine net" into a glass jar of sea water—a confused mass of elegant medusæ and glistening ctenophores, swift-darting sagittas and snake-like tomopteris, copepods and schizopods, the pelagic larvæ of worms and echinoderms. The important stimulus and instruction of the founder of planktonic investigation has exercised a constant influence on my entire later life, and has given me a lasting interest in this branch of biology.*

Two years later (in August and September, 1856), while at Würtzburg, I accepted the invitation of my honored teacher, A. Kölliker, to accompany him to Nizza, and, under his excellent guidance, became acquainted with the zoölogical treasures of the Mediterranean. In company with Heinrich Müller and K. Kupffer, we investigated especially the rich pelagic animal life of the beautiful bay of Villafranca. There, for the first time, I met those wonderful forms of the pelagic fauna which belong to the classes of the siphonophores, pteropods, and heteropods. I also there first saw living polycyttaria, acanthometra, and polycystina, those phantasmic forms of radiolaria, in the study of which I spent so many later years.

Johannes Müller, who was at this time at Nizza, and had already begun his special investigation of this latter order, called my attention to the many and important questions which the natural history of these enigmatical microscopical organisms present. These valuable suggestions resulted some years later in my going to Italy and spending an entire year in pelagic fishing on the Mediterranean coast. Dur-

---

* When at Helgoland, investigating the wonders of the plankton with the microscope, Johannes Müller, pleased with the care and patience with which his zealous students tried to study the charming forms of medusæ and ctenophores, spoke to me the ever-memorable words, "There you can do much; and as soon as you have entered into this pelagic wonderland you will see that you can not leave it."

ing the summer of 1859, at Naples and at Capri, I endeavored to gain as wide a knowledge as possible of the marine fauna. In the following winter, at Messina, I devoted my entire attention to the investigation of the radiolaria, and thus obtained the material which forms the basis of my monograph of this class (1862). Daily boat trips in the harbor of Messina made me acquainted with all the forms in the pelagic fauna which make this classic spot, in consequence of the combination of uncommonly favorable conditions, far richer for planktonic study and investigation than any other point on the Mediterranean (3, pp. v, 25, 166, 170).

For a full generation, since that time, the study of plankton has remained my most pleasant occupation, and I have hardly let a year pass without going to the seacoast and, by means of the pelagic net, getting new material for work. Various inducements were offered to me in addition; on the one hand the radiolaria, on the other the siphonophores and medusæ, to which I had already given some attention while at Nizza in 1864. The results of these studies are given in my monographs of these two classes (1879 and 1888). In the course of these three decades I have by degrees become acquainted with the entire coast of the Mediterranean and its fauna. I have already made reference, in the preface to my "System of Medusæ," p. XVI, to the places where I have studied this subject. In addition to the Mediterranean I have continued my planktonic studies on the west coast of Norway (1869); on the Atlantic coast of France (1878); on the British coast (1876 and 1879); at the Canary Islands (1866–67); in the Red Sea (1873), and in the Indian Ocean (1881–82).

By far my richest results and my deepest insight into the biology of the plankton were vouchsafed me during a three months' residence at Puerto del Arrecife, the seaport of the Canary island Lanzarote (in December, 1886, and in January and February, 1887). The pelagic fauna in this part of the Atlantic is so rich in genera and species; the fabulous wealth of life in the wonderful "animal roads" or Zain currents (18, p. 309) is, every day, so great, and the opportunities for investigation on the spot are so favorable that Lanzarote afforded me greater advantages for planktonic study than all the other places ever visited by me (excepting perhaps Messina). Every day the pelagic net brought to me and to my companions (Prof. Richard Greeff and my two students, N. Miklucho-Maclay and H. Fol) such quantities of valuable tow-stuff (*Auftrieb*) that we were able to work up only a very small part of it. At that time I concentrated my chief interest on the medusæ and siphonophores, and the larger part of the new material which is worked up in my monographs of these two classes was collected at Lanzarote. All my observations "On the Development of the Siphonophores" (1869) were made there.

The excursion to the coral reefs of the Red Sea (1873), which is recounted in my "Arabic Corals," and the trip to Ceylon, about which I have written in my "Indian Journal" (*Indische Reisebriefe*, 1882), were extremely valuable to me, because I thereby gained an insight into the wonders of the Indian fauna and flora. On the journey from Suez to Bombay (in November, 1881), as well as on the return from Colombo to Aden (in March, 1882), I was able to make interesting observations on the pelagic fauna of the Indian Ocean, as well as during a six weeks' stay at Belligam and in the pelagic excursions which I made from there. I obtained thereby a living picture of the oceanic and neritic fauna of the Indo-Pacific region, which differs in so many respects from that of the Atlantic-Mediterranean region. The special results of my experience there are, with the kind consent of Dr. John Murray, for the most part embraced in my report on the Radiolaria (1887), and on the Siphonophora (1888), which form parts XVIII and XXVIII of the *Challenger* Report. These two monographic reports also contain many observations on plankton, which I had made in earlier journeys and had not yet published.

The extensive experience which I had gained through my own observations of living plankton during a period of three decades was well filled out by the investigation of the large and well-preserved planktonic collections placed at my disposal from two different sources by Capt. Heinrich Rabbe, of Bremen, and by the *Challenger* directors of Edinburgh. Capt. Rabbe, with very great liberality, turned over to me the valuable collection of pelagic animals which he had obtained on three different trips (with the ship *Joseph Haydn*, of Bremen) in the Atlantic, Indian, and Pacific oceans, and which he had carefully preserved according to my directions and by approved methods. This extraordinarily rich and valuable material, contained in numerous bottles, embraced planktonic samples from the most diverse localities of the three oceans, chiefly in the southern hemisphere. Like the much more extensive collection of the *Challenger*, it gives (though to a smaller degree) a complete summary of the complexity of the composition of the plankton and the difference in its constituents. Rabbe's collection supplements that of the *Challenger* in a most welcome manner, since the course of the *Challenger* was southward from the Indian Ocean through the Antarctic region, and between the Cape of Good Hope and Melbourne was always south of 40° south latitude. The course of the *Joseph Haydn*, on the other hand, on the repeated voyages through the Indian Ocean, was much more northerly, and between Madagascar, the Cocos Islands, and Sumatra included a number of points where the pelagic net obtained a very rich and peculiarly constituted capture. I hope to be able to publish soon in detail the special results which I have obtained by investigation of Rabbe's plankton collection, with the aid of the carefully kept journal which Capt. Rabbe made of his observations. The discoveries of new radiolaria, medusæ, and siphonophores which I owe to these are already embraced in my monographs on these three classes in the *Challenger* Report, and in the preface I have expressed to Capt. Rabbe my sincere thanks for his very valuable aid.

Of all expeditions which have been sent out for investigating the biology of the ocean, that of the *Challenger* was, without doubt, the greatest and the most fruitful, and I recognize it with additional gratitude since I was permitted for twelve years to take part in working up its wonderful material. When, after the return of the expedition, I was honored by its leader, Sir Wyville Thompson, by being summoned to work up the extensive collection of radiolaria, I believed, after a hasty survey of the treasures, that I could complete their investigation in the course of three to five years; but the further I proceeded in the investigation the greater seemed the assemblage of new forms (4, p. XV), and it was a whole decade before the report on the radiolaria (part XVIII) was completed. Three other reports were also then finished—on deep-sea horny sponges (part LXXXII), on the deep-sea medusæ (part XII), and on the siphonophores (part XXVIII) collected by the *Challenger*. The comparative study of these extremely rich planktonic treasures was highly interesting and instructive, not only on account of the daily additions to the number of new forms of organisms in these classes, but also because my general ideas on the formation, composition, and importance of the plankton were enriched and extended. I am sincerely thankful for the liberality with which Sir Wyville Thompson, and after his untimely death (1882) his successor, Dr. John Murray, placed these at my entire disposal.

A record of the 168 stations of observations of the *Challenger* expedition, whose soundings, plankton results, and surface preparations I have been able to investigate, has been given in § 240 of the report on the radiolaria (4, p. CLX). The number of the bottles containing plankton (from all parts of the ocean) in alcohol amounts to more than a hundred, and in addition there are a great number of wonderful preparations which Dr. John Murray finished at the different observation stations, stained with carmine and mounted in Canada balsam. A single such preparation (for example, from station 271) contains often 20 to 30 and sometimes over 50 new species. Since the material for these preparations was taken with the tow net, not only from the surface of all parts of the sea traveled by the *Challenger*, but also from zones of different depths, they make important disclosures in morphology as well as in physiology and chorology. To the study of these station preparations I am indebted for many new discoveries. I have been able to examine over a thousand (4, p. 16).

If I here refer to the development and extension of my own plankton studies, it is because I feel compelled to make the following brief summary of results. I am not now in a position to give the proofs in detail, and must defer the thorough establishment of the most weighty series of observations for a later and more detailed work. But since, to my regret, I am compelled to decidedly contradict the far-reaching assertions made by Hensen (22), it is only to justify and prove these that I refer to my extended experience of many years. I believe I do not err in the assumption that among living naturalists I am one of those who by extensive investigation on the spot have become most thoroughly acquainted with the conditions of the plankton and have worked deepest into these intricate problems of marine biology. If I had not for so many years had these continually in mind, and at each new visit to the sea begun them anew, I would not dare to defend with such determination the assertions expressed in the following pages.

III.—CHOROLOGICAL TERMINOLOGY.

The science of the distribution and division of organic life in the sea (marine chorology) has in the last decade made astonishing progress. Still this new branch of biology stands far behind the closely related terrestrial chorology, the topography and geography of land-dwelling organisms. We have as yet no single work which treats distinctly and comprehensively of the chorology of marine plants and animals in a manner similar to Griesbach's "Vegetation of the Earth" (1872) for the land plants, and Wallace's "Geographical Distribution of Animals" (1876) for the land animals.

How much there is still to be done is shown by the fact that not one of the simplest fundamental conceptions of marine chorology has yet been established. For example, the most important conception of one subject, that of the pelagic fauna and flora, is now employed in three different senses. Originally, and through several decades, this term was used only in the sense in which Johannes Müller used it, for animals and plants which are found swimming at the surface of the sea. Then the term was extended to all the different animals and plants which are found at the surface of fresh-water basins. It was so used, for example, by A. Weismann in his lecture upon "the animal life at the sea-bottom" (1877), in which he "distinguishes the animal world living on the shore from the 'pelagic or oceanic company living in the open sea.'" To a third quite different meaning has the conception of the pelagic living world been widened by Chun (1887), who extends it from the surface of the ocean down to the greatest depths (15, p. 45). In this sense the conception of the pelagic organisms practically agrees with the "plankton" of Hensen.

Errors have already arisen from the varied use of such a fundamental conception, and it seems necessary to attempt to clear this up, and to establish at least the most important fundamental conception of marine chorology. In the use of words I will, as far as possible, conform to the usage of the better authors.

MARINE FLORA AND FAUNA.

Since the old mooted question about "the limits of the animal and vegetable kingdom" comes anew into the foreground in the planktonic studies, a few words must first be devoted to its consideration. In the plankton, those organisms (for the most part microscopic) which stand on the boundary line and which may be regarded as examples of a neutral "Protista realm," play a conspicuous part—the unicellular diatoms and murracytes, dictyochea and palmellaria, thalamophora and radiolaria, dinoflagellata and cystoflagellata. Since it is still asserted that for replies to this boundary question we need new researches, "more exact observations and experiments," I must here express the opposing belief, that the desired answer is not to be obtained by this empirical and inductive method, but only by the philosophic and deductive method of more logical definite conception (*logischer Begriff-Bestimmung*). Either we must use as a definite distinction between the two great organic realms the physiological antithesis of assimilation, and consider as "plants" all "reducing organisms" (with chemical-synthetic functions) and as "animals" all "oxidizing organisms" (with chemical-analytical functions) or we may lay greater weight on the morphological differences of bodily structure and place the unicellular "*Protista*" (without tissues) over against the multicellular *Histona* (with tissues).[*]

For the problem before us, and with more particular reference to the important questions of the fundamental food supply (*Urnahrung*) and the cycle of matter in the sea (*Stoffwechsel des Meeres*), it is here more suitable to employ the first method. I regard the diatoms, murracytes, and dinoflagellates as *Protophytes*, the thalamophores, radiolarians, and cystoflagellates as *Protozoa*.

For a term to designate the totality of the marine flora and fauna, the expression *halobios* seems to be suitable, in opposition to *limnobios* (the organic world of fresh water) and to *geobios* (as the totality of the land-dwelling or terrestrial plant and animal world). The term *bios* was applied by the father of natural history, Aristotle, "to the whole world of living" as opposed to the lifeless forms, the *abion*. The term biology should be used only in this comprehensive sense, for the whole organic natural science, as opposed to the inorganic, the abiology. In this sense, zoölogy and botany on the one side, and morphology and physiology on the other, are only subordinate parts of biology, the general science of organisms. But if (as is frequently done to-day even in Germany) the term biology is used in a much narrower sense, instead of *œcology*, this narrowing leads to misunderstandings. I mention

---

[*] *Protista* and *Histona* may both again be divided into two groups, on the ground of the different assimilation, into an animal and a vegetable group, the *Protista* into *Protophyta* and *Protozoa*, the *Histona* into *Metaphyta* and *Metazoa*. Compare my "Natural History of Creation" (*Natürliche Schöpfungsgeschichte*), 8th edition, 1889, pp. 420 and 453.

this here because in planktology the interesting and complex vital relations of pelagic organisms, their manner of life and economy, are very often called biological instead of œcological problems.*

PLANKTON AND BENTHOS.

If under the term *Halobios* we embrace the totality of all organisms living in the sea, then these, in œcological relation, fall into two great chief groups, *benthos* and *plankton*. I give the term *benthos*† (in opposition to *plankton*) to all the non-swimming organisms of the sea, and to all animals and plants which remain upon the sea bottom either fixed (sessile) or capable of freely changing their place by creeping or running (vagrant). The great œcological differences in the entire mode of life, and consequently in form, which exist between the benthonic and planktonic organisms, justify this intelligible distinction, though here as elsewhere a sharp limit is not to be drawn. The *benthos* can itself be divided into littoral and abyssal. The *littoral-benthos* embraces the sessile and vagrant marine animals of the coast, as well as all the plants fixed to the sea-bottom. The *abyssal-benthos*, on the other hand, comprises all the fixed or creeping (but not the swimming) animals of the deep sea. Although as a whole the morphological character of the *benthos*, corresponding to the physiological peculiarities of the mode of life, is very different from that of the *plankton*, still these two chief groups of the *halobios* stand in manifold and intimate correlation to one another. In part these relations are only phylogenetic, but also in part at the present day of an ontogenetic nature, as, for example, the alternation of generations of the benthonic polyps and the planktonic medusæ. The adaptation of marine organisms to the mode of life and the organization conditioned thereby may in both chief groups be primary or secondary. These and other relations, as, well as the general characteristics of the pelagic fauna and flora, have already been thoroughly considered by Fuchs (12) and Moseley (7).

PLANKTON AND NEKTON.

The term *plankton* may be used in a wider and in a narrower sense; either we understand it as embracing all organisms swimming in the sea, those floating passively and those actively swimming; or we may exclude these latter. Hensen comprehends under *plankton* "everything which is in the water, whether near the surface or far down, whether dead or living." The distinction is, whether the animals are driven involuntarily with the water or whether they display a certain degree of independence of this impetus. Fishes in the form of eggs

---

* The terms biology and œcology are not interchangeable, because the latter only forms a part of physiology. Comp. my "Generelle Morphologie," 1866, Bd. I, p. 8, 21; Bd. II, p. 286; also my "Ueber Etwickelungsgang und Aufgabe der Zoölogie," Jena. Zeitsch. fur Med. u. Nat., Bd. v, 1870.

† βένθος, the bottom of the ocean; hence the organisms living there.

and young belong in the highest degree to the plankton, but not when mature animals. The copepods, although lively swimmers, are tossed about involuntarily by the water, and, therefore, must be reckoned in the plankton (9, p. 1). If, with Hensen, we thus limit the conception of *plankton*, then we must distinguish the *actively* swimming *nekton* from the *passively* driven *plankton*. The term thus loses its firm hold, and becomes dependent on quite variable conditions; upon the changing force of the current in which the animal is driven, by the momentary energy of voluntary swimming movements, etc. A pelagic fish or copepod, which is borne along by a strong current, belongs to the plankton; if he can make a little progress across this current, and if, besides this, he can voluntarily and independently define his course, then he belongs also to the nekton. It therefore seems to me advisable, as preliminary, to regard the term plankton in the wider sense, in opposition to benthos.

Still, for the chief theme which Hensen has set up in his plankton studies, for the physiological investigation of the cycle of matter in the sea (*Stoffwechsel des Meeres*), this limitation of the plankton conception will not hold; for a single large fish which daily devours hundreds of pteropods or thousands of copepods exerts a greater influence on the economy of the sea than the hundreds of small animals which belong to the plankton. I will return to this in speaking of the vertebrates of the plankton. If with Hensen we could, on practical grounds, separate those animals of the plankton which are carried involuntarily from those following their own voluntary swimming movements (independent of the current), we might distinguish the former as *ploteric*,* the latter as *necteric*.*

HALIPLANKTON AND LIMNOPLANKTON.

Although the swimming population of fresh water shows far less variety and peculiarity than that of the sea, still among the former as among the latter similar conditions are developed. Already the study begins to take a joyous flight to the pelagic animals of the mountain lakes, etc. Therefore, it will be necessary here also to fix limits, as has been already done for the marine fauna; but since the term "pelagic" should only be used for marine animals, it becomes advisable to designate as *limnetic* the so-called "pelagic" animals of fresh water. Among these we can again distinguish *autolimnetic* (living only at the surface), *zonolimnetic* (limited to certain depths), and *bathylimnetic* (dwellers in the deep waters). The totality of the swimming and floating population of the fresh water may be called *limnoplankton*, as opposed to the marine *haliplankton* (9, p. 1), which we here briefly call *plankton*.

---

* Πλωτήρ = drifting; νηκτής = swimming.

OCEANIC AND NERITIC* PLANKTON.

The manifold differences which the character of the plankton shows according to its distribution in the sea, lead first, with reference to its horizontal extension, to a distinction between oceanic and neritic plankton. *Oceanic plankton* is that of the open ocean, exclusive of the swimming *bios* of the coast. The region of oceanic plankton may from a zoölogical point of view be divided into five great provinces: (1) the Arctic Ocean; (2) the Atlantic; (3) the Indian; (4) the Pacific; (5) the Antarctic. In each of these five great provinces the characteristic genera of the plankton are apparent through the different species, even if the differences in general are not so significant as in the different provinces of the neritic and still more of the littoral fauna.

The neritic plankton embraces the swimming fauna and flora of the coast regions of the continents as well as the archipelagos and islands. This is in its composition essentially different from the oceanic plankton, and is quantitatively as well as qualitatively richer. For along the coast there develop, partly under protection of the littoral *bios*, or in genetic relation with it, numerous swimming animal and vegetable forms which do not generally occur in the open ocean, or there quickly die; but the floating organisms of the latter may be driven by currents or storms to the coast and there mingled with the neritic plankton. Aside from this the richness of the neritic plankton in genera and species is much greater than that of the oceanic. The complicated and manifold relations of the latter to the former, as well as the relations of both to the benthos (littoral as well as abyssal), have been but little investigated and contain a fund of interesting problems. One could designate the neritic plankton also as "littoral plankton" if it were not better to limit the conception of the littoral bios to the non-swimming organisms of the coast, the vagrant and sessile forms.

PELAGIC, ZONARY, AND BATHYBIC PLANKTON.

I keep the original meaning of the *pelagic plankton* as given forty-five years ago by Johannes Müller, and used since by the great majority of authors. I also limit the meaning of the pelagic fauna and flora to those actively swimming or passively floating animals and plants, which are taken swimming at the surface of the sea, no matter whether they are found here alone or also at a variable depth below the surface. These are the *superficial* and *interzonary* organisms of Chun (15, p. 54). On the other hand, I distinguish the *zonary* and *bathybic* organisms; I call *zonary plankton* those organisms which occur only in zones of definite depths of the ocean, and above this (at the surface of the sea) or below (at the sea bottom) are only found occasionally, as for example many phæodaria and crustacea; also the deep-sea siphonophores dis-

---
* Νηρίτης, son of Nereus.

covered by Chierchia, which were taken by him in great numbers and in great vertical and horizontal extension, but never higher than 1,000 meters below the surface and never deeper than 1,000 meters above the sea bottom (8, p. 85). The deepest part of this zonary fauna forms the *bathybic plankton* (or the profound tow-stuff, *Auftrieb*), i. e., animals of the deep sea, which only hover over the bottom but never touch it, whether they stand in definite relation to the abyssal benthos or not. One might also call them "*abyssal plankton*," if it were not more practicable to limit the term "abyssal" to the (vagrant and sessile) benthos of the deep sea. To the bathybic plankton belong many phæodaria, some medusæ and siphonophores, many deep-sea crustacea, *Tomopteris euchæta*, *Megalocereus abyssorum*, etc. (15, pp. 55–57).

In each of these vertical parts of the plankton, distinctions may be made which apply to the *horizontal* distribution. We may also distinguish oceanic and neritic forms in the pelagic fauna as in the zonary and bathybic fauna.

AUTOPELAGIC, BATHYPELAGIC, AND SPANIPELAGIC PLANKTON.

If, following the old custom, we limit the term "pelagic *bios*" to those organisms which, at some time, swim or float at the *surface* of the sea— if we do not with Chun (15, p. 45) extend this term to the zonary and bathybic animals—it still is necessary to further distinguish by different terms those forms of life which constantly, temporarily, or only exceptionally live at the surface of the sea. I suggest for these the terms autopelagic, bathypelagic, and spanipelagic. *Autopelagic* are those animals and plants which are constantly found only at the surface (or in stormy weather at slight depths below it), the "superficial" of Chun (15, pp. 45, 60). To this "constant superficial fauna" belong, for example, many polycyttaria (most sphærozoids), many medusæ (*e. g.*, *Eucopidæ*), and many siphonophores (*e. g.*, *Forskalidæ*); further, the lobate ctenophores (*Eucharis*, *Bolina*), particular species of *Sagitta* (*e. g.*, *bipunctata*), and many copepods (*e. g.*, *Pontellina*, 15, p. 27).

I call *bathypelagic* all those organisms which occur not merely at the surface, but also extend down into the depths, and often fill the deep layers of the ocean in not less astonishing multitudes than the surface layers. Chun designates such bathypelagic animals as "interzonary pelagic animals" (15, p. 45). Here belongs properly the chief mass of the plankton; for through the agreeing researches of Murray (5, 6), Moseley (7), Chierchia (8), and Chun (15, 16), as well as from my own wide experience, it becomes highly probable that the great number of pelagic animals and plants only pass a part of their lives at the surface; swimming at different depths during the other part. Among the bathypelagic animals there are farther to be distinguished: (*a*) *Nyctipelagic*, which arise to the surface only at night, living in the depths during the day; very many medusæ, siphonophores, pyrosoma, most pteropods, and heteropods, very many crustacea, etc.; (*b*) *Chimopelagic*, which appear at the surface only in winter and in summer are hidden in the depths—radiolaria, medusæ, siphonophores, ctenophores, a part of the pteropods and heteropods, many crustacea, etc.; (*c*) *Allopelagic*, which perform irregular vertical wanderings, sometimes appearing at the surface, sometimes in the depths, independently of the changes of temperature, which condition the change of abode of the nyctipelagic and chimopelagic animals; the final cause of these wanderings ought to be found in different œcological conditions, as of reproduction, of ontogeny, of food supply, etc.

Finally one may call *spanipelagic* those animals which always live in the ocean depths (zonary or bathybic), and come to the surface only exceptionally and rarely. This does not apply to a few deep-sea animals which once every year ascend to the surface, but only for a short time, for a few weeks or perhaps for a single day, *e. g.*, Athorybia and Physophora among the siphonophores, Charybdea and Periphylla among the medusæ. The final cause of this remarkable spanipelagic mode of life must lie chiefly in the conditions of reproduction and ontogeny. These animals must be much more numerous than present appearances show.

HOLOPLANKTONIC AND MEROPLANKTONIC ORGANISMS.

Numerous organisms pass their whole life and whole cycle of development hovering in the ocean, while with others this is not the case. These rather pass a part of their life in the benthos, either vagrant or sessile. The first group we call *holoplanktonic*, and the second *meroplanktonic*. To the holoplanktonic organisms, which have no relation whatever to the benthos, belong the greater part of the diatoms and oscillaria, all murracytes and peridinea; further all radiolaria, many globigerina, the hypogenetic medusæ (without alternation of generations), all siphonophores and ctenophores, all chætognathæ, pteropods, the copelata, pyrosoma, and thalidia, etc. Among these we find "purely pelagic, zonary, or bathybic" forms.

The *meroplanktonic* organisms, on the other hand, which are found swimming in the sea only for a part of their life, passing the other part vagrant or sessile in the benthos (either littoral or abyssal), are represented by the following groups: A part of the diatoms and oscillaria, the planktonic *fucoids*, the metagenetic medusæ (*Craspedota* with hydroid nurse, *Acraspeda* with scyphistoma nurse), some turbellarians and annelids, etc; further, the "pelagic larvæ" of hydroids and corals, many helminths and echinoderms, acephala and gasteropods, etc.

IV.—SUMMARY OF THE PLANKTONIC ORGANISMS.

A.—PROTOPHYTES OF THE PLANKTON.

The *unicellular plants* (*Protophyta**) have very great importance in the physiology of the plankton and the cycle of matter in the sea (*Stoffwechsel des Meeres*), for they furnish by far the greater part of the fundamental food (*Urnahrung*). The inconceivable amount of food which the countless myriads of swimming marine animals consume daily is chiefly derived, directly or indirectly, from the planktonic flora, and in this the unicellular protophytes are of much greater importance than the multicellular metaphytes. Nevertheless the natural history of these small plants has thus far been very much neglected. As yet no botanist has attempted to consider the planktonic flora in general, and its relation to the planktonic fauna. Only that single class, so rich in forms, the diatoms, has been thoroughly investigated and systematically worked up; as regards the other groups, not a single attempt at systemization has been made; and many simple forms of great importance have lately been recognized for the first time as unicellular plants. I must, therefore, limit myself here to a brief enumeration of the most important groups of the plankton flora. Its general extent and quantitative development have in my opinion hitherto been much undervalued, and with reference to the cycle of matter in the sea (*Stoffwechsel des Meeres*) deserve a thorough consideration. I find masses of various protophytes everywhere in the plankton, and suspect that they have been neglected chiefly because of their small size and inconspicuous form. Many of these, indeed, have been regarded as protozoa or as eggs of planktonic metazoa.

As a foundation for a most important province of botany, the classification of the protophytes, we must keep in the foreground the following considerations: (1) The kind of reproduction, whether by simple division (*Schizophyta*) into two, four, or many parts, or by formation of motile swarm-spores, *Mastigophyta;* (2) the constitution of the phytochroms, of yellow, red, or brown pigment, which is distributed in the protoplasm of the cell (usually in the form of granules), and has great significance in assimilation (chlorophyll, diatomin, erethrin, phæodin, etc.); (3) the morphological and chemical constitution of the *cell-membrane* (cellulose, siliceous, capsular, or bivalvular, etc.). So long as we hold to the present view of the vegetable physiologists, that for the fundamental process of vegetal assimilation, for the synthesis of protoplasm and amylum, the presence of the vegetal pigment matter is necessary, we can regard as true protophytes only such unicellular organisms as are provided with such a phytochrom, but we will have to

---

*The separation of the *Protophyta* from the *Metaphyta* is as justifiable as that of the *Protozoa* from the *Metazoa*. The latter form tissues, the former do not. (Compare Naturl. Schöpfungsgeschichte, VIII Aufl., 1889, pp. 420–453.)

include here a great number of protista, which have hitherto been reckoned as protozoa, *e. g.*, the *Murracyteæ*, *Dictyocheæ*, *Peridineæ*. As characteristic and important protophytes of the plankton I here mention seven groups: (1) *Chromaceæ*, (2) *Calcocyteæ*, (3) *Murracyteæ*, (4) *Diatomeæ*, (5) *Xanthelleæ*, (6) *Dictyocheæ*, (7) *Peridineæ*.

1. *Chromaceæ* (30, p. 452).—In this lowest vegetable group is probably to be placed a number of small "unicellular algæ" of simplest form, which occur in great abundance in the plankton, but on account of their minute size and simple spherical shape have for the most part been overlooked, or possibly regarded as germ cells of other organisms. They may here be provisionally distinguished as *Procytella primordialis*. The diameter of the spherical cells in the smaller forms is only about .001 to .005 mm., in the larger .008 to .012 mm, seldom more. Usually each cell contains only one phytochrom granule of greenish color, sometimes approaching a yellow or red, sometimes a blue or brown. Whether there is also a diminutive nucleus is doubtful. Increase takes place simply by division into two or four parts, and appears to go on with excessive rapidity, but swarm spores do not appear to be formed. Hundreds or thousands of such green spheres may be united in a mass of jelly. The decision whether these simplest *Chromaceæ* belong to the *Chlorcocceæ* or *Protococceæ*, or to some other primitive protophytic group, must be left to the botanist for further investigation, as well as the question whether these diminutive *Procytellæ* are actually true nucleated cells or only unnucleated cytodes. For our plankton studies these are of interest only so far as they develop in astonishing quantities in many (the colder) regions of the ocean, like the diatoms; and with the latter form a great part of the fundamental food (*Urnahrung*). Over wide areas the sea is often colored brown or green, and they form the chief food (described as *Protococcus marinus*) of inconceivable myriads of copepods, as Kükenthal has mentioned in his "Contributions on the Fauna of Spitzbergen."

2. *Calcocyteæ*.—In the eighth edition of the "*Naturliche Schopfungsgeschichte*" (30, p. 437) I have designated as *Calcocyteæ* or "unicellular calcareous algæ" those important minute organisms which, as "*Coccosphæra, Cyathosphæra*, and *Rhabdosphæra*, play a great rôle in oceanic life. They are found abundantly in the plankton of the tropical and subtropical seas, less abundantly in colder zones, and are never absent where pelagic *Thalamophora* occur in great numbers. Like the latter, they are bathypelagic. The ball of protoplasm which completely fills the interior of the small calcareous-shelled plastid seems, when stained red with carmine or brown with iodine, to be unnucleated, and therefore a cytode. The beautiful calcareous plates which compose the shell (*Coccolitha, Cyatholitha, Rhabdolitha*), and which in the *Rhabdosphæra* bear a radial spine, fall apart after death and are found in great numbers in all parts of the warmer oceans and in the globigerina ooze of the bottom. Murray (5, p. 533; 6, p. 939) and Wyville Thompson (14, I, p. 222)

were the first to demonstrate the wide distribution and innumerable abundance of this unicellular calcareous alga, and I agree with them in the supposition that these play a significant part in the biology of the ocean and in the formation of its globigerina ooze.

3. *Murracyteæ.*—Under this name I may here refer to the very important but hitherto neglected group of planktonic protophytes, which were first discovered by John Murray and described under the name *Pyrocystis* (5, p. 533, plate XXI; 6, pp. 935–938). These "unicellular algæ" are transparent vesicles, from 0.5 to 1 or 1.5 millimeters in diameter, and spherical, oval, or spindle-shaped in form. Their simple continuous cell membrane is very thin and fragile, like glass. It is stained blue by iodine and sulphuric acid, and seems to contain a small quantity of siliceous earth. The contents of the vesicle is a vacuolated cell, whose protoplasmic network contains many yellow granules of diatomin. The spherical form (*Pyrocystis noctiluca* Murray) is very similar in size and form to the common *Noctiluca miliaris* and probably is very often mistaken for it. I saw these thirty years ago (1860) at Messina, and later (1866) at Lanzarote, in the Canary Islands.

When John Murray published in 1876 the first figures and careful description, he at first placed them with the diatoms, but later (6, p. 935) he has, with justice, separated them. He there says of *Pyrocystis noctiluca*:

This organism is everywhere present, often in enormous masses, at the surface of the tropical and subtropical oceans, where the temperature is not more than 20° to 21° C., and the specific gravity of the oceanic water is not diminished by the presence of coast and river water. *Pyrocystis* shines very brightly; the light comes from the nucleus and is *the chief source of the diffuse phosphorescence of the equatorial oceans* in calm weather.

Since these unicellular vegetable organisms do not have the characteristic bivalve shell or siliceous case of the diatoms, but their cell membrane forms a completely closed capsule, they can not be reckoned with the latter, but must be regarded as representatives of a different group of protophytes, for which I propose the name *Murracyteæ* or "glass bladders" (*Murra*, a name given by the Romans to a glasslike mineral—fluospar (?)—from which costly articles are made.)*

---

*In the Atlantic and Indian oceans I have seen great masses of *Murracyteæ*, and have distinguished many species, which may be regarded as representatives of four genera: (1) *Pyrocystis noctiluca* Murray; spherical. (2) *Photocystis ellipsoides* Hkl; ellipsoid. (3) *Murracystis fusiformis* Hkl (*Pyrocystis fusiformis* Murray); spindle-shaped. (4) *Nectocystis murrayana* Hkl; cylindrical. The *Murracytes* multiply, as it appears, only by simple division (commonly into two parts, less frequently into four). After the nucleus, lying eccentrically or against the cell wall, has divided, there follows division of the soft cell body, which is separated from the firm capsulelike membrane by a wide space (filled with a jelly). Then the membrane bursts, and around the two halves or four tetrads there is immediately formed a new covering. Considered phylogenetically, the *Murracytes* appear as very old oceanic *Protophytes* of very simple structure. Perhaps they ought to be regarded as the ancestral form of the diatoms, for the bivalvular shell of the latter could have arisen by a simple halving of the capsule of the former.

4. *Diatomeæ.*—The inconceivable quantities in which the diatoms populate the whole ocean and the extraordinary importance which they possess as one of the most important constituents of the "fundamental food supply" (*Urnahrung*) in the cycle of matter in the sea has been considered so many times that it is sufficient here to point to the comparatively recent accounts of Murray (5, p. 533; 6, p. 737, etc.), Fuchs (12, p. 49), Castracane (6, p. 930), and Hensen (9, p. 80). Earlier the chief attention was paid to the benthonic diatoms which everywhere cover the seacoast and the shallow depths of the sea bottom in astonishing quantities; in part fixed on stalks, in part slowly moving among the forests of seaweed and the fixed animal banks (*festsitzenden Thierbanken*) of the coast. The importance of the planktonic diatoms was recognized much later, those abounding in the open ocean as well as in the coast waters furnishing one of the most important sources of food for the pelagic animals. The *oceanic diatoms*, which often cover the surface of the open sea as a thick layer of slime, form another flora, very insufficiently studied and characterized by many forms of colossal size (several millimeters in diameter), peculiarly regular in form, and with extremely thin-walled siliceous shells (species of *Ethmodiscus, Coscinodiscus, Rhizosolenia*, etc., discovered in such numbers by the *Challenger*). The *neritic diatoms*, on the other hand, which, swimming free in no small numbers, populate the coast waters, are less in diameter and with thicker walls, and stand on the whole between the oceanic and littoral forms. The absolute and relative quantity of the planktonic diatoms seems to increase gradually from the equator towards both poles.

In the tropical zone the pelagic diatoms are much less developed than in the temperate zone, and here again much less than in the polar zone. Wide stretches of the Arctic Ocean are often changed by inconceivable masses of diatoms into a thick dark slime, the "black water," which forms the feeding-ground of whales. The pteropods and crustaceans, upon which these cetaceans live, feed upon this diatom slime, the "black water" of the Arctic voyager. Not less wonderful are the vast masses of diatoms which fill the Antarctic Ocean south of the fiftieth degree of latitude, and whose siliceous shells, sinking to the bottom after the death of the organism, form the diatom ooze (*Challenger*, stations 152–157). The tow nets here were quickly filled with such masses of diatoms (for the most part composed of *Chætoceros*) that these when dried in the oven formed a thick matted felt (6, p. 920).

5. *Xanthelleæ.*—A highly important share in the cycle of matter in the sea belongs to the remarkable *xanthelleæ* or "yellow cells," which live in *symbiosis* in the bodies of many marine animals, in the plankton as well as in the benthos. I first proved that these "yellow cells," which were observed by Huxley (1851) and by Johannes Müller (1858) in the calymma of radiolarians, were "undoubted cells," and also described their structure and increase by division (3, p. 84), and later (1870) showed that they constantly contained amylum (4, § 90). But Cien-

kowski first advanced the view that the yellow cells are independent unicellular organisms, parasitic algæ, which for a time live in the bodies of the radiolarians, but after the death of the latter come forth and multiply by division. This supposition was confirmed experimentally by Karl Brandt (24, p. 65) and Patrick Geddes, who explained further the nature of their symbiosis, and finally showed the wide distribution of the *xanthelleæ* in the bodies of numerous marine animals, as well as their production of zoöspores (*Zooxanthella, Philozoön*). Whether these are ontogenetically connected with certain "yellow unicellular algæ" which live free in the plankton, remains to be farther investigated. Perhaps also in this group belong the *Xanthidea* which were described by Hensen (9, p. 79) and Möbius (10, p. 124) as species of *Xanthidium* and as "spiny cystids," spherical cells which reach 1 millimeter in diameter, contain yellow diatomin granules, and multiply by division. Their thick hyaline shell, which seems to consist of slightly silicified cellulose, armed with simple or star-shaped radial spines, is characteristic. I find these *Xanthideæ* very numerous in the oceanic plankton. Perhaps the siliceous-shelled *Xanthidia*, which Ehrenberg has found so abundantly as fossils, also belong here.

6. *Dictyocheæ.*—The ornamented latticed cases of the *Dictyochidæ*, formed of hollow siliceous spicules, are often found in great numbers in the plankton, pelagic as well as zonary. Although these have long been known, both living and as fossils, to microscopists, two very different views as to their true nature are entertained.*

In a preliminary contribution "On the Structure of *Distephanus* (*Dictyocha*) *speculum*" Zoöl. Anzeiger, No. 334, one of my earlier students, Adolf Borgert, briefly showed that each single case contains an independent ciliated cell. He therefore considered it a new group of *Flagellata* (or *Mastigophora*), for which he proposed the term *Silicoflagellata*. The "twin parts" described by me (4, p. 1549) he regarded as a double case which had arisen through the conjugation of two individual *flagellata*. To my mind this new interpretation seems to have very considerable probability, although I do not regard it as settled that the ciliated cells are the swarm-spores of the *Phæodarium*. In case

---

* Ehrenberg, who in 1838 and 1841 first described the ornamented siliceous skeletons of *Dictyocha* and *Mesocena*, called them diatoms and distinguished no less than 50 species of them, some living, some fossil. Later, at Messina (1859), I noticed, inclosed within the ornamented hat-shaped latticed shell a small cell, and on that account referred it to the *Radiolaria*, with reference particularly to the similar siliceous skeletons of some *Nassellaria* (*Acanthodesmida*). Twenty years later R. Hertwig found a spherical *Phæodarium*, the surface of whose calymma was covered with numerous *Dictyocha* little hats (*Dictyocha-Hütchen*), and he therefore believed that they must belong to this legion. He compares the single siliceous little hats (*Hütchen*) with the scattered spicules of the *Sphærozoida*. In my *Challenger* report (4, p. 1558) I agreed with this interpretation; so much the more when I myself saw numerous similar *Phæcystina* (*Dictyocha stapedia*) living among a similar *Phæodaria* in Ceylon, and found specimens in several bottles of the *Challenger* collections, especially from Station 144, from the Cape of Good Hope (4, p. 1561, pl. 101, Figs. 10–12).

the greenish-yellow pigment granules in the protoplasm of the *Dictyochidæ* are chlorophyll or phytochrom, they must be placed with "unicellular algæ." If, as I believe, the supposition of Borgert is correct, then the masses of *Dictyochidæ* shells found so abundantly in the calymma of *Phæodariæ* can be regarded only as the empty shells of *Silicoflagellata*, which the skeletonless *Phæodina* has taken in as food. This supposition is much more probable since these, together with siliceous scales of diatoms and tintinnoids, have been found in great numbers in the calymma of other radiolarians. This case would then be analogous to two similar appearances which I myself have previously described, *Myxobrachia pluteus* (4, p. 22) and *Dalcaromma calcarea* (4, p. 70, § 102).

7. *Peridineæ* (*Dinoflagellata* or *Dinocytea*, earlier *Cilioflagellata*).—This group of *Flagellata* (or *Mastigophora*) earlier placed with the *Infusoria*, has lately, with more certainty, been recognized as a protophytic group with vegetable metabolism. They are represented in the plankton by numerous and, in part, remarkable and beautiful forms, a part of which have been lately figured by Stein under the name *Arthrodele flagellata*. Many such forms occur in the neritic, fewer in the oceanic plankton, and often in such masses that they take a great part in the formation of the fundamental food supply. Hensen correctly points out the great importance of these *Protista*, of whose quantity he attempted to give a conception by counting (9, p. 71). Many of these participate in a prominent way in the marine population (*Ceratium*, *Prorocentrum*, etc.). John Murray very often found chains of *Ceratium tripus* (each composed of eight cells) floating in the plankton of the open ocean, without ciliary movements, while the ciliated single cells inhabited the neritic plankton in vast numbers close to the shore. Sometimes these crowds of *Peridineæ*, like the diatoms, appeared so abundantly as to fill the tow net with a yellow slime (6, p. 934).

B.—METAPHYTES OF THE PLANKTON.

The only class of metaphytes which occurs in the plankton are the algæ. The great majority of this class, so rich in forms, belong to the littoral benthos; only a few forms have adopted the pelagic mode of life, and of these only two, from their great abundance, are of any considerable importance in the oceanic fundamental food supply, the *Oscillatoriæ* which live in the depths, and the *Sargassa* which grow at the surface. A third group, the *Halosphæreæ*, is much less abundant and important, but of considerable interest in many relations.*

---

* The *Oscillatoriæ* must be regarded as true algæ, since their characteristic "jointed threads" ("*Glieder-faden*") form an actual *Thallus*, and indeed a thread-like thallus, as in the *Confervæ*. But on the same grounds also we must regard as algæ the *Volvocineæ* and *Halosphæreæ* with spherical thallus; they are also multicellular *Metaphytes*, which show the simplest form of tissue (*Histones*, 30, p. 420). The foregoing prototypes, on the other hand, have no tissue, since the entire organism is only a simple cell (*Protista*, 30, p. 453).

1. *Halosphæreæ.*—Under the name *Halosphæra viridis*, Schmitz (1879) first described a new genus of green algæ from the Mediterranean, which appear floating in the plankton of the Gulf of Naples in great numbers from the middle of January until the middle of April. They form swimming hollow spheres, from 0.55 to 0.62 mm. in diameter, whose thin cellulose wall is covered within by a single layer of chlorophyll containing cells analogous to the blastoderm of the metazoic egg. Each of these epithelial cells divides later into several daughter cells, each of which forms four cone-shaped swarm-spores with two ciliated cells. I have known this green ball for thirty years. In February, 1860, I found them numerous in the plankton of Messina. I observed a second kind in February, 1867, at Lanzarote, in the Canary Islands. The hollow spheres found in the Atlantic are twice as large, and reach a diameter of 1 to 1.2 mm. They have pear-shaped swarm-spores. I named them *Halosphæra blastula*. Morphologically these hollow spherical algæ are of great interest, since they are directly comparable to the blastula (or blastosphere stage) of the metazoic embryo. As the latter is to be regarded as the simplest type of the metazoon, so *Halosphæra* (like *Volvox*) can be looked upon as the primitive ancestral form of the *Metaphyta* (4, p. 499). Hensen has lately found numerous living specimens of *Halosphæra viridis* in five hauls from a depth of 1,000 to 2,000 meters (10, p. 521). The light of the bathybic luminiferous animals may possibly be sufficient for their metabolic activity.

2. *Oscillatoriæ.*—Like the diatoms in the cold regions of the ocean, the oscillatoriæ (*Trichodesmium* and its allies) are found in the warm regions in inconceivable quantities. It is very certain that the latter, as well as the former, belong to the most important source of the "fundamental food supply." Ehrenberg in 1823 observed in the Red Sea, at Tur, such large quantities of *Trichodesmium erythræum* that the water along the shore was colored blood-red by them. Möbius has recently carefully described the same thing anew, and has (quite correctly) traced from it the name of the Red Sea (26, p. 7). Later, I myself found just as great numbers as these in the Indian Ocean at Maledira and Ceylon (25, p. 225). In Rabbe's collections are several bottles of plankton (from the Indian and Pacific oceans) entirely filled with them.* The *Challenger* found great quantities of *Trichodesmium* in the Arafura Sea and Celebes Sea (6, p. 545, 607), and also in the Guinea stream (6, p. 218); and between St. Thomas and the Bermudas (6, p. 136) wide stretches of the sea were colored by it dark red or yellowish brown. Murray found it only in the superficial, never in the deeper layers of the ocean.

3. *Sargasseæ.*—The higher algæ are represented in the planktonic flora only by a single group, the *Sargasseæ*, and these again are com-

---

*In the collection of *Radiolaria*, which may be purchased from the famulus Franz Pohle, at Jena, preparation No. 5, from Madagascar, contains many flakes of this *Oscillatoria*.

monly only of a single species, *Sargassum bacciferum*; but this has the greatest importance, since, as is known, it alone forms the floating sargasso banks, which cover such extensive portions of the ocean. Besides this very important species, other fucoids are found floating in the ocean, especially species of *Fucus* (*F. vesiculosus*, *F. nodosus*, and others). Still they never appear in such masses as the familiar "berry weed." The floating sargasso banks are well known to have their characteristic animal life, which Wyville Thompson accurately described and fittingly termed nomadic (14, II, pp. 9, 339).

This remarkable sargasso fauna bears the same character in both the Atlantic and the Pacific oceans and consists partly of benthonic animals, which live sessile or creeping on the sargasso weed, partly of planktonic organisms which swim among the weeds; the latter are more neritic than oceanic. Hensen has lately described this fauna as remarkably poor, and could only find 10 species of animals in it (9, p. 246). The *Challenger* found more than five times as many species in this same Atlantic sargasso, namely, 55 (6, p. 136). It is obvious that the remarkable negative results of Hensen on this as on other planktonic questions can have no value against the positive results of other investigators.

C.—Protozoa of the Plankton.

The two great chief groups of unicellular animals, *Rhizopoda* and *Infusoria*, occur in the ocean in very different proportions, in the reverse condition to that in fresh water.

The *Infusoria* (*Flagellata* and *Ciliata*), which chiefly form the protozoic fauna in the latter, are indeed represented in the sea by a great number of species, but the most belong to the littoral benthos, and only a few swimming species occur in such quantities that they are of importance in the plankton, the *Noctilucidæ* among the *Flagellata*, the *Tintinnoidæ* among the *Ciliata*. Much greater is the wealth of the ocean in *Rhizopoda*, calcareous-shelled *Thalamophora* and siliceous-shelled *Radiolaria*. The accumulated masses of these shells form the most important sediment of the ocean, while their unicellular soft bodies constitute the chief food supply for many planktonic animals.

Infusoria.—As is known, the *Infusoria* do not play so great a rôle in the life of the ocean as in that of the fresh water. It is true that a great number of *Flagellata* and *Ciliata* occur in the neritic or littoral fauna, but neither on account of the number of individuals nor the richness of forms are they elsewhere of importance, and only a few small groups extend out into the open sea. It seems as if these tender and for the most part uncovered *Protozoa* are not suited for the contest which the wild "struggle for existence" offers here. The armored rhizopods take their place. Still two small and very peculiar groups of *Infusoria* are found in very great numbers in the plankton, and sometimes in such quantities as to form the chief bulk; the *Noctiluca* among the *Flagellata*, and the *Tintinna* among the *Ciliata*. Both groups, and particularly the *Noctilucidæ*, belong to the neritic plankton. They occur in the oceanic only where the coast water flows in (6, pp. 679, 750, 933).

The common *Noctiluca miliaris* and some related species sometimes cover the surface of the coast waters in such masses as to form a thick reddish-yellow slime, often like "tomato soup," and at night are brightly luminous. The *Tintinnoidæ* (*Tintinnus, Dictyocysta, Codonella*) appear in smaller quantities, but often in great numbers. Some forms of these elegant *Ciliata* are oceanic.

Thalamophora (*Foraminifera*).—The *Thalamophora*, often and very properly called *Foraminifera*, were once generally regarded as benthonic. New observations first showed that a part of these are planktonic, and through the comprehensive series of observations by the *Challenger* the abundant occurrence of these pelagic *Foraminifera* and their great part in the formation of that most important sediment, the *Globigerina* ooze, was first established. All these *Thalamophora* of the plankton belong to the peculiar perforated *Polythalamia*, to the family of the *Globigerinidæ*; only *Orbulina* (provided it is independent) to the *Monothalamia*. The number of their genera (8–10) and species (20–25) is relatively small, but the number of individuals is inconceivably great. By far the most important and numerous belong to the genera *Globigerina, Orbulina*, and *Pulvinulina*; after these *Sphæroidina* and *Pullenia*. They occur everywhere in the open ocean in numberless myriads. J. Murray could often from a boat scoop up thick masses of them with a glass, and never fished with the tow net in 200 fathoms without obtaining some (5, p. 534). A few forms (*Hastigerina* and *Cymbalopora*) show more local increase in numbers, while others are rare everywhere (*Chilostomella, Candeina*). In the equatorial countercurrents of the Western Pacific, between the equator and the Caroline Islands, the *Challenger* found "great banks of pelagic foraminifera. On one day an unheard-of quantity of *Pulvinulina* was taken in the tow nets; on the following day they were entirely absent, and *Pullenia* was extraordinarily abundant." These important observations by Murray I can confirm from my own experience in the Atlantic and Indian oceans* (comp. 3, pp. 166, 188).

---

*The important relations of these pelagic *Polythalamia* to the rest of the fauna of the plankton on the one side, as well as its importance in the formation of the "*Globigerina* ooze" on the other, has been expressly stated by Murray (6, p. 919). I agree completely with him in the view that these oceanic *Globigerinidæ* are true pelagic rhizopods, which in part are found swimming only at the surface or at slight depths (autopelagic), in part at zones of different depths (zonary), but they are not benthonic. The enormous sediment of "*Globigerina* ooze" is composed of the sunken calcareous shells of the dead pelagic animals. On the other hand, the benthonic thalamophores, living partly abyssal, on the bottom of the deep sea, partly littoral, creeping among the forests of seaweed on the coasts, are of other species and genera. They develop a much greater variety of form. The neritic thalamophores found swimming in the coast waters are in part again characterized by various forms.

Radiolaria.—No class of organisms has remained so long unknown to us, and by the brilliant discoveries of the last decade has been suddenly placed in so clear a light, as the *Radiolaria* (comp. 4, § 251–260). For half a century we knew next to nothing of these wonderful rhizopods; to-day they appear as one of the most important planktonic classes.* These, the most varied in form of all the unicellular organisms, form a *purely oceanic class*, and live and swim in all seas, especially in the warmer ones. Numerous species are also found near the coasts, yet these are not distinguishable from those of the open sea. They constitute no separate neritic fauna.

Vast crowds of *Radiolaria* occur at the surface of the ocean, as well as at different depths. Long ago Johannes Müller remarked:

> It is a great phenomenon that *Acanthometra* can be taken daily by thousands in a calm sea and independently of storms; and that of many species of *Polycystina*, hundreds of individuals were seen during my last residence at the seashore (2, p. 25).

I have tried myself, on the hundreds of voyages to different coasts which I have made since 1856, to thoroughly study the natural history of the *Radiolaria*. The incomparable collections of the *Challenger* afforded me by far the richest material for observation. The results obtained therefrom are embodied in the report (1887). Among other references to the conditions of the plankton there mentioned, it brought up the following propositions: (1) *Radiolaria* occur abundantly in all seas which contain a medium amount of salt, and which do not (like the Baltic) receive a strong influx of fresh water. (2) In the colder seas only a few species occur (chiefly *Acantharia*), but immense quantities of individuals; towards the equator the variety in form gradually increases (horizontal distribution, comp. 4, § 226–231). (3) The chief groups of *Radiolaria* are distributed unequally in the five *bathyzones* or girdles of depth of the open ocean. The subclass *Porulosa* (the two legions of *Spumellaria* and *Acantharia*) inhabit especially the two upper zones. On the other hand, the subclass *Osculosa* (*Nasselaria*

---

*After Ehrenberg, in 1847, had described the siliceous shells of some hundred species from the Barbados, we obtained in 1858 the first description of their organization through Johannes Müller. In the work with which this great master closed his renowned life he described 50 species which he had observed alive in the Mediterranean Sea (2). When in continuation of this I devoted a winter's residence in Messina to their further investigation, I was able in 1862, in the monograph consequent thereupon, to distinguish 144 new species, in all 113 genera and 15 families (3). But this rich *Radiolaria* fauna of Messina still gave no promise of the immense quantities of these delicately ornamented creatures peopling the open ocean, and whose variously formed siliceous shells, sinking to the bottom after death, formed that wonderful sediment, the "*Radiolaria ooze.*" This was first discovered thirteen years later by the *Challenger*. The investigation of the fabulous radiolarian treasures (chiefly from the Pacific) which this expedition brought home has led to the discrimination of 20 orders, 85 families, 739 genera, and 4,318 species (4, § 256). Further study of the *Radiolaria* slime of the deep sea will bring to light many new forms from this inexhaustibly rich mine.

and *Phæodaria*) move in the three lower zones (vertical distribution, 4, § 232-239). The dependence of their appearance upon the various conditions of life has been investigated by Brandt (24, p. 102).

D.—COELENTERATES OF THE PLANKTON.

The ancestral group of the *cœlenterates* has important significance and manifold interest for the natural history of the plankton; still this applies in very varied degrees to the different principal groups of this numerous circle (comp. 30, p. 522). The great class of the *sponges*, which belongs exclusively to the benthos, has never acquired a pelagic habit of life. The phylum of the *platodes* also needs no further reference here. We know, to be sure, a small number of pelagic turbellarians and trematodes. Arnold Lang, in his monograph on the sea-planarians or polyclads (1884, p. 629), mentions as "purely pelagic" or oceanic 8 species and 4 genera (*Planocera, Stylochus, Leptoplana, Planaria*). Parasitic trematodes are occasionally found as "pelagic parasites" in medusæ, siphonophores, and ctenophores; but these trematodes and turbellarians are usually found only individually; they never appear in such quantities as are characteristic of the majority of the plankton animals. Much more important for us is the third type of the cœlenterates, the diversified chief group of the nettle animals or *Cnidaria* (30, p. 524).

Cnidaria.—With reference to the mode of life and the form conditioned thereby, one may divide the whole group of *Cnidaria* into two great principal divisions, polyps and acalephs, which since the time of Cuvier have lain at the foundation of the older systems. The polyps (in the sense of the older zoölogists) embrace all nettle animals, which are fixed to the bottom of the sea, hydropolyps as well as scyphopolyps (*Anthozoa*). They belong exclusively to the benthos. Only a few forms have acquired the pelagic mode of life (*Minyadæ, Arachanactis*, larvæ of *Actiniæ, Cerinthidæ*, and some other corals). The second principal division of the nettle animals, the *Acalepha*, embraces, in the sense of their first investigator Eschscholtz (1829), the three classes of medusæ, siphonophores, and ctenophores; all swimming marine animals, which, from their richness in forms, their general distribution in the ocean, and their abundant occurrence, possess much importance for plankton study. Since the above-mentioned pelagic polyps (*Minyadæ*, etc.) on the whole are rare, and never appear in great quantities, we need make no further reference to them here. Much more important are the *Acalephs*, which offer a fund of interesting problems for plankton study. Commonly, all these animals are roughly termed "pelagic," but a new consideration shows us that they are so in a very different sense, and that the distinction which we have made above in reference to their chorological terminology here finds its complete justification. We will first consider the medusæ, then the siphonophores and ctenophores.

Medusæ.—The great interest which I have felt in this wonderful class of animals since my first acquaintance with living medusæ, in 1854, and which has been increased by my numerous sea voyages, led me to the monographing of them (1879). I immediately gained thereby a number of definite chorological and œcological ideas, which have been of permanent influence in the further course of my plankton studies. By it was definitely fixed the knowledge that the whole race of the medusæ is *polyphyletic*, and that on the one side the *Craspedota* (or *Hydromedusæ*) have arisen independently from the *Hydropolyps*, just as on the other side the *Acraspedota* (or *Scyphomedusæ*) from the *Scyphopolyps*. In both analogous cases the transition to the pelagic, free-swimming mode of life has led to the formation, from a lower, sessile, very simply organized benthic animal, of a much higher planktonic metazoön, with differentiated tissues and organs—a fact which is of great significance for our general understanding of the phylogeny of tissues.

I have in that monograph broadly distinguished two principal forms of ontogeny or individual developmental history among the medusæ, *metagenesis* and *hypogenesis*. Of these I regard metagenesis, the alternation of generations with polyps, as the primary or *palingenetic* form; on the other hand, *hypogenesis*, the "direct development" without alternation of generations, as the secondary abbreviated or *cenogenetic* form. This distinction is of great importance in the chorology, in so far as the great majority of the *oceanic* medusæ are *hypogenetic;* the *neritic*, on the other hand, are *metagenic*. To the oceanic medusæ in the widest sense I refer the *Trachylinæ* (*Trachymedusæ* and *Narcomedusæ*) among the *Craspedota;* to the neritic, the *Leptolinæ* (*Anthomedusæ* and *Leptomedusæ*: comp. 29, p. 233). While the former have lost their relation to the benthonic polyps, the latter have retained it through heredity. The same seems to obtain also for the majority of the *Acraspedota*, namely the *Discomedusæ*. Among these there are only a few oceanic genera with *hypogenesis*, *e. g.*, *Pelagia*. The development of the smaller but very important acraspedote orders, which I have distinguished as *Stauromedusæ*, *Peromedusæ*, and *Cubomedusæ*, is, I am sorry to say, as yet quite unknown. The first is to be regarded as neritic and metagenic; the two latter, on the other hand, oceanic and hypogenic. That the majority of the large *Discomedusæ* are neritic and not oceanic is shown from their limited local distribution.

Although ten years ago the *Medusæ* were generally held to be purely pelagic animals, it has now been found that a certain (perhaps considerable) part of them are zonary or bathybic. Among the 18 deep-sea *medusæ* which I have described in part XII of the *Challenger* Report (1881) there are, however, some forms which occur also at the surface, and a few which perhaps were accidentally taken in the tow net while drawing it up. But others are certainly true deep-sea dwellers, as the *Pectyllidæ* among the *Craspedota*, the *Periphyllidæ* and *Atollidæ* among the *Acraspedota*. Some *Medusæ* have partly or entirely given up the swimming mode of life, as *Polyclonia*, *Cephea*, and other *Rhizostoma*, which lie with the back towards the sea bottom, the many-mouthed bunch of tentacles directed upwards. The *Lucernaridæ* have completely passed over to the benthos. Many *Medusæ* are spanipelagic, rise to the surface only during a few months (for the purpose of reproduction?), and pass the greater part of the year in the depths; thus in the Mediterranean the beautiful *Cotylorrhiza tuberculata*, *Charybdea marsupialis*, *Tima flavilabris*, and *Olindias mülleri*. These bathybic forms are sometimes brought up in great numbers with the bottom net (19, p. 122). Many cling with their tentacles to *Algæ* and other objects (20, p. 341).

The immense swarms in which the *Medusæ* sometimes appear, millions crowded thickly together, are known to all seafaring naturalists. Thus in Arctic waters, *Codonium princeps*, *Hippocrene superciliaris;* in the North Sea, *Tiara pileata*, *Aglantha digitalis;* in the Mediterranean, *Liriantha mucronata*, *Rhopalonema velatum;* in the tropics, *Cytæis nigritina;* in the Antarctic Ocean, *Hippocrene mocloriana* and others. Hensen (9, p. 65) in the North Sea found a swarm of *Aglantha*, the number of which he estimated at twenty-three and one-half billions. The extent of the multitude was so great that "the thought of approximately estimating the animals in this swarm must be given up." In such cases the whole sea for a few days, or even weeks, seems everywhere full of *Medusæ;* and then again weeks, or even months, may pass without finding an individual. The *uncertainty* of appearance, the "capriciousness of these brilliant beauties," in other words the dependence upon many different, and for the most part unknown causes, is in this interesting animal group remarkably impressed upon us. I will, therefore, in another place, refer to it on the ground of my own experience.

Siphonophores.—What I have said above concerning the unequal distribution of the medusæ applies also to their wonderful descendants, the purely *oceanic* class of the siphonophores. This highly interesting class was, up to a few years ago, also regarded as purely pelagic; but of these, too, it is now known that they are in great part bathypelagic, in part also zonary and bathybic. The new and very peculiar group of the *Auronectæ* (*Stephalidæ* and *Rhodalidæ*), taken by the *Challenger* at a depth of 200 to 600 fathoms, is described in my "Report of the Siphonophores of H. M. S. *Challenger*" (1888, p. 296). The *Bathyphysa* taken by Studer, and some of the *Rhizophysidæ* (*Aurophysa*, *Linophysa*) captured by the *Gazelle*, were taken at a depth of 600 to 1,600 fathoms (l. c.). But that such deep-sea siphonophores (probably mostly *Rhizophysidæ*) inhabited the ocean in great masses was first shown by Chierchia (8, p. 84–86). Previously, in numerous soundings which the *Vettor Pisani* had made in the Atlantic and Pacific oceans, the line of the deep-sea lead when drawn up was found to be wound around with the torn-off stinging tentacles of great siphonophores. By means of the new closible net invented by Palumbo, he was enabled to bring up the entire animals from definite depths. From these experiments Chierchia concluded "that certain characteristic species of siphonophores live in great numbers at certain depths, from 1,000 meters above the bottom upwards, the strongest and most resistant in the depths, the weaker higher up" (8, p. 86). Other siphonophores, which belong to the forms most numerous at the surface, extend down to considerable depths, as *Diphyes sieboldii* (15, p. 12). The larvæ of *Hippopodius luteus*, which are very numerous in winter and spring, have quite disappeared in summer, and, according to Chun, live in greater depths, even to 1,200 meters (15, p. 14). Other forms are spanipelagic and come to the surface only for a short time, only a few weeks in the year, like so many *Physonectæ*. From these and other grounds the participation of the siphonophores in the plankton, like that of their ancestors, the *Hydromedusæ*, is extremely irregular, and their appearance at the surface of the sea is subject to the most remarkable changes.

Ctenophores.—This *Cnidarian* class also, like the preceding, is purely oceanic, not neritic. They also show the same phenomena of pelagic distribution as the *Siphonophores* and *Medusæ*, frequent appearance in great swarms, sudden disappearance for long periods, unaccountable irregularity in their participation in plankton formation. The tables which Schmidtlein has given on the basis of three years' observations, on their periodical appearance in the Gulf of Naples, are very instructive for all three classes of the planktonic *Cnidaria* (19, p. 120). The ctenophores also, up to a short time ago, were regarded as autopelagic animals; but of them also it has been discovered that they extend in abundance to various, somewhat definite depths. Chun, in his monograph of the ctenophores of Naples (1880, p. 236–238) has pointed out that these most tender of all pelagic animals have just as definite vertical as horizontal migrations. Many ctenophores, which in the spring are found as larvæ at the surface, later sink, pass the summer in the cooler depths, and rise to the surface in the autumn in crowds, as mature animals. The irregularity of their appearance is also mentioned by Graeffe (20, p. 361).

E.—HELMINTHS OF THE PLANKTON.

The race of the helminths or "worms" (the cross of suffering for systematic zoölogy) obtains a more natural unity and more logical definition, if one removes therefrom the platodes and annelids, placing the former with the cœlenterates, the latter with the articulates. The justice of this limitation and also the grounds for regarding the worms as the common ancestral group of the higher animals, I have set forth already in the "Gastrea Theory" (1873), and many times at later opportunities, last in the eighth edition of my "Natural History of Creation" (1889, p. 540). There remain then as helminths, in the narrower sense, four divisions with about 12 classes, namely, (1) the *Rotatoriæ*

(*Trochosphæra, Ichthydina, Rotifera*); (2) the *Strongylariæ* (*Nematoda, Acanthocephala, Chætognatha*); (3) the *Rhynchocœla* (*Nemertina, Enteropneusta*), and (4) the *Prosopygiæ* (*Bryozoa, Brachiopoda, Phoroneæ, Sipunculeæ*). The larvæ of many of these worms have acquired the pelagic mode of life, but most of them are too small and too scattered in the plankton to be of any considerable importance in its composition.

Chætognatha.—In its mature condition only a single class of helminths plays an independent and indeed an important rôle in the plankton—the small and peculiar class of arrow-worms or *Chætognatha* (*Sagitta, Spadella*, etc.). These, together with the copepods, salpæ, pteropods, and radiolarians belong to the most substantial, most generally distributed, and usually unfailing constituents of the plankton. Hensen (9, p. 59) has made some calculations of the immense numbers in which they appear. He reckons them in the "perennial plankton," yet does not find "everywhere the regularity which one might expect." He is astonished at the "highly remarkable variations" in their numbers, and finds this very unequal distribution very puzzling (9, p. 60). Chun has lately shown that the troops of *Sagitta* not only populate the surface of the sea, but also "in common with the *Radiolaria, Tomopteridæ, Diphyidæ, Crustacea*, constitute the most numerous and most constant inhabitants of the greater depths. In countless multitudes they are taken in the open as well as in the closible net, from 100 meters down to 1,300 meters" (15, p. 17). It seems that *Sagitta*, as a whole purely *oceanic*, is represented by pelagic as well as zonary and bathybic species.

F.—MOLLUSKS OF THE PLANKTON.

The race of mollusks play a very important rôle in the plankton. Although the great majority of the genera and species belong to the benthos, yet there are a few families which have become adapted to the pelagic mode of life, of great importance on account of the great swarms in which they often appear. The three chief classes which we distinguish in this race (30, p. 546) live very differently. The *Acephala*, entirely benthonic, can take part only as swarming larvæ in the composition of the plankton; so also the swimming larvæ of many meroplanktonic *Gastropoda*. Of these latter only a very few genera have adopted completely the pelagic mode of life, like *Ianthina* among the prosobranchs, *Glaucus* and *Phyllirrhæ* among the opisthobranchs.

Pteropods and Heteropods.—These two groups of snails are holoplanktonic, chiefly nyctipelagic animals, which come to the surface of the sea, preferably during the night, in vast numbers (14, pp. 121-125). Chun has lately discovered that many of them are found at considerable depths (15, p. 36). Some kinds of pteropods (*e. g., Spirialis*) seem to belong to the zonary and bathybic fauna. The heteropods are on the whole of less importance. They occur in great swarms less frequently and only in certain parts of the warmer seas. The pteropods on the other hand surpass the former, not only by a great diversity of genera and species, but particularly from their enormous development in all parts of the ocean. *Clio* and *Limacina* are known to occur in the Arctic and Antarctic ocean in schools so vast as to form the chief food supply of the whales; the swarms of *Creseis*, *Hyalea*, and others which appear in the seas of the warmer and temperate zones, are also so considerable that these fluttering "sea butterflies (*Farfalle di mare*)" often play a very important part in the "cycle of matter in the sea" (*Stoffwechsel des Meeres*"). The irregularity of the distribution and phenomena is also shown by the fact that Hensen, during his plankton expedition through the North Sea (July and August, 1887), completely missed the pteropods (9, p. 59; 10, p. 116). On the other hand, when in August, 1879, I fished at Scoury, on the northwest coast of Scotland, we found such immense quantities of *Limacina* (during the forenoon in still weather) that these pteropods certainly formed more than nine-tenths of the entire plankton, and with a bucket we could scoop up many thousands. The mass of the swarm had the same density for a depth of two fathoms and for more than a square kilometer in horizontal extent.

Cephalopods.—Although entirely swimming animals, these highly developed mollusks for the most part do not fall under the term plankton, if with Hensen we limit this to those "animals floating involuntarily in the sea" (9, p. 1). They must then be included in the "nekton;" but naturally it depends in some cases entirely on the strength of the current whether the small cephalopods should be included in the former or in the latter. In any case this highest developed class of mollusks is of very great importance in the physiology of the plankton, the question of the "cycle of matter in the sea." On the one hand they daily consume vast masses of pteropods, crustacea, sagitta, medusæ, and other planktonic animals; on the other, they furnish the most important food for fishes and cetaceans. From recent investigations it is found that the cephalopods are partly pelagic, partly zonary or bathybic (*Spirula*, *Nautilus*, etc.). Characteristic small, transparent *Decolenæ* (*Loligopsidæ*) are known as partly pelagic, partly bathybic species (15, p. 36). The same is true also of some *Octolenæ* (*Philonexidæ*). Young forms of cephalopods are captured swimming in the plankton at the surface as well as in the depths.

G.—ECHINODERMS OF THE PLANKTON.

The rayed animals in their significance in the plankton, as also in many other morphological and physiological relations, show highly peculiar and varied conditions. Although all echinoderms are without exception purely marine animals, and no single form of this great group inhabits fresh water, still not a single species has completely adopted the planktonic life. Not a single echinoderm in its full-grown and sexually mature condition can be called pelagic. The few forms which temporarily swim about (*Comatulidæ*) belong only to the neritic fauna and do not occur in the ocean. They also are found in such limited numbers that they are without importance for the plankton.

Much more important for us are the free-swimming echinoderm larvæ, which often play a great part in the neritic plankton. Indeed they are classical objects in the history of plankton investigation; for to their study their discoverer, Johannes Müller, forty-five years ago first applied the method of "pelagic fishery with the fine net," which soon led to such remarkable and brilliant results. The distribution and number of the larval rayed animals is naturally dependent upon that of their benthonic parents; but in addition also partly upon chorological, partly œcological causes. According to Sir Wyville Thompson (14, II, pp. 217–245; 6, p. 379), the remarkable metamorphosis, discovered and described in a masterly way by Müller, is the rule only among the littoral forms, chiefly in the temperate and warm zones; on the other hand, it is the exception in the case of the majority, for star animals of the deep sea and cold zones, in the Arctic as well as in the Antarctic, develop directly. Therefore, great troops of pelagic larvæ of these animals occur commonly only in the *neritic plankton* of the temperate and warm zones, not in the open ocean. They seem to visit the depths (below 100 meters) very seldom (15, p. 17). Besides, their appearance is naturally connected with the time of year of this development; often only during a few months (9, p. 62). The variation in the constitution of the "periodic plankton" is here very remarkable.

H.—ARTICULATES OF THE PLANKTON.

Of the three chief divisions which we distinguish in the group of articulated animals (30, p. 570) two, the *Annelids* and *Tracheates*, take no part in the constitution of the plankton. Both are represented only by a few pelagic genera, and these have a limited distribution. Much greater in importance is the third chief division, the *Crustacea*. It is the only animal class which is never lacking in the tow-net collections (or only very exceptionally), and which commonly appears in such numbers that their predominant position in the animal world of the sea is evident at the first glance. This applies as well to the oceanic as to the neritic fauna, to the littoral as to the abyssal benthos.

Annelids.—The great mass of this group, so rich in forms, belongs to the benthos, and is represented in the abyssal as well as in the littoral fauna by numerous creeping and sessile forms. Only very few ringed animals have acquired the pelagic mode of life and have assumed the characteristic hyaline condition of the oceanic glasslike animals, the swimming *Tomopteridæ* and *Alciopidæ*. Both families are represented in the plankton only by a few genera and species, and as a rule their number of individuals is not very considerable. Chun has lately shown by means of the closible net that both forms, *Tomopteris* as well as *Alciope*, are represented in the different depths, from 500 to 1,300 meters, by peculiar zonary and bathybic species, which are distinguishable from the pelagic species of the surface by characteristic marks. "The wealth in such *Alciopidæ* (and *Tomopteridæ*) at all depths of 100 meters or over is very surprising, and it requires a careful scrutiny, for the beautiful transparent worms often press actively by dozens in serpentine course through the crowd of other forms in the dishes" (15, p. 24).

Crustacea.—In their general œcological importance, in their universal distribution over all parts of the ocean, and especially in their incomprehensible fertility and the abundance of their appearance conditioned thereby, the *Crustacea* surpass all other classes of animals. In the physiology of the plankton the first rank in the animal kingdom belongs to them, as to diatoms in the vegetable kingdom. On the whole, in the organic life of the ocean they have the same predominant importance as the insects for the fauna and flora of the land. In a similar way, as the complicated "struggle for existence" has called up for the latter a quantity of remarkable œcological relations and morphological differences conditioned thereby within the insect class, so has the same occurred in the ocean within the crustacean class. Meanwhile the numerous orders and families of this class, so rich in forms, participate in very different degrees in the constitution of the plankton. The order of copepods by far surpasses all other orders. Next to these follow the ostracods and schizopods, then the phyllopods, amphipods and decapods. The other orders of crustaceans participate in the constitution of the plankton in a much less degree—part of them very little. It is to be added that larvæ of all orders may appear in great numbers therein. Thus, for example, the pelagic larvæ of the sessile benthonic cirripeds often appear in the neritic plankton so numerously that they constitute four-fifths to nine-tenths or even more of the entire mass.

The chorology of the *Crustacea* offers to the plankton investigator one of the most important and interesting fields of work, the elaboration of which has yet scarcely been begun. The same applies also to the geography and topography of the oceanic and neritic *Crustacea*, both in their horizontal and vertical distribution, to their relations to the benthonic *Crustacea* as well as to the marine fauna and flora in general. As a very important result of the recent discoveries, particularly of the *Challenger*, the fact must here as elsewhere be brought up that in the different groups of *Crustacea* (just as in the *Radiolaria*) the vertical divisions of the planktonic fauna can be very plainly distinguished. Pelagic, zonary, and bathybic forms are found here in quite definite relations.

Copepoda.—As the *Crustacea* are on the whole the most important and influential among the planktonic animals in their œcological relations, so are the copepods among the *Crustacea*. Only one who has seen with his own eyes can gain a conception of the innumerable masses in which these small crustaceans crowd the surface of the ocean as well as the zones of different depths. For days the ship may sail through wide stretches of ocean whose surface always remains covered with the same yellowish or reddish "animal mush," composed in by far the greater part of copepods. In the journal which I kept in the winter of 1866–67, at Lanzarote, in the Canary Islands, of the varying constitution of the plankton, for many days there is only the remark: "almost pure buckets of copepods," or "the collection consisted almost entirely of *Crustacea*, by far the greater part of copepods." That these small crustaceans form the chief food supply for many of the most important food-fishes (*e. g.*, the herring) has long been known. In the Arctic as well as the Antarctic Ocean *Calanus finmarchicus* and a few related species form in general the chief bulk of the plankton, and furnish food for pteropods and cephalopods, for the divers and penguins, for many fishes and whales. On the voyage from Japan to Honolulu the *Challenger* sailed through wide stretches of the North Pacific Ocean which were covered with red and white patches, caused by great accumulations of two species of small copepods, the red being *Calanus propinquus* (8, p. 758). In many other regions, from the Polar Circle to the Equator, the ship passed through white bands many miles wide, composed solely of copepods (8, p. 843). That their appearance is *very irregular* and dependent on many conditions is true of this very important group of plankton animals as for all others. For two days the *Challenger* went through thick shoals of *Corycaeus pellucidus*. For the next three days the copepods had entirely disappeared.

Hensen has made statistical statements upon the appearance of the copepods of the North and Baltic seas (9, p. 45). Chun has lately shown that this order plays a highly significant rôle, not only at the surface, but also at considerable depths (600 to 1,300 meters), (15, p. 25). "Their abundance and richness in forms in greater depths is absolutely astonishing. Larval forms of species sessile or living upon the bottom mingle in confusion with the young forms and sexually mature stages of eupelagic species. Many species hitherto regarded as varieties are numerously represented in the depths." On the other hand, the order seems to be very poorly represented at very great depths. The *Challenger* found only one very characteristic deep-sea species in 2,200 fathoms—*Pontostratioides abyssicolla* (8, p. 845). Some genera never leave the surface and are autopelagic, *e. g.*, *Pontellina* (15, p. 27).

Ostracoda.—The ostracods are, next to the copepods, the most important *Crustacea* of the plankton, and are represented at the surface as well as in different depths by masses of many species. In the œcology of the ocean they play a similar rôle, as do the near-related cladocerans (*Daphnidæ*) in the fresh water. The *Challenger* collected 221 species of ostracods. Of these 52 were found below 500 fathoms, 19 below 1,500, and 8 below 2,000 fathoms in depth. Many ostracods, like many copepods and other crustaceans, belong to the most important luminous animals of the ocean. On my journey to Ceylon (in the beginning of November, 1881), as well as on the return trip (middle of March, 1882), I admired as never before the oceanic light in its splendor. "The whole ocean, so far as the eye could reach, was a continuous shimmering sea of light." Microscopical investigation of the water showed that the luminous animals were for the most part small *Crustacea* (*Ostracoda*), to a less extent *Medusæ, Salpæ*, worms," etc. (25, pp. 42, 372). Chierchia, three years later, in the same region and in the same month, saw the same brilliant phenomenon: "The most brilliant emerald-green light was produced by an infinitude of ostracods" (8, p. 108).

Schizopoda.—Not less important in the planktonic life than the ostracods (sometimes even more important) are the schizopods. They also occur in wide stretches in immense swarms at the surface, as well as in greater and lesser depths. They also play a great rôle in the cycle of matter in the sea (*Stoffwechsel des Meeres*); on the one side since they devour great quantities of protozoa and planktonic larvæ, and on the other because they serve as food for the cephalopods and fishes. Many schizopods, like many ostracods and copepods, belong to the most brilliantly luminous animals, and, like the latter, furnish very interesting problems for the bathygraphy of the plankton. G. O. Sars, who has worked up the rich material collected by the *Challenger*, distinguished 57 species, and found that 32 of these lived only at the surface, 6 from 32 to 300 fathoms, and 4 extended down below 2,000 fathoms (as far as 2,740 fathoms), (6, p. 739). Chun also has discovered in the Mediterranean a number of new zonary and bathybic schizopods very different from the pelagic varieties of the surface, *Stylochiron, Arachnomysis*, etc. (15, p. 30).

The phyllopods (*Daphnidæ*), the amphipods (*Phronimidæ, Hyperidæ*), and the decapods (*Miersidæ, Sergestidæ*) are indeed represented in the plankton by a number of interesting forms, partly oceanic, partly neritic; and some of these occasionally appear in considerable quantities. But as a whole they are of far less importance than the copepods, ostracods, and schizopods. The same applies also to the other groups of *Crustacea*, although many of them in their larval state take a great part in the constitution of the plankton. Also in regard to these multiformed and often abundant *pelagic crustacean larvæ*, as well as for the mature crustacean animals, the advancing plankton study has still to establish and explain a fund of facts; namely, in relation to their pelagic, zonary, and bathybic distribution; their migrations, and the relations in which this planktonic fauna stands to the benthic fauna.

Insecta.—That important branch the *Tracheata*, the most numerous in forms of all the principal divisions of the animal kingdom, has in the sea no representatives whatever. The *Protracheata, Myriapoda*, and *Arachnidæ* are exclusively inhabitants of the land and in small part of the fresh water, except the pycnogonids or pantopods (in case these really belong to the *Arachnidæ*). Among the *Insecta* there is only a single small group of true marine animals, the family of the *Halobatidæ*. These small insects, belonging to the *Hemiptera*, have completely acquired a pelagic mode of life, and run about in the tropical ocean just as our "water-runner" (*Hydrometra*) on the surface of fresh water.

Both of the genera belonging there (*Halobates* and *Halobatodes*, with about a dozen species) are limited to the tropical and subtropical zone. The *Challenger* found them in the Atlantic between 35° north latitude and 20° south latitude; in the Pacific between 37° north latitude and 23° south latitude. I myself observed *Halobates* numerously in the Indian Ocean, and on one day in crowds in the neighborhood of Belligam. Although they can dive, they never go into the depths.

J.—TUNICATES OF THE PLANKTON.

The tribe of mantle animals falls into two chief divisions, according to their mode of life. The *ascidians* belong to the *benthos;* all other *tunicates* to the *plankton*. The *Copelata* (or *Appendicularidæ*) are morphologically the oldest branch of the stem, and are to be regarded as the nearest of the now living relatives of the *Prochordiniæ*, the hypothetical common ancestor of the tunicates and vertebrates (30, p. 605). The near relationship of the *Copelata* and the ascidian larva makes it very probable that the whole class of ascidians has sprung from the primarily pelagic *Copelata*, and has diverged from this through the acquirement of a sessile mode of life. The *Lucidiæ* or *Pyrosomidæ*, on the other hand, are probably secondarily pelagic animals, and sprang from the *Cælocormidæ*, a benthonic synascidian group. The *Thalidiæ* (the *Doliolidæ* as well as the *Salpidæ*) are to be regarded as primarily pelagic animals. These conditions are doubly interesting, because the tunicates in an exemplary manner demonstrate the peculiarities which the transition on one side to a sessile mode of life in the benthos (in case of the ascidians), and on the other to a free-swimming mode of life in the plankton (in the case of all other tunicates), has brought about. All the latter are transparent and luminous fragile animals, poor in genera and species, but rich in numbers of individuals. The ascidians, on the other hand, fastened to the bottom, in part littoral on the coast, in part abyssal in the deep sea, are much richer in genera and species, in many ways adapted to the manifold local conditions of the bottom, and mostly opaque. The few hyaline forms (*e. g., Clavellina*) may be regarded as the remnant of the old ascidian branch, which diverged from the pelagic *Copelata*.

All planktonic tunicates are exquisite *oceanic* animals and all may appear in immense swarms of astonishing extent. Murray (6, pp. 170, 521, 738, etc.) and Chierchia (8, pp. 32, 53, 75, etc.) met with great swarms of *Appendicularia, Pyrosoma, Doliolum,* and *Salpa* in the middle of the open ocean, both in the Atlantic and Pacific, particularly in the equatorial zone. I observed the same in the Indian Ocean, between Ceylon and Aden. Further, I have whole bottles full of closely pressed *Thalidiæ*, which Captain Rabbe collected in the middle of the Atlantic, Pacific, and Indian oceans, far removed from all coasts. In many log books also these swimming and luminous crowds of *Salpa* and *Pyrosoma* on the open sea, far from all coasts, are spoken of. On the other hand we know of no *neritic* tunicates, no other forms of swimming mantled animals which are found only on the coasts, except the omnipresent ascidian larva.

Lately Chun has established the interesting fact that the planktonic tunicates occur in numbers not only at the surface and in slight depths, but also during the summer extend down into greater depths (15, pp. 32, 42). He discovered further in the Mediterranean new *Copelata*, which are only zonary or bathybic, never coming to the surface and characterized by peculiar organization as well as difference in size (*Megalocercus abyssorum*, 3 centimeters long, 15, p. 40).

The small, delicate *Copelata* and *Doliola*, from their small size, are naturally more difficult to see than the large luminous *Salpæ* and *Pyrosoma*. Whoever has carefully examined great quantities of oceanic plankton can readily testify that the former also occur almost everywhere and occasionally take an important part in the constitution of the mixed plankton. Among the *Salpæ* there are for example the smaller species which form extensive swimming shoals. From the three-year observations of Schmidtlein it is learned that the salpas belong to the perennial plankton and are numerous throughout the whole year (19, p. 123).

K.—VERTEBRATES OF THE PLANKTON.

The vertebrates of the sea are in their mature condition for the most part too large and have too powerful voluntary movements to be reckoned in the true plankton in Hensen's sense, as "animals carried involuntarily with the water." The sea fishes, as well as the aquatic birds and mammals of the sea, overcome more or less easily the impetus of the currents, and thereby prove their independence by voluntary movements, which is not commonly the case with the floating invertebrate animals of the plankton. Meanwhile I have already shown above that this limitation of the *plankton* against the *nekton* is very arbitrary and at any moment may be changed in favor of the latter through diminution of the strength of the current. For the chief point of Hensen's plankton investigation, for the question of the "cycle of matter in the sea," the vertebrates are of greatest importance, since they, the largest of the rapacious animals of the sea, daily consume the greatest quantity of plankton, no matter whether directly or indirectly. A single sea fish of medium size may daily consume hundreds of pteropods and thousands of crustacea, and in case of the giant cetaceans this quantity may be increased ten or a hundred fold. In a comprehensive consideration of the plankton conditions, and particularly in its physiological, œcological, and chorological discussion, a thorough investigation of the vertebrates swimming in the sea, the marine fishes, the aquatic birds, seals, and cetaceans, is not to be undertaken. We can then turn from it here, since it has no further relation to the purpose of this plankton study. We can here in Hensen's sense (9, p. 1) provisionally limit ourselves to the vertebrates of the sea "carried involuntarily with the water," and as such (apart from a few small pelagic fishes) only the pelagic eggs, young brood, and larvæ of the marine fishes come into consideration. Some few teleosts (*Scopelidæ, Trichiuridæ, et al.*) occur sometimes in schools in the plankton and are partly autopelagic, partly bathypelagic. The remarkable *Leptocephalidæ* are possibly planktonic larvæ (of *Murænoidæ*), which never become sexually mature (7, p. 562).

Fish eggs.—The planktonic fish eggs, found in great numbers at the surface of the sea, as well as the young fish escaped from them, play without doubt a great rôle in the natural history of the sea. Hensen, whose planktonic investigation started from this point, had thereupon "based the hope to obtain a far more definite conclusion upon the supply of certain species of fishes than had hitherto seemed to be possible" (9, p. 39). But the assumption from which he starts is wholly untenable. Hensen says (*loc. cit.*):

It is scarcely to be doubted that an opinion upon the relative wealth of various kinds of fish in the Baltic or in any other part of the ocean whatever can be obtained through the determination of the quantity of eggs in the area under consideration.

Brandt also characterizes this proposition as very lucid and weighty (23, p. 517).

This standard proposition of Hensen and Brandt, from which a series of very important and complicated computations are to be made, was disposed of in a brilliant manner thirty years ago by Charles Darwin. In the third chapter of his epoch-making "Origin of Species," treating of the "Struggle for Existence," Darwin, under the head of Malthus' theory of population, speaks of the conditions and results of individual increase, the geometric relation of their increase, and the nature of the hindrances to increase. He points out that "*in all cases* the average number of individuals of any species of plant or animal depends only indirectly on the number of seeds or eggs, but directly on the conditions of existence under which they develop." Striking examples of these facts are everywhere at hand, and I myself have mentioned a number of them in my "Natural History of Creation" (30, p. 143). Still, to draw a few examples from the life of the plankton, I recall in this connection many pelagic animals; *e. g.*, crustacea and medusæ. Many small medusæ, which belong to the most numerous animals of the pelagic fauna (*e. g., Obelia* and *Lirope*) produce relatively few eggs; as also copepods, the commonest of all planktonic animals. Incomparably greater is the number of eggs produced by a single large medusa or decapod, which belongs to the rarer species. So, from the number of pelagic fish eggs *not the slightest conclusion* can be drawn as to the number of fish which develop from them and reach maturity. The major portion of the planktonic fish eggs and young are early consumed as food by other animals.

V.—COMPOSITION OF THE PLANKTON.

The composition of the plankton is in qualitative as well as quantitative relations very irregular, and the distribution of the same in place and time in the ocean also very unequal. These two axioms apply to the oceanic as well as to the neritic plankton. In both these important axioms, which in my opinion must form the starting-point and the foundation for the *œcology and chorology of the plankton,* are embodied the concordant fundamental conceptions of all those naturalists who have hitherto studied carefully for a long time the natural history of the pelagic fauna and flora.

The surprise was general when Prof. Hensen this year advanced an entirely opposite opinion, "that in the ocean the plankton was distributed so equally that from a few hauls a correct estimate could be made of the condition in a very much greater area of the sea" (22, p. 243). He says himself that the plankton expedition of Kiel, directed by him, started on this "*purely theoretical view*," and that it had "*full results* because this hypothesis was proven far more completely than could have been hoped" (22, p. 244).*

These highly remarkable opinions of Hensen, contradictory to all previous conceptions, demand the most thorough investigation; for if they are true, then all naturalists who many years previously, and in the most extensive compass, have studied the composition and distribution of the plankton are completely in error and have arrived at entirely false conclusions. If, on the other hand, these propositions of Hensen are false, then his entire plankton theory based thereon falls, and all his painstaking computations (on which in the last six years he has spent 17,000 hours, which he wishes to have number the individuals distributed in the plankton) are utterly worthless.

In the first place, the *empirical basis* upon which Hensen founded his assumptions must be proved, "starting from a purely theoretical point of view." The plankton expedition of Kiel was 93 days at sea, and in the months of late summer (July 15 to November 7) which, as is known, offer in the northern hemisphere the most unfavorable time of all for pelagic fishery (28, p. 16, 18). Hensen himself says that it bore the "character of a trial trip" (22, p. 10), and his companion Brandt names it a "reconnaissance" upon which they had come to investigate rapidly

---

* Hensen speaks of this in the following terms: "Hitherto it was the prevailing view that the inhabitants of the sea were distributed in schools, and that one, according to luck and chance, according to wind, current, and season, sometimes came upon thick masses, sometimes upon uninhabited parts. This in fact applies only in a certain degree for the harbors. For the open sea our knowledge teaches that normally regular distribution obtains there, which changes in thickness and ingredients only within wide zones corresponding to the climatic conditions. In any case one must seek the variation from such condition according to the cause which has produced it, and the occurrence of inequality is not to be taken as the given starting-point for relative investigation" (22, p. 244).

in succession as great areas as possible" (23, p. 525). In a more remarkable way he adds: "Thereby has resulted the furnishing of a fixed basis for a thorough quantitative and qualitative analysis of marine organisms." According to my view such "fixed basis" was obtained long ago, particularly by the widely extended investigations of the *Challenger* expedition (from January, 1873, to May, 1876), fitted out with all appliances. This embraced a period of forty months, and included "the whole expanse of the ocean." Their experience ought to lay claim to much greater value than that of the *National*, whose voyage of three months took in only a part of the Atlantic, and was in addition trammeled by bad weather, accidents to the ship, early loss of the large vertical nets, and other misfortunes in the carrying out of their plans. It is hardly conceivable how an "exact investigator," from so incomplete and fragmentary experience, can derive the "fixed basis" for new and far-reaching views, which stand in remarkable contradiction to all previous experience.

It would here lead too far, if, from the numerous old and new narratives of voyages, I should collect the observations of seafarers upon the remarkable inequality of the sea population, the different fauna and flora of the regions of currents, the alternation of immense swimming swarms of animals and almost uninhabited areas of sea. It is sufficient to point out the two works in which the most extensive and thorough knowledge up to this time is collected, the "Narrative of the Cruise of H. M. S. *Challenger*," edited by John Murray (6), and the "Collezioni della R. Corvetta *Vettor Pisani*" (8), published by Chierchia. Since the general chorological and œcological results in these two principal works agree fully with my own views gained from thirty years' experience, I pass immediately to a general exposition of these latter, reserving their proof for a later special work.

A.—POLYMIXIC AND MONOTONIC PLANKTON.

The constitution of the plankton of swimming plants and animals of different classes is exceedingly manifold. In this regard I distinguish first two principal forms, polymixic and monotonic plankton.*

The "mixed tow-stuff (*Auftrieb*), or the *polymixic plankton*," is composed of organisms of different species and classes in such a way that no one form or group of forms composes more than the one-half of the whole volume. The "*simple* tow-stuff, on the other hand, or the *monotonic plankton*," shows a very homogeneous composition, while a single group of organisms, a single species or a single genus, or even a single family or order, forms very predominantly the chief mass of the capture, at least the greater part of the entire volume of the plankton, often two-thirds or three-fourths of it, sometimes even more. Under this monotonic plankton one may again distinguish *prevalent plankton*, when the predominant group forms up to three-fourths of the total volume,

---

* Πολέμικτος = much mixed, complex; μονότονος = of a single form, simple.

and *uniform plankton* when this exceeds three-fourths and forms almost the whole mass.

In general the mixed plankton is more abundant than the simple, since as a rule the circumstances of the "struggle for existence" condition and vary in many ways the constitution of the planktonic flora and fauna. Still there are numerous exceptions to this rule, and at many points in the ocean (especially in the zoöcurrents) there occurs locally a development so numerous, and an accumulation of a single form or group of forms in such swarms, that these in the haul of the pelagic net form more than one-half the entire volume. This *monotonic plankton* appears in very different definite forms; for the difference of climate, the season, the oceanic currents, the neritic relation, etc., determine significant differences in the quantitative development of the plankton organisms, which simultaneously appear in vast numbers in a definite region. I will next briefly go over the single forms of the monotonic plankton known to me, passing over, however, the consideration of the extremely manifold composition of the *polymixic plankton*, since I am reserving that as well as a contribution of a number of mixture-tables for a later work.

1. *Monotonic Protophytic Plankton.*—Of the seven groups of pelagic *Protophytes*, at least three, the *Diatoms*, *Murracytes*, and *Peridineæ*, appear in such quantities in the ocean that they alone may constitute the larger part of the collection of the pelagic nets. The most important and most common is the *monotonic diatom-plankton*, particularly in brackish and coast waters. The siliceous-shelled unicellular *Protophytes* which compose this belong, often predominantly or almost entirely, to a single species or genus, as *Synedre* in the colder, *Chætoceros* in the warmer seas. The colossal masses of Arctic and Antarctic diatoms, which form the "black-water," the feeding-ground of whales, have been mentioned above. In the warmer tropical and subtropical parts of the ocean such accumulations of diatoms seldom or never occur. Here their place is taken by the *monotonic murracyte-plankton*, composed of immense swarms of nyctipelagic *Pyrocystidæ*. Less frequent is the *monotonic peridineæ-plankton*. Although these *Dinoflagellata* take a very significant part in the composition, especially of the neritic plankton, yet they do not often occur in such quantities as to form the greater part of the volume of the capture.

2. *Monotonic Metaphytic-Plankton.*—Among the pelagic *Metaphytes* there are only two forms, the *Oscillatoriæ* and the *Sargasseæ*, which appear so numerously that they form the greater part of the pelagic tow-stuff. The *monotonic oscillatoriæ-plankton*, as a rule formed of swimming bundles of fibers of a single species of *Trichodesmium*, appears in many regions of the tropical ocean in such masses that the quantity of the pelagic fauna is diminished on that account. The *monotonic sargassum-plankton*, formed of "swimming banks" of a single fucoid, *Sargassum bacciferum*, is the characteristic massive form of organic life in the *Halistasa* of the "Sargasso Sea."

3. *Monotonic Protozoic-Plankton.*—Among the unicellular *Protozoa*, three different groups, the *Noctiluca*, *Globigerina*, and *Radiolaria*, appear pelagically in such quantities that they form the greater part of the volume of the plankton. The *monotonic noctiluca-plankton* is neritic, and is composed almost exclusively of milliards of the common *Noctiluca miliaris*. It forms the reddish-yellow covering of slime upon the surface of the coast seas, and in the ocean always points out the littoral currents. On the other hand, the widely distributed *monotonic globigerina-plankton* is purely oceanic, the point of origin of the *globigerina* ooze of the deep sea. In different regions of the ocean it is composed of different genera of the above-mentioned pelagic thalamophores. Much more manifold is the *monotonic radiolaria-plankton*, also oceanic. Of these, one can distinguish the three following modifications:*

(1) *Polycyttaria-Plankton*, sometimes composed only of *Collozoum*, sometimes of *Sphærozoum*, sometimes of *Collosphæra*, most often of a mixture of these three forms; in the warmer seas, partly pelagic, partly zonary; very abundant.

(2) *Acantharia-Plankton*, commonly formed of milliards of a single or of a few species of *Acanthometron* (in the colder seas, *e. g.*, on the east and west coast of South America, south of 40° S. lat.; also north of 50° N. lat. on the coast of Shetland, Faroë-Orkney, and Norway); partly autopelagic, partly bathypelagic.

(3) *Phæodaria-Plankton*, zonary and bathybic, mostly composed of the larger species of *Aulosphæridæ* and *Sagosphæridæ*, *Cælodendridæ* and *Cælographidæ* (*e. g.*, *Cæloplegina murrayanum* from the Faroë-Orkney Channel, 4, p. 1757).

4. *Monotonic Cnidaria-Plankton.*—In the group of nettle animals there are numerous forms of medusæ, siphonophores, and ctenophores, which appear in immense schools. The *monotone medusa-plankton* is chiefly neritic, composed of very different local forms on the different coasts. Of the larger *Acraspedota*, in the warmer seas *Rhizostoma* (*Pilemidæ*, *Crambessidæ*) particularly occur; in the colder, *Semostoma* (*Aurelidæ*, *Cyanidæ*), which in schools fill the littoral bays and currents. Of the oceanic *Scyphomedusæ*, *Pelagia* seems to form similar schools. Among the *Craspedota*, monotonic medusa-plankton is especially formed of neritic *Cordonidæ*, *Margelidæ*, and *Eucopidæ*, of oceanic *Æquoridæ*, *Liriopidæ*, and *Trachynemidæ*. *Monotonic siphonophora-plankton* occurs only in the warmer seas, although *Diphyidea* are found abundantly in all parts of the ocean. The remarkable blue troops of the pelagic *Physalidæ*, *Porpididæ*, and *Velellidæ* have for a long time

---

*Radiolarian-plankton is contained in 13 preparations of the Radiolaria collection, which I have collected (1890) and which can be bought through the famulus Franz Pohle at Jena; 8 of these preparations contain polycyttaria-plankton, 2 acantharia-plankton, and 3 phæodaria-plankton. This collection (of 34 microscopical preparations) embraces in addition 17 preparations of the radiolarian-ooze of the deep sea, and 4 preparations of deep sea horny-sponges, whose pseudo-skeleton is composed of radiolarian slime. (Challenger Report, part LXXXII.)

in the tropical and subtropical seas attracted the attention of seafarers by their immense numbers as well as by the irregularity of their sudden appearance and disappearance. Rarer is a purely *physonectic* plankton chiefly composed of *Forskalia;* I have observed such repeatedly at Lanzarote. At that same place also occurred frequently a *monotonic ctenophora-plankton*. These delicate nettle animals also, as is well known, like the *Medusæ* and *Siphonophores,* appear in such closely packed crowds that there is scarcely room between them for other pelagic animals. Not infrequently the great accumulation of a single species of ctenophore imparts to the plankton a very remarkable character, and this is true in all oceans, in the cold as well as in the warm and temperate zones. More often it happens that the monotonic cnidaria-plankton is composed of several species of *Medusæ, Siphonophores,* and *Ctenophores,* while other classes of animals take only a very limited share in its constitution.

5. *Monotonic Sagittidæ-Plankton.*—The only form of monotonic plankton which the branch of *Helminthes* furnishes is made up by the class of the *Chætognatha,* various species of the genera *Sagitta* and *Spadella.* Although purely oceanic according to their mode of life, yet they occur numerously in the neritic tow-stuff (*Auftrieb*). Sometimes only a single species of these genera, sometimes several species close together, appear in such swarms as to make up more than half of the entire plankton. These phenomena have been observed in the colder as well as in the warmer seas. In the former the plankton is composed of the smaller, in the latter of the larger species. These forms occur also in the deep sea, and indeed the zonary *sagittidæ-plankton* is composed of different species from the pelagic.

6. *Monotonic Pteropoda-Plankton.*—Astonishing masses of oceanic pteropods are very widely distributed in all parts of the ocean, and in part are formed of characteristic genera and species in the different zones. The immense schools of *Clio borealis* and *Limacina arctica,* which inhabit the northern seas and (as "whale-food") furnish the chief food supply for many cetaceans, sea-birds, fishes, and cephalopods, have long been known. But no less immense are other swarms of pteropods, composed of different genera and species, which populate the seas of the temperate and tropical zones. These have often escaped the notice of seafarers, because most species are nyctipelagic. Of the immense quantities of these floating snails, direct evidence is furnished by the accumulated calcareous shells, which in many stretches of ocean (especially in the tropical zone) thickly cover the bottom at depths between 500 and 1,500 fathoms. Often the greater part of this "pteropod-ooze" is formed solely of them (6, pp. 126, 922). At Messina as well as at Lanzarote I found the pteropod-plankton often mixed with considerable numbers of heteropods. Still the latter never form the greater part of the volume.

7. *Monotonic Crustacea-Plankton.*—As the crustaceans surpass all other classes of the animals of the plankton in quantitative development, so they form monotonic plankton far more often than all other classes. Most commonly this simple crustacean-plankton is composed of copepods, not infrequently entirely of a single species (6, pp. 758, 843). Next to this I have more frequently found monotonic *ostracoda-plankton;* next *schizopoda-plankton.* Sometimes also there are in these two orders only numberless individuals of a single species, sometimes of many different species, which compose the monotonic plankton, often almost exclusively, and at other times mixed with additions of other *Crustacea, Sagitta, Salpa,* etc. The other above-mentioned orders of crustaceans, which also take a considerable part in the constitution of the plankton, the decapods, amphipods, and phyllopods, I have never found in such quantities that they formed more than half of the mass of tow-stuff. On the contrary, such quantities of *crustacean-larvæ* of one species (*e. g.,* of *Lepas* and other cirripeds) occasionally appear that they predominantly determine the character of the plankton.

8. *Monotonic Tunicata-Plankton.*—Next to the monotonic forms of plankton, which are composed of *Crustacea* and *Cnidaria*, that of the *Tunicata* is most numerous. Quite preponderant in quantity are the *Thalidiæ* or *Salpaceæ* (*Salpa* and *Salpella*), and among these, especially the smaller species (*Salpa democratica-mucronata, S. runcinata-fusiformis,* and related species). I have often taken such *monotonic salpa-plankton* in the Mediterranean, in the Atlantic and Indian oceans, and have received the same also through Capt. Rabbe from different parts of the Pacific Ocean. Masses of *Doliolum* and of *Copelata* (*Appendicularia, Vexillaria,* etc.) are also commonly mixed with this in greater or less quantities. Still these planktonic tunicates, on account of their small size, recede before the *Salpa.* I know of no instance where they have by themselves formed a monotonic plankton. But this is the case with the nyctipelagic *pyrosoma.* The *Challenger* and the *Vettor Pisani* in the tropics, on dark nights, met with quantities of *monotonic pyrosoma-plankton* in the middle of the Atlantic and Pacific. By day not a single one of these "cones of fire" was to be seen, and as soon as the moon arose they went into the depths (8, pp. 32, 34).

9. *Monotonic Fish-Plankton.*—If, with Hensen, we limit the term plankton to the *halobios* floating passively in the sea, we can designate as "monotonic fish-plankton" only the schools of very young and small fishes, which often appear abundantly in the currents, occasionally so compact that very few other pelagic animals can find room between them. If one wishes to extend the term still farther, and wipe out the sharp distinction between *plankton* and *nekton,* all those sea fishes (oceanic as well as neritic) which appear in schools, and which play so significant an œcological rôle in the cycle of matter in the sea (*e. g., Scopelidæ, Clupeidæ, Leptocephalidæ, Scomberoidæ*) will in general belong here (12, p. 51).

B.—Temporal Planktonic Differences.

The first and most remarkable phenomenon, known to every seafaring planktologist, is the varying constitution of the plankton and the variable mingling of its constituents. The remarkable differences of composition apply *qualitatively* and *quantitatively* to the *oceanic* as well as to the *neritic* plankton. They are just as important in the comparison of different places during the same time as at different times in one and the same place. We can therefore distinguish local and temporal variations, and will first of all consider the latter.

To obtain a complete and more certain survey of the temporary variations of plankton composition, there would be needed especially an unbroken series of observations, which had been carried on at one and the same place at least for the space of a full year—still better for several successive years—to obtain from the yearly and monthly oscillations a general average. Such complete *series of observations*, comparable to the meteorological (with which they stand in direct causal connection), have not hitherto been made. They belong to the most important tasks of the zoölogical stations now everywhere springing up.*

Meanwhile, a general conception of the considerable size of the yearly and monthly oscillations can be obtained from a comparative summary based upon the important series of observations extending over three years, which Schmidtlein has given upon the appearance of the larger pelagic animals in the Gulf of Naples, during 1875–77 (19, p. 120). The contributions of Graeffe upon the occurrence and time of appearance of marine animals in the Gulf of Trieste are also worthy of notice in this connection (20).

The considerable temporal variations which underlie the appearance of the pelagic organisms and which determine such great differences in the plankton composition, relative to quality and quantity, may be divided into four groups: (1) yearly, (2) monthly, (3) weekly, (4) hourly variations. Their causes are manifold, partly meteorological, partly biological. They are comparable to corresponding temporal oscillations of the terrestrial flora and fauna, on one side depending upon climatic conditions and meteorological processes, and on the other upon the changing mode of life, especially upon the conditions of reproduction and development. As the annual development of most terrestrial plants is connected with definite time conditions, as the period of budding and leaf development, of their blossoming and fructification, has

* My own extensive experience, I am sorry to say, is in this regard very insufficient, since I have never worked at a zoölogical station, and since usually I was only so fortunate as to go to the seacoast for a few months (or even only for a few weeks) during the academic vacation. Only once have I had the opportunity to extend my plankton studies at one and the same place to a half year (from October, 1859, to April, 1860, at Messina, 3, p. v, 166), and three times have I carried them on for three months at the same place—in the summer of 1859 at Naples, in the winter of 1866–67 at Lanzarote, and in the winter of 1881–82 in Ceylon.

become adapted to the meteorological conditions, the time of year and other conditions of life in the "struggle for existence," so also the annual development of most marine animals is governed by definite, inherited habits. With them also the influence of meteorological variations on the one side, of œcological relations on the other, are of the greatest importance for the periodical appearance. Most organisms appear in the plankton only periodically, and only very few can be reckoned as belonging to the "perennial plankton" in Hensen's sense (9, p. 1). This investigator also attaches great importance to the *temporal* "highly remarkable variations" in the plankton composition (9, pp. 29, 59); he explains it in part by "periods of famine" (p. 53).

Yearly oscillations.—The plankton literature has hitherto contained only a few reliable statements upon the yearly variations, which underlie the appearance of the pelagic animals and plants. Still there are a few contributions of high merit, extending over a series of years, namely those of Schmidtlein from Naples (19) and of Graeffe from Trieste (20). Even the first glance at the tables, those of the former relating to the appearance of the pelagic animals in the Gulf of Naples, shows us how remarkably different was the action of the majority of these in several successive years. As there are good and bad wine and fruit years, so there are rich and barren plankton years. But Schmidtlein correctly remarks that extensive observations extending through a long series of years are demanded to gain a deeper insight into the meaning of these yearly and monthly variations shown in the tables. The same view is also held by Chun, who, in his monograph of the ctenophores of the Gulf of Naples (p. 236), points out how very different was the number of these in five successive years.

Graeffe, resting on the basis of his observations for many years, says of *Cotylorhiza tuberculata*, that this beautiful acaleph has not for many years been found in the Adriatic, in other years only individually, but not at all rarely (yet always only in the three months of July, August, and September). Just as variable is the occurrence—"*according to the year*"—of *Umbrosa lobata* and other medusæ. Of the six species of ctenophores of the Gulf of Trieste, only one appears every year, the five others only now and then. Not only do the *quantities of individuals*, but also the "*time of appearance* of pelagic animals change according to the meteorological conditions of the time of year" (20, v, p. 361). I myself can fully establish this proposition on the ground of observations which I have made in the course of many years of medusa studies. Many of these "capricious beauties" occur in one and the same place on the Mediterranean coast (*e. g.*, in Portofino, in Villafranca), numerously in the first year, rarely in the second, and not at all in the third. When, in April, 1873, I fished in the Gulf of Smyrna, it was full of swarms of the great pelagic *Chrysaora hyoscella*. In April, 1887, when for the second time I sought the same gulf, I could not find a single individual of that beautiful medusa, but instead the gulf was filled by crowds of a new, hitherto undescribed, large medusa, *Drymonema cordelia*. Thousands of these *Cyaneidæ* lay cast upon the beach at Cordelio.*

Monthly oscillations.—The time of year is of just as great importance for the appearance of very many pelagic animals as for the flowering and fruit formation of land plants. Many of the larger planktonic animals, *Medusæ*, *Siphonophores*, *Ctenophores*, *Heteropods*, *Pyrosoma*, etc., appear only in one month or during a few months of the year. They form Hensen's "periodic plankton." In the Mediterranean many pelagic animals are numerous in the winter, while in the summer they are entirely wanting. This "periodical appearance of pelagic animals" has long been known and often mentioned; but not so the important fact that these ethoral periods themselves show considerable variations. For this the tables of Schmidtlein (19) and the notes of Graeffe (20) give important points of support. Especially the *Disconectæ* and other *Siphonophores*† behave very irregularly. The cause of the monthly variation lies on the one side in the conditions of reproduction and development; on the other in the varying temperature of the season, as Chun has lately shown (15, 16).

Daily oscillations.—Every naturalist who has observed and fished pelagic animals and plants in the sea for a long time, knows how unlike their appearance is on different days in the same period of the year or in the same month, when one may daily hope to find them. As a rule, the weather, and particularly the *wind*, conditions the remarkable difference of appearance. In long-continuing calms the surface of the sea becomes covered with swarms of various pelagic creatures. In long bands, smooth as oil, the most wonderful zoöcurrents appear. But as soon as a fresh wind stirs up lively waves, the majority sink into the quiet depths, and if a more violent storm churns up the deeper layers, all life vanishes from the surface for days. Many animals of the plankton (especially oceanic) are very susceptible to the influence of fresh water, and therefore disappear during violent rains. Warm sunshine entices the one to the surface, while it drives the other into the depths. This influence of the weather upon the quality and quantity of the planktonic composition is so well known that it is not necessary to give examples. Hensen (9) has even gone over his work many times, without thinking how the above endangers his "exact methods" and made their results illusionary.

---

* *Drymonema cordelia*, whose milk-white umbrella reaches half a meter in diameter, I will describe hereafter. It differs in the formation of the gonads and oval tentacles, as in several other points, from the Adriatic species, which I have described as *Drymonema victoria* (=*dalmatinum*) (II, 29).

† Of the *Disconectæ* (*Porpita* and *Velella*) Chun during a 7 months' residence at the Canary Islands (1887-88) could find not a single specimen. According to him they should appear first in midsummer (July to September). On the other hand I saw at Lanzarote an isolated swarm of these *Disconectæ* in midwinter, in February, 1867.

Hourly oscillations.—Many pelagic animals appear at the surface of the sea *only at a definite hour* of the day, some in the morning, others at noon, still others towards evening. During the remainder of the day not a single individual of the species is to be found. Agassiz, thirty years ago, brought forward noticeable examples of this from the class of *Medusæ*, and I can from my own experience adduce a number of other examples. But many other pelagic animals also (*e. g.*, *Siphonophores*, *Heteropods*) come to the surface only for a few hours. We have long known that the swarms of the nyctipelagic *Pteropods*, *Pyrosoma* and many *Crustacea*, come to the surface only during the night and flee the light of day. Other groups act just reversely. But the late extensive observations, especially of Murray (6), Chierchia (8), and Chun (15) have taught us how great is the extent and importance of those hourly variations. That these are of great influence upon the composition of the plankton, and that this accordingly is very different at different times of day, needs no repetition. But we must allude once more to how all these temporal oscillations must be taken into consideration, if the *equality of plankton distribution* is to be proved by observation and estimation. In point of fact they all seem to tend to very remarkable *inequality*.

C.—Climatic Plankton Differences.

The numerous contributions which earlier and later observers have made upon the appearance of the swarms of the pelagic animals in different regions of the ocean, agree in pointing out the differences among them, corresponding to the climatic zones. Thus the Arctic oceans are characterized by masses of monotonic plankton of *Diatom*, *Beroidæ*, *Copepod*, and *Pteropod* groups, swarms which are often composed of milliards of single species. In the oceanic regions of the temperate zone we meet monotonic plankton of the *Fucoid*, *Noctiluca*, *Medusa*, *Ctenophore*, *Salpa*, *Schizopod*, etc., classes, sometimes composed of one, sometimes of several species. In the tropical ocean immense banks of monotonic plankton appear, in which the *Murracytes*, *Oscillatoriæ*, *Physalia*, *Pyrosoma*, *Ostracoda*, determine the character of the swimming oceanic population. Although these facts have long been known, up to this time no attempt has been made to arrange them chorologically or to define more closely the characteristic features of the plankton in the climatic zones. Yet I believe, partly upon the ground of the accounts referred to above (particularly of the *Challenger* and of the *Vettor Pisani*), partly on the ground of my own comparative investigations (of the *Challenger* as well as of the Rabbe collections), that even now an important proposition can be formulated.

The quantity of the plankton is little dependent upon the climatic differences of the zones, the quality very dependent; especially in this way, that the number of component species diminishes from the equator towards both poles. This proposition corresponds, on the whole, with the conditions which the climatic differences show in the terrestrial fauna and flora. Here as there the explanation of the facts is above all to be sought in the influence of the sun, that "all-powerful creator," which in the tropical zone conditions a much more lively interaction of the natural forces than in the polar zones. The "cycle of matter in the sea" (*Stoffwechsel des Meeres*) is no less influenced by the perpendicular rays of the sun than is the terrestrial fauna and flora; and as in the tropics the quantity and the complexity of the terrestrial organic living forms is by far most highly developed, so is it also the case with the marine forms.

Hensen places himself in remarkable opposition to this hitherto accepted view when in his account of the results of the *National* expedition he surprises us with the following statement:

Although we have found plankton everywhere, the amount of it under and near the tropics was relatively small, namely on an average 8 times less than in the north near the Banks of Newfoundland. Each one of these hauls contained upwards of a hundred different forms; but the poverty of the quantity is still a remarkably apparent established fact (22, p. 245).

In the notable account which E. du Bois-Reymond (on January 23, 1890) laid before the Berlin Academy upon the results of the *National* expedition, it was said concerning its scientific results that a complete account could not be given for three years, but then he added:

Only one chief result may here be assumed beforehand. Contrary to all expectations, established upon a theoretical basis, the quantity of plankton in the tropical waters is shown to be surprisingly small (21, p. 87).

Since Hensen with this "chief result" of the *National* expedition stands in strong opposition to the familiar experience of the *Challenger*, of the *Vettor Pisani*, and of many other expeditions, we must first of all again examine the *empirical foundations* upon which his assertions rest. For these he admits that he regards as such only the results of his "*trial trip*" *through a part of the Atlantic ocean*, in which the *residence in the tropics* embraced *scarcely two months*. The results which he here draws from his plankton fisheries, which obviously turned out remarkably poorly as a result of accidental conditions, may contradict the results which were set up by the *Challenger* and the *Vettor Pisani* during a residence in the tropics of altogether four years, in different parts of three great oceans. It is not indeed saying too much, if we declare this kind of conclusion by Hensen as hasty, and the "exact method" which he wishes to establish by *computation* as useless.

My own comparative study of the rich planktonic collections which Murray and Rabbe have brought in from the different parts of the three great oceans, has convinced me that the tropical ocean is not only qualitatively much richer (by the variety and number of planktonic species and genera) than the oceans of the temperate and cold zones, but that it also does not fall behind the latter quantitatively (in the abundant distribution and vast accumulations of individuals). To be sure, one ought not to take into consideration merely the *surface* of the tropical ocean (although this also is often extremely densely populated), but also the deeper zonary regions. For in the tropical zone there are numerous nyctipelagic organisms, which by day shun the glow of the perpendicular rays of the sun and betake themselves into the cooler, more or less deep layers of water; but at night these bathypelagic animals and plants appear at the surface in such immense crowds that they are not surpassed in quantity by the "immeasurable swarms" of pelagic organisms in the temperate and cold zones.

During my trip through the tropical region of the Indian Ocean, as well on the way to Ceylon (from Bombay) as on the return (from Socotora), I daily wondered at the great richness of pelagic life on the mirrored surface. At night the "whole ocean, as far as the eye could see, was a continuous shimmering sea of light" (25, p. 52). The luminous water, which at night we scooped up directly from the surface with buckets, showed a confused mass of nyctipelagic luminous animals (*Ostracods, Salpa, Pyrosoma, Medusæ, Pyrocystæ*), so closely packed that in a dark night we could plainly read the print in a book by the brightness of their pelagic light. The crowded mass of individuals was not less considerable than I have so often found in the Mediterranean in the currents of Messina. What quantities of food the plankton must here furnish to the larger animals was shown by the vast schools of great medusæ and flying-fish, which for days accompanied our vessel; and this mass covered large areas of the open Indian Ocean, midway between Aden and Ceylon. Just such plankton masses I have received through the kindness of Capt. Rabbe from other tropical parts of the Indian Ocean, between Madagascar and the Cocos Islands, and between these and the Sunda Archipelago. I encountered a wonderfully rich and thick planktonic mass in a pelagic current of the southwest monsoon drift, 50 nautical miles south of Dondra Head, the southern point of Ceylon.* I have referred to the richness of this in my "Indian Journal" (25, p. 275).

That the *tropical zone of the Atlantic Ocean* also possesses a vast wealth of plankton is shown by many older accounts, but especially from the experience of the *Challenger*. In the middle of the Atlantic, between Cape Verde and Brazil, Murray observed colossal masses of pelagic animals; and if by day they were scarce at the surface, he continually found them below the surface, in depths of 50 to 100 fathoms and more (6, pp. 195, 218, 276, etc.); at night they ascended to the surface and filled the sea far and wide with a brilliant glow (pp. 170, 195, etc.). "*On the whole cruise along the Guinea and equatorial currents, the pelagic life was exceedingly rich and varied,* in the quantities of individuals as well as of species, *much more than anywhere else in the northern or southern part of the Atlantic Ocean.* The greatest quantities were seen in the Guinea current during calms, *when the sea literally swarmed*

---

*A part of the new species of pelagic animals which I found in this astonishingly rich oceanic current are described in my "Reports on the Siphonophora and Radiolaria of H. M. S. *Challenger.*"

with life" (p. 218). This astonishing wealth of plankton was observed in the whole breadth of the Atlantic tropical zone in August and September, 1873; but it was not less than that passed by the *Challenger* on her return in March and April, 1876, in the eastern part of the same region, between Tristan d'Acunha and Cape Verde. "When the water was calm, an extraordinary superabundance of pelagic life appeared at the surface. *Oscillatoriœ* covered the sea for miles, and vast quantities of *Radiolaria* (*Collozoun*) filled the nets" (p. 930). With those and other accounts by the *Challenger*, those of the *Vettor Pisani* quite agree. "*The zone of equatorial calms is out of all proportion rich in organic life.* Sometimes the water seems coagulated, jelly-like, even to the touch. It is impossible to describe the quantities of variously colored forms" (8, p. 31). Chierchia enthusiastically describes the wonderful spectacle which the luminous ocean furnishes at night—"a sea of light which extends to the whole horizon" (pp. 32, 53, etc.). The numerous plankton samples which I myself have investigated from the Atlantic tropical zone show for the most part an extraordinarily rich composition, particularly those between Ascension and the Canary Islands (*Challenger* stations 345 to 353), above all the two equatorial stations 347 and 348, which, like the *Canary currents*, which I studied for three months at Lanzarote, whose fabulous wealth I have already mentioned, also belong to the region of the *tropical* trades-drift.

The quantity and wealth of forms of the plankton in the *tropical zone of the Pacific Ocean* is not less than in the tropical region of the Atlantic and Indian oceans. In the most diverse parts of this region the *Challenger* sailed through "thick banks of pelagic animals." Between the New Hebrides and New Guinea "the surface of the water and its deeper levels swarmed with life. All the common tropical forms were found in great abundance. The list of genera of animals was about the same as in the Atlantic tropical region (pp. 218, 219), but it showed *considerable difference in the relative abundance of species*" (6, p. 521). Among the Philippines the water showed "a quite uncommon quantity and variety of oceanic surface animals" (p. 662). On the voyage from the Admiralty Islands to Japan the oceanic "fauna and flora of the surface was everywhere *especially rich and varied.* In the neighborhood of the equatorial countercurrents, between the equator and the Carolines, pelagic foraminifera and mollusks were taken in such quantities in the surface net that they surpassed all earlier observations," etc. (p. 738). On the voyage through the *central part of the tropical* Pacific, from Honolulu to Tahiti, between 20° N. lat., and 20° S. lat., "the catch of the tow net was everywhere very rich. *The superabundance of organic life in the equatorial current and countercurrent is very noticeable, as well with reference to the number of species as of individuals*" (p. 776). From this wonderfully rich region, which of all parts of the tropical ocean *is farthest removed from all continents,* came *the absolutely richest* plankton samples which I have ever studied, those which the *Challenger* brought from her stations 262–280. My astonishment was great when I first saw these planktonic masses, in the autumn of 1876; but it grew boundless when a year later I studied preparations taken from them and found in them hundreds of new species of pelagic animals.

The wonderfully rich *Radiolaria* ooze which the *Challenger* brought up at the central Pacific stations 263–274 (from depths of 2,000 to 3,000 fathoms) is only the siliceous remains of that planktonic mass, from which all organic constituents have vanished and the calcareous shells for the most part dissolved by the carbonic acid of the deep currents.* The numerous surface preparations which Murray finished upon the spot on this remarkable voyage of planktonic discovery through the central Pacific, and mounted in Canada balsam, are *absolutely the richest plankton preparations* which I have ever studied, especially those of stations 266–274, between 11° N. lat. and 7° S. lat. The richest of all stations is 271, lying almost under the equator (0° 33′ S. lat., 152° 56′ W. long.). I have since shown these preparations for microscopical studies to many colleagues and friends, and they have always expressed the liveliest astonishment over the new "wonder-world" concealed in them. They are jokingly called the "mira-preparations" (comp. 4, §§ 228–235).

The wonderful plankton wealth of the tropical Pacific is as well established by the manifold observations of Chierchia: "*The quantity and quality of the organisms which inhabit the tropical regions of the sea surpass all conception*" (8, p. 75). Inconceivable quantities of pelagic animals of all groups were seen in the middle of the tropical Pacific, between Callao and Hawaii, between Honolulu and Hongkong, not only at the surface, but in the most various depths up to 4,000 meters. The quantity of deep-sea siphonophores was here so enormous that the sounding lead was never drawn up without its being surrounded with torn-off tentacles (p. 85). During the forty days' voyage from Peru to Hawaii the pelagic fishery at the surface as well as in the depths brought to light "such a quantity of different organisms that it must seem almost impossible to one who did not follow the work with his own eyes" (8, p. 88). Similarly, in the Chinese sea and in the Sunda Archipelago immense masses of plankton were encountered.

It is my intention here to bring together the most general impressions of the relative planktonic wealth of the various oceanic regions, which I have gained from a comparative study of many thousand planktonic preparations. The pelagic fauna and flora of the tropical zone is richer in different forms of life than that of the temperate zone, and this again is richer than that of the cold zone of the ocean. This is true of the oceanic as well as of the neritic plankton. Everywhere the neritic plankton is more varied than the oceanic. The wealth of

---

*Of this *Radiolaria* ooze there are 16 samples (embracing about 1,000 different species) contained in the "Radiolarian collection" (1890) above mentioned. The 8 richest of these (Nos. 20–27) belong to the *tropical central Pacific* (stations 265–274).

individuals can in none of these regions be called absolutely greater than in the others, since the quantitative development is very dependent upon local and temporal conditions and, according to time and place, is on the whole extremely irregular. Estimation of individuals can in this relation prove nothing.

D.—CURRENTIC PLANKTONIC DIFFERENCES.

By far the most important of all the causes which determine the changing and irregular distribution of the plankton in the sea are the *marine currents*. The fundamental importance of these *currents* for all planktonic studies is generally recognized and has lately been mentioned many times and explained by Murray (6) and Chierchia (8). Even the zoölogists of the plankton expedition of Kiel have not been able to close themselves to this intelligence. Brandt calls special attention to "the importance of the marine currents as a means of, and limit to, the distribution of the planktonic organisms," so that in the various Atlantic currents numerous forms continually appear which were wanting in the regions previously traveled" (23, p. 518). Thus, Hensen mentions the "extraordinarily large plankton catches, which were transported by various currents."

I learned thirty years ago to recognize the great importance of the marine currents and their direct influence upon the composition of the plankton, when at Messina I went out almost daily in the boat for six months to secure the rich pelagic treasures of the strait (3, p. 172). The periodical strong marine current, which there is known to the Messinese under the name of the *current* or the *Rema*, enters the harbor twice daily and brings to it inexhaustible treasures of pelagic animals which since the time of Johannes Müller have aroused the wonder and desire for investigation of all naturalists tarrying there. Not less important did I find later the planktonic importance of the local marine currents (at Lanzarote), when the "Zain" current of the Canary Sea in like manner brought with it an extraordinary wealth of pelagic animals. My companion on the trip, Richard Greeff, has very vividly described these marine currents as "animal roads" (18, p. 307). During my numerous pelagic journeys on the Mediterranean it was always my first care to investigate the conditions of the currents, and on the most different parts of its coast (from Gibraltar to the Bosporus, from Corfu to Rhodos, from Nizza to Tunis, I have always been convinced of the determining influence which they exerted upon the composition and distribution of the plankton.

Although the fundamental importance of the marine currents for the diverse questions of oceanography are now generally recognized, still very little has been done to follow out in detail their significance for planktology. It seems to me, we must here, with reference to our theme, particularly distinguish (1) *halicurrents* (the great oceanic currents); (2) the *bathycurrents* (the manifold deep currents or undercurrents);

(3) the *nerocurrents* (the littoral currents or local coast currents); and (4) the *zoöcurrents* (the local planktonic streams or very crowded animal roads).

Halicurrents or ocean streams.—The unequal distribution of plankton in the ocean is in great part the direct result of the oceanic currents. In general the proposition is recognized as true that the great ocean streams, which we briefly designate as *halicurrents*, effect a greater accumulation of swimming organisms and thereby are richer in plankton than the *halistasa* or "still streams," the extensive regions which are inclosed by them and relatively free from currents. For a long time the richness in plankton which characterizes the Gulf Stream on the east coast of North America, the Falkland Stream on the east coast of South America, and the Guinea Stream on the west coast of Central Africa, has been known. Less understood and investigated than these Atlantic streams, but also very rich in varied plankton, are the great streams of the Indian and Pacific oceans, the Monsoon Stream on the south coast of Asia, the Mozambique Stream on the east coast of South Africa, the Black Stream of Japan, the Peru Stream on the west coast of South America, etc.

It is very difficult, from the numerous scattered accounts of the pelagic fauna and flora of these great ocean currents, to form a general picture of them, but it is now possible to draw from them the conclusion that generally the plankton of the *halicurrents*, qualitatively as well as quantitatively, is richer than the plankton of the *halistasa*, or the great oceanic sea basins around which flow the great streams and counter streams, and which meet the first glance on every recent map of the marine currents.*

In defending this proposition I rely especially upon the rich experience of the two most important plankton expeditions, of the *Challenger* (6) and of the *Vettor Pisani* (8), and also upon my own comparative study of several hundred plankton samples, which were collected in part by Murray, in part by Capt. Rabbe, in the most diverse parts of three great oceans. The planktonic wealth of the great halicurrents is most remarkable at the place where they are narrowest, when the masses of swimming animals and plants are most closely pressed together. Highly remarkable here is the opposition which the rich pelagic fauna and flora of the stream forms in qualitative and quantitative relation to the sparse population of the immediately adjacent halistase. As the temperature and often even the color of the sea

---

* The systematic biological investigation of the *halistasa* seems to me to form one of the nearest and most pressing problems of planktology, and also of oceanography. Apart from the smaller and little investigated Arctic and Antarctic regions, in all five great areas of quiet water ought to be distinguished, namely: (1) the North Atlantic *halistase* (with the Sargasso Sea); (2) the South Atlantic (between Benguela and Brazil streams); (3) the Indian (between Madagascar and Australia); (4) the North Pacific (between California and China), and (5) the South Pacific *halistase* (between Chili and Tahiti).

water in two adjacent regions is remarkably different and often sharply contrasted, so also is the constitution of their animal and vegetable world. Thus Murray observed a strong contrast between the cool green coast streams and the warmer deep-blue ocean water when the *Challenger* neared the coast of Chili, between Juan Fernandez and Valparaiso, and correspondingly there occurred a sudden change of pelagic fauna, for the oceanic globigerina disappeared and the neritic diatoms, infusoria, and hydromedusæ appeared in greater abundance (6, p. 833).

This change was very remarkable when the *Challenger* (at station 240, June 21, 1875) left the warm "black stream" of Japan and entered the cold area of quiet water adjacent on the south (about 35° N. lat., 153° E. long.). Great polymixic masses of plankton, dwellers in the first area, were here killed by the sudden change of temperature and replaced by the monotonic copepodan fauna of the cold halistase (16, p. 758). Also, later, on the voyage through the Japan stream, the planktonic contents of the tow net plainly showed the proximity of two different currents. "In the cold streams there appeared a greater mass of small diatoms, noctiluca, and hydromedusæ than in the warmer streams where the richer pelagic animal world (*Radiolaria, Globigerina*) remained the same which the *Challenger* observed from the Admiralty Islands to Japan." Many similar cases occurred during the voyage, when proximity to the coast or the presence of coast currents was indicated by the contents of the tow net (6, p. 750).

Observations upon the plankton richness of the oceanic currents, similar to those of Wyville Thompson and Murray on the *Challenger* (6) were made by Palumbo and Chierchia on the *Vettor Pisani*. The latter calls attention especially to the great importance of these and the great accumulation of pelagic animals in limited regions of currents.

It is a fact, that generally on a voyage through the ocean *great quantities of individuals of one species are found pressed together in relatively small spaces*, and this is true of organisms which, on account of their small size, are not capable of extensive movements. In addition, it is also a fact that when the ship is in the midst of the great oceanic currents, the pelagic fishery gives the most brilliant results (8, p. 109). *It is quite certain that the investigation of the distribution of the pelagic organisms can not progress unless accompanied by a parallel study of the currents, the temperature, and the density of the water* (8, pp. 109, 110).

Even the participators in the *National* expedition of Kiel could not avoid noticing the great irregularity of planktonic distribution in the ocean and the importance of the oceanic currents in this respect. During the voyage it was noticed that in different Atlantic currents numerous forms appeared continually which were absent in the regions previously traversed:

The conditions are much more complicated (!) than we had hitherto supposed (23, p. 518).

But it is worthy of notice how Hensen, the leader of this plankton expedition, has noticed this abundant accumulation of pelagic organisms in single regions of currents, and has twisted it in favor of his theory of the *regular distribution of the plankton:*

The tests of the volume of the plankton show that, five times in the north, once north of Ascension, *extraordinarily large catches* (!) were made. These must have been caused by various *currents* in this region, and can therefore be left *out of consideration* (9, p. 249).

It seems to me that Hensen would have done better to take into consideration these and other facts observed by him relative to the unequal plankton distribution before he built up his fundamental, certainly *adequate*, theory of the equality of the same. This was to be expected, since he himself in his first oceanic plankton studies (1887) observed many "*remarkable inequalities*," and his own tables furnish proof of this. While he many times mentions the immense swarms of *Medusæ* and declares this "quite superabundant accumulation to be mysterious," he adds: "such places must be avoided in this fishery" (9, pp. 27, 65). When Hensen later, in comparing the different catches of copepods (one of the most important planktonic constituents), finds that the distribution of the plankton in the ocean is *very irregular* and that the constitution of this seems to very strongly contradict his general conceptions of natural life (9, p. 52), he holds it to be best that these catches, which are of "such a different kind, should be excluded from consideration" (pp. 51, 53). Also, in the case of *Sagitta*, which Hensen reckons with the copepods as belonging to the uniform perennial plankton, he finds "throughout not the equality which one might expect, but much more remarkable variations" (p. 59).

These "surprising inequalities," "variations even to tenfold," he finds in case of the *Daphnidæ* (pp. 54, 56) and *Hyperidæ* (p. 57), the pelagic larvæ of snails and mussels (pp. 57, 58), *Appendicularia* and *Salpa* (pp. 63, 64), the *Medusæ* and *Ctenophores* (64, 65), the *Tintinnoids* (p. 68), the *Peridiniæ* (p. 71), and even in the *Diatoms* (p. 82)—in brief, in all groups of pelagic organisms which by the numerous production of individuals are of importance for the plankton and upon which Hensen employs his painstaking method of calculation by quantitative planktonic analysis. If one freely "sets apart from consideration" all these cases of remarkable inequality (because they do not fall in with the theoretically preconceived ideas of the equality of planktonic composition), then finally the latter must be proved by counting.

Bathycurrents or deep streams.—Through recent investigations, particularly of Englishmen (Carpenter, Wyville Thompson, John Murray, *et al.*), we have become acquainted with the great importance of the submarine currents or deep streams. It has been demonstrated that the *epicurrents*, or the surface streams, furnish us no evidence relative to the understreams to be found below them, which we name *bathycurrents*. These undercurrents may in different depths of the ocean have a quite different constitution, direction, and force from the overcurrents. This is as true of the great oceanic as of the local coast currents. If the more accurate study of marine currents is a very difficult subject and great hindrances lie, as they do, in the way of exact determinations, the same applies especially to the deep currents. New ways and means must first be found for pressing into the dark labyrinth of very complicated physical transactions. Now we can only say that the bathycurrents are of great importance for the *irregular constitution and distribution of the plankton*. Since the time when, through the discoveries of Murray (1875), Chierchia (1885), and Chun (1887), we learned to recognize the existence and importance of the zonary and bathybic fauna, and particularly, through Chun, the *vertical migration of the bathypelagic animals*, the complicated conditions of the submarine currents must evidently have exerted an extraordinary significance for planktology. Although we have hitherto known so little about this subject, yet two points stand out clearly: First, that these are of great influence upon the local and temporal oscillations of planktonic composition; second, that it is an untenable illusion if Hensen and Brandt believe that, by means of their perfect-working vertical plankton net, "a column of water whose height and base area can be accurately determined has been completely filtered" (23, p. 515); for one can never certainly know what considerable changes in the plankton of this column of water one or more undercurrents have caused during the drawing up of the vertical net.

Nerocurrents or coast streams.—While the halicurrents or the great ocean streams are influenced in the first place by the winds and stand in immediate connection with the air currents of our atmosphere, it is only partly the case with the local coast currents, for here a number of local causes, which are to be sought in the climatic and geographical condition of the neighboring coast, work together. In the case of coasts which are much indented, in archipelagos with numerous islands, etc., the study of the littoral currents becomes a very complicated problem. The physical and geological natural condition of the coast mountains and of the beach, the number and force of the incoming rivers, the quality and quantity of the coast flora, etc., are here important factors. The fishermen, pilots, etc., are very well acquainted with these local coast currents, which we will briefly call *nerocurrents*, and are usually to be trusted in the details. Scientifically these currents should be studied more closely in smaller part and less quantity. For planktology they are of very high interest and not less important than the oceanic currents.

Next, the above-intimated *reciprocal relations of the neritic and oceanic plankton* are to be taken into consideration. He who for a long time has carried on the pelagic fishery at a definite point on the coast knows how very much the result of this is influenced by the natural condition of the coast, by the course and the extent of the coast currents. Straits like those of Messina and Gibraltar, harbors like those of Villafranca and Portofino, furnish uncommonly rich plankton results, because in consequence of the littoral currents a mass of swimming animals and plants are collected together in a limited space. The volume of this planktonic mass thus heaped up is often ten or many times greater than that in the immediately adjacent parts of the sea. On the contrary, the planktonic mass is extraordinarily poor in pelagic animals and plants, where by the emptying of great floods a quantity of fresh water is brought into the sea and its saltness diminished. Johannes Müller pointed out how very much the result of pelagic fishery was influenced thereby. Again, on the other hand, the rivers day by day bring into the sea a quantity of organic substances which serve as food for the benthonic organisms, and since the benthos again stands in manifold reciprocal relation to the plankton, since the meroplanktonic animals (like the medusæ, the pelagic larvæ of worms, echinoderms, etc.) are the means of a considerable interchange between the two, so is it easily understood how the distribution of the holoplanktonic animals is also influenced thereby and how irregular becomes the composition of the plankton.

Zoöcurrents, or planktonic streams.—Among the most noteworthy and important phenomena of marine biology is the great accumulation of swimming bodies which form long and narrow bands of thickened plankton. All naturalists who have worked at the seashore for a long period and have followed the irregular appearance of the pelagic organisms know these peculiar streams, which the Italian fishermen call by the name "*correnti*." Carl Vogt, in 1848, pointed out their great importance for pelagic fishery (17, p. 303). For their scientific designation and their distinction from the other marine currents I propose the term *Zoöcurrents* or *Zoörema*.*

The pelagic animals and plants are so numerous and so closely packed in these zoöcurrents as to resemble somewhat the human population in the busiest street of a great commercial city. But millions and millions of small creatures from all the above-mentioned groups of planktonic organisms are crowded confusedly together, and furnish a spectacle of whose charm a conception can be formed only by seeing it. If one directly scoops up a portion of this motley crowd with a tumbler, not infrequently "the greater part of the contents of the glass (an actual living animal broth) is composed of the volume of animals, the smaller of the volume of water" (3, p. 171). From a distance these "crowded sea-animal streets" are usually discernible from the smoothness which the surface of the sea presents, while close beside it the surface is more or less rippled. Often one can follow such an "oil-like animal stream," which usually has a breadth of 5 to 10 meters, for more than a kilometer without finding any diminution of the thick crowd of animals in it, while on both sides of it, right and left, the sea is almost vacant, or shows only a few scattered stragglers. At Messina, as at Lanzarote, the phenomena of the zoöcurrents were especially pronounced. My companion

---

*Rema (used in Messina) is from the Greek ρεύμα = current; comp. 3, p. 172 note.

on the trip, Richard Greeff, has described the Canary animal streams so vividly that I will here give his description verbatim:

Our gaze was directed to the highly peculiar long and narrow currents, which are of very especial importance for pelagic fishery with fine nets. If one looks at the calm sea, especially from an elevation over a wide expanse of water, here and there are seen strongly marked shining streaks, which intersect the surface as long narrow bands. Their course and place of appearance seem to be continually changing and irregular. Sometimes they are numerous, sometimes only few or entirely absent; to-day they appear here, to-morrow there; some have one direction, others the opposite or crossing the first. Occasionally they run along close to one another and unite in a single stream. If one approaches this streak it becomes evident that here in fact a current prevails different from the movements of the surrounding water, and that thereby is brought about the smooth band-like appearance. They give the impression of streams cutting through the rest of the ocean, with their own channel and banks, which, notwithstanding the great variations in the time and place of their appearance, yet during their existence, which is often brief, show a certain independence.

If one comes upon such streams, which are not too far distant from the coast, he sees that all the smaller, lighter objects which formerly scattered over the surface, floated about or cast upon the shore, were drawn into it. Pieces of wood and cork, straw, algae, and tangle torn loose from the bottom, all in motley procession are carried along in this current. But in addition (and this is for us the most important phenomenon) all the animals belonging in the region of these currents are drawn in and fill it, often in such great quantities that one is tempted to believe it is not merely the mechanical influence of the narrow stream which has brought about such an accumulation of animals, but that the latter voluntarily seek out these smooth, quiet streams, perhaps in connection with certain vital expressions. A trip upon such a pelagic animal road furnishes a fund of very interesting observations. We can lean over the edge of the boat and review the countless brightly colored sea-dwellers, sometimes passing by singly, so that we can inspect them in their unique peculiarities, sometimes in such thickly massed hordes that they seem to form an unbroken layer of animals for a few feet below the surface. Yet these animal roads, where one meets them in the sea, will always form the most certain and richest mine for the so-called pelagic fauna, although naturally, from their changeableness and their dependence upon a calm sea, they can never be definitely counted upon. Likewise, the origin of these noticeable streams and their significance in the natural history of the sea is still almost completely dark, in spite of the fact that they can be observed in almost all seas and under favorable circumstances daily, and also are known to the fishermen of Arrecife under the name Zain (18, p. 307).

Although the zoöcurrents seem to occur in the most diverse parts of the ocean, and have often aroused the astonishment of observers, yet a recent investigation of them is wanting. What I know about them from my own experience and from the contributions of others is essentially the following:

The zoöcurrents occur in the open ocean as well as in the coast regions, particularly in the region of those nerocurrents which run in straits between islands or along indented coasts. They are dependent upon the weather, especially the wind, and appear as a rule only during calms. Although in the case of the neritic zoöcurrents the local course is more or less constant, still it is subject to daily (or even hourly) variations. Their breadth is usually between 5 and 10 meters, but sometimes 20 to 30 meters or more; their length is sometimes only a few hundred meters, and at others several kilometers. Oceanic animal streams reach much greater extension. Their constitution is sometimes polymixic, sometimes monotonic, often changing from day to day. Highly remarkable is the sharp boundary of the smooth, thickly populated animal roads, especially if the less inhabited and plankton-poor water on both sides is rippled by the wind. What combination of causes produces this vast accumulation is still quite unknown; certainly wind and weather play a rôle in it; often, also, the ebb and flow of the tide and other local conditions of the regions, especially local currents. As whirlwinds on land drive together the scattered masses of dust and smaller objects and raise a column of dust upwards, so may the submarine whirlwind press closely together the bathypelagic planktonic masses and carry them upward to the surface. But probably, also, in the same connection, complicated œcological conditions come into play, e, g., sudden simultaneous development of quantities of eggs of one species of animal. A new study of the zoöcurrents is one of the most urgent problems of planktology.

VI.—METHODS OF PLANKTOLOGY.

The new aspects and methods which three years ago were introduced by Prof. Hensen into planktology, and of which I have already spoken, have for their main purpose the *quantitative analysis of the plankton*, *i. e.*, the most exact determination possible of the quantity of organic substance which the swimming organisms of the sea produce. To solve this subject and come nearer to the question connected with it of the "cycle of matter in the sea," Hensen devised a new mathematical method which aims chiefly at *the counting of the individuals of animals and plants* which populate the ocean. This new method we can briefly term *the oceanic population statistics* of Hensen. The high value which this indefatigable physiologist attributes to his new arithmetical method is shown by the special mention which he makes of it in his first contribution (9, pp. 2–33), from the wonderful patience with which he counted for months the single *Diatoms, Peridineæ, Infusoria, Crustacea*, and other pelagic individuals in a single haul of the Müller net, and from the long tables of numbers, the numerical protocols, and records of captures which he has appended to his first plankton volume which appeared in 1887.

Any ordinary pelagic haul with the Müller net or tow net brings up thousands of living beings from the sea; under most favorable circumstances hundreds of thousands and millions of individuals.* How much labor and time was involved in the counting of these organisms (for the greater part microscopic) is shown from the fact that "even the counting of one Baltic Sea catch, which is pretty uniform in its composition, required eight full days, reckoning eight working hours to the

---

* In a small catch, which filtered scarcely 2 cubic meters of Baltic Sea water, were found 5,700,000 organisms, including 5,000,000 microscopic peridineæ, 630,000 diatoms, 80,000 copepods and 70,000 other animals (23, p. 516).

day" (23, p. 516). Meanwhile Brandt, explaining the "highly original procedure" of Hensen ("turning attention to attacking a problem, the solution of which no one had ever thought of"), remarks, with reference to the foregoing quantitative analysis of the Atlantic plankton expedition of the *National* (1889), "that the very much more manifold ocean catches will consume presumably twice as much time, and since on the plankton voyage at least 120 such catches were made, then the working out of these (quite apart from the preliminary preparations) will fully occupy an investigator for 120 × 14 days, or about 6 years" (23, p. 516).*

Opinions respecting the significance and the value of the oceanic population statistics of Hensen are very different. E. du Bois-Reymond, in his paper before the Berlin Academy (21, p. 83),† attributes to it extraordinary importance, "wherefore the uncommon sacrifice made for it was justified." According to his opinion, the plankton expedition of the *National*, arranged for this purpose, within its definite limits, from the novelty and beauty of its well-described task, assumes a unique place, and the Humboldt fund ought to be proud at having been among the first to contribute to its execution" (21, p. 87). On the ground of this honorable recognition, as well as of the great hopes which the naturalist of Kiel himself based upon the results of the *National* expedition, numerous notices have appeared in German newspapers, disseminating the view that an entirely new field of scientific investigation had been thereby actually entered upon, and that a further extension of it was of great importance. I am sorry to say that I can not agree with this very favorable conception.

DISTRIBUTION OF THE PLANKTON.

The foundation upon which the entire planktonic conception and computation of Hensen rests is the view "that in the ocean the plankton must be regularly distributed; that from a few catches very safe estimates can be made upon the condition of very great areas of the sea" (22, p. 243). As Hensen himself says, he started with this "*purely theoretical view*," and he believes that a completely successful result is to be had, because these theoretical premises have been more fully

---

*According to this, the unfortunate plankton counter would in these 120 catches have to count for over 17,000 hours. How such an arithmetical Danaidæ work can be carried through without ruin of mind and body I can not conceive.

†In the introduction to this noteworthy paper Du Bois-Reymond says that since 1882 Hensen "had been mindful that, especially on the surface of the sea, there was found a more unequally numerous population of minutest living forms than had previously been supposed" (21, p. 83). This remark needs correction, because many times in the celebrated log book of the *National* plankton expedition this has been overlooked, and therefore it has wrongly been inferred that Hensen eight years ago was the first to discover the *existence* and *abundance* of the pelagic fauna and flora. In fact, for forty-five years they have been the object of wonder and study for numerous naturalists.

established than could have been hoped. I have already shown that this fundamental premise is entirely wrong. *The mass of plankton in the ocean is not perennial and constant, but of highly variable and oscillating size.* The biological composition is very diverse, dependent upon *temporal* variations—year, season, weather, time of day, upon *climatic* conditions and especially upon the complicated *currentic* conditions of the streams of the sea, of the oceanic and littoral currents, the deep currents, and the local zoöcurrents.

A comprehensive and fair estimation of all these œcological conditions must *a priori* lead to the conviction *that the distribution of the plankton in the ocean must be extremely irregular*, and we find this "purely theoretical view completely established" *a posteriori* by the comparative consideration and comparison of all the earlier abovementioned observations. These can not be regarded as refuted by the opposing view of Hensen; for the empirical basis of the latter is, in regard to its time and place, *much too scanty and incomplete.*

One might perhaps object that the *technical methods* of plankton capture which Hensen employed gave more complete results than the methods hitherto used; but this is not the case. The recent description which Hensen gives of his technical methods for obtaining plankton (or pelagic fishery) is very praiseworthy (9, pp. 3 to 14). The construction of the net (material, structure of the net, size of filtration), the management of the catch and of the craft, are there carefully described. The advance of the new technique there realized may indeed serve to carry on the pelagic fishery or plankton capture more productively and more completely than was possible with the previous simpler technical apparatus of planktology; but I can not find that one of the proposed improvements of this pelagic technique shows a great advance in principle and is at all comparable to the great advance which Palumbo and Chierchia made in 1884 by the invention of the closible net. Besides, I can not understand how the new "plankton net" constructed by Hensen could give more accurate results than the simple "Müller net" hitherto employed, and the "tow net" used by the *Challenger.* Such a vertical net will always bring up only a part of the plankton contained in the volume of water going through it, and by no means, as Hensen and Brandt believe, is a column of water whose height and base area can be measured with sufficient accuracy *perfectly* filtered. In this supposition the incalculable disturbances by conditions of currents, especially of concealed deep streams, are left out of account, as already mentioned. Besides, Chierchia has lately shown how unreliable and little productive is the fishery with the vertical net on account of the considerable horizontal swimming movements of the pelagic animals (8, p. 79). At any rate, the improvements Hensen has introduced into the technical methods of plankton capture are not so important that the remarkable difference between his and the earlier results can thereby be explained.

OCEANIC POPULATION—STATISTICS.

Statistics in general is known to be a very dangerous science, because it is commonly employed to find from a number of incomplete observations the approximate average of a great many. Since the results are given in numbers, they arouse the deceptive appearance of mathematical accuracy. This is especially true of the complicated biological and sociological conditions, whose total phenomenon is conditioned by the coöperation of numerous different factors, and is, therefore, very variable according to time and place. Such a highly complicated condition, as I believe I have shown, is the composition of the plankton. If, as Hensen actually wishes, this were to be sufficiently analyzed by counting the individuals, and oceanic population statistics were thereby to be made, then this would only be possible by the formation of numerous statistical tables, which should give results in figures of the plankton fishery quantitatively in at least a hundred different parts of the ocean, and in each of these at least during ten different periods of the year.

A single "reconnoitering voyage" on the ocean, a single "trial trip," limited in time and place, like the three-months Atlantic voyage of the *National* expedition, can furnish only a single contribution to this subject. But it can in no way, as Brandt thinks, offer "firm foundations" for the solution of this and that "thorough analysis" (23, p. 525). If, also, after six years the 120 catches should actually be counted through (after a labor of more than 17,000 hours), if by statistical arrangement of this numerical protocol, by rational reckoning of their results, a serviceable conception of the quantity of individuals of the oceanic region investigated should be obtained, then at best this one computation would give us an *approximate* conception of the conditions of population of a *very small part* of the ocean; but from it by no means can we, as the investigator of Kiel wishes, arrive at conclusions bearing upon the whole ocean; for that purpose hundreds of similar computations must be made, including the most diverse regions and based upon continuous series of observations during whole years. The zoölogical stations would be the best observatories to carry out *complete series of observations* of this character, not such trial trips as the three-months voyage of the *National*.*

---

* In my opinion the results of the *National* expedition of Kiel would have been quite different if it had been carried out in the three months from January to March, instead of from July to October. On the whole, the volume of planktonic catch, at least in the North Atlantic Ocean, would have more than doubled; in some places it would have been increased many fold. Its constitution would have been entirely different. If the expedition had by accident fallen in with a zoöcurrent, and its voyage had continued in it for a few miles, the contents of the nets would have certainly been a hundredfold, possibly a thousandfold, greater.

COUNTING OF INDIVIDUALS.

Since the new method of oceanic population statistics introduced by Hensen seeks its peculiar basis in the counting of the individuals which compose the plankton, and since it finds in this "counting the only basis upon which a judgment can rest" (9, p. 26), then we must examine more critically this cardinal point of his method, upon which he lays the greatest stress. The counting of the single organic individuals, which compose the mass of the plankton, is in itself, quite apart from its eventual value, an extremely difficult and doubtful task. Hensen himself has not concealed a part of this great difficulty, and attempts to partly allay the doubts which arise against his whole method.* But in fact these are much greater and more dangerous than he is inclined to admit.

WHAT IS AN ORGANIC INDIVIDUAL?

This simple question, as is known, is extremely difficult to answer. If one does not accept all the grades of physiological and morphological individuality, which I have distinguished in the third book of my "*Generelle Morphologie*," 1866, there are at least three distinct chief grades to be kept apart: (1) The cell (or plastid); (2) the person (or bud); (3) the cormus (or colony).† Only among the *Protista* (*Protophyta* and *Protozoa*) is the actual individual represented by a single cell; on the other hand, among the *Histona* (*Metaphyta* and *Metazoa*), by

---

* The fourth part of the "Methodik" in the plankton volume of Hensen, which treats of "the work on land," (*a*) Determination of the volume, (*b*) the counting (9, pp. 15-30), is especially worthy of reading, not only because it gives the deepest insight into the error of his method, but also into his very peculiar conception of a general biological problem.

†The swimming animals and plants which compose the plankton should in this respect be arranged under the following heads: (*a*) *Protophyta*—among the *Chromaceæ, Calcocyteæ, Murracyteæ, Xanthelleæ, Dictyocheæ,* and *Peridineæ*, all single cells are to be counted; among the diatoms partly the latter, partly the cenobia or cell aggregates. (*b*) *Metaphyta*—among the *Halosphæra* are to be counted the spherical *Thalli;* among the *Oscillatoriæ* the single, thread-like *Thalli;* among the *Sargassa* the cormus as well as its buds; but the cells which constitute each thallus and bud are also peculiar. (*c*) *Protozoa*—the *Infusoria* (*Noctiluca* and *Tintinna*) as well as the rhizopods (*Thalamophora* and *Radiolaria*), are all to be counted as unicellular individuals, but among the *Polycyttaria*, besides the *Cænobia* (colonies of *Collozoidæ, Spharozoidæ,* and *Collosphæridæ*). (*d*) *Cælenterata*—among the *Medusæ* and *Ctenophores*, as also among the pelagic *Anthozoa* and *Turbellaria* the single persons are to be counted; among the *Siphonophores* these as well as the single colonies; for each person (or each medusom) of a cormus is here equivalent to a medusa. (*e*) *Tunicata*—among the *Copelata, Doliolum,* and the generations of solitary *Salpas*, the single persons are to be counted; on the other hand, among the *Pyrosoma* and the *Salpa* chains, the single cormi as well as the persons which compose them. (*f–k*) In all the remaining groups of planktonic animals, in the case of sagitta, mollusks, echinoderm larvæ, articulates, and fishes, not merely the persons are to be counted, but also the cells which make up each of these metazoa.

the higher unit of the person or of the colony, which is composed of many cells. If we actually wish to carry out exactly the method, held by Hensen as indispensable, of counting the individuals, and wish to obtain useful results for his statistical work, then nothing remains except a counting of all single cells which live in the sea. For only the single cells, as the "organic elementary individual," can form the natural arithmetical unit of such statistical calculations and the computations based thereon. If Hensen in his long " numerical protocols and comparisons of captures" (9, pp. XI-XXXIII) places close to one another as counted individuals—as coördinated categories—the unicellular radiolaria, the *cormi* of siphonophores and tunicates, the persons of medusæ, ctenophores, echinoderms, and crustacea, the eggs and persons of fishes, then he places together vastly incommensurable bulks of quite different individual value. These can only be comparable for his purpose if all single cells are counted. But since each fish and each whale in the ocean daily destroys milliards of these planktonic organisms, so, in order to gain an "exact" insight into the "cycle of matter in the sea," the cell milliards which compose the bodies of these gigantic animals must be counted and placed in the reckoning.

ECONOMIC YIELD OF THE OCEAN.

Hensen holds the quantitative determinations of the plankton not only as of the highest importance in theoretical interest to science, but also in practical interest to national economy. He thinks "that we will be able to invent correct modes of action in the interest of the fisheries,* only if we are in position to form a judgment upon the productive possibilities of the sea" (9, p. 2). Accordingly he regards it as the most pressing problem to determine the economic yield of the ocean in the same way as the farmer determines the useful yield of his fields and meadows, the yearly production of grass and grain. By the counting of the planktonic individuals which Hensen has carried on for a long time for a small part of the Baltic Sea, he thinks he has become convinced that the "entire production of the Baltic in organic substance is only a little inferior to the yield of grass upon an equally large area of meadow land."

The farmer determines the yield of his meadows, garden, and field by quantity and weight, not by counting the individuals. If instead of this he wished to introduce Hensen's new exact method of deter-

---

*How the practical interests of the fisheries can be advanced by quantitative plankton analysis I am not able to understand. The most important modes of action which we can employ for the increase of the fish production of the ocean— artificial propagation, increase and protection of the fry, increase of their food supply, destruction of the predaceous fishes, etc.—are entirely independent of the numerical tables which Hensen's enumeration of individuals gives. That the number of swimming fish eggs furnishes no safe conclusion upon the number of mature fish has been pointed out above.

mination, he must count all the potatoes, kernels of grain, grapes, cherries, etc., and not only that, he must also count the blades of grass of his meadow, even every individual weed which grows among the grain of his field and the useful plants of his garden; for these also, regarded from the physiological point of view, belong to the "total production" of the ground. And what would be gained by all these immense countings? Just as little as with the "desolate figures" in Hensen's long numerical protocols.*

VOLUME AND WEIGHT OF THE PLANKTON.

If one actually regards the determination of the planktonic yield as a highly important subject, and believes that this can be solved by a certain number of quantitative plankton analyses, then this goal can be reached in the simplest way by determination of the volume and weight of each planktonic catch. Hensen himself naturally first trod this nearest way; but he thinks that it is not accurate enough and encounters difficulties (9, p. 15). In his opinion, "an accurate analysis of the plankton, on account of the great variety of its parts, can only be obtained by counting; he quite forgets that such a counting of individuals also possesses only an approximate and relative value, not a complete and absolute one; farther, that from the counting of the different individuals no more certain measure for the economic value of the whole diversely constituted planktonic catch is furnished; finally that the counting of one catch is of highest value as a single factor of a great computation, which is made from thousands of different factors.

The only thorough method of determining the yield, in planktology as in economy, is the determination of the useful substance according to mass and weight and subsequent chemical analysis. In fact, the determination of the planktonic volume, as of the weight, just as the qualitative and quantitative chemical analysis of the plankton, is possible up to a certain degree. The difficulties are less than Hensen believes. It seems odd that the latter has not mentioned these two simplest methods in a single place in his comprehensive volume (9, p. 15), but hastily casts them aside and replaces them with the quite useless "counting of individuals," a Danaide task of many years.

---

* While Hensen is going over the counting of the single constituent parts of the plankton, he calls special attention to the fact "that in spite of the apparently" desolate figures, in almost every single case certain results of general interest have come out, though the opportunity is not offered to show them in a comparison.

CYCLE OF MATTER IN THE OCEAN (*Stoffwechsel des Meeres.*)

The many and great questions which the mighty cycle of matter in the ocean furnishes to biology, the questions of the source of the fundamental food supply, of the reciprocal trophic relations of the marine flora and fauna, of the conditions of support of the benthonic and planktonic organisms, etc., have, within the last twenty years, since the beginning of the epoch-making deep-sea investigation (13), been much discussed and have received very different answers (11). Hensen has also devoted considerable attention to this, and particularly emphasizes the physiological importance of the fundamental food supply (*Urnahrung*). He believes this complicated question can be solved especially by *quantitative determination of the fundamental food supply.*

I have already shown why this method of quantitative plankton analysis must be regarded as useless. Even assuming that it were possible and practicable, I can not understand how it could lead to a definite solution of this question. On the other hand, I might here point to one side of the oceanic cycle of matter whose further pursuit seems very profitable. The two chief sources of the "oceanic fundamental food supply" have already been correctly recognized by Möbius (11), Wyville Thompson (13, 14), Murray (6), and others: First, the vast terrigenous masses of organic and particularly vegetable substances, which are daily brought by the rivers to the sea; secondly, the immense quantities of plant food which the marine flora itself furnishes. Of the latter we previously had in mind chiefly the benthonic littoral flora, the mighty forests of algæ, meadows of *Zostera*, etc., which grow in the coast waters. Only in recent times have we learned to value the astonishing quantity of vegetable food which the planktonic flora produces, the *Fucoids* of the Sargasso Sea on the one side, the *Oscillatoriæ* and the microscopic *Diatoms* and *Peridineæ* on the other. But the smaller groups of pelagic *Protophytes*, which I have mentioned above, the *Chromaceæ*, *Murracyteæ*, *Xanthelleæ*, *Dictyocheæ*, etc., also play an important rôle. The great importance which devolves upon the small symbiotic *Xanthelleæ*, has been especially emphasized by Brandt (24), Moseley (7), and Geddes. Evidently their multiplication is extremely rapid, and if each second milliard of such *Protophytes* were eaten by small animals, new milliards would take their places. Whether or not the number of these milliards is shown to us by the quantitative planktonic analysis seems to me wholly indifferent. More important for the understanding of their physiological importance would be the ascertainment of the rapidity of the increase.

The importance of these *Protophytes* and of the *Protozoa* living upon them has lately been particularly described by Chun (28, pp. 10, 13). He has also rightly emphasized the extraordinary importance which the *vertical migration* of the bathypelagic animals has for the support of the deep-sea animals. They are to a great extent the under workmen, who constantly bring the provision transports into the deep sea (15, pp. 49, 57). Thither, in addition, come the immense quantities of marine plant and animal corpses, which daily sink into the depths and are borne away by currents. Thither comes the constant "rain" of the corpses of zonary *Protozoa* (especially *Globigerina* and *Radiolaria*), which uninterruptedly pour down through all the zones of depth into the deepest abysses, and whose shells form the most abundant sediment of the deep sea, the calcareous *Globigerina* ooze and the siliceous *Radiolaria* ooze. In general, it seems to me that the daily supply of food materials which the decaying corpses of numberless marine organisms furnish to the others, is much more important than is commonly supposed.* How much food would a single dead whale alone furnish?

But especially important and not sufficiently valued in this regard, it seems to me, is the trophic importance of the benthos for the plankton. Immense quantities of littoral benthos are daily carried out into the ocean by the currents. Here they soon disappear, since they serve as food for the organisms of the plankton. If one weighs all these complicated reciprocal relations, he obtains without counting a sufficient general conception of the "cycle of the organic material in the marine world."

COMPARATIVE AND EXACT METHODS.

The farther the two great branches of biology, namely, morphology and physiology, have developed into higher planes during the last decade, so much farther have the methods of investigation in both sciences diverged from one another. In morphology the high worth of comparative or declarant methods has always been justly more recognized, since the general phenomena of structure (*e. g.*, in ontogeny and systemization) have been in great part removed from exact investigation, and comprise historical problems, the solution of which we can strive for only indirectly (by way of comparative anatomy and phylogenetic speculation). In physiology, on the other hand, we constantly strive to employ the exact or mathematical methods, which have the advantage of relative accuracy and which enable us to trace back the general phenomena of vital activity directly to physical (particularly to chemical) processes. Plainly it must be the endeavor of all sciences (of morphology also) to find and retain as much as possible this exact mode of investigation. But it is to be regretted that among most branches of science (and particularly the biological ones) this is not possible, because the empirical foundations are much too incomplete and

---

* Hensen values this source of food very slightly, because "only a very few animals live upon dead matter," and explains it in this way, "that material in a state of foul putrefaction requires a stronger digestive power than the organization of the lower animals can produce" (9, p. 2). I must contradict both ideas. The sponges live chiefly upon decaying organisms, as do also many *Protozoa*, *Helminths*, **Crustacea**, etc.

the problems in hand much too complicated. Mathematical treatment of these does more harm than good, because it gives a deceptive semblance of accuracy, which in fact is not attainable.* A part of physiology also embraces such subjects as are with difficulty, or even not at all, accessible to exact definition, and to these also belong the chorology and œcology of the plankton.

The fundamental fault of Hensen's plankton theory in my opinion lies in the fact that he regards a highly complicated problem of biology as a relatively simple one, that he regards its many oscillating parts as proportionally constant bulks, and that he believes that a knowledge of these can be reached by the exact method of mathematical counting and computation. This error is partly excusable from the circumstance that the physiology of to-day, in its one-sided pursuit of exact research, has lost sight of many general problems which are not suited for exact special investigation. This is shown especially in the case of the most important question of our present theory of development, the species problem. The discussions which Hensen gives upon the nature of the species, upon systemization, Darwinism, and the descent theory, in many places in his plankton volume (pp. 19, 41, 73, etc.) are among the most peculiar which the volume contains. They deserve the special attention of the systematist. The "actual species" is for him a physiological conception, while, as is known, all distinction of species has hitherto been reached by morphological means.†

In my Report on the *Radiolaria* of H. M. S. *Challenger* I have attempted to point out how the extremely manifold forms of this most numerous class (739 genera and 4,318 species) are on the one hand distinguished as species by morphological characters, and yet on the other hand may be regarded as modifications of 85 family types, or as descendants of 20 ancestral orders, and these again as derived from one common simple ancestral form (*Actissa*, 4, § 158). Hensen on the other hand is of the opinion that therein is to be found "a strong opposing proof against the independence of the species" (9, p. 100). He hopes "to lighten the systematic difficulties by the help of computation" (p. 75). Through his systematic plankton investigations he has reached

---

* A familiar and very instructive example of this perverted employment of exact methods in morphology is furnished by the familiar "Mechanical theory of development" of His, which I have examined in my anthropogeny (3d edition, p. 53, 655) as well as in my paper upon *Ziele and Wege der Entwickelungsgeschichte* (Jena, 1875).

† Since of late the physiological importance of the "species" conception has often been emphasized and the "system of the future" by the way of "comparative physiology" has been pointed out, it must here be considered that up to this time not one of these systematic physiologists has given even a hint how this new system of description of species can be practically carried out. What Hensen has said about it (9, pp. 41, 73, 100) is just as worthless as the earlier discussions by Poléjáeff, which have been critically considered in my Report on the Deep-Sea Keratosa (*Challenger*, Zoölogy, vol. XXXII, part 82, pp. 82–85).

the conviction that "the more accurately the investigation has been made, so much the more plain becomes the distinction of species" (9, p. 100). On the other side I, like Charles Darwin, through many years of comparative and systematic work, have arrived at the opposite conclusion: "*The more accurately the systematic investigations are made, the greater the number of individuals of a species compared, the intenser the study of individual variation, by so much more impossible becomes the distinction of actual species, so much more arbitrary the subjective limits of their extent, so much stronger the conviction of the truth of the Theory of Descent.*"*

PLANKTOLOGICAL PROBLEMS.

The wonderful world of organic life, which fills the vast oceans, offers a fund of very interesting subjects. Without question, it is one of the most attractive and profitable fields of biology. If we consider that the greater part of this field has been open to us scarcely fifty years, and if we wonder at the new discoveries which the *Challenger* expedition alone has brought to light, then we ought to count upon a brilliant future for planktology.

Above all we ought to cherish the hope that our German *National* expedition, the first great German undertaking in the field, may promote many planktonic problems, and that the six naturalists who, under such favorable conditions and with such important instruments, studied the oceanic plankton for ninety-three days and in 400 hauls of the net were able to obtain a rich collection of pelagic organisms, will by their careful working up of these enrich our knowledge many fold. However, the preliminary contributions of Hensen (22) and Brandt (23) give us no means of passing judgment upon the matter now. Among the results which the former has briefly given to the Berlin Academy few require consideration; but for this the difference of our general point of view is to blame. Thus, for example, I have attempted to explain the remarkable "similarity to water of the pelagic fauna," the transparency of the colorless glassy animals, in 1866, in my General Morphology (II, p. 242), according to Darwin's Theory of Selection, by natural selection of like colors (30, p. 248). Hensen, on the other hand,

---

* F. Heincke has briefly, in his careful "Investigations upon the Stickleback," given expression to the same conviction in the following words: "All the conclusions here deduced by me are simply and solely founded upon the comparison of very many individuals of living species, or, in other words, upon the study of *individual variation*. I am convinced that in essentials the study of embryology will confirm my theory. It will be a proof of this, that he who wishes accurately to describe related species, and races of a species, and to study their genealogical relation to one another, must begin by *comparing a very great number of individuals from different localities* accurately and methodically. He will then soon see *that proofs of the theory of descent by this means are found in great numbers at all times*, if only one does not spare the pains to trace them out." (Ofversigt af K. V. Akad. Forh. Stockholm, 1889, No. 6, p. 410.) This view of Heincke is shared by every experienced and unbiased systematist.

regards hunger as the cause of this, and the "tendency to explore a relatively great bulk of water." In general, according to his view, " many larger pelagic animals bear the outspoken character of unfavorable conditions of life, of a life of hunger."

Regarding the *appearance* of many pelagic animals *in swarms*, Hensen explains "that the young do not float, but swim freely. In consequence of this, the mother animals drive them away, and if the larvæ finally rise to the surface the former can not enter into competition with them" (22, p. 252). The accumulation of numbers of *Physalia* in great swarms stands, according to his view, in correlation with the mode of movement. The animals which are capable of no independent movement of progression must remain rather closely crowded together, in order to be able to reproduce *bisexually;* those carried too far away must perish." On the other hand it is to be noted that the *Physalia* is not, as Hensen assumes, *gonochoristic*, but always *hermaphroditic*.*

The above-mentioned phenomena, the similarity to water of the pelagic fauna, the periodic appearance of many pelagic organisms in swarms, their abundant accumulation in the zoöcurrents (p. 85), particularly their relation to the currents, are only a few of the greater problems which planktology furnishes for human investigative energy. For these, as for so many other fields of biology, Charles Darwin, by the establishment of the descent theory, has opened to us the way to a knowledge of causes. We must study the complicated reciprocal relations of the organisms crowded together in the struggle for existence, the interaction of heredity and variation, in order to learn to understand the life of the plankton. But in these plankton studies, as well in physiological as in morphological questions, we must use that method which Johannes Müller, the discoverer of this field, always employed in a manner worthy of imitation: simultaneous "observation and reflection."

---

* The cormi of all *Physalidæ* are monœcious, their cormidia monoclinic. Each single branch of the racemose gonodendron is monostylic, and bears one female and several male medusoids. The facts were brought out thirty-five years ago by Huxley. (Compare my Report on the Siphonophoræ: Zoölogy of the *Challenger*, vol. XXVIII, pp. 347, 356.)

LITERATURE.

1. JOHANNES MÜLLER, 1845–1855. Ueber die Larven und die Metamorphose der Echinodermen. Abhandlungen der Berliner Akademie der Wissenschaften.
2. ———, 1858. Ueber die Thassicollen, Polycystinen und Acanthometren des Mittelmeeres. *Idem.*
3. ERNST HAECKEL, 1862. Monographie der Radiolarien. Uebersicht der Verbreitung, pp. 166–193.
4. ———, 1887. Report on the Radiolaria collected by H. M. S. Challenger during the years 1873–1876. Chronological section. §§ 226–240. (Deutsch in der "Allgemeinen Naturgeschichte der Radiolarien." 1887, pp. 123–137.)
5. JOHN MURRAY, 1876. Preliminary report on some surface organisms examined on board H. M. S. Challenger, and their relation to ocean deposits. Proceed. Roy. Soc., vol. XXIV, pp. 532–537.
6. ———, 1885. Narrative of cruise of H. M. S. Challenger, with a general account of the scientific results of the expedition (1873–1876), vol. I, II.
7. H. N. MOSELEY, 1882. Pelagic Life. Address at the Southampton Meet. Brit. Assoc., Nature, vol. XXVI, No. 675, p. 559.
8. GAETANO CHIERCHIA, 1885. Collezioni per Studi di Scienze Naturali, fatte nel Viaggio intorno al mondo dalla R. corvetta Vettor Pisani. Anni 1882–1885.
9. VICTOR HENSEN, 1887. Ueber die Bestimmung des Planktons, oder des im Meere treibenden Materials an Pflanzen und Thieren. V. Bericht der Commission zur wissensch. Unters. der Deutschen Meere in Kiel.
10. K. MÖBIUS, 1887. Systematische Darstellung der Thiere des Plankton in der westl. Ostsee und auf einer Fahrt von Kiel in den Atlantischen Ocean bis jenseit der Hebriden. (V. *Idem*).
11. ———, 1871. Wo kommt die Nahrung für die Tiefseethiere her? Zeitsch. für wissensch. Zool., 21. Bd., p. 294.
12. TH. FUCHS, 1882. Ueber die pelagische Flora und Fauna. Verhandl. d. k. k. Geolog. Reichsanstalt in Wien, 4. Febr., 1882, pp. 49–55.
13. WYVILLE THOMPSON, 1873. The Depths of the Sea. An account of the general results of the dredging cruises of H. M. S. S. Porcupine and Lightning.
14. ———, 1877. The Atlantic. A preliminary account of the general results of the exploring voyage of H. M. S. Challenger.
15. CARL CHUN, 1888. Die pelagische Thierwelt in grösseren Meerestiefen und ihre Beziehungen zu der Oberflächen-Fauna. Bibliotheca zoologica, Heft I.
16. ———, 1889. Bericht über eine nach den Canarischen Inseln im Winter 1887–88 ausgeführte Reise. Sitzungsberichte der Berliner Akademie der Wiss., p. 519.
17. CARL VOGT, 1848. Ocean und Mittelmeer, p. 303.
18. RICHARD GREEFF, 1868. Reise nach den Canarischen Inseln. "Die Meeresströmungen als Thierstrassen," pp. 307–309.
19. R. SCHMIDTLEIN, 1879. Vergleichende Uebersicht über das Erscheinen grösserer pelagischer Thiere während der Jahre 1875–1877. Mittheil der zoolog. Station Neapel, Bd I, p. 119.
20. EDWARD GRAEFFE, 1881–88. Uebersicht der Seethier-Fauna des Golfes von Triest, nebst Notizen über Vorkommen, Lebensweise, Erscheinungs- und Fortpflanzungs-Zeit. Arbeiten d. zool. Station Triest.
21. E. DU BOIS-REYMOND, 1890. Bericht über die Humboldt-Stiftung und die Kieler Plankton-Expedition des National. Sitzungsberichte der Berliner Akademie d. Wissensch. vom 23. Januar 1890, pp. 83–87.

22. Victor Hensen, 1890. Einige Ergebnisse der Plankton-Expedition der Humboldt-Stiftung. Sitzungsberichte der Berliner Akademie der Wissenschaften vom 13. März 1890, pp. 243-253.
23. Karl Brandt, 1889. Ueber die biologischen Untersuchungen der Plankton-Expedition. Verhandl. der Gesellschaft für Erdkunde zu Berlin, vom 7. Dec. 1889, p. 515.
24. ———, 1885. Die coloniebildenden Radiolarien (Sphaerotseu) des Golfes von Neapel.
25. Ernst Haeckel, 1882. Indische Reisebriefe. II. Aufl.
26. Karl Möbius, 1880. Beiträge zur Meeres-Fauna der Insel Mauritius und der Seychellen, 1880.
27. Carl Chun, 1886. Ueber die geographische Verbreitung der pelagisch lebenden Seethiere. Zoolog. Anzeiger, Nr. 214, 215.
28. ———, 1890. Die pelagische Thierwelt in grossen Tiefen. Verhandl. d. Gesellsch. deutsch. Naturf. u. Aerzte, Bremen, 1890.
29. Ernst Haeckel, 1879. Monographie der Medusen. I. Bd. Das System der Medusen. II. Bd. Der Organismus der Medusen.
30. ———, 1889. Natürliche Schöpfungsgeschichte. Achte Auflage.

VICTOR HENSEN
AND THE DEVELOPMENT OF
SAMPLING METHODS IN ECOLOGY

John Lussenhop

Victor Hensen and the Development of Sampling Methods in Ecology

JOHN LUSSENHOP

Department of Biological Sciences
University of Illinois at Chicago Circle
Box 4348, Chicago, Illinois 60680

INTRODUCTION

Nineteenth-century ecological problems drew investigators from many biological disciplines. Often social needs defined ecological problems and may have helped focus the investigator's attention on specific goals. For example, the ecological community concept was first stated by a natural historian who was trying to improve the oyster fisheries.[1]

Another aspect of nineteenth-century ecology is that significant research problems and appropriate methodology had not been agreed upon, and this resulted in disputes between scientists of different backgrounds working on similar ecological problems. Quantitative ecological research exemplifies these features of nineteenth-century ecology. The first statistical estimates of the size of a population of organisms in nature were made of fish eggs and plankton in the Baltic Sea by the physiologist Victor Hensen, who was trying to improve the fisheries. Hensen's conclusions were significant contributions to ecology, and his work stimulated many students. Yet Hensen could not convince many of his contemporaries of the validity of his results. Hensen's efforts to justify his sampling methods illustrate the degree to which ecology was limited by lack of methodology.

This discussion traces the source and development of Hensen's statistical methods. It illustrates how the techniques Hensen used limited his concept of local variation in animal populations and also limited his ability to respond to criticism. The limitation imposed on Hensen's work were faced by many nineteenth-century ecologists. When ecologists first attempted to estimate the numbers of organisms present in large areas, there was no statistical sampling methodology that could be adopted from other disciplines, and biologists had little experience with the extent of variability of natural populations or organisms. As a

---

1. Karl Mobius, "Die Auster und die Austernwirtschaft" (Berlin, 1877), transl. J. H. Rice in *Rep. U.S. Commnr. Fish*, (1880), 683–751.

result, although much quantitative sampling apparatus was invented by nineteenth-century ecologists, the samples taken with nets, dredges, and quadrats were not treated statistically until the 1930's.

The first sampling of highly variable natural populations which might have provided the methodology needed by ecologists was of human populations. In order to obtain data, seventeenth- and eighteenth-century statisticians counted households, births, marriages, and deaths, or used tax lists as a sample on the basis of which they estimated the number of people.[2] By the beginning of the nineteenth-century most governments had established regular censuses by complete enumeration. Demographers did not contribute to development of a sampling methodology which ecologists could have used but instead relied on the enumerations of government statistical bureaus. Only in the twentieth-century have government statistical bureaus and demographers returned to an interest in samples rather than complete enumeration.[3]

SAMPLING METHODS

Early naturalists who made estimates of the numbers of organisms in natural populations did so in the absence of any generally accepted procedures. Three characteristics of currently accepted sampling procedures have been distinguished by S. S. Wilks,[4] and may be used to analyze the characteristics of nineteenth-century sampling methods. Sampling must be (1) operationally defined, (2) reproducible, and (3) valid. The development of a sampling procedure to meet these criteria entails (1) a well-defined sequence of steps leading from proper deployment of a specific kind of collecting gear to the procedure for calculating the estimate. The collecting gear and statistical procedure must ensure that (2) the result of different investigators following the same steps show a reasonable agreement, and (3) the resulting estimate should be a true or valid description of the population in nature.

---

2. Harold Westergaard, *Contributions to the History of Statistics* (New York: Agathon, 1968), chap. 8.

3. Frederick Stephan, "History of the Uses of Modern Sampling Procedures," *J. Am. Stat. Assn., 43* (1948), 12–39. Recent use of sampling in demography is described by C. L. Chiang, *Introduction to Stachastic Processes in Biostatistics* (New York: John Wiley, 1968).

4. S. S. Wilks, "Some Aspects of Quantification in Science," *Isis, 52* (1961), 135–142.

Victor Hensen and the Development of Sampling Methods in Ecology

The naturalists who first made estimates of the numbers of organisms in natural populations did use operationally well-defined procedures. And although their results were often reproducible, they were not valid because of incorrect assumptions about how organisms are arranged or patterned in their habitats. For example, William Scoresby estimated the number of organisms in an algal bloom by dipping out a bucketful of the bloom and counting the number of organisms in a measured, small volume of water under a microscope.[5] Scoresby estimated the size of the whole bloom and multiplied the number he had observed in a small volume by the total volume. Scoresby's method tacitly assumed that the algae within the bloom were distributed at approximately uniform distances from one another, and that this pattern was not changed in the collecting bucket. If this were true, Scoresby's estimate would be reproducible and valid. However, an alternative assumption might be that organisms are spaced randomly in the bloom, and that this pattern is not affected by collection. In this case, one would expect each sample to contain a different number of organisms, and a number of samples would be required to characterize the population. Scoresby's estimate would not be reproducible and might not be valid if the organisms were patterned randomly.

The 1802 estimate Laplace made of the French population is significant in that he introduced a method for checking the reproducibility of his estimate.[6] His procedure may be divided into the steps outlined by Wilks. A well-defined method was used for choosing localities in thirty departments for which the number of inhabitants and births in the preceding three years was obtained. Laplace assumed that the number of men and women in the fraction of France he sampled were a simple proportion of the number in the entire country. He estimated reproducibility of his sample by noting that the numbers in his sample were so large that reasonable errors would result in values with a small percent of error distributed symmetrically on either side of his estimate. Laplace did not discuss the validity of his sample — he felt it was determined by choosing representative areas as samples.

These examples suggest that the development of sampling procedures has been impeded by slowness in developing methods for judging the

---

5. William Scoresby, *Journal of a Voyage to the Northern Whale-Fishery: including Researches and Discoveries on the Eastern Coast of West Greenland, made in the summer of 1822 in the Ship Baffin of Liverpool* (Edinburgh: Archibald Constable, 1823), pp. 353–356.

6. Pierre S. Laplace, *Theorie analytique des probabilities*, 3rd ed. (Paris: 1847).

validity and reproducibility of a sampling procedure. One might suppose that a biologist could demonstrate the validity of a sampling procedure by completely enumerating a portion of the population, after which reproducibility could be easily evaluated. However, particularly when organisms are sampled from water or soil, their pattern can be approximated only by taking many contiguous samples – and even then only a small area can be sampled in this way. Instead, collections from the biological populations are compared with collections one would obtain from a model or theoretical population with a known pattern. After a model population which fits the sample sufficiently closely is found, reproducibility can be measured, and often adjustments in the sampling procedure can be made as well. The following sections trace Hensen's attempts to demonstrate the validity of his sampling method, culminating in his construction of a model population with which to compare his results.

HENSEN'S BACKGROUND

Victor Hensen's background, his teachers, and his choice of subjects suggest that he chose research in the most rapidly developing areas of his time.[7] Hensen grew up on the Baltic Coast in the city of Schleswig, and went to Würzburg and Berlin to study medicine. His teachers at Würzburg were Albert Kölliker, Rudolf Virchow, and Franz Leydig. At the end of two years working there under the guidance of the biochemist Johann Scherer, Hensen was able to publish an account of the isolation of glycogen only a month after Claude Bernard.[8] Hensen wrote his medical dissertation at Kiel on the chemistry of the urine of epileptics.[9]

Hensen was interested in marine life from the time he was a student, and occasionally a marine invertebrate was the subject of one of his largely physiological early papers. This interest in marine life is under-

---

7. Important biographical sources are: Rudiger Porep, "Der Physiologe und Planktonforscher Victor Hensen," *Kieler Beiträge zur Geschichte der Medizin und Pharmazie, 9* (1970), 1–147, and Karl Brandt, "Victor Hensen und die Meeresforschung," *Wissenschaftliche Meeresuntersuchungen der Kommission zur wissenschaftlichen Untersuchungen der deutschen Meere, Abteilung* (Kiel, n.f. *20* (1925), 51–103.

8. V. Hensen, "Ueber Zuckerbildung in der Leber," *Virchows' Arch. Patt. and Physiol., 11* (1857), 395–398.

9. V. Hensen, "De Urinae excretione in epilepsia," *Schriften der Universität zu Kiel, 6* (1860).

Victor Hensen and the Development of Sampling Methods in Ecology

standable because Hensen grew up close to the Baltic, and his teacher, Kölliker, transmitted to students the enthusiasm for the minute marine invertebrates his own teacher, Johannes Müller, had first collected. As an undergraduate, Hensen and his friend Carl Semper used a stipend to pay for a collecting trip to the Mediterranean at Trieste.

The opportunity for Hensen's appointment as Professor Ordinarius and Director of the Physiological Institute at Kiel was the Prussian annexation of Schleswig-Holstein in October 1864. C. L. Pannum, the Danish Director of the Physiological Institute, returned to Copenhagen six months after the Peace of Vienna, and Hensen was chosen to replace him. Prussian influence had begun to transform Kiel into an important port, and the university into a center of excellence in medicine and oceanography.[10] Although published sources are not entirely adequate for an understanding of Hensen's motivation, the pattern of his publications suggests an initial strengthening of the Physiological Institute, and then a developing interest in marine research. We are reminded of Hensen's descriptive and experimental work at the Physiological Institute by terminology stemming from his work — the H line (Hensen's line) of muscle,[11] and Hensen's node of the avian embryo — as well as by his discovery of the role of the ciliary muscle in visual accommodation.[12]

Hensen obtained the resources for the establishment of a fisheries research program at the University of Kiel from a government commission which he had a part in forming, and locating at Kiel. In the first election after annexation of Schleswig-Holstein, Hensen (and his former teacher, Rudolf Virchow) were elected to the Prussian Landtag. Hensen spent only four months of 1867–1868 in Berlin as an active Landtag member before he was made Professor Ordinarius and returned to the university. While a Landtag member, he worked actively for establishment of a government fisheries research program. Hensen believed that such a program at the university would improve the economic basis of coastal fishing villages and the province as a whole.[13] Sixteen years after

10. Geert Seelig, *Eine deutsche Jugend, Erinnerungen an Kiel und den Schwanenweg* (Hamburg: Alster Verlag, 1922), p. 167.
11. V. Hensen, "Ueber ein neues Strukturverhaltnis der quergestreiften Muskelfaser," in *Arbeiten aus dem Kieler physiolischen Institut, 1868* (Kiel, 1869), pp. 1–36, 172–176.
12. V. Hensen and Carl Voelckers, *Experimentaluntersuchung über den Mechanismus der Accomodation*, (Kiel: Schwers, 1868).
13. Rudolf Illing, "Die Entwicklung der Seefischerei an der Nordsee Küste Schleswig-Holsteins." *Zeitschrift der Gesellschaft für Schleswig-Holsteinische Geschichte, 53* (1923), 135. The economic problems in Schleswig-Holstein before

establishment of the commission, Hensen summarized his motivation and the reasons for his increasing interest in plankton:

> Pursuit of this source of nutrient material [plankton] is the work to which the course of my research is tending. In the Prussian Landtag in 1867 during the planning of scientific research for the study of oceans in the interest of fisheries, I remember thinking that the only correct part to take in the interest of fisheries would be to study the possibility of judging the productivity of the ocean. As a member of the commission I have above all sought to strengthen and expand fishing on the German coasts.[14]

DEVELOPMENT OF HENSEN'S SAMPLING METHODS

Procedure

The Prussian government established the Kommission zur Wissenschaftlichen Untersuchung der deutschen Meere in 1870. Hensen was a Commission member, and, as a participant in the Commission's summer expeditions from 1871 to 1885, developed sampling methods for pelagic fish eggs and for plankton. Georg Sars had shown that the eggs of codfish float freely,[15] and Hensen extended this observation to plaice and flounder. Hensen was interested in these observations because they suggested the possibility of estimating the size of the parental fish population on the basis of an egg sample. Later in this period, Hensen became increasingly interested in the small organisms the fish feed upon. He introduced the term "plankton" for these organisms, and concentrated on sampling their numbers because he felt that the productivity of fisheries depended on the size of the plankton population.[16]

Hensen's first task in developing a quantative sampling procedure was devising a way of estimating the volume of water which had been sampled. For this purpose he initiated the use of nets made of uni-

---

unification with Prussia and the growth of the fishing industry after unification are discussed in this article.

14. V. Hensen, "Ueber die Bestimmung der Planktons oder des im Meer treibenden Materials an Pflanzen und Tieren," *Bericht der Commission zur Wissenschaftlichen Untersuchungen der Deutschen Meere*, 5 (1884), 2.

15. Georg O. Sars, "Om Vintertorskens *(Gadus morrhua)* forplantning og udvikling," *Forh. Vidensk Selsk. Krist.*, 8 (1886), 237–249; transl. H. Jacobson in *Rep. U.S. Commnr. Fish.*, 1873–74, 1874–75., 213–222.

16. V. Hensen, "Bestimmung der Planktons," p. 2.

Victor Hensen and the Development of Sampling Methods in Ecology

formly woven silk netting used by millers to separate different grades of flour.[17] Nets do not filter as much water at the end of a haul as at the beginning because the meshes gradually become occluded with plankton and detritus which are not swept back into the collecting canister. Hensen simulated net hauls in laboratory tanks and developed tables of coefficients to estimate the volume of water actually filtered while collecting.

All of the fish eggs in a net haul were counted, but when Hensen began to study plankton, there were often too many organisms in the sample to count. Hence the number of organisms in a small subsample of the original sample were counted in the laboratory.[18] Hensen's procedure involved sampling the well-shaken sample with a piston pipette which he developed to eliminate the poor sampling of larger organisms due to the round opening of ordinary pipettes. He estimated the number of organisms in the entire sample by using the theory developed by the Jena physicist Ernst Abbe for estimating the number of red blood cells.[19] Abbe had independently derived the Poisson distribution in order to calculate the reproducibility of his estimates by calculating the expected variation in the number of red blood cells counted in equivalent volumes of blood. He did not, however, discuss the validity of his estimate by testing the agreement between the predications of the Poisson distribution and the number of red blood corpuscles in counts of many samples. Applying these methods, Hensen felt that the plankton in a well-shaken sample were distributed uniformly *(Gleichmässig)*, using the same word he used for plankton patterning in nature.[20] He considered differences between the numbers of organisms in subsamples from the same sample to be random *(zutreffende Vertheilung)*. The reproducibility of subsampling could be calculated by using Abbe's Poisson distribution. On this basis Hensen was able to calculate not only the average number of organisms per unit volume of

---

17. Ibid.
18. Hensen's laboratory procedures are evalutated by K. F. Wiborg, "Estimates of Numbers in the Laboratory," in J. H. Fraser and J. Corlett, ed., "Contributions to a Symposium on Zooplankton Production," *Rapp. P. V. Réun. Cons. Perm. int. Explor. Mer., 153* (1962), 74–77.
19. Ernst Abbe, "Ueber Blutkorper-Zahlung," *Jena. Z. Naturw., 12* (1878), Suppliment-Heft, Sitzungsberichte, pp. xcvii-cv. Abbe wrote a section in V. Hensen "Methodik der Untersuchungen bei der Plankton Expedition," *Ergebnisse der Plankton Expedition* (Kiel, 1895), I, B, 166–169.
20. V. Hensen, "Methodik der Untersuchungen," p. 144.

water, but also how many subsamples to count for a desired degree of reproducibility.[21]

Hensen had few precursors in the problem of sampling organisms in nature. Furthermore, his only previous experience was reported in a minor portion of a study devoted to chemical analysis of earthworm feces and to the importance of earthworm burrows to root growth.[22] Hensen was not explicit about his sampling technique, but he believed that earthworm holes were found at approximately equal distances from one another, and that earthworms effected uniform distribution of humus. The study is a good example of Hensen's readiness to extrapolate his results over large areas. Hensen counted the number of earthworm holes in a garden of known area, and on this basis calculated the number of earthworms and the amount of their excrement which he expected to find in much larger areas. Charles Darwin recognized Hensen's earthworm study as the only source of quantitative data on earthworms, and he also did not hesitate to extrapolate from Hensen's data.[23]

At sea Hensen could not directly observe the organisms he sampled, as he had observed earthworm burrows. He understood that the problem at sea might be different from laboratory subsampling or earthworm sampling because the organisms he was sampling might be patterned differently. He also knew that both the reproducibility and the validity of his estimate depended on knowing this pattern.

Validity

The Commission's first four summer expeditions covered the Baltic and the two connecting arms of the North Sea, the Kattegat and the Skagerrak. Hensen collected fish eggs with vertical net hauls, and characterized the average number for the area by dividing the total number of eggs by the total volume of water filtered.[24] He believed that the number of eggs of a given species in each sample differed from the true number for the area by a small error due to the condition of the sea and

21. Ibid., p. 164.
22. V. Hensen, "Die thatigkeit des Regenworms *(Lumbricus terrestris L.)* für die Fruchtbarkeit der Erdbodems," *Z. Wiss. Zool.,* 28 (1877), 354–364.
23. Charles Darwin. *The Formation of Vegetable Mould through the action of Worms, with Observations on their Habits* (London: John Murray, 1881).
24. V. Hensen, "Ueber das Vorkommen und die Menge der Eier einiger Ostseefische...," *Bericht der Commission zur Wissenschaftlichen Untersuchungen der Deutschen Meere,* 5 (1884), 297–313.

the facility with which the net haul was made. On this assumption, his estimates were valid samples of the actual egg density, and only relatively few samples were necessary for a reproducible description of egg density.

Becoming aware of the need to test his assumption, he wrote in the fourth Commission Report (1884):

> ... eggs will distribute themselves approximately uniformly, each isolated on the ocean surface if they remain long enough. I must admit that I had taken this sentence as self-evident, and accordingly undertook to collect experimental data.[25]

The experiments consisted of observations in Kiel Harbor on distances traveled by fish eggs over three days, and the distance three glass balls traveled in ten minutes.[26] He felt they took up uniform distributions, but acknowledged that in more turbulent water, such as in a river mouth or in the Kattegat, uniformity should not be expected.

The distribution of freely floating organisms was discussed again by Hensen in a report on plankton collected on the Commission's North Sea Expedition during the summer of 1884.[27] He had become increasingly interested in the food chain on which fish and the commercial fisheries depended, and wished to demonstrate uniform distribution as a basis for sampling plankton as well as fish eggs. Using ten silvered glass balls, he watched their movement in an agitated sea for twenty-four hours. But it was difficult to decide what kind of a distribution the balls took up, hence after this occasion Hensen began to use statistics to investigate plankton distributions.

Hensen reasoned that if the differences between the number of organisms in a given species in paired net hauls were dure to random differences in the collecting conditions and not due to the larger differences to be expected if the organisms were localized in small patches, then the average difference between the numbers of individuals in each net haul ought to be less than the median difference. For this purpose Hensen used the number of individuals of plankton species collected in sixty-four pairs of net hauls in the Baltic and Kattegat. He calculated the geometric mean for the difference between each pair, and found

---

25. V. Hensen, "Verkommen und Menge der Eier," p. 310.
26. Ibid., p. 311.
27. V. Hensen, "Bestimmung der Plankton," pp. 22–29.

that the average geometric mean for differences between pairs was less than the median difference between pairs.[28]

Hensen considered this a satisfactory demonstration that the errors involved were small, and that vertical net hauls gave a valid representation of the true population numbers. However, Hensen's test did not involve a comparison of his results with an independent standard from a population on spaced uniformly, and so it is more properly a test of reproducibility.

After the North Sea Expedition, Hensen came to feel that if one was to understand fisheries in the North Sea, a plankton survey of the whole North Atlantic was necessary. Hensen was able to obtain 105,600 marks from the Kaiser, the Berlin Academy, and the German Fishing Union. He chartered a boat, and with five other biologists spent 115 days, July 15 to November 7, 1889, collecting from Greenland to Newfoundland, south to Bermuda, the Cape Verde Islands, and Ascension Island, west to Balem, and north diagonally across the Atlantic. Hensen's results, particularly the conclusion that the numbers of plankton were greater in the north Atlantic than in the tropical Atlantic, aroused considerable interest and criticism.

Controversy over the Validity of Hensen's Results

Criticisms of Hensen's methods and conclusions began after the Plankton Expedition, and stimulated Hensen to elaborate his descriptive statistical methods further. After the preliminary results of the expedition were published,[29] the physiologist Emil Du Bois-Reymond reported favorably on the expedition to the Berlin Academy, and said Hensen's methods were "convincing."[30] Du Bois-Reymond's acceptance of Hensen's methods contrasts with the reaction of many naturalists who were critical of the validity of Hensen's data, partly because of their field experience, but perhaps because of their attitudes toward statistics. A good example is Ernst Haeckel. Although Haeckel had developed the concept of ecology on the basis of Darwin's *Origin of Species*, he was critical of Hensen's conclusions, mainly because of his distrust of statistics.[31]

---

28. Ibid., p. 24.

29. V. Hensen, "Einige ergebnisse der Plankton-Expedition der Humboldt-Stiftung," *Sber. preuss. Akad. Wiss. 14* (1890), 243–253.

30. Emil Du Bois-Reymond, "Berichte des Curatoriums der Humboldt-Stiftung fur Naturforschung und Reisen," *Sber. preuss. Akad. Wiss., 14* (1890), 83–87.

31. This paradox is discussed by Robert Stauffer, "Haeckel, Darwin and Ecology," *Q. Rev. Biol. 32* (1957), 138–144.

Victor Hensen and the Development of Sampling Methods in Ecology

Haeckel wrote that although the expedition had been afforded more money than any other in German history, the results were "remarkably negative."[32] Haeckel questioned Hensen's assumption of basis uniformity in plankton populations and cited the great aggregations of plankton he had observed (but not sampled) in the Indian Ocean.[33] Haeckel did not believe that the North Atlantic supported greater summer plankton populations than the Tropical Atlantic, as the Plankton Expedition's results showed. He contended that Hensen's collection methods were not as efficient as those of the British Challenger Expedition (December 1872 – May 1876). Haeckel compared the Plankton Expedition's Sargasso Sea collections with those of the Challenger Expedition because biologists believed that the Sargasso Sea was the most uniform habitat in the Atlantic.[34] Yet, Haeckel claimed, the Plankton Expedition had netted fewer species there than the Challenger Expedition, and variation in the numbers collected was too great for them to be patterned uniformly. Finally, Haeckel was skeptical of Hensen's earlier attempts to estimate the number of fish with collections of pelagic eggs. Hensen had begun an extensive program of egg sampling at stations set up on the Baltic coast by the Commission. By sampling eggs at stations around the Baltic, Hensen wished to estimate the number of mature fish available to the fisheries. Haeckel pointed out that Darwin had concluded that most of the reproductive effort of a species is lost; an adult population estimate determined by counting eggs would be too large.[35] Hensen responded by pointing out that he was estimating the size of the parental populations, not the size of the populations which would develop from the eggs.[36] However, the basic controversy was over the different attitudes toward statistics held by the naturalist Haeckel and the experimentalist Hensen.

Haeckel viewed statistics as giving "a deceptive semblance of accuracy when it is not attainable," and felt the result was to regard "a highly

---

32. Ernst Haeckel, "Plankton Studien," *Jena. Z. Naturw.*, 25 (1891), 232–336, transl. G. W. Field in *Rep. U.S. Commnr. Fish.*, 1889–1891, 565–641.

33. Ibid., p. 321.

34. This reasoning is discussed by William Herdman, "Variation in Successive Vertical Plankton Hauls at Port Erin," *Proc. Trans. Lpool biol. Soc.*, 35 (1920), 161.

35. Haeckel, "Plankton Studien," pp. 287–288. Charles Darwin, *On the Origin of Species* (London: John Murray, 1859), chap. 3–4.

36. V. Hensen, *Die Plankton-Expedition und Haeckel's Darwinismus* (Kiel: Lipsius and Tischer, 1891), p. 22. Hensen does not consider the possibility of predation on eggs.

complicated problem of biology as a relatively simple one."[37] In his reply to Haeckel, Hensen only used more graphic statistical methods, methods which, although Hensen meant them to demonstrate the validity of his method, only illustrated reproducibility. Taking the volume of plankton in twenty-six hauls made between Bermuda and the Cape Verde Islands, Hensen considered that about half of the deviations from the mean of all of the samples should fall under the central half of the normal curve of error. Plotting his data under the normal curve, Hensen showed that if two quite large deviations were discarded, the rest were symmetrically distributed under the curve, and that twelve of the twenty-four observations lay under the central half of the curve.[38]

Hensen understood that not all of this variation was likely to be due to variation in the plankton populations. Part of it was undoubtedly attributable to collecting techniques and the condition of the sea. Just how much variation was due to variation in the plankton populations was important to determine if Hensen were to be able to claim that a few samples were a reproducible and valid characterization of large areas of ocean. Frans Schütt, a member of the Plankton Expedition, developed a method for separating the variability in the catch due to sampling equipment and conditions from the variability due to the patterning of the organisms. Using Hensen's sixty-four paired North Sea hauls as an ideal, he assumed that deviations between the pairs were mostly due to error caused by the collecting method because the second of each pair was taken immediately after the first in very nearly the same area.[39] These hauls had an average error (or standard deviation) of about 20 percent of the mean plankton volume. Schütt, and Hensen as well, assumed that error due to collecting conditions and gear

---

37. Haeckel, "Plankton Studien," pp. 287–288.

38. Hensen, *Plankton-Expedition*, p. 66. This was standard practice, as evidence by Pearson's criticism of it: "In the current text-books of the theory of errors it is customary to give various series of actual errors of observations to compare them with theory by means of a table or distribution based on the normal curve, or graphically by means of a plotted frequency diagram, and on the basis of these comparisons to assert that an experimental foundation has been established to the normal law of errors." See Karl Pearson, "On the Criterion That a Given System of Deviations from the Probable in the Case of a Correlated System of Variables Is Such That It Can Be Reasonably Supposed to Have Arisen from Random Sampling," *Lond. Edinb. Dubl. Phil. Mag.*, 5th ser., *50* (1900), 171.

39. Franz Schütt, *Analytische Plankton-Studien. Ziele, Methoden und Anfangs-Resultate der quantitativ-analytischen Planktonforschung* (Kiel: Lipsius and Tischer, 1891), pp. 56–57.

could be separated from variability in plankton populations by subtracting 20 percent from the observed error for single-net hauls made in other areas. Even for the relatively uniform Sargasso Sea, subtracting 20 percent left a variability in the plankton volume of about 16 percent of the mean volume.

Hensen's Independent Test of Validity for Plankton Samples

Haeckel's criticism of the validity of Hensen's Plankton Expedition results came before the complete expedition results were published. By the time Hensen's volume in the expedition results appeared as the final volume in the series, he was able to explain the 16 percent difference in plankton volume observed under even the most ideal collecting conditions by comparing his results with an independent sample from a uniform population. Hensen calculated the amount of variation to be expected from all possible ways of sampling an idealized uniform population of organisms, and then compared the variability of this independent standard with that of his collections.

Hensen felt that the 16 percent population variation could be due to the way in which a net passed through a plankton population. When writing the summary volume of the Plankton Expedition in 1911, Hensen explained variability of the plankton caches in this way and used geometrical probability to calculate frequency distributions which would explain the variation.[40] He conceptualized plankton populations as independent particles occupying the center of a hexagonal column.[41] He began by considering the catch made by a net with a diameter equal to one hexagon as it is pulled vertically through the idealized tiers of uniformly spaced plankton. With net-center passing through the center of a hexagon, one plankter could be caught. But

---

40. V. Hensen, "Das Leben in Ozean nach Zahlung seiner Bewohner," *Ergebnisse der Plankton Expedition* (Kiel, 1911), V.O. 22–29.

41. The closest or most economical packing of equally spaced points on a plane is at the vertices of adjacent equilateral triangles. The midpoints of the six equal vertices connecting any point with its neighbors define the edges of the hexagon enclosing a point. H. S. M. Coxiter, *Introduction to Geometry* (New York: John Wiley, 1961), and D'Arcy Thompson, *On Growth and Form* (Cambridge: Cambridge University Press, 1963), vol. II, discuss the theory and history of biological applications of this type of spacing. Frank Egerton has called by attention to the suggested use of irregular hexagons in botanical surveys by W. H. Coleman, "On the Geographical Distribution of British Plants," *The Phytologist*, 3 (1848), 217.

were the net center to pass near the sides of a hexagonal column, it would overlap enough into a neighbouring column so that two plankters might be caught. If the net-center passed through the corner of the hexagon, no plankton would be caught.

Hensen accommodated plankton populations of different densities by considering all possible catches resulting from passing a net through equally spaced individuals separated by distances from 1 to 1/20 of the net diameter. For each plankton density, Hensen moved an imaginary net-center over the surface of the central hexagon, and mapped the areas of this hexagon for which different numbers of individuals in the surrounding hexagons would be caught. The fraction of the central hexagon area through which net-center must pass in order to capture a specific number of plankton gave the probability of capturing that number of a uniform population. Initially Hensen measured areas within the central hexagon with a planimeter, but in his last paper on plankton distribution,[42] he described a trigonometric formula for calculating the areas. He emphasized that a table giving the probabilities of each possible catch size for a given density was needed as an independent standard for judging the validity of his sample. "The extent of non-uniformity entering the distribution will only be measurable when the size of the deviation from a uniform distribution can be measured."[43]

If the validity of the sample was accepted, then reproducibility of a plankton sample could be specified with confidence limits. As a method of calculating confidence limits, Hensen calculated the largest and smallest catch for densities up to one hundred organisms per net. On this basis, if less than fifty individuals were caught in a series of net hauls, the average number of individuals would not be expected to differ more than 33 percent from the true value, and a series of net hauls containing more than fifty individuals should not differ from the true value by more than 25 percent.[44]

Hensen was the first biologist to discuss and suggest a solution for the validity and reproducibility of biological samples from nature. However, by simply using the mean of his sample to find a model uniform population of equal density in his table, he did not use the information on variability of his samples. The result was that his confidence limits

---

42. V. Hensen, "Zur Feststellung der unregelmassigkeiten in der Verteilung der Plankton," *Wiss. Meeresunters.*, n.f. *14* (1912), 193–203.
43. Ibid., p. 193.
44. Ibid., p. 200.

were so broad as to be useless in deciding whether collections were made in the same population (reproducibility) or whether they were from a uniform population at all (validity). Results of more recent statistical analyses of vertical Hensen net hauls suggest that much more than 16 percent on the variability of single-net hauls is due to local differences in plankton abundance; Hensen must have often confused local dense patches of plankton with high regional abundance. Nevertheless, the bias toward overestimation in his method did not prevent Hensen from discovering the major pattern in plankton distribution in the Atlantic, and his sampling method for fish eggs was used until about 1920.[45]

Hensen's Influence on other Biologists

Early twentieth-century oceanographers recognized the need for sampling methods, but they did not all agree that organisms have about the same pattern and density over large areas of the ocean. C. G. J. Peterson accepted Hensen's sampling method and the uniform distribution. When Peterson invented a dredge for collecting from the ocean bottom he commented:

> I believe, now that we have found a method [for making quantitative collections of organisms], and I consider it as an extension of Hensen's excellent investigations, that quantitative benthos investigations can now also be included under the study of the metabolism of the sea.[46]

H. H. Grau, writing on pelagic plant life in Murry's and Hjort's *The Depths of the Ocean*, felt that Hensen and his school had shown that the "distribution of the pelagic plants, at any rate, is extremely regular."[47] Researchers at the Liverpool Biological Station considered Hensen's methods so important that on two occasions a biologist from Liverpool,

---

45. An evaluation of Hensen's contribution to sampling and a comparison of his results with current work can be found in D. H. Cushing, "Patchiness," in J. H. Fraser and J. Corless, eds., "Contributions to a Symposium on Zooplankton Production," *Rapp. P. −v. Réun. Cons. int. Explor. Mer.*, *153* (1962), 152–163.
46. C. G. J. Peterson, "Valuation of the Sea. 1. Animal Life of the Sea Bottom, Its Food and Quantity," *Rep. Dan. biol. Stn.*, *20* (1911), 47.
47. H. H. Grau, "Pelagic Plant Life," in John Murray and John Hjort, *The Depths of the Ocean* (London: Macmillan, 1912), p. 360.

visited Kiel.[48] However, although the laboratory director at Liverpool, W. A. Herdman, believed in the value of sampling, he did not accept the idea of extensive uniformity in populations of organisms, and warned that "if your observations are liable to be affected by any accidental factor which does not apply to the entire area, then your results may be so erroneous as to be useless."[49]

Hensen's example stimulated other biologists to quantify their studies. Charles Kofoid found that there was so much particulate matter in the Illinois River that the filtering coefficients Hensen advocated using in order to adjust for the progressive clogging of the nets were not sufficient. Kofoid's reaction was to quantify the use of a pump for collecting plankton.[50] Charles Adams, writing a summary of ecology in 1913, mentioned that Hensen's methods were

> extended to the sea bottom by [C. G. J.] Peterson, to fresh water by many students, to land animals mainly by [Fredrich] Dahl, and in recent years to plants by [Frederick E.] Clements and others.[51]

Fredrich Dahl, who was a member of the Plankton Expedition, was the first to sample benthic organisms with quadrats,[52] and later made quantitative observations of the ability of insects to find localized food sources.[53] The American limnologist C.D. Marsh was convinced of the value of Hensen's quantitative approach. However, he had collected plankton from the same spot in a lake for a number of years, often many times a day, and had found such large variation in the number of individuals belonging to the same species that he did not accept

---

48. Reports of these visits are given by J. T. Jenkens, "The Methods and Results of the German Plankton Investigations with Special Reference to the Hensen Nets," *Proc. Trans. Lpool. biol. Soc., 15* (1901), 278–341, and W. J. Dakin, "Methods of Plankton Research," *Proc. Trans. Lpool. biol. Soc., 22* (1908), 500–553.

49. W. A. Herdman, "Some Problems of the Sea," *Proc. Trans. Lpool. biol. Soc., 21* (1907), 10.

50. Charles Kofoid, "On Some Sources of Error in the Plankton Method," *Science, 6* (1897), 829–832.

51. Charles C. Adams, *Guide to the Study of Animal Ecology* (New York: Macmillan, 1913), p. 61.

52. Fredrich Dahl, "Untersuchungen uber die Thierwelt der Unterelbe," *Bericht der Commission zur Wissenschaftlichen Untersuchungen der Deutschen Meere, 6* (1893), 149–185.

53. Friedrich Dahl. "Experimental-Statistische Ethologie," *Verh. dt. zool. Ges., 8* (1898), 121–132.

Hensen's notion of uniform patterning.[54] The sources of Clements' first use of quadrats is not clear, unfortunately, for he makes no mention of the origin of his method.[55]

CONCLUSIONS

Why was Hensen unsuccesful in the quantification of ecological sampling? No aspect of plankton research itself seems to have hindered quantification; both collecting methods and taxonomy were sufficiently advanced. The reason is probably that at the time he began sampling, Hensen had to devise his own statistical methods for expressing the reproducibility and validity of samples. Hensen might have succeeded in this if he had overcome prevalent nineteenth-century attitudes toward randomness.

The statistical literature of medicine and physics with which Hensen was probably familiar gave methods for expressing reproducibility and for comparing differences between means of different sets of observations. For example, a student of Poisson writing on medical statistics advocated using Poisson's limit (standard error $2\sqrt{2}$) to test the difference between two means.[56] Other authors suggested that differences between means were most meaningful if very large numbers of observations were used.[57] In his laboratory subsampling, Hensen used the propable error as a limit about means. In this and other ways, he seems most indebted to the physicist Ernst Abbe for statistical methods.[58] However, all the methodology available to Hensen had been developed for situations in which errors are a property of the measurement or sampling process, and not of the phenomena themselves. The available methods for measuring reproducibility were based on the assumption that differences from the average were small and that they tended to accumulate about the mean in a bell-shaped pattern. Hensen constantly reinvestigated the distribution of plankton numbers about the average

---

54. Early work on the patterning of plankton in lakes is reviewed by C. Dwight Marsh, "On the Limnetic Crustacea of Green Lake," *Trans. Wis. Acad. Sci. Arts Lett.*, 11 (1897), 179–224.

55. Roscoe Pound and Frederick Clements, "A Method of Determining the Abundance of Secondary Species," *Minn. bot. Stud.*, 2 (1898), 19–24.

56. Jules Gavarret, *Principes généraux de statistique médicale* (Paris: Bechet jeune et Labé, 1840).

57. F. Oesterlen, *Handbuch der medicinischen Statistik* (Tübingen: H. Laupp, 1865).

58. See Wiborg, "Estimates of Numbers."

using a different method each time. Westergaard points out that medical statisticians did not make such investigations with their biological data.[59]

To a considerable extent, biological sampling problems forced development of theory because samples afforded the only information on a pattern in water or soil which could not be directly observed. The sampling methods of Laplace and the late nineteenth-century government statisticians contrasted strongly with Hensen's because, either through subjective knowledge of the population sampled or through censuses, they attempted to choose representative or typical samples.[60] The high reproducibility and validity of representative sampling is attained by knowing more about a population than a biologist can ordinarily know. The uncertain reproducibility and validity of biological sampling spurred the development of formal sampling theory.

A formal sampling theory developed only after change in the general intellectual attitude toward randomness, which was reflected in nineteenth-century statistics.[61] The nineteenth-century attitude that randomness is not part of nature changed in the twentieth century to a view of randomness as a property of nature.[62] The physicist's incorporation of randomness into physical models in response to this intellectual change late in the nineteenth-century is discussed by Bork.[63] In biology, the change was initiated by the attention Darwin focused on morphological variation. The English biometricians – Francis Galton and W. F. R. Weldon, for example – were prominent in developing methods for the analysis of biological variation.[64] Most pertinent to the development of

59. Westergaard, *History of Statistics*, p. 150.

60. Late nineteenth-century offical sampling is reviewed by A. N. Kiaer, "Die reprasentative Untersuchungs-methode," *Allgemeines statistiches Archiv, 5* (1899), 1–22. Kiaer mentions that sampling has previously been used in geology, oceanography, and meterology, but gives no references.

61. The first application of sampling theory to plankton sampling was made by H. J. Buchanan-Wollaston, "The Philosophic Basis of Statistical Analysis," *J. Cons. perm. int. Explor. Mer., 10–11* (1935–36), 249–265, and by W. E. Ricker, "Statistical Treatment of Sampling Processes Useful in the Enumeration of Plankton," *Arch. Hydrobiol., 31* (1937), 68–84.

62. This change is discussed by Ronald A. Fisher, "The Expansion of Statistics," *Jl. R. statis. Soc. A, 116* (1953), 1–6.

63. Alfred M. Bork, "Randomness and the Twentieth Century," *The Antioch Review, 27* (1967), 40–61.

64. For a nineteenth-century summary, see John T. Merz, *A History of European Thought in the Nineteenth Century* (London: William Blackwood, 1904–1912) II, 608–626.

sampling theory was Karl Pearson's use of frequency distributions as models of biological variation. In ecology, quantification was brought about by Ronald Fisher more than by anyone else; he incorporated randomness into sampling plans and built upon the methods developed earlier for analysis of individual variation. Fisher's use of random sampling allowed comparison between the sample collections and the collections expected from a model population of known patterning (calculated with a frequency distribution). This is a much more efficient method of determining the validity of a sample than Hensen's comparison of collections with a model uniform collection. Intellectual background and accumulated biological information caused Fisher to find variability where Hensen had seen uniformity.

In summary, Victor Hensen became interested in fisheries research because of the economic importance of fishing to Germany. Hensen had considerable understanding of the prerequisites for valid sampling, but the value of his quantitative approach was limited by the general preconceptions shared by most nineteenth-century biologists. Through Hensen's efforts many other biologists were stimulated to undertake quantitative samples, even though the statistical methods for analyzing variation among populations developed only after methods for analyzing variation among individuals had been developed.

Ackowledgements

I am very grateful to Robert C. Stauffer, Department of the History of Science, University of Wisconsin (Madison), for guiding me to this topic; to John C. Neess, Zoology Department, University of Wisconsin (Madison), for discussions on statistics; and to Frank N. Egerton, University of Wisconsin (Parkside), for many suggestions on background and style. Portions of this paper were presented at the History of Science Society Annual Meeting in New York, December 28, 1971.

UEBER DIE BEFISCHUNG DER DEUTSCHEN KÜSTEN

Victor Hensen

Ueber die
Befischung der deutschen Küsten.

Bearbeitet

von

Dr. HENSEN.

(Separatabdruck aus dem II. Jahresberichte der Kommission zur Untersuchung der deutschen Meere in Kiel.)

Berlin.
Wiegandt, Hempel & Parey.
1874.

Einleitung.

Die Untersuchung des Meeres stösst, sobald man sich dem Studium des Lebens der Fische zu nähern sucht, auf erhebliche Schwierigkeiten, welche ein direktes Vorgehen sehr hemmen.

Der allgemeine Gesichtspunkt, von dem die Wissenschaft ausgehen kann, ist der, das Dunkel, welches einen Einblick in die Bahnen, denen das Material an Kohlenstoff und Stickstoff im Meere folgt und in denen als besonderer Wendepunkt die Fische sich erweisen dürften, zu lichten. Ohne Zweifel bilden die wissenschaftlich gesicherten Thatsachen, welche auf dem Gebiet der Meeresuntersuchungen gewonnen sind, auch in der hier ins Auge gefassten Richtung ein sehr werthvolles, grundlegendes Material, jedoch wird es etwa in ähnlichem Maasse schwierig oder unmöglich sein, sich aus diesem Materiale ein Bild über die Laufbahn des Stoffes im Meere zu verschaffen, wie es z. B. in der Physiologie der Thiere möglich wird, die Resultate der Forschung über einzelne Functionen des Körpers zu einem Gesammtbilde von dessen Stoffumsatz zu gestalten. Dort musste ein besonderer Weg eingeschlagen werden, der in steter Anlehnung an die anderweit erworbenen Kenntnisse, von dem Studium der Anfangsmaterialien und Endproducte ausgehend, so nach innen vorzudringen sucht, wie die Detailforschung nach aussen heraus die Thatsachen zu einem Gesammtbilde zusammen zu fügen bestrebt ist.

Das eben gewählte Beispiel unterscheidet sich jedoch von der hier in Rede stehenden Forschung erheblich. Es handelt sich nicht um den Stoffwechsel eines Thieres, dessen Nahrung seiner Herstammung nach, dessen Excretionen ihren weiteren Schicksalen nach als genügend bekannt und durchsichtig angesehen werden, sondern das Thier soll als Maassstab dienen für den Strom, in welchem der organische Stoff des Meeres kreist, und von welchem es einen Theil bildet.

Da man bei derartigen Forschungen zunächst die Punkte zu suchen hat, von denen sich ein Ausgang gewinnen lässt, würde zu überlegen sein, wo derselbe zu finden ist. Bei dem bisher gewonnenen Material der Meeresuntersuchungen sind es mehr die qualitativen Verhältnisse, denen sich die Aufmerksamkeit zuwandte, denn selbst die Strömungen, die Vertheilung der Temperatur, des Salzgehaltes und der Gase geben, obgleich auf messender Basis gewonnen, für Massenberechnungen theils keine sicheren Anhaltspunkte, theils würde z. B. mit der Kenntniss der Kohlensäure- und Sauerstoffmasse des Meeres für die Schicksale des Kohlenstoffs nur in bedingter Weise etwas gewonnen. Wir können uns die Bahnen der Stoffe des Meeres in den belebten Körpern in ähnlicher Weise als einen schliesslich zu seinem Ausgangspunkt zurückkehrenden Strom denken, wie dies für dieselben Materien in Luft und Erde erlaubt ist, wo vom Anorganischen durch Pflanzen und verschiedene Thiere hindurch wieder Anorganisches sich bildet. Dann ist also hier wie dort das Anorganische eine Abtheilung des Stromes, an welche folgende sich anreiht, diese zweite Abtheilung zu wählen und zu fassen ist eben die Aufgabe, welche vorliegt.

Es ist schon angedeutet worden, dass die Fische diese zweite Abtheilung sind. Nicht als ob alles Material um seinen Kreislauf zu vollenden in das Blut der Fische gelangen müsse; im Gegentheil, es wird der viel grössere Theil des Materials als Moder von Pflanzen und Thieren der gänzlichen Auflösung zu anorganischer Materie verfallen, sondern weil der vollendetste und ausgedehnteste Kreislauf der Materie durch die Fische, oder allgemeiner gesprochen, durch die Wirbelthiere des Meeres hindurchgeht.

Im Interesse besserer Verständigung wird die Vergleichung des Stoffwechsels im Meere mit einem Strom noch näher auszuführen sein. Denken wir uns eine beschränkte heisse Fläche am Boden des Meeres. Sie wird das auf ihr lastende Wasser erwärmen (beleben) und dasselbe in einer ihrer Ausdehnung an Breite entsprechenden Strom in die Höhe treiben. Doch bald kühlen sich an den Rändern die in mannigfachen Wirbeln fliessenden Wassertheilchen ab, mischen sich mit den benachbarten kalten Wassermassen und hören in ihrer aufwärtssteigenden Bewegung auf (sterben ab). An die Oberfläche kommt schliesslich nur ein schmaler Strom warmen Wassers, die ganze bewegte oder belebte Masse bewegt sich also innerhalb eines kegelförmigen, mit abgestumpfter Spitze versehenen, Stromes. Das kalte Meer, die unorganische Materie, umgiebt diesen Strom und wird in seine Basis hineingezogen; in dieser finden sich die niedersten organisirten Wesen, in grösster Menge abkühlend (absterbend), ehe sie weit fortgeschritten sind. Die Spitze des Stromes repräsentirt die höchsten Thierformen. Die Theilchen in der Mitte des Kegels können verschiedene Bahnen durchlaufen, sie können nach oben steigen, können aber auch durch Wirbel nach unten gezogen oder in die kalte unorganische Peripherie gestossen werden.

Die Theilchen an der Spitze verfolgen nur einen Weg, sie verbreiten sich in die umgebende Masse und kühlen ab, sie sind deshalb diejenige Abtheilung des Stromes, welche sich an den anorganischen Zustand sicher anreihen lässt.

Dass die Fische wirklich als die abgeplattete Spitze unseres Stoffwechselkegels zu betrachten sind, ist schon wegen ihres anatomischen Baues nicht zweifelhaft. Dass sie in anorganisches Material übergehen, ergiebt sich daraus, dass ihre Excretionen, von denen allerdings nur die Kohlensäureausathmung bis jetzt näher untersucht ist, dem anorganischen Material nicht minder nahe stehen, wie die Excrete der niederen Thiere, wenigstens soweit wir nach Erfahrungen an den niederen Thieren des Festlandes schliessen dürfen. Man könnte einwenden, dass die durch die Vögel und von den Menschen und Säugethieren verzehrten Fische nicht zu anorganischem Material des Meeres werden, sondern zu einem grossen Theil der Luft und dem Lande übergeben werden, jedoch dagegen ist geltend zu machen, dass die Menge an Substanz, um die es sich handeln kann, eine relativ geringe sein wird und dass vom Lande aus wohl mehr Material ins Meer hineinkommt, wie auf dem genannten Wege demselben entnommen wird. Man könnte ferner einwenden, dass die abgestorbenen Fische häufig von niederen Wesen werden assimilirt werden und also nicht unmittelbar sich zu anorganischer Substanz umwandeln. Es werden jedoch die Fische vorzugsweise die Beute anderer Fische und in beschränkter Weise diejenige der Vögel und Säugethiere. Die Cephalopoden und vielleicht einige Krebse betheiligen sich zwar auch am Fang der Fische, jedoch ist nicht anzunehmen, dass ein erheblicher Theil ihrer Nahrung aus dieser Classe entnommen werde. Die Zahl der Fische, welche durch Krankheiten absterben und in Folge dessen niederen Thieren zur Beute anheimfallen können, ist an sich nicht unbedeutend, denn man braucht keine sehr grosse Strecke des Meeresstrandes abzugehen, um auf die Reste eines oder des anderen Fisches zu stossen, während man beim Durchwandern der Felder nur sehr selten auf die Leichen unverletzter Landthiere stösst. Jedoch der Strand bildet den Sammelpunkt aller auf dem Wasser schwimmenden todten Theile und wenn man annehmen kann, dass die Mehrzahl der abgestorbenen Fische zuerst an der Oberfläche treibe, (ich habe darüber keine ausreichenden Erfahrungen) wird die Menge der an Krankheit gestorbenen und nicht alsbald von Möven u. s. w. verzehrten Fische doch eine relativ verschwindende sein müssen.

Man kann dem Gesagten zu Folge also die Fische für den Gipfel des Stoffwechselstroms im Meere erklären, durch sie hindurch wandert derjenige Theil des Kohlenstoffs und Stickstoffs desselben, welcher durchschnittlich die ausgedehnteste Wanderung innerhalb belebter Formen vollzieht ehe er zu der anorganischen Beschaffenheit, welche er ursprünglich hatte, wieder zurückkehrt.

Diese Masse an Kohlenstoff und Stickstoff wird gemessen durch einen Cylinder, den wir uns aus der Masse unseres Stromeskegels, dieselbe als gleich dicht vorausgesetzt, herausschneiden.

Der Querschnitt dieses Cylinders muss dann der Masse der in dem betrachteten Moment vorhandenen Fische entsprechen, die die Endscheibe von der Höhe Eins bilden. Es würde also genügen, diesen und die Gesammthöhe des Cylinders zu kennen, um ein genaues Bild des bezüglichen Stoffwechsels zu erhalten.

Somit hätten wir also die Aufgabe, welche die Statik des Meeresstoffwechsels sich zu stellen hat, präcisirt.

Diese Aufgabe ist für das Gebiet des ganzen Meeres nicht zu lösen, jedoch man braucht das nicht sehr zu bedauern. Da die Wissenschaft ihre Ziele nicht erstrebt, um in ihnen den Ruhepunkt einer Vollendung, die überhaupt nirgends erreichbar ist, zu finden, sondern nur um eine Basis zur Vervollkommnung der weiteren Arbeiten zu haben, so genügt es, auf die allmälige Annäherung an ein richtiges Ziel hinzuarbeiten, wobei man dann überzeugt sein kann mit jeder Annäherung nicht nur auf mancherlei Wissenswerthes zu stossen, sondern auch die Bedingungen, welche vorwärts zu führen vermögen in ganz anderer Vollkommenheit zu lernen, als man sie im Anfang des Versuchs übersehen konnte.

Betrachten wir die Aufgabe, eine Messung der Fischmenge des Meeres auszuführen, eingehender, so zeigt sich, dass die Bestrebung einer annähernden Schätzung unter bestimmten Einschränkungen nicht so sehr aussichtslos ist. Offenbar ist keine im Meer vorkommende Classe von Organismen der statistischen Forschung so zugänglich, wie die Fische, denn überall wo Menschen an den Küsten wohnen, bilden die Fische ein Object der Nachstellung und somit auch der Beobachtung.

Mit einfacher Beachtung des Fischfangs werden wir nicht viel weiter kommen, es wird jedenfalls eines durchgeführten Beobachtungssystems bedürfen. Ein solches ist für einen Ocean jedoch in keinem Falle zu machen, weil die Ausdehnung desselben zu kolossal ist und seine Küsten nicht ausreichend von civilisirten Völkern besetzt sind. Man wird sich also mit kleineren Meerestheilen zu beschäftigen haben und besten Falles mit Hülfe der dort gemachten Erfahrungen sich an grössere Meere heranwagen können.

Wo ist ein entsprechendes Meer zu finden und welche Bedingungen müssen an dasselbe gestellt werden? Wir suchen offenbar einen Ocean im Kleinen, die Bedingungen sind also: grosse Abgeschlossenheit des Stoffwechsels in sich, genügende Mannigfaltigkeit des Terrains, genügende Tiefe, entsprechender Salzgehalt, entsprechende Temperaturen. Hier würde das kaspische Meer in erster Linie zu nennen sein, allein es enthält sein Wasser, wie ich Hrn. KLÖDENS Erdkunde*) entnehme, viel mehr Bittersalz, wie das atlantische Meer, aber nur $1/3$ von dessen Kochsalz.

Ich kann zur Zeit nicht darüber urtheilen, ob dies Verhältniss einen wesentlichen Unterschied in der Entwicklung des Lebens bewirkt. Aus der mir nur nach Citaten bekannten Schrift: Notice sur les Pêcheries et la chasse aux Phoques dans la Mer blanche, l'Ocean glacial et la Mer Caspienne, par ALEXANDER SCHULTZ, Conseiller d'état. Petersbourg 1873, ergiebt sich eine Fischerei von jährlich 260 Millionen Kilo, das ist pro Quadratmeile 30,900 Kilo. Zum Vergleich erwähne ich, dass der Ort Hela in der Ostsee pro Quadratmeile 164,000 Kilo im Jahr 1874 fischte, es liegen jedoch vor ihm bis zur Mitte der Ostsee noch mindestens 10 Meilen unbefischte Fläche. Doch dies nur beiläufig. Ohne Zweifel wird uns das caspische Meer ein sehr schönes Vergleichsobjekt werden können.

Das Mittelmeer ist schon zu gross, um an umfassende Beobachtungen denken zu lassen. Kleinere Gewässer, wie z. B. die Haffe der Ostsee oder noch kleinere Buchten, wie z. B. die Schlei, haben meist ein zu brakisches Wasser und entbehren zu sehr der Selbstständigkeit. Es wird nämlich den in ihnen in reicher Zahl lebenden Thieren durch die Brut der dort laichenden Bewohner der Ostsee ein sehr reiches Nahrungsmaterial zugeführt.

Die Nordsee kann als ein genügend abgeschlossenes Areal nicht betrachtet werden, da die Communication mit dem atlantischen Meer an zwei fast entgegengesetzten Stellen geschieht und da die nördliche von beiden eine so grosse Breite und Tiefe hat.

Die Ostsee ist dagegen unter den europäischen Meerestheilen weit günstiger gestellt. Allerdings ist die Communication mit der Nordsee, respective dem vorliegenden Kattegat, eine mehrfache, und wie dies namentlich die Studien von Hrn. Dr. A. MEYER [2]) ergeben haben, eine lebhafte, aber die Verbindungswege sind nur schmal und ziemlich flach. Dass auch grosse Massen von Thieren diese Wege passiren, darf als unzweifelhaft betrachtet werden, aber es bleibt vorläufig die Frage offen, ob die Ostsee auf diesem Wege mehr an thierischem Material empfange oder verliere. Man muss jedoch denken, dass diese Frage ihre Erledigung finden werde, wenn sich die Aufmerksamkeit in Dänemark und Schweden derselben zuwendet; die Entscheidung würde schon einen Fortschritt unserer Kunde über den Stoffwechsel des Meeres bringen. Jedenfalls möchte ich diese beschränkten Communicationsöffnungen zwischen Ostsee und Kattegat nicht als ein Hinderniss für den Erfolg eines Studiums des Fischreichthums der Ostsee gelten lassen.

In Bezug auf Tiefe, Salzgehalt und Temperaturwechsel ist ein erheblicher Unterschied vorhanden, zwischen dem Ocean und der Ostsee. Es kann nicht verkannt werden, dass diese Umstände sehr ungünstig auf die Mannigfaltigkeit der organisirten Welt einwirken, denn es verlieren sich nicht nur eine ganze Reihe von Formen vollständig, je weiter man vom Kattegat aus in die Ostsee vordringt, sondern der Rest zeigt schlechtere Entwicklung und tritt schliesslich gegen Süsswasser-Formen zurück [3]). Damit ist eine Unähnlichkeit zwischen Ostsee und dem offenen Meere hervorgebracht, die eine Vergleichung beider mindestens sehr erschwert. Man kann zur Zeit nicht einmal sagen, ob der Stoffumsatz in der Ostsee durch jene Verhältnisse beschleunigt oder verlangsamt wird, noch sich ein ausreichendes Bild darüber machen, welche Modificationen derselbe erleidet. In Wahrheit haben wir ja weder von den Vorgängen in dem Ocean, noch von denen in der Ostsee befriedigende Kunde. Daher bleibt es, trotz der erwähnten Bedenken, immer sehr wünschenswerth, dass wir uns wenigstens über die Orte, von denen wir eine nähere Kunde erhalten können, unterrichten. Ich glaube behaupten zu können, dass diese Möglichkeit für die Ostsee vorliegt [4]).

Für das Studium wird, wie schon erwähnt, eine systematische Forschung nothwendig werden.

Eine Registrirung des gesammten Fischfangs der Ostsee bis zu genügender Genauigkeit wäre wohl nicht unmöglich. Durch die Registrirung des Eisenbahntransports und des Imports durch die Berichte der Handelskammern und eine ungefähre Ermittelung der Quote, welche auf den Lokalconsum kommt, ist die Grösse des Fangs zu bestimmen. Dies Verfahren wäre jedoch höchstens als Controlle zu empfehlen [5]). Derartige Auf-

---

[1]) Handbuch der Erdkunde, Thl. 1, Seite 584.

[2]) Untersuchungen über physikalische Verhältnisse des westlichen Theiles der Ostsee. Kiel 1871. S. 22 u. f.

[3]) Die Expedition zur physikalisch-chemischen und biologischen Untersuchung der Ostsee im Sommer 1871. S. 138 u. f.

[4]) Das adriatische Meer wäre für solches Stadium vielleicht mehr geeignet, wie die Ostsee, jedoch ist dessen Mündung in das Mittelmeer viel breiter, wie die Mündung der Ostsee.

[5]) Durch Hrn. Dr. J. WITTMACK ist in seinem Werk: Beiträge zur Fischerei-Statistik des deutschen Reichs, sowie eines Theils von Oesterreich, Ungarn und der Schweiz. Circular d. deutschen Fischereivereins No. 1, 1875, eine Uebersicht des Fischtransports auf den Eisenbahnen und der Ein- und Ausfuhr von Fischen gegeben. Es ist jedoch sehr schwierig, das Material in der jetzigen Form zu der gedachten

nahmen haben nämlich, wie ich aus Erfahrung weiss, das Missliche, dass es sehr schwierig ist, den Import von dem Fang des in's Auge gefassten Rayons zu trennen, dass ebenso der Export aus den Häfen sich der Controlle entzieht und dass schliesslich der Fang des Binnenlandes nicht abzusondern ist. Ausserdem fördern die so gewonnenen Daten nur wenig das Studium des Meeres, denn man erreicht vielleicht die Kunde des Fangs während grösserer Perioden, aber man kann nur Unbestimmtes über den Ort des Fangs und die damit verknüpften Umstände erfahren; doch ist gerade Letzteres zunächst das Wichtigere für diejenige Kunde, welche zu erwerben wir bestrebt sein müssen.

Es muss ein anderes Verfahren gefunden werden. Eine directe Registrirung des Fangs ist aus leicht ersichtlichen Gründen nicht möglich, dieselbe kann aber durch die Combination zweier anderweiten Massnahmen ersetzt werden. Die eine derselben ist: eine statistische Aufnahme der Küstenbevölkerung, derselben ist die vorliegende Arbeit gewidmet, die andere: die Einrichtung von passend gewählten und vertheilten Beobachtungsstationen an der Küste.

Wenn ich ein solches Vorgehen empfehle, so geschieht dies auf Grund von Erfahrungen, welche ich gemacht habe, denn nicht nur liegt eine Aufnahme der Küstenbevölkerung des deutschen Reichs nunmehr fertig vor, sondern es sind bereits seit mehreren Jahren Beobachtungen über die tägliche Fischerei gemacht worden. Obgleich die Resultate dieses Versuchs sich keineswegs ganz übersehen lassen, können doch schon einige Andeutungen über das, was erreichbar scheint, gegeben werden.

Der Durchschnitt des monatlichen oder mehrmonatlichen Fangs pr. Fischerboot oder pr. Fischer wird für die einzelnen Orte Unterschiede ergeben, welche (unter den oben gemachten Voraussetzungen einer richtigen Wahl des Beobachtungsorts), es gestatten werden, eine Curve zu construiren, deren Höhe den durchschnittlichen Betrag der Fischerei pr. Boot oder Fischer an jedem Theil der Küste angiebt. Dann wird die Kenntniss der Lage der einzelnen Fischerorte längs der Küste und der dort verwandten Böte resp. Fischer erlauben, die Grösse des gesammten Fangs zu schätzen. Die Genauigkeit einer solchen Schätzung hängt natürlich sehr von der Vollkommenheit der Curve ab und es ist klar, dass es sehr schwierig sein wird, namentlich an den Bodden und den Mündungen der Ströme und Haffe die Curve zu ziehen, jedoch darf man vielleicht erwarten, auf diese Weise die Grösse des Gesammtfanges mit der Zeit wohl ebenso genau zu schätzen, wie dies auf Grundlage des zuerst besprochenen Verfahrens geschehen sein würde. Ein wesentlicher Vortheil dieses Verfahrens besteht darin, dass die Arbeiten in eng geschlossenem Kreise unter unmittelbarster Direction und Verantwortlichkeit derjenigen, welche mit wissenschaftlichem Sinn sich derselben annehmen wollen, könnten ausgeführt werden und somit, ich sage nicht vollkommen, aber doch angenähert einen streng wissenschaftlichen Charakter annehmen würden, ohne dabei im mindesten den praktischen Bedürfnissen den Rücken zu kehren.

Eine einfachere Methode der Bestimmung der Grösse des jährlichen Fangs würde eine Feststellung der Einnahme eines Fischers und daraus, des für seine Erhaltung nothwendigen Fangs sein. Man würde auf diese Weise zu einer Maximal- und Minimal-Zahl kommen. Es hat jedoch eine solche Berechnung ihre grossen Schwierigkeiten, weil der Verdienst keineswegs überall die gleichen Procente des Fangs ausmacht und nicht nur von Ort zu Ort, sondern auch für die einzelnen Fischarten sehr veränderlich ist. Aber selbst davon abgesehen, würde uns eine solche Bestimmung nicht weiter führen, als die einfache Darstellung der Vertheilung der Fischer dies thut. Es sind nicht die statistischen Daten in irgend einer Gestalt, welche als wissenschaftliches Endziel von uns hingestellt worden sind, sondern es sind die Beziehungen, welche diese Zahlen mit dem Leben im Meere verbinden, die wir zu verfolgen haben. In dieser Richtung leistet die Methode der Beobachtungsstationen weit mehr. Ich erwarte sogar, dass wir durch sie viel eher in wichtige Beziehungen des Lebens der Fische und anderer Thiere zu dem Gesammtstoffwechsel des Meeres und in andere interessante Verhältnisse werden eingeführt werden, als wir in die Lage kommen würden, mit Erfolg Berechnungen anzustellen.

Man ist bisher bei der Beantwortung scheinbar einfacher die Fischerei betreffender Fragen in erhebliche Schwierigkeiten gekommen, und zwar allein aus dem Grunde, weil keine genügenden Registrirungen des Fanges, sondern nur ältere Angaben über auffallenden Fischreichthum vorlagen. So z. B. erwies es sich als sehr schwierig, zu entscheiden, ob die Befischung des Meeres Einfluss auf den Fischreichthum desselben habe. Die englische Parlaments-Commission für die Untersuchung dieser Frage, kam[1], in Folge einer ausgedehnten Nachfrage bei den Fischern, deren Ergebniss sich kaum auf 1500 Quartseiten hat unterbringen lassen, zu dem Resultat, dass die Fischereierträge eher im Zunehmen als im Abnehmen begriffen waren.

Hr. Spencer F. Baird[2] kommt in Folge seiner mehrjährigen ausgedehnten Untersuchungen der ameri-

---

Bestimmung zu verwenden. In dem Bericht ist ein solcher Versuch nicht gemacht worden, um so mehr musste ich davon absehen. Sollte es nicht möglich sein, die Eisenbahndirektionen zu veranlassen, sich Kunde davon zu verschaffen, welche Grösse des deutschen Fischfangs aus dem Fischtransport ihrer Bahn sich direkt ergiebt. Die einzelnen Aufgabestellen werden in der Regel wissen oder leicht ermitteln können, ob die transportirte Waare ein Fang aus der Umgegend ist oder nicht.

[1] Report of the commissioners appointed to inquire into the Seafisching of the united Kingdom 1866.
[2] Report of the condition of the sea fischeries of the south coast of New-England. Washington 1873. Pg. XXXVIII.

kanischen Küste (Neu-England) zu dem bestimmten Nachweis, dass die Küstenfische in den letzten Jahren erheblich an Zahl abgenommen haben, aber, sagt er, da man Grund hat anzunehmen, dass Stenotomus argyrops und einige andere, den Hauptfang abgebende Fische, zu verschiedenen Zeiten in mehr oder weniger ausgedehntem Maasse verschwunden waren, können wir dessen nicht sicher sein, dass die Vernichtung durch die ausgedehnte Fischerei wirklich bewirkt worden ist. Hr. BAIRD hat jedoch gleichzeitig durch ausgedehnte faunistische Untersuchungen eine festere Grundlage, als dies die Erfahrungen der Fischer zu sein scheinen, für die weitere Behandlung der einschlägigen Fragen gelegt, und auch die englische Commission betont scharf, dass eingehendere Untersuchungen, namentlich statistischer Art, für jede sichere Entscheidung auf dem Gebiete der Fischerei nothwendig seien.

Letztere Commission berichtet,[1]) dass der Acker guten, gut bebauten Landes im Jahre 300 Centner Fleisch bringe, dieselbe Strecke geeigneten Meeresgrundes aber jede Woche und zwar das ganze Jahr hindurch ebensoviel an Fischgewicht liefere. Hier ist also eine Angabe über den Ertrag des Meeres, aber offenbar kann eine solche Angabe nicht als richtiges Maass für seine Leistungen gelten, sondern muss meines Erachtens als Folge eines Zusammenströmens der Thiere an gewissen Orten angesehen werden.

Der Beweis, dass dem wirklich so sei, ist jedoch zur Zeit nicht beizubringen, aber eine allseitige regelmässige Beobachtung der Fischerei, wie sie hier in Anregung gebracht wird, dürfte bald zu sicheren Entscheidungen führen.

Durch die täglichen Angaben der Beobachtungsstationen über die Fischerei eines Complexes von Fischern können uns die Erfahrungen der Fischer in solcher Form gegeben werden, dass wir dieselben selbstständig zu extrahiren vermögen. Das wird um so vollständiger geschehen, je zutreffender die Fragen gestellt worden sind und man wird mit der Zeit mehr und mehr lernen, richtig zu fragen. Durch die Combination der verschiedenen Beobachtungen werden wir in die Lage versetzt, Gesetzmässigkeiten in dem Verhalten der Fische zu entdecken und festzustellen und werden dann, wenn es gilt dieselben weiter zu verfolgen, mit Aussicht auf raschen Erfolg selbstthätig einzugreifen vermögen.

Alle diese Umstände sind es, welche es wohl der Mühe werth erscheinen lassen, auf die Einrichtung von Beobachtungsstationen rings an den Küsten der Ostsee hinzuwirken. Man könnte wohl im bottnischen Meerbusen, dort wo keine Salzwasserfische mehr vorkommen, damit aufhören, dann würden die zu besetzenden Küstenstrecken keine zu grosse Länge haben, da sie sich unter vier Staaten vertheilen.[2])

Es braucht kaum gesagt zu werden, dass auch dann noch eine grosse Reihe von Schwierigkeiten bleiben. Abgesehen davon, dass es nicht leicht sein wird, die passenden Stationen und Beobachter überall zu gewinnen, treffen unsere Beobachtungen eine ganze Reihe von Fischen überhaupt nicht. Der grössere Theil dieser Fische sind kleinere Formen und bilden die Nahrung der Marktfische, ein anderer Theil kommt sparsam vor oder wird nicht geschätzt, während er doch für die wissenschaftlichen Zwecke von Bedeutung ist. Diesem Mangel ist nicht abzuhelfen. Es ist fraglich, ob der Vorzug der Beobachtung in geschlossenem Becken ein sehr erheblicher ist. An der Westküste Schwedens ist seit längerer Zeit der Fang sorgfältig registrirt und diese Arbeiten haben gute Erfolge gehabt, ebenso wird gewiss die Verfolgung des Fangs der Schiffe in freier Nordsee, wie sie von Hrn. V. FREDEN ausgeführt wird, sehr förderlich sein, aber es spricht doch sehr Vieles zu Gunsten einer durchgeführten Beobachtung der Ostseeküsten.

Eingehender wird dieser Gegenstand in dem nächsten Bericht zur Sprache kommen, hier sollten nur die allgemeinen Grundzüge, nach denen diesseits verfahren worden ist, dargelegt werden. Es sollte gezeigt werden, wie wir die vorliegende Arbeit aufgefasst zu sehen wünschen und der Verfasser, welcher sie in diesem Sinne aufgefasst hat, bedurfte dieser Darlegung, um vor Anderen und vor sich selbst den Zeitaufwand zu rechtfertigen, welchen die anscheinend etwas sterile und in Wirklichkeit etwas gleichförmige statistische Bearbeitung der Küstenbefischung erfordert.

Das Verfahren der statistischen Aufnahme.

Auf den Antrag der Commission genehmigte Se. Excellenz Herr V. SELCHOW, damals Minister für die Landwirthschaft in Preussen, nachdem er an die Behörden eine betreffende Anfrage hatte ergehen lassen, dass durch die Regierungen eine statistische Aufnahme der Fischer an den preussischen Küsten ins Werk gesetzt werde. Die Aufnahme fand im Jahre 1872 statt. Derselben haben sich dann die Staaten Lübeck, Hamburg und

---

[1]) L. c. Vol. I. Pg. XVII.
[2]) Ich glaube, dass 30 Stationen ausreichen dürften, der Beobachter ist mit 25 Thlr. gut bezahlt, macht 750 Thlr. Dazu kämen dann Porto und Druckkosten, jedoch die ganze Summe auf die vier Staaten vertheilt ist so gering, dass sie kein Hinderniss für die Einleitung der Beobachtungen abgeben kann.

Bremen, sowie Mecklenburg-Schwerin und Oldenburg[1]) auf desfallsiges Ansuchen angeschlossen, so dass durch dies geneigte Entgegenkommen eine Statistik für die gesammte Meeresküste des deutschen Reichs zu Stande gebracht ist. Die von Helgoland gegebenen Daten verdanken wir unserem dortigen Beobachter, Herrn PARKINSON. Die Formulare, nach denen die Aufnahme geschehen ist, hatten folgenden Inhalt.

Aufschrift: Zusammenstellung der Fischer, welche die Fischerei auf Salzwasser, in Haffen, auf Brakwasser und in den Mündungen grosser Ströme betreiben.

Kreis: Ortschaft:

Die Rubriken waren: a) Zahl der Fischer, welche die Fischerei als Hauptgewerbe betreiben. b) Zahl der Gehülfen, welche von den Fischern für die Fischerei gehalten werden. c) Zahl derjenigen, welche die Fischerei als Nebengewerbe betreiben und dadurch wenigstens für ein Zwölftheil des Jahres ihren Unterhalt haben. d) Welche Fische werden hauptsächlich gefangen von den Fischern in: Rubrik a, Rubrik c. e) Wieviel Fahrzeuge sind für die Fischerei in Gebrauch? Welcher Art sind die Fahrzeuge? f) Ausdehnung des Bezirks, auf welchem gefischt wird — in See hinaus und wie weit? wie weit hin an den Küsten? g) Bemerkungen und Name des Ortsvorstehers oder des Sachkundigen, welcher diese Fragen beantwortet.

Zu Rubrik a) war die Anmerkung gemacht: Hierher sind diejenigen Personen zu rechnen, welche mehr als die Hälfte ihrer Zeit auf den Fischfang verwenden, oder mehr als die Hälfte ihrer Einnahme daraus beziehen.

Zu Rubrik b) war bemerkt: Wenn der Fischer zum Rudern, Ziehen der Netze, Räuchern, Einsalzen oder sonst zu einer die Fischerei betreffenden Beschäftigung eine Person männlichen oder weiblichen Geschlechts, die älter als 14 Jahre ist, gebraucht, so hat er einen Gehülfen, braucht er mehrere, so hat er so viel mehr. Die Gehülfen der Fischer in Rubrik c) werden nicht in Rubrik b) angegeben, sondern, wenn ihr Verdienst dabei mehr wie den Erwerb eines Monats ausmacht, unter c) angeführt, sonst ganz weggelassen.

Zur Erläuterung ist Folgendes zu bemerken. Obgleich die Absicht war, nur die eigentliche Küstenfischerei zu berücksichtigen, mussten doch Haffe und Mündungen der Ströme mit aufgenommen werden, weil die Aussonderung derselben soweit nöthig besser nachträglich geschah. Die Definition eines Fischers muss gegeben werden, weil sehr häufig die Fischer ein Nebengewerbe treiben und dasselbe gelegentlich solcher Aufnahmen vorschützen. Die Gelegenheitsfischer machen einen grossen Bruchtheil aller Fischer aus und fischen theilweise gerade dann, wenn der Fischfang am ergiebigsten ist. Deshalb mussten sie mit berücksichtigt werden. Wenn diese Leute den Erwerb für ein bis sechs Monate aus der Fischerei gewinnen, darf man sie auch wohl als wirkliche Fischer bezeichnen, um so mehr, als alle diejenigen, welche sich gelegentlich ihre Mahlzeiten mit der Angel oder im Winter mit dem Stecheisen erwerben, ganz aus der Rechnung fallen. Dass alle, welche nur indirect sich bei der Fischerei betheiligen, durch Räuchern, Einsalzen u. s. w., nach der Anmerkung mit aufgeführt werden sollten, hat sich als ein Missgriff erwiesen. Der Anforderung ist, wie aus den Bögen deutlich genug hervorging und auch in der Tabelle aus der Anzahl der Böte zu entnehmen ist, wenig entsprochen worden, nur an wenigen Orten sind Frauen aufgezählt.[2])

Die Frage: welcher Art sind die Fahrzeuge? war nicht glücklich, es wird besser sein, die in Betracht kommenden Arten von Fahrzeugen namentlich aufzuführen, also Rubriken für z. B. Kähne von 8 bis 18 Fuss, von 18 bis 30 Fuss u. s. w. und eine Rubrik für nicht namhaft gemachte Fahrzeuge einzurichten. Die Rubrik: Ausdehnung des Bezirks, auf welchem gefischt wird, ist durchgehends richtig verstanden, doch würde anstatt der Frage: »in See hinaus und wie weit« besser zu setzen sein: »wie weit pflegt sich der Fischer von der Küste zu entfernen? (geographische Meilen.) Die Stellen, auf welchen vorzugsweise gefischt wird, werden die Fischer kaum angeben, und wenn dies geschähe, würden sie oft schwer auf den Karten ausfindig zu machen sein.

Die namentliche Unterschrift ward verlangt, um Jemanden zu haben, der sich für die Beantwortung verantwortlich fühlt.

Die Beantwortung der Fragebögen.

Die ca. 700 Fragebogen, welche einkamen, waren in sehr verschiedener Weise ausgefüllt. Aus einzelnen Kreisen, z. B. aus Rügen, einigen Theilen der Haffe war durch Fischereibeamte, die sog. Fischkieper, die Ant-

---

[1]) Lübeck, Hamburg, Bremen und Helgoland 1873, Mecklenburg und Oldenburg 1874.

[2]) Ich habe, es es zu ermitteln war, die Anzahl der Frauen angegeben, aber namentlich an der Westküste, wo häufig einzelstehende Frauen die Granatfischerei betreiben, werden einige mehr sein. Für die Ostsee, wo häufig angegeben ist, so und so viel Mann Gehülfen oder auch die Namen, wo sehr oft Gehülfen ganz fehlen oder die Zahl nach die Frauen der Fischer sich ausschliessen, sind an den meisten Orten die Frauen nicht mitgezählt, ob mir einzelne Fälle entgangen sind, kann ich leider nicht sagen.

Die Statistik hält sich übrigens so vorwiegend an die Zahl der Fahrzeuge, dass sie von geringer Bedeutung ist, auch hat es nach dem darin festgehaltenen Standpunkt kein Bedenken, die wirklich fischenden Frauen als Gehülfen zu rechnen. Die grosse Anzahl Frauen auf Usedom sind durch meine Schuld hineingekommen, sie waren alle auf einem Bogen, der von Dorf zu Dorf gewandert war, eingetragen, der erste Ort hatte geschrieben: die 23 Frauen der Fischer sind ihre Gehülfen und alle anderen Orte machten dann eine entsprechende Bemerkung. Erst nach dem Druck der Karte erkannte ich, dass in diesem Fall die Frauen besser weggelassen worden wären.

Es würde erforderlich gewesen sein, auf den Fragebogen den Frauen eine besondere Rubrik zu widmen.

wort mit Sorgsamkeit gegeben, und da Nachfragen, welche ich wegen einiger mir auffallenden Verhältnisse anstellte, befriedigende Bestätigungen ergaben, glaube ich, dass dieselben durchaus verlässlich geantwortet haben Eine Reihe Bögen sind auf den Landrathsämtern auf Grund von Vernehmungen der betreffenden Ortsvorstände ausgefüllt worden; noch andere sind von den Bürgermeistern der Städte, andere von sachverständigen Fischhändlern, sehr viele direct von den Ortsvorstehern mit oder ohne Hülfe der Fischerälterleute ausgefüllt, einige wenige von Gensd'armen.

Die Aufnahme ist dem Gesagten zu Folge als eine im Ganzen sorgfältige zu bezeichnen, jedoch wäre es verkehrt, wenn man von ihr eine so hohe Genauigkeit erwarten wollte, wie z. B. von allgemeinen Volkszählungen. Die Staatsbeamten, welche die Aufnahme vermittelten, haben darüber gewacht, dass keine groben Fehler und Unterlassungen eintraten, aber auf eine Richtigstellung der Details im Sinne der Frage konnten sie sich nicht einlassen. In der That hat es auch keine Bedeutung, ob die Aufnahme vollständig richtige Zahlen ergiebt, ein Fehler von etwa 10 %, der wohl vorhanden sein dürfte, ist schon deshalb nicht zu vermeiden, weil die Zahl der Gelegenheitsfischer eine wechselnde ist und davon abhängt, ob die Schiffahrt den Vorrath an Matrosen mehr oder weniger vollständig absorbirt. Es kam darauf an, die Aufnahme möglichst gleichmässig geschehen zu lassen und dies ist meines Erachtens durch dieselbe erreicht. Wir werden die Angaben in dem nächstfolgenden Abschnitt noch genauer prüfen.

Im Einzelnen ist hier nur hervorzuheben, dass eine kleine Lücke am westlichen Ufer des Peenestroms zu vermuthen ist und dass das kleine Gebiet von Mecklenburg-Strelitz an der Trave gleichfalls noch einige Gelegenheitsfischer bergen könnte, da unser Ansuchen auf specielle Nachfrage abgelehnt ward. Dagegen sind manche Orte mit so unbedeutender Fischerei aufgenommen, die z. B. ohne Boot sind und nur auf dem Eise fischen, dass man zweifelhaft sein kann, ob sie überall in diese Aufnahme gehören. Dadurch findet eine theilweise Compensation der Fehler statt.

Die Grenze zwischen Gelegenheitsfischern und eigentlichen Fischern ist keinenfalls eine genügend scharfe. Die Angaben über die Art der Fahrzeuge sind häufig unterblieben. Nachfragen haben ergeben, dass in solchen Fällen die Fahrzeuge entweder kleine Kähne oder kleine Kielboote zu sein pflegen.[*])

Die Frage: wie weit in See hinaus, ist meistens so aufgefasst, dass die äusserste Grenze, bis zu der hinausgefischt wird, angegeben wurde, in der Regel werden sich die Fischer dichter am Lande halten.

Bearbeitung des Materials.

Das eingegangene Material machte, um ausgenützt zu werden, eine Bearbeitung in dem Sinne erforderlich, dass eine möglichst unmittelbare Anschauung der Verhältnisse daraus resultire. In den Angaben über die Ausdehnung des Fischereibezirks war eine Grundlage zur Entwicklung solcher Anschauung gegeben, denn es konnte für jeden Fischerort der Bezirk mit genügender Genauigkeit umgrenzt werden. Dies ward zunächst ausgeführt; das dabei eingeschlagene Verfahren erläutert sich am besten durch einige Beispiele. Auf Karte I ist einer der nördlichsten Orte: Karkelbeck. Dasselbe giebt an, es fische an der Küste eine Meile nach Norden, ¼ Meilen nach Süden und bis auf 6 Meilen ins Meer hinaus. Es kann demnach der Bezirk dieses Orts im Ganzen nicht zweifelhaft sein. Er wird die Form eines Rechtecks haben von 1¾ und 6 Meilen Seite, doch wird man mit Fug und Recht die freien Ecken der Figur etwas abrunden, denn dass bis in diese Ecken hinein gefischt werde, ist unwahrscheinlich. Das weiter südlich gelegene Schwarzort giebt an, es fische an der Küste von Schwarzort bis Brail und ins Meer hinaus ¼ Meile; es kann nicht zweifelhaft sein, dass sein Bezirk so umgrenzt ist, wie auf Karte I geschehen, nämlich durch die punktirte Linie, welche ein wenig nördlich von Brail ins Wasser führt und ein wenig nördlich Schwarzort wieder aus demselben herausgeht. Bommels-Vitte giebt an, es fische an der Küste von Polangen (in Russland dicht an der Grenze) bis Nidden und ins Meer hinaus bis auf 4 Meilen. Der grosse mit punktirter Linie umgrenzte Bezirk auf Karte I, welcher bei Nidden beginnt und bis zur russischen Grenze sich erstreckt, wird also der Ausdehnung der Fläche entsprechen, auf welcher von Bommels-Vitte aus die Fischerei betrieben wird. Ein Blick auf die Karte wird lehren, dass bei den so gemachten Abgrenzungen Manches willkührlich ist, aber andererseits muss zugegeben werden, dass die Bezirke, innerhalb deren die Fischerei sich bewegt, überhaupt keine scharfen Grenzen haben. Dennoch haben die Bezirke eine reelle Existenz und sind also einer Darstellung fähig, eine solche würde aber nutzlos erschwert oder gar unmöglich gemacht werden, wenn man die Grenzen so unbestimmt halten wollte, wie sie in Wirklichkeit sind. Durch die Abgrenzung wird zunächst nur ausgesagt, es sei dort der Bezirk innerhalb dessen gefischt werde und der Ort fische nicht über die angegebene Grenze hinaus.

In der angedeuteten Weise ist für alle Küstenorte der Fischereibezirk gebildet worden. Es stellte sich dabei sehr bald heraus, dass durch das Ineinandergreifen benachbarter Bezirke die Darstellung häufig sehr com-

---
*) Nähere Angaben über die Beschaffenheit der Fahrzeuge, welche in der Ostsee gebraucht werden, findet man im Circular des tschen Fischereivereins 1873, No. 1, S. 137.

plicirt werde. Es hat kein grosses Interesse den Fischereibezirk jedes einzelnen Orts erkennen z können, dies war nur für die weitere Ausführung der Arbeit erforderlich, aber da beschlossen war, eine graphisch Darstellung zu geben, schien es doch erwünscht, die erworbene Kunde so vollständig wie möglich darin zu geben Es handelt sich dabei um zweierlei, nämlich den Fischereibezirk eines Orts zu umgrenzen und ersichtlic zu machen, zu welchem Ort jeder Bezirk gehöre.

In einfachster Weise geschieht die Darstellung dadurch, dass der Bezirk durch eine in sich geschlossen Linie abgegrenzt wird und dass von dem zugehörigen Ort ein Verbindungsstrich an die nächste Grenze de Bezirks geführt und in dieser mit einer Marke, etwa einem Kreise beendet wird. So ist es Karte I bei Mc neraggen (fälschlich Melmeraggen gedruckt) geschehen. In der Regel musste eine andere Art der Darstellun benutzt werden. Dabei bildet die Küste den einen Theil der Grenze des Bezirks und von ihr aus gehe punktirte Linien in's Meer, welche die Umgrenzung vollenden. Die Verbindung des so entstandenen Bezirk mit dem zugehörigen Ort wird dadurch ersichtlich gemacht, dass von dem betreffenden Ort aus an die Küst nach Nord und Süd resp. Ost und West eine ausgezogene (zuweilen mit nach der Küste zielenden Pfeilen ve sehene) Linie geführt wird, welche die Grenze des Fischereibezirks an den beiden Stellen trifft, wo die im Mee verlaufende punktirte Linie beginnt. In anderen Worten: um den Bezirk eines Orts zu finden, hat man d von ihm ausgezogenen Linien bis an die Küste hin zu verfolgen, die beiden so aufgefundenen Küstenpunk zeigen sich durch eine punktirte Linie in See verlaufende Linie verbunden. Diese Linie und die Küste bilden d Grenze des gesuchten Bezirks.

Es ist z. B. auf Karte I leicht, den Fischereibezirk von Memel, Nidden und Schwarzort zu erkenne Derjenige von Süderspitze scheint dagegen zweideutig zu sein. Er wird nämlich von dem Bezirk Melneragge welcher bis Schwarzort herabreicht, zweimal geschnitten, man kann folglich die Küstenpunkte von Schwarzo durch drei verschiedene Seelinien mit einander verbinden, eine dicht am Lande verlaufende, die zweite $^3/_4$, d dritte $1^3/_4$ Meilen in See. Hier genügt die einfache Regel, welche bei der Zeichnung überall verfolgt word ist: Die Linien, welche einen Bezirk abgrenzen, bilden niemals scharfe Winkel, sonde biegen sich stets im Bogen, dagegen sind alle Schnittpunkte mit anderen Linien möglich im rechten Winkel ausgeführt. Nach dieser Regel*) kann es nicht zweifelhaft sein, dass der Bezir Süderspitze $1^3/_4$ Meilen in See hinausgeht.

Das eingeschlagene Verfahren erwies sich nicht überall als ausreichend, es mussten Markirungen d Seegrenzen, hin und wieder auch der Landlinien vorgenommen werden. Dabei ist mit wenigen Ausnahm die Regel innegehalten, dass Markirungen nur an solchen Bezirken sich finden, welche mehreren Orten geme schaftlich sind. So haben Karte I Immersatt und Uszeneiten einen gemeinschaftlichen Bezirk, die Landlini beider führen auf denselben Küstenpunkt, die Seelinie ist markirt. Solche Markirungen wurden namentli für Usedom — Wollin und Rügen nöthig. Hier kommt es häufig vor, dass Orte und Ortcomplexe eine Bezir grenze theilweise gemeinsam mit anderen haben, so dass die markirten Linien streckenweise verschmelzen u sich nachher wieder zu sondern. So läuft auf Karte VI die einfache Seelinie kl. Divenow in die marki Linie von Berg- und Ost-Divenow aus, so haben Aalbeck, Heringsdorf, Neuhof und Neukrug theils gemeinscha liche, theils gesonderte Grenze und ihre Grenze läuft wiederum z. Thl. gemeinschaftlich mit der Linie Kar hagen, Hammelstall, Zempin u. s. w. Die gewählte Darstellungsweise wird häufig in den Musterzeitungen benu und ich habe daher angenommen, dass es, wenn erforderlich, gelingen werde sich auch in den verwickelter Theilen der vorliegenden Darstellung zurecht zu finden.

Zuweilen kam es vor, dass zwei Orte gemeinschaftliche Angaben gemacht haben, d. h. die Fische gemeinschaftlich betreiben. Dann sind die betreffenden Orte durch einen Strich mit einander verbunden u haben nur einen Bezirk, so Karte I Brail und Perwelk, Alt-Pilkoppen und Neu-Pilkoppen.

In anderen Fällen fischt derselbe Ort in zwei getrennten Meerestheilen z. B. an der Küste und in ein Haff oder Bodden. Die einzelnen Bezirke in den Haffen waren nicht darzustellen. Tritt dort der erwäh Fall ein, so erkennt man dies an den doppelten Zahlen, welche bei dem betreffenden Ort stehen und von de die eine die Ostseefischerei die andere die Hafffischerei angiebt, so Karte I bei Nidden und Pilkoppen; Karte bei Saager, Ziegenort u. s. w. bei letzteren führt eine Linie von den Orten zu ihren Bezirken in der Ostsee. anderen Fällen konnten die Bezirke in beiden Gewässern gezeichnet werden, dann führen von dem betreffend Ort an die in Betracht kommenden Bezirke 2 (Plogshagen u. s. w. auf Hiddensoe) oder 4 Linien (Breege Rügen, Ahrenshoop auf dem Dars) an die betreffenden Punkte.

Auch der Fall tritt ein, dass in einem Ort einige Fischer nur in kleinem, andere in weit grösser Bezirk fischen. In solchem Falle kommen auf einen Ort zwei Bezirke und dem entsprechend 4 Küstenpun Dies findet statt bei Travemünde und Dahme, Karte VIII.

Endlich fischen zwei Orte, Uckermünde Karte VI, und Möltenort Karte IX, theilweise in so gross Bezirk, dass derselbe nicht verzeichnet werden konnte, dies ist auf den betreffenden Karten eingeschrieben.

*) Auf Karte VI und VII in dem Punkt eine Meile nördlich von Greifswalder Oye mussten einige Linien rechtwinklig gebogen wer

Es wird nur selten erforderlich sein, sich über den Bezirk bestimmter Orte zu orientiren. Dann wird in der Regel die Karte deutlich genug sein, namentlich wenn man die in der Tabelle enthaltene Angabe über die Bezirksgrösse zu Hülfe nimmt. Es ist jedoch nicht zu verkennen, dass an einigen Stellen, so im nördlichen Greifswalder Bodden, im Alsensund die Orientirung schwierig wird; bei dem gewählten Maassstab und Steindruck war es mir nicht möglich, die Darstellung deutlicher zu geben.

Nachdem die Bezirke verzeichnet waren, wurden sie mit dem Polarplanimeter einzeln vermessen und die Messungen in der nachfolgenden Tabelle verzeichnet.*) Dann wurden die Strecken gemessen, welche von den einzelnen Kreisen und den Provinzen aus befischt werden. Diese Maasse sind also keine Summirungen der Bezirke, welche wegen des Ineinandergreifens zu falschen Resultaten geführt hätten, sondern direkte Bestimmungen. Endlich sind die Küstenlängen der Provinzen ohne Berücksichtigung der Bodden und kleinen Buchten gemessen, wobei Rügen und Alsen als mit breiter Basis aufsitzende Halbinseln betrachtet wurden, und die Halbinsel Hela sowie die Inseln Poel, Fehmarn und Aaroe nicht gerechnet wurden. Will man nämlich einen ungefähren Einblick in das Verhältniss der Fischereibevölkerung zu der vorliegenden Meeresfläche haben, so dürfen die kleinen Sunde zwischen den Inseln ebensowenig in Betracht gezogen werden wie die Flussmündungen und Häfen. Misslich bleibt das Verfahren immer, aber ich wüsste kein besseres an seine Stelle zu setzen.

Für den Bericht über das weitere Verfahren müssen einige zu gebrauchende Namen festgestellt werden. Durch die Ausdehnung eines Bezirks ist die Extensität der Fischerei bestimmt, man wird also von einem Orte sagen, seine Fischerei sei extensiv, wenn der zugehörige Bezirk gross ist, und man wird von dem Meere sagen, es sei ein Küstentheil desselben mehr oder weniger extensiv befischt, je nachdem der dort liegende Complex von Bezirken weit von der Küste ins Meer geht oder nicht.

Die Intensität des Fischereibetriebes (welche nicht völlig sich deckt mit dem Ertrage der Fischerei) ist bestimmt durch die Menge der Fischer auf der Flächeneinheit und durch die Vollkommenheit, mit welcher ihre Kraft für die Fischerei ausgenutzt wird. Diese Vollkommenheit ist aber ein Produkt aus der Güte des Arbeitsgeräths (durch welches die menschliche Kraft in mehr oder weniger hohem Grade für den Fischfang ausgenutzt wird) und aus der Zeitdauer, während welcher gefischt wird.

Wenn ich zur Erläuterung dem Gange des Berichts ein wenig vorgreifen darf, so würde auf Karte I Melneraggen intensiv aber wenig extensiv fischen im Vergleich zu Bommels-Vitte, welches extensiv aber weniger intensiv fischt. Abgesehen davon, dass Bommels-Vitte eine grössere Dichte der Fischereibevölkerung (177 Fischer mit 84 Böten) als Melneraggen (130 Fischer mit 55 Böten) hat, ist sein Betrieb offenbar besser entwickelt. Karkelbeck hat einen auffallend extensiven Fischereibetrieb, aber es hat zu wenig Fischer und Böte, um den Betrieb bedeutend zu nennen. Neufähr, Karte III, hat einen auffallend intensiven Betrieb, aber er ist sehr wenig extensiv, so dass auch er im Vergleich zu Bommels-Vitte die Bezeichnung »gut entwickelt und bedeutend« nicht verdient. Aber die Fischerei ist in Bezug auf den Ertrag in Neufähr vielleicht bedeutender wie in Bommels-Vitte. Aus den Karten ist also besten Falls nur die Bedeutsamkeit des Fischereibetriebes, die nicht unbedingt, sondern nur bis zu einem gewissen Grade sich mit der Grösse des Fischereiertrages deckt, zu ersehen. Es ist jedoch überhaupt nicht gewiss, ob die Bedeutsamkeit des Betriebes sich ersehen lässt, denn diese resultirt aus der In- und Extensität und es kann allerdings bezweifelt werden, ob die Intensität sich aus den Karten genügend erkennen lasse.

Es würde erfreulich sein, wenn die Intensität aus dem vorhandenen Material abgeleitet werden könnte, doch weder die Arbeitszeit noch die Vollkommenheit des Geräths lassen sich daraus mit Bestimmtheit entnehmen, wir können daher nur versuchen, uns dem gewünschten Ziele möglichst anzunähern.

Am nächsten scheint es zu liegen, als Maass der Intensität die Anzahl Fischer zu nehmen, welche auf die Einheit der befischten Fläche kommt. Es würde aber damit stillschweigend vorausgesetzt werden, dass jeder Fischer durchschnittlich dieselbe Zeit und gleich passendes Geräth für den Fischfang verwende, und dies ist durchaus unstatthaft. Die grössere Anzahl der Gelegenheitsfischer fischt jedenfalls verhältnissmässig wenig und schlecht; die eigentlichen Fischer allein, die ja in der That, soweit sie nicht auch Landwirthschaft treiben, ihre Zeit möglichst auf den Fischfang verwenden, geben keinen Anhalt für die wahre Befischung, weil ganze Meeresstrecken nur von Gelegenheitsfischern befischt werden.

Wir müssen uns daher an die Mittel zur Befischung wenden. Mittel sind einestheils das Geräth, anderentheils die Fahrzeuge. In Bezug auf das Geräth kann gesagt werden, dass dasselbe, obgleich an sich keineswegs vollkommen, doch so leicht, den Traditionen des Orts gemäss, beschafft wird, dass es als ausreichend und praktisch gelten kann, oder doch mindestens, dass es zurücktritt gegen die Bedeutung, welche das Fahrzeug für den Küstenfischer hat und stetig nach der Güte desselben richtet. Die Anzahl der Böte scheint mir das beste Merkmal für die Intensität des Fischereibetriebes zu sein, denn eine Fischerei an der See ohne Böte ist nur an beschränkten Stellen möglich, dass man diese als Bedingung für die Entwicklung eines nennenswerthen Betriebes betrachten kann. Ferner kann man m. E. sagen, dass dort, wo Böte für den Fischfang vorhanden sind, sie auch entsprechend für den Fischfang benutzt werden. Dies gilt um so mehr, je vollkommener die Böte sind,

---

*) Bei dieser Arbeit ist Hr. Chr. Jenssen behülflich gewesen.

mehr also noch für hoffentlich kommende Zeiten, als für die gegenwärtige, aber auch jetzt schon bildet das Boot an der Küste einen Besitz, der nicht gerne unbenutzt gelassen wird. Wenn das Gesagte richtig ist, wird die Anzahl der Böte vollkommener der auf die Fischerei verwandten Zeit proportional sein als dies die Anzahl der Fischer ist. Dabei findet doch eine bestimmte Beziehung zwischen der Anzahl der Böte und der Anzahl der gewerbsmässig Fischenden statt, denn wo die Böte verhältnissmässig zahlreich sind, finden sich die gewerbsmässigen Fischer und ihre Gehülfen zahlreicher vertreten wie die Gelegenheitsfischer, oder ist doch die Fischerei, soweit ich Kunde gewinnen konnte, factisch intensiver.

Dagegen ist nicht zu verkennen, dass die Böte keineswegs einander gleichwerthig sind. Die Seetüchtigkeit der Böte (und der Mannschaft) findet in der Regel einen Ausdruck in der Grösse des Bezirks, resp. dessen Erstreckung in's Meer hinein, die Stärke der Bemannung wird daraus nicht ersichtlich.

Die Bemannung besteht in der Regel aus zwei bis drei Mann, jedoch es kommen Fälle vor, wo deren 12 aber oft auch nur 1 Mann im Boote sind, ja zuweilen hat ein Fischer 2 Böte.

Es bilden demnach auch die Böte kein Maass von wünschenswerther Genauigkeit, aber theils aus den angeführten Gründen, theils nach dem Eindruck, welchen mir das vorliegende Material machte, habe ich ihnen den Vorzug gegeben. Man hätte an den Figuren, welche die einzelnen Böte repräsentiren, noch die Stärke der Besatzung irgendwie markiren können und dies Verfahren wird sich vielleicht für den Fall empfehlen, dass derartige Darstellungen sich wiederholen sollten, für den eigentlichen Zweck, die Bedeutsamkeit des Fischereibetriebes an den einzelnen Küstentheilen dem Auge wahrnehmbar zu machen, wird dadurch wenig erreicht. Eine anderweite Darstellung, etwa eine Theilung der grösseren Böte in die entsprechende Anzahl kleinerer, würde den Uebelstand gehabt haben, statt der Darstellung eines factischen Verhältnisses etwas Imaginäres zu schaffen und damit vielleicht den Werth des Ganzen in Frage zu stellen. Wenn auch für mich die Untersuchung des Fischereibetriebs und dessen Beziehung zur Fischerei das hervorragendere Interesse hatte, so konnte es doch nicht erlaubt sein, den sicheren Gewinn für Gegenwart und Zukunft aus dem Auge zu lassen, der darin besteht, dass die Kunde eines thatsächlichen aber nicht auf einfache Art zu erforschenden, für die Fischerei nicht unwichtigen Verhaltens in übersichtlicher Weise niedergelegt ist.

Das Verfahren der Darstellung ist einfach dieses, dass die Böte, welche auf den einzelnen Ortsbezirk kommen, in diesem gleichmässig vertheilt werden. Das Resultat solcher Vertheilung ist dann natürlich, dass an Stellen, wo viele Orte gemeinsam fischen, die Bootfiguren dichter liegen, als an den Nachbarstellen, wo wenige Orte fischen.

In Wirklichkeit musste etwas anders verfahren werden. Es wurden die einzelnen kleinen Bezirke, welche durch die Grenzlinien der ineinandergreifenden Ortsbezirke entstehen, einzeln ausgemessen, die Böte, welche auf diese Fläche kamen berechnet, summirt und dann eingetragen. Der Grund für das Verfahren liegt in de Schwierigkeit, die Figuren der Böte von vornherein gleichmässig in den oft grossen und unregelmässigen Bezirke zu vertheilen.

Es sind die Namen der Fischerorte und einiger grösseren Städte, die s   ld sie keine Fischer beher bergen, in Klammer stehen, eingetragen und jedem Ort die Gesammtzahl der Fisc.   über dem Strich, die Zah der Böte unter dem Strich beigeschrieben, so dass man sich stets über das Verhält.   zwischen Fischern un Böten orientiren kann.

Der Fischereibetrieb in der Nordsee ist nicht graphisch dargestellt worden, obg.   h die Zeichnunge ganz so wie für die Ostsee durchgeführt worden sind. Es geben die Fahrzeuge nämlich keinen rauchbaren Mas stab mit ab, da theils mit Schiffen, theils mit offenen Böten theils gewerbsmässig theils ohne Fahrzeuge geb_cht wird. D Schiffe haben natürlich sehr grosse Bezirke, die Böte finden sich vorzugsweise in den Flüssen und die Fischere im Ganzen, namentlich aber an den Küsten ist so unbedeutend, dass sie die Darstellung nicht rechtfertigen kan

Die gewonnenen numerischen Resultate sind in der nachfolgenden Tabelle dargelegt. Man findet derselben auch die Fische verzeichnet, welche von den Orten der verschiedenen Kreise angegeben werde Nachdem die sorgfältige Arbeit des Hrn. Dr. WITTMACK*) erschienen ist, dürfte ein Eingehen auf diese Angabe unnöthig geworden sein.

Im Allgemeinen muss ich bemerken, dass man an Tabellen und Karten nicht einen solchen Maassst legen darf, wie an streng wissenschaftliche Arbeiten. Die einzelnen Fragebögen zu prüfen und zu diskutire wie eine Urkunde oder ein Naturobjekt würde die Arbeit sehr erschwert haben und würde ungerechtferti gewesen sein. Ich habe die möglichste Treue obwalten lassen, aber ich habe z. B. zweimal bei ganz unbede tenden Orten, wo die Bezirksangabe fehlte, dieselbe nach Analogie der Nachbarbezirke ergänzt und Abänderung eintreten lassen, wo die Bezirke auf kleine Strecken gar zu complicirt in einander griffen. Ich glaube erst we diese Ungenauigkeiten von Wichtigkeit werden, wird es an der Zeit sein, strengere kritische Arbeiten zu schaffe vorerst ist es zweifelhaft, ob auch nur so eingehende Arbeiten wie diese schon an der Zeit sind.

---

*) l. c.

Statistische Tabelle.

Provinz Preussen.

Regierungs-Bezirk Königsberg.

1) Kreis Memel.

Ortschaften	Eigentliche Fischer	Gehülfen derselben	Gelegenheitsfischer	Sa. der Fischer	Zahl der Fahrzeuge	Grosse des Fischereibezirks (☐Meilen¹)	Zahl der Böte pr.☐Meil.	Bemerkungen
Immersatt	—	—	5	5	3	1,32	2,27	Hauptfang: Stör, Aal, Hering, Dorsch, Stint, Neunauge, Flunder, Lachs, Hecht, Zander, Zärthe, Barsch, Kaulbarsch, Brassen, Plötz.
Uszeneiten	—	—	2	2	1	1,32	0,76	
Karkelbeck	—	—	15	15	3	13,91	0,22	
Mellnerraggen	55	75	—	130	55	2,17	27,15	
Bommels-Vitte	42	115	20	177	84	28,56	2,94	
Memel	—	—	6	6	6	0,03	200	
Süderspitze	4	2	1	7	5	2,29	2,06	
Schwarzort	37	14	—	51	15	0,61	25,02	6 Bradderkähne, 20 kleine Kähne.
Brail & Perwelk	16	48	—	64	8	0,69	18,37	
Nidden	30	90	18	138	18	0,30	60,0	¹) Gesammtfläche des betischten Bezirks.
Ortschaften 10. Summa	184	344	67	595	213	34,80¹)	6,12²)	²) Die Zahl der Böte auf die Gesammtfläche verrechnet.

2) Kr. Fischhausen.

Ortschaften								
Pillkoppen	19	36	—	48	6	0,42	14,29	
Rossitten	—	—	8	8	4	0,45	8,81	
Sarkau	20	60	4	84	10	0,54	18,52	
Kranz	12	—	12	24	12	1,85	6,48	
Eiseln	—	—	2	2	1	0,15	6,67	
Alknicken	—	—	4	4	1	0,59	1,70	
Rautau	—	—	7	7	2	0,44	4,55	
Neu-Kuhren	—	—	13	13	4	0,75	5,34	
Loppehnen	—	—	2	2	2	1,00	2,00	
Rauschen	—	—	9	9	4	0,21	19,05	
Gr. Kuhren	—	—	24	24	5	1,47	3,40	
Kl. Kuhren	—	—	11	11	3	1,47	2,04	
Gr. Dirschkeim	—	—	5	5	1	1,96	0,51	Hauptfang: Stör, Hering, Dorsch, Flunder, Steinbutte, Lachs.
Kraxtepellen	10	10	—	20	8	5,56	1,44	
Palmnicken	5	5	—	10	1	10,90	0,09	
Sorgenau	6	18	10	34	6	10,90	0,55	
Nodens	—	—	4	4	1	5,56	0,18	
Rothenen	8	—	4	12	6	5,56	1,08	
Loch-städt	2	6	—	8	4	0,05	80,0	
Neuhäuser	—	—	2	2	1	0,04	25,0	
Alt-Pillau	30	30	—	60	30	0,61	49,18	
Wogram	3	3	—	6	3	0,61	4,92	
Camstigal	17	17	—	34	17	0,61	27,87	
Neutief	9	11	—	20	5	4,58	1,09	
Alttief	2	2	—	4	1	1,96	0,51	
Grenzhaus	2	2	—	4	1	1,95	0,51	
Ortschaften 26. Summa	138	200	121	459	139	31,30	4,44	

Regierungs-Bezirk Danzig.

3) Kreis Danzig.

Ortschaften								
Narmeln	12	12	—	24	12	7,21	1,66	Flache Böte (Kähne).
Neukrug	14	15	—	29	15	7,21	2,08	do.
Vöglers	12	11	1	24	14	7,45	1,88	do.
Kahlberg	9	—	5	14	5	0,13	38,46	do.
Liep	15	—	—	15	14	0,15	93,33	do.
Probernau	23	—	—	23	10	6,09	1,64	
Vogelsang	19	—	27	46	30	14,14	2,17	
Bodenwinkel	96	36	6	138	58	14,14	4,20	
Stutthof	24	—	—	24	8	14,14	0,58	Hauptfang: Stör, Aal, Hering, Dorsch, Stint, Neunauge, Flunder, Steinbutte, Lachs, Meerforelle (Lachsforelle), Zärthe, Brassen, Schleihe, Güster.
Stegen	—	—	3	3	2	0,19	10,53	
Junkeracker	—	—	2	2	1	0,02	50,00	
Pasewark	—	—	9	9	3	1,25	2,40	
Schiewenhorst	1	6	4	11	3	0,17	17,65	
Schnackenburg	—	—	8	8	5	0,23	21,74	
Kronenhof	—	—	4	4	4	0,23	39,12	
Wordel	1	1	—	2	1	0,01	100	
Bohnsack	43	40	48	131	92	5,86	15,70	16 Sicken oder Sinken (?) und 60 Kähne.
Neufähr	6	40	98	144	110	0,37	297,0	Flache Böte (Kähne).
Krakau	2	4	8	14	6	0,70	8,57	4 Kähne.
Heubude	2	10	16	28	16	0,11	145,5	8 Kähne.
Weichselmünde	46	40	11	97	30	9,44	31,78	
Brösen	15	7	2	24	7	8,57	0,82	
Glettkau	15	15	7	37	16	8,57	1,87	
Ortschaften 23. Summa	355	237	259	851	452	28,9	15,6	

*) Geograp...

4) Kreis Neustadt.

Ortschaften	Eigentliche Fischer	Gehülfen derselben	Gelegenheitsfischer	Sa. der Fischer	Zahl der Fahrzeuge	Grösse des Fischereibezirks □Meilen pr. □Meil.	Zahl der Böte	Bemerkungen
Karlikau-Schmierau	17	18	—	35	21	9,07	2,32	13 Kähne.
Zoppot	21	26	—	47	70	9,07	0,07	18 Kähne.
Koliebke	3	2	—	5	2	5,34	0,37	Kähne.
Hoch-Redlau	5	8	—	13	5	5,34	0,94	Kähne.
Gdingen	4	9	—	13	5	8,67	0,58	Kähne.
Oxhöft	10	24	9	43	15	8,67	1,73	
Mechlinken	2	1	6	9	4	5,11	0,78	Hauptfang: Aal, Hering, Breitling, Dorsch, Flunder,
Rewa	—	—	32	32	5	5,11	0,98	Steinbutt, Lachs, Barsch.
Oslanin	—	—	2	2	2	0,39	5,13	
Putzig	1	1	—	2	2	0,19	10,53	
Schwarzau	—	—	26	26	—	0,26	—	
Grossendorf	52	11	22	85	22	0,62	35,48	Böte werden von Grossendorf gemiethet.
Hela	77	69	—	146	48	1,28	29,78	
Danziger Heisternest	79	69	—	148	51	2,74	18,61	
Putziger Heisternest	36	36	2	74	39	2,15	18,14	
Kussfeld	58	47	4	109	63	1,40	54,0	
Ceynowa	71	33	—	104	79	1,27	62,20	
Podscharnin	—	—	2	2	1	0,016	62,50	
Chlappau	—	—	32	32	5	0,54	9,26	
Tupadel	—	—	41	41	4	0,50	8,00	
Ostrow	—	—	20	20	2	0,87	2,30	
Karwen	—	—	30	30	5	0,60	8,33	
Karwenbruch	—	—	25	25	5	0,08	62,5	
Wiedow mit Neuhof	—	—	5	5	1	0,08	12,5	
Odergau	—	—	3	3	—	0,16	—	Betheiligen sich am Dembecker Netz.
Zarnowitz	—	—	15	15	3	0,16	18,75	
Lübkau	—	—	4	4	—	0,16	—	Betheiligen sich am Dembecker Netz.
Ortschaften 27. Summa	436	354	280	1070	418	23,0	18,3	

Provinz Pommern.
Reg.-Bezirk Coeslin.

1) Kreis Lauenburg.

Ortschaften								
Wittenberg	—	—	11	11	5	0,67	71,43	Hauptfang: Aal, Hering, Breitling, Dorsch, Neunauge, Flunder, Lachs, Meerforelle (Lachsforelle).
Lübtow	1	—	38	39	6	2,76	2,7L	
Koppalin	1	—	18	19	6	7,66	0,78	3 Kähne.
Sassin	—	—	8	8	1	0,21	4,76	Kahn.
Leba	38	29	22	89	27	14,21	1,90	8 Kähne.
Ortschaften 5. Summa	40	29	97	166	45	18,0	2,5	

2) Kreis Stolp.

Giesebitz	10	—	17	27	10	0,30	33,33	6 Kähne.
Fuchsberg	27	—	—	27	10	0,30	33,33	6 Kähne.
Zemminer Klucken	6	—	—	6	1	0,30	3,33	
Selessen	15	10	—	25	8	0,30	2,67	4 Kähne.
Schmolsin	28	4	4	36	12	0,30	40,0	8 Kähne.
Holzkathen	—	—	44	44	5	0,08	62,5	Hauptfang: Stör, Aal, Hering, Breitling, Neunaugen,
Gross-Garde	—	—	74	74	6	0,08	75,0	Dorsch, Stint, Flunder, Lachs, Brassen, Kaulbarsch, Uckelei, Karauschen, Hecht, Plötz (Weissfisch), Zander, Barsch, Schleihe.
Gross-Rowe	6	28	30	64	9	1,81	4,97	
Klein-Machmin	—	—	18	18	5	0,76	6,58	
Stolpmünde	16	9	—	25	9	4,63	1,94	2 Frauen unter den Gehülfen.
Görshagen	—	—	4	4	2	1,59	9,34	
Vietzke	26	23	—	49	14	5,95	2,35	
Ortschaften 12. Summa	134	74	191	399	91	12,98	7	

3) Kreis Schlawe.

Saleske Strand	8	—	—	8	4	0,20	20,0	Hauptfang: Aal, Tobiasfisch (Sandaal), Hering, Breitling, Dorsch, Neunauge, Flunder, Lachs, Brassen, Hecht, Zander.
Krolower Strand	40	20	—	60	20	3,54	5,65	
Vitte	—	—	22	22	18	0,97	18,56	
Rügenwaldermünde	4	—	12	16	4	0,03	133,03	
Neuwasser	31	20	—	51	2	8,12	0,24	
Dankerort	8	4	—	12	2	8,12	0,49	
Ortschaften 6. Summa	91	44	34	169	50	10,92	4,6	

4) Kreis (Fürstenthum) Cammin.

Ortschaften	Eigentliche Fischer	Gehülfen derselben	Gelegenheitsfischer	Sa. der Fischer	Zahl der Fahrzeuge	Grösse des Fischereibezirks ☐Meilen pr.☐Meil.	Zahl der Böte	Bemerkungen.
Laase	19	28	—	47	23	1,40	16,39	19 Böte, 30 Fuss Länge, 12 Fuss Breite. 4 Lachsböte
Coesliner Deep	24	69	—	93	45	3,7	12,45	von 40 bis 50 Fuss Länge u. 20 Fuss Breite.
Nest	33	55	4	92	41	3,34	12,27	
Mölln	6	—	—	6	6	1,59	3,77	15 Böte, 15 Fuss Länge, 3½ Fuss Breite.
Bauerhufen	—	—	16	16	5	0,22	22,73	
Sohrenbohm	—	—	8	8	2	0,31	6,45	Hauptfang: Stör, Aal, Tobiasfisch, Hering, Breitling,
Funkenhagen	16	16	—	32	4	0,45	8,89	Dorsch, Stint, Neunauge, Flunder, Schollen,
Pleushagen und Altenhagen	—	—	2	2	2	0,13	15,38	Lachs, Brassen, Kaulbarsch, Uckelei, Hecht,
Henkenhagen	—	—	14	14	2	0,12	16,67	Plötz, Zander, Speitsch (mir unbekannter Name),
Colbergermünde	30	16	17	63	28	22,61	1,24	Rodang (ebenso)
Gribow	25	—	8	33	15	12,40	1,21	24 Fuss Länge und 3 Fuss Breite.
Colberger Deep	—	—	23	23	11	9,07	1,21	
Ortschaften 12. Summa	153	184	92	429	184	34,0	5,4	

Reg.-Bez. Stettin.

5) Kr. Greiffenberg.

Camp	11	11	—	22	11	0,29	37,93	
Treptower Deep	36	—	24	60	16	10,83	1,48	3-bis 4gängige Böte.
Kl. Horst	24	—	—	24	6	5,92	1,01	
Revahl	4	4	—	8	4	2,51	1,59	Hauptfang: Aal, Hering, Dorsch, Flunder, Lachs,
Pustchow	1	2	—	3	2	6,54	3,06	Hecht, Plötz, Barsch.
Ortschaften 5. Summa	76	17	24	117	39	19,0	2,1	

6) Kreis Kammin.

								Hauptfang: Stör, Aal, Hering, Dorsch, Stint, Flunder,
Kl. Dievenow	24	24	—	48	16	5,55	2,88	Lachs, Kaulbarsch, Zander, Plötz.
Berg-Dievenow	16	9	—	25	10	12,45	0,80	3gängige Böte.
Ost-Dievenow	4	5	—	9	3	8,76	0,34	do.
Sager	2	4	—	6	2	0,55	3,6	do.
Ortschaften 4. Summa	66	75	—	141	51	13,0	3,9	? 1 Quase.

7) Kreis Usedom-Wollin.
(a. Insel Wollin.)

West-Dievenow	15	41	—	56	14	8,76	1,60	Hauptfang: Aal, Goldfisch (Alose?), Hering, Flunder,
Heidebrink	3	3	—	6	2	8,76	0,23	Lachs, Plötz, Zander, Barsch.
Neuendorf	20	—	—	20	10	2,12	4,72	
Warnow	12	—	—	12	6	2,31	2,60	
Misdroy	31	42	17	90	35	5,57	6,28	5 Garnböte.
Osternothhafen	30	30	4	64	17	3,28	5,18	
Ostswine	1	—	2	3	2	3,28	0,61	
Ins. Woll. 7 Ortsch. Sum.	112	116	23	251	86			

(b. Insel Usedom.)

Westswine	12	—	—	12	5	3,28	1,52	
Svinemünde	10	42	23	75	30	0,24	157,4	20 Lieger.
Aalbek a/A.	73	69	—	142	48	51,67	} 1,61	6- bis 7gängige Böte.
Aalbek k/A.	56	38	—	94	118	51,67		do.
Heringsdorf	—	—	12	12	5	28,3	0,18	4- bis 5gängige Böte.
Neuhof und Neukrug	—	—	18	18	5	7,92	0,63	Drei 6- bis 7gängige Böte.
Banzin	—	—	28	28	12	4,10	2,93	4- bis 5gängige Böte.
Neu-Sallenthin	—	—	5	5	2	1,19	} 4,20	
Alt-Sallenthin	—	—	6	6	3	1,19		
Sellin	—	—	10	10	5	1,19	4,20	
Benz	—	—	2	2	1	1,19	0,84	Auf der Karte einzutragen vergessen.
Stoben	—	—	2	2	1	1,19	0,84	
Uckeritz	60	60	—	120	15	12,64	1,19	Von Uckeritz bis Peenemünde sind 280 Frauen als Ge-
Loddin	26	22	6	54	12	2,08	5,77	hülfen aufgezählt.
Coserow	48	48	—	96	14	11,66	1,20	
Zempin	42	42	—	84	12	9,67	1,25	
Zinnowitz	1	—	12	13	13	0,21	62,0	
Hammelstall	23	23	3	49	8	8,35	0,96	
Carlshagen	56	56	5	117	33	8,35	3,96	Hauptfang: Stör, Aal, Goldfisch, Hering, Dorsch,
Peenemünde	37	37	6	80	14	5,70	2,46	Flunder, Steinbutte, Karauschen, Hecht, Plötz,
Ins. Used. 20 Ortsch. Sa.	444	437	138	1019	356			Zander, Barsch, Schleihe, Brassen, Zärthe.

| Ortsch. 27. Sa. im Kreise | 556 | 553 | 161 | 1270 | 442 | 63,0 | 7,0 | |

8) Kreis Uckermünde.

Ortschaften	Eigentliche Fischer	Gehülfen derselben	Gelegenheitsfischer	Sa. der Fischer	Zahl der Fahrzeuge	Grösse des Fischereibezirks ☐Meilen pr.☐Meil.	Zahl der Böte	Bemerkungen
Ziegenort	1	1	—	2	1	0,55	1,82	Hauptfang: Aal, Goldbutte.
Uckermünde	2	—	—	2	2	—	—	Bis an die Mecklenburgische u. Schleswig-Holsteinische Küste, also kein genauer Bezirk.
Ortschaften 2. Summa	3	1	—	4	3			

9) Kreis Anclam.

Fähre	1	1	—	1	1	0,55	1,82	
Ortschaft 1. Summa	1	1	—	1	1	0,55	1,82	

Reg.-Bez. Stralsund.

10) Kr. Greifswald.

Hollendorf	12	—	—	12	5	5,35	0,94	Hauptfang: Aal, Hering, Hornhecht, Flunder, Hecht, Plötz, Barsch.
Croeslin	38	12	4	54	15	5,35	2,81	Segelböte mit Grosssegel, Fock und Klüver.
Freest	43	10	—	53	27	5,35	5,04	do.
Spandowerhagen	23	4	—	27	10	5,35	1,87	
Greifswalder Oie	—	—	9	9	9	0,93	9,68	Ruderböte.
Insel Ruden	—	—	4	4	10	0,37	27,1	1 Segelboot.
Freesendorf	2	2	—	4	2	5,35	0,37	Segelböte.
Lubmin	37	—	—	37	18	5,35	3,36	do.
Vierow	24	—	—	24	13	5,35	2,43	do.
Lossin	2	2	—	4	2	5,35	0,37	do.
Ludwigsburg	2	2	—	4	2	5,35	0,37	do.
Wiek	54	30	22	106	38	0,85	44,8	do.
Greifswald	21	4	2	27	15	0,85	17,7	13 Segelböte.
Leist	1	3	—	4	2	0,03	66,6	
Ortschaften 14. Summa	259	69	41	369	168	5,35	31,4	

11) Kreis Grimmen.

Karrendorf	—	—	2	2	2	0,03	66,6	Hauptfang: Aal, Hering, Hornhecht, Flunder, Hecht, Plötz, Barsch.
Gristow	16	—	1	17	4	0,85	4,71	
Kalkvitz	8	—	—	8	2	0,85	2,36	
Stahlbrode	14	5	8	27	14	4,24	3,30	Segelboote auf Bodden gebaut mit Schwertern,
Gr. Miltzow	1	1	—	2	1	7,93	0,13	Zeeseboot, 28′ lang, 9′ breit, 5 Fuss hoch, 2½′ Tiefgang.
Niederhof	1	1	—	2	1	0,15	6,66	
Neuhof	2	2	—	4	2	0.15	13,3	Jolle, 15′ lang, 7′ breit, 3′ hoch, 1′ Tiefgang.
Ortschaften 7. Summa	42	9	11	62	26	7,03	3,3	

12) Kr. (Ins.) Rügen.

Drigge	1	2	—	3	1	1,2	0,83	Segelboot, 17 Fuss lang, 3 Fuss breit, 1 Fuss Tiefgang.
Wampen	1	1	—	2	1	1,2	0,83	
Rügenhof	—	—	1	1	1	0,35	2,86	
Kl. Kubitz	—	—	1	1	1	0,95	1,05	
Lieschow	—	—	16	16	10	0,95	10,5	
Mursewiek	3	—	4	7	5	2,30	2,12	1 Zeeseboot auf Bodden gebaut mit Schwertern u. 1 Mast.
Gingst	—	—	4	4	4	1,35	2,96	Polten, 10 Fuss lang.
Büschow, Tankow u. Markow	—	—	5	5	4	1,32	3,03	
Waase	2	—	4	6	4	1,32	3,03	1 Zeeseboot.
Wusse	—	—	1	1	1	2,48	0,40	
Freesenort	1	2	—	3	3	2,48	1,21	1 Zeeseboot.
Suhrendorf	—	—	5	5	1	1,67	0,60	
Heidekathen	—	—	2	2	1	1,67	0,60	
Trent	—	—	5	5	4	3,40	1,16	Polten.
Schaprode	—	—	25	25	7	1,35	5,19	2 Polten.
Hiddensee Fähre	—	—	2	2	1	1,67	0,6	
Plogshagen a/H.	25	16	—	41	30	1,67	18,0	4 Zeeseboote, 7 Garnböote, 1 Schwert, 2 Masten.
Neuendorf a/H.	28	15	—	43	27	1,67	16,2	3 Zeeseboote, 7 Garnboote.
Vitte a/H.	59	10	—	69	48	11,37	28,7	5 Zeeseboote, 10 Garnboote.
Grieben a/H.	3	3	—	6	6	1,67	3,60	3 Zeeseboote.
Breetz	—	—	5	5	3	0,09	33,3	
Vieregge	—	—	10	10	3	0,39	7,69	
Neuenkirchen	1	—	13	14	2	1,03	1,95	
Sylvin	1	1	—	2	1	0,10	20	
Banzelvitz	1	1	—	2	1	0,10	10	
Ralswiek	3	1	2	6	4	1,55	2,58	2 Zeeseboote.
Stedar	1	1	—	2	2	1,55	1,29	
Buschvitz	6	2	—	8	6	0,50	12	Hauptfang: Stör, Aal, Goldfisch, Hering, Hornhecht, Dorsch, Flunder, Steinbutte, Lachs, Makrelen, Brassen, Kaulbarsch, Hecht, Plötz, Zander, Barsch, Schleihe.
Zittvitz	6	—	3	9	8	1,55	5,16	
Bergen	5	2	—	7	2	1,55	1,29	
Zirzevitz	1	—	1	2	1	0,50	2	
Lietzow	7	7	14	28	12	1,57	7,64	
Polchow	9	8	4	21	10	1,57	6,33	3 Zeeseboote.
Breege	23	4	137	164	24	1,72	14	8 Zeeseboote.

12) Kreis (Insel) Rügen. (Fortsetzung.)

Ortschaften	Eigentliche Fischer	Gehülfen derselben	Gelegenheitsfischer	Sa. der Fischer	Zahl der Fahrzeuge	Grösse des Fischereibezirks ☐Meilen	Zahl der Böte pr. ☐Meil.	Bemerkungen.
Cammin	—	—	2	2	1	0,25	4	
Wittower Fähre	1	1	1	3	2	1,67	1,2	1 Zeesenboot.
Wiek a/R.	5	3	45	53	17	1,67	1,02	3 Zeesenboote, 2 auf Bodden, 1 auf Kiel.
Kuhle	—	—	1	1	1	1,67	0,6	
Wittower Posthaus	—	—	—	1	2	1,67	1,2	1 Zeesenboot.
Dranske	12	—	—	12	14	0,65	21,5	5 Zeesenboote.
Arcona und Puttgarten	—	—	5	5	3	0,16	18,8	
Vitte	11	—	—	11	4	0,94	4,25	
Goor	—	—	3	3	2	0,04	50	
Nobbin	—	—	2	2	1	0,06	16,7	
Glowe	9	—	—	9	5	4,90	1,02	
Bisdamitz	1	—	2	3	1	2,50	0,4	
Blandow	1	1	—	2	1	2,50	0,4	
Lohme	10	2	1	13	6	2,50	2,4	
Sassnitz	24	1	6	31	8	5,04	1,59	
Crampas	6	1	2	9	3	3,02	0,99	
Neu-Mukram	2	1	1	4	2	0,32	6,25	
Lubkow	1	—	1	2	1	0,23	4,35	
Trips	1	—	1	2	1	0,23	4,35	
Binz	25	—	2	27	18	1,50	1,2	
Sellin	9	—	—	9	4	1,10	3,62	
Goehren	23	7	—	30	17	2,70	6,3	
Lobbe	16	6	—	22	21	2,70	7,8	
Thiessow	13	8	17	38	24	5,62	4,27	
Kl. Zicker	9	5	3	17	11	5,62	1,96	
Gr. Zicker	28	11	8	47	16	5,62	2,85	
Gager	22	10	4	36	14	5,62	2,51	
Kleinhagen	9	8	2	19	10	5,62	1,78	
Mariendorf	9	—	—	9	8	2,83	2,83	
Alt-Reddewitz	11	6	9	26	25	2,83	8,84	
Baabe	19	5	1	25	11	5,62	1,96	
Moritzdorf	9	—	—	9	4	5,62	0,71	
Altensien	—	—	1	1	—	—	—	Fischen nur zu Eis.
Seedorf	10	—	14	24	12	0,05	240	
Neuensien	—	—	1	1	—	—	—	Nur zu Eis.
Preetz	—	—	2	2	—	—	—	do.
Sandort	1	1	—	2	2	0,46	4,35	
Neu-Reddewitz	15	2	4	21	24	5,62	4,27	
Stresow	4	2	9	15	17	9,32	1,83	
Muglitz	3	1	—	4	5	3,68	1,36	
Freetz	5	1	2	8	8	3,68	2,18	
Vilmnitz	5	—	6	11	3	3,68	0,82	
Beuchow	—	—	2	2	—	—	—	Nur zu Eis.
Lonvitz	—	—	1	1	—	—	—	do.
Lauterbach	1	1	11	13	7	0,85	8,24	
Neuendorf bei Puttb.	17	—	12	29	22	3,68	5,98	
Putbus	—	—	10	10	—	—	—	Nur zu Eis.
Wreechen	13	9	—	22	8	0,85	9,41	
Zehnmorgen	—	—	2	2	—	—	—	Nur zu Eis.
Casnevitz	—	—	3	3	—	—	—	do.
Neukamp	23	6	5	34	34	3,68	9,24	
Krakvitz	2	—	—	2	2	0,85	2,36	
Lanschvitz	—	—	2	2	—	—	—	Nur zu Eis.
Garz	—	—	32	32	—	—	—	do.
Gr. Schoritz	1	3	—	4	2	0,04	50	
Zudar	1	3	3	7	4	0,04	100	
Poppelvitz a/Z.	3	3	—	6	3	10,14	0,3	
Pritzwald	1	1	—	2	2	9,32	0,22	
Grabow	1	1	—	2	1	10,14	0,1	
Glevitzer Fähre	—	—	2	2	1	6,53	0,15	
Glevitz a/Z.	2	2	—	4	2	6,53	0,31	Auf Bodden gebaut mit Schwertern und 1 Mast.
Tannenort	1	3	—	4	2	0,4	5	
Prossnitz	1	1	—	2	1	0,97	1,04	
Ortschaften 97. Summa	584	192	507	1283	660	37,35	15,0	

13) Kr. Franzburg.								Hauptfang: Stör, Aal, Hering, Dorsch, Schollen, Flunder, Lachs, Brassen, Kaulbarsch, Hecht, Plötz, Jolle. | Zander, Barsch, Schleihe.
Devin	1	1	—	2	1	0,92	1,09	
Stralsund, Stadt	149	75	45	269	96	3,38	28,4	7 Zeesekähne, 40' lang, 14' breit, 7' hoch, 4' Tiefge. 68 Zeeseb., 28' lg., 9' br. 21 Jollen, 24' lg., 7' br. Beide Letzteren 5' hoch, 2½' Tiefgang.
Parow	1	1	—	2	1	1,54	0,63	
Solkendorf	1	1	—	2	1	7,93	0,13	Zeseboot, 30' lang, 10' breit.
Barhövt	2	2	—	4	2	7,93	0,25	do. 28' „ 9' „
Zarrenzin	1	1	—	2	1	7,93	0,13	do.
Wend. Langendorf	2	2	—	4	3	2,35	1,28	2 do. do.
Kinnbackenhagen	—	—	3	3	2	0,60	3,33	
Barth	19	19	—	38	33	2,09	15,8	17 Zeesenboote, 28' lang, 9' breit.

13) Kreis Franzburg. (Fortsetzung.)

Ortschaften	Eigentliche Fischer	Gehülfen derselben	Gelegenheitsfischer	Sa. der Fischer	Zahl der Fahrzeuge	Grösse des Fischereibezirks ☐Meilen	Zahl der Böte pr. ☐Meil.	Bemerkungen.
Pruchten	33	31	—	64	39	2,09	18,7	2 Polte, 33 Zeeseboote.
Bresewitz	1	1	—	2	6	2,09	2,87	4 Zeeseböte.
Bedstaedt	6	6	—	12	6	2,09	2,87	do.
Fuhlendorf	1	1	—	2	1	2,09	0,48	do.
Michaelsdorf	—	—	6	6	6	0,25	24	
Neuendorf b. Saal u. b. Heide	—	—	4	4	4	1,78	2,25	
Saal	—	—	1	1	1	1,14	0,88	Polte.
Langendamm	—	—	3	3	2	1,14	1,76	
Born a/Dars	9	—	—	9	5	1,29	3,88	
Wick a/Dars	2	—	2	4	2	0,54	3,71	
Sundische Wiese	—	—	3	3	1	0,56	1,79	
Pramort	—	—	6	6	6	0,22	2,73	
Zingst	16	—	—	16	2	0,12	1,67	
Prerow	17	—	—	17	3	0,04	7,50	
Darserort	—	—	2	2	1	0,21	4,77	
Ahrenshoop	10	2	—	12	3	1,96	1,53	
Ortschaften 25. Summa	274	146	75	495	228	13,27	17,2	

Staat Grossherzogthum Mecklenburg-Schwerin.

Dom.-Amt Ribnitz.								
Althagen und Niehagen	17	9	3	29	17	0,9	3,33 u. 1,47	Davon 14 Böte in Bodden mit 0,95 ☐M.
Wustrow	—	—	9	9	5	0,9	5,55	
Stadt Ribnitz.								
Körckwitz	22	—	—	22	22	0,25	88	
Ribnitz	11	—	—	11	11	0,25	44	Böte mit losem Kiel.
Stadt Rostock.								
Hinrichshagen	} 24	—	—	24	3	0,23	13	3 Fischervereine à 8 Personen 3 Monate fischend. Warnemünder Jollen, 22 Fuss lang, 8 Fuss breit, 4 Fuss tief.
Rowershagen								
Torfbrücke								
Rostock	41	11	—	52	58	0,24	241,7	52 Kähne.
Warnemünde	56	—	30	86	32	3,35	9,06	Warnemünder Jollen.
Rethwisch	2	—	—	2	1	3,94	0,25	Halbböte.
Boergerende	10	—	—	10	7	3,94	1,77	do.
Amt Rackow.								Hauptfang: Hering, Dorsch, Plattfische, Lachs, Hornhecht, Stint, Uckelei, Sandart, Hecht, Barsch, Alander (Cyprinus Idus), Aalquabben, Kaulbarsch, Brassen, Schleihe, Plötz, Rothauge, Garnelen.
Brunshaupten	7	—	—	7	3	5,45	0,55	
Ahrensee	26	5	—	31	13	1,24	10,5	
Alt-Gaarz	6	2	3	11	14	0,49	28,6	6 Zeeseböte.
Dom.-Amt Redentin.								
Boiensdorf	3	2	—	5	3	0,76	3,95	1 Zeeseboot.
Redentin	4	—	—	4	4	2,39	1,67	
Fischkathen	3	—	8	11	11	2,39	4,6	
Amt Poel.								
Gollwitz Vorwerk, Malchow	4	—	4	8	6	2,39	2,51	
Fährdorf-Niendorf	7	—	3	10	7	2,39	2,92	1 Zeeseboot.
Kirchdorf	20	5	25	50	20	2,39	1,84	5 Zeeseböte.
Wangern mit Vorwangern	} 30	—	6	36	10	3,13	3,19	
Einhusen								
Weitendorf								
Brandenhusen								
Timmendorf	5	—	7	12	8	2,39	3,35	
Wismar	21	21	8	50	25	2,39	10,5	21 Zeeseböte, ausserdem fischen im Winter unbeschäftigte Matrosen mit Angeln.
Beckerwitz	1	—	—	1	1	0,78	1,28	
Domanial-Amt Grevesmühlen.								
Boltenhagen	—	—	3	3	1	1,06	0,95	
Tarnewitz	8	—	8	16	6	2,38	2,52	Darunter 7 Frauen.
Warnkenhagen	6	9	—	15	6	0,53	11,3	Darunter 6 Frauen.
Dassow	13	5	2	20	14	0,26	53,8	9 Kähne.
Ortschaften 35. Summa	347	77	111	535	298			

Fürstenthum Lübeck.

Lübeck	11	2	—	13	11	0,65	13,4	Hauptfang: Hering, Breitlinge, Dorsch, Aal, Bütt, Garnelen, Hecht, Barsch, Brassen.
Gothmund	17	12	—	29	17	1,61	10,5	
Schlutop	42	42	—	84	54	1,61	33,6	
Travemünde	24	2	42	68	32	10,43	1,63 u. 12,4	15 Böte nur 1,21 ☐M.
Ortschaften 4. Summa	94	58	42	194	114	11,00		

Grossherzogl. oldenburgischer Antheil des Fürstenthums Lübeck.

Ortschaften	Eigentliche Fischer	Gehülfen derselben	Gelegenheitsfischer	Sa. der Fischer	Zahl der Fahrzeuge	Grösse des Fischereibezirks ☐Meilen pr. ☐Meil.	Zahl der Böte	Bemerkungen
Niendorf	10	16	—	26	10	1,58	6,33	Hauptfang: Dorsch, Bütt, Hering, Breitling, Aal,
Scharbeutz	—	—	2	2	1	0,72	1,39	Makrelen, Lachs, Meerforelle.
Haffkrug	24	39	—	63	27	2,95	9,15	
Ortschaften 3. Summa	34	55	2	91	38	3,27	11,6	

Provinz Schleswig-Holstein. (Ostküste.)

1) Kreis Oldenburg.

Wintershagen	2	—	—	2	1	1,08	0,93	
Oevelgönne	5	3	2	10	8	2,25	3,56	
Neustadt	44	2	15	61	59	2,25	26,22	3 Zeesenböte à 2 Last und 49 Kähne à ½ Last.
Grömitz	9	—	4	13	8	1,09	7,34	
Kellenhusen	1	—	5	6	4	0,79	5,06	
Dahme	10	—	5	15	9	13,32	0,08 u.14,3	1 Boot 13,32 ☐M., die übrigen 0,56. 4 gedeckte Böte.
Siggen	2	4	—	6	2	0,19	10,52	Darunter zwei Frauen.
Godderstorff	1	2	—	3	2	0,92	2,17	1 Kahn.
Lochrstorff	4	—	9	13	14	0,74	18,92	Zum Theil Kähne.
Wulfen	2	—	7	9	4	0,15	26,67	
Sahrensdorf	1	1	—	2	1	0,31	3,23	
Burg a/F.	6	4	5	15	6	2,43	2,47	1 Zeesenboot.
Gahlendorf	1	2	—	3	3	1,72	1,74	
Preesen	—	—	2	2	1	0,47	2,13	
Puttgarden a/F.	—	—	1	1	1	0,10	10	
Danschendorf	—	—	1	1	1	0,18	5,56	Hauptfang: Aal, Hering, Hornhecht, Dorsch, Schollen,
Orth	6	—	1	7	7	0,40	17,99	Goldbutte, Lachs, Quappen (Blennius), Barsch,
Lemkenhafen	2	—	8	10	7	0,20	35,0	Garnelen, Brassen.
Albertsdorf	1	—	2	3	1	0,16	6,25	
Strucamp	5	2	1	8	6	0,60	10	1 gedecktes Boot.
Heiligenhafen	9	—	7	16	16	1,33	12,03	do.
Ortschaften 21. Summa	111	20	75	206	161	18,9	8,5	

2) Kr. Plön u. Kiel.

Schlendorf	—	—	1	1	1	0,11	9,09	Kahn.
Neudorf	8	1	6	15	13	3,40	3,82	
Waterneverstorff	3	3	3	9	9	1,48	7,66	
Herrschaft Hessenstein:								
Rethkuhl								
Hubertsberg	8	—	—	8	7	3,57	1,96	3 Kähne.
Grünberg								
Todendorf								
Stackendorfer Strand	5	—	—	5	2	1,04	1,92	
Schönberger Strand	6	—	—	6	2	1,04	1,92	Fang: Dorsch, Plattfische, Aal, Hering, Breitling,
Wendtorff	5	—	—	5	4	6,84	0,58	Hornhecht, Barsch, Raapfen, Makrelen, Lachs,
Stein	12	—	6	18	6	6,84	0,88	ferner Garnelen.
Laboe	7	—	5	12	3	1,34	1,78	
Möltenort	16	14	13	43	17	0,46	3,70	3 gedeckte Boote.
Wellingdorf	7	9	2	18	15	21,68	1,93	9 Kähne. } Kr. Kiel.
Ellerbeck	47	27	2	76	85	21,68	4,94	54 Kähne.
Ortschaften 12. Summa	124	54	38	216	164	21,05	7,5	

3) Kr. Eckernförde.

Am Strande bei Braunberg	1	1	—	2	1	0,25	4	Hauptfang: Aal, Hering, Breitling, Hornhecht, Dorsch,
Eckhof	1	—	—	1	1	0,12	8,33	Flunder, Goldbutt, Lachs, Makrelen, Brassen,
Noer	1	1	—	2	1	0,16	6,25	Hecht, Zander, Schleihe.
Eckernförde	120	42	164	326	170	16,49	10,31	Eine Anzahl Quasen.
Borbye	2	—	7	9	4	1,42	2,87	1 Zeesenboot.
Mohrberg	1	3	—	4	1	1,42	0,70	Kahn.
Ludwigsburg	4	1	—	5	3	0,16	4,87	
Grünholz	—	—	1	1	1	0,57	1,75	
Olpenitz	1	—	—	1	1	0,01	100	
Ortschaften 9. Summa	131	48	172	351	183	16,50	11,1	

4) Kreis Schleswig.

Stadt Schleswig	95	96	—	191	60	1,55	38,71	Kähne, 22 bis 38 Fuss Länge.
Cappeln	3	—	4	7	7	0,41	17,07	Fang: Aal, Hering, Hornhecht, Dorsch, Goldbutte.
Ortschaften 2. Summa	98	96	4	198	67	1,55	43,2	Brassen, Schnäpel, Hecht, Plötz, Barsch.

5) Kreis Flensburg.

Ortschaften	Eigentliche Fischer	Gehülfen derselben	Gelegenheitsfischer	Sa. der Fischer	Zahl der Fahrzeuge	Grösse des Fischereibezirks □ Meilen	Zahl der Böte pr. □ Meil.	Bemerkungen.
Mehlby und Luklos	1	—	—	1	1	0,53	1,89	
Rabelsund	1	—	—	1	2	0,03	150	
Maasholm	21	—	11	32	26	0,77	63,41	Böte und Kähne.
Oeher Drecht	3	—	—	3	3	0,20	15,0	
Düttebüll	6	—	7	13	9	0,43	20,93	
Wackerballig	3	—	3	6	5	0,20	25	1 Kahn.
Ohrfeldt	3	—	—	3	5	0,52	8,06	2 Kähne.
Steinberghaff	4	—	1	5	4	1,60	2,50	
Gut Oestergaard	2	—	—	2	1	1,60	0,63	
Steinbergholz	2	—	2	4	2	1,60	1,25	
Habernis	—	—	1	1	1	0,50	2	Hauptfang: Aal, Hering, Dorsch, Goldbutt, Flunder,
Neukirchen	1	—	—	1	1	0,43	2,33	Steinbutt, Makrelen, Lachs, Hornhecht, Garnelen,
Dollerupholz	—	—	1	1	1	0,06	16,66	Taschenkrebse (Carcinus maenas).
Westerholz	—	—	2	2	2	0,67	2,99	
Langballigholz	5	—	2	7	5	0,67	7,46	
Bockholmwick	—	—	4	4	4	0,25	16	
Hollnis	1	—	1	2	2	1,97	1,02	1 Quase.
Twedterholz	2	—	1	3	3	0,05	60	
Flensburg	18	30	31	79	31	5,03	6,16	4 Quasen.
Duburg	1	—	—	1	1	4,80	0,21	
Nichuus	2	—	—	2	2	0,33	6,06	
Collund	—	—	3	3	3	0,64	4,69	
Hönsnap	—	—	16	16	8	0,33	24,8	
Ortschaften 22. Summa	76	30	86	191	123	6,47	19,0	

6) Kr. Apenrade. (I.)

Randershof	1	—	2	3	1	0,04	25,86	Hauptfang: Aal, Hering, Dorsch.
Beken	10	10	1	21	15	0,58	25,86	
Rinkenis	10	3	6	19	14	0,58	24,14	
Treppe	9	—	8	17	10	0,58	17,24	
Alnør, Nalmaybro	—	—	12	12	7	0,58	12,07	
Atzbüll	—	—	1	1	1	0,58	1,72	
Ortschaften 6. Latus	30	13	30	73	48			

7) Kr. Sonderburg.

Ekensund	2	—	24	26	14	3,41	4,10	Hauptfang: Aal, Hering, Dorsch, Goldbutte, Lachs,
Iller	1	1	10	12	6	1,78	3,37	Makrelen.
Gammelgab	—	—	2	2	2	2,55	0,78	
Schelde	3	—	1	4	4	0,48	8,33	
Broacker	8	3	6	17	12	0,11	106,8	
Schmoel	1	—	—	1	1	0,11	9,08	
Düppel	—	—	2	2	1	0,57	1,75	
Sonderburg	22	—	15	37	40	2,96	13,5	
Höruphaf	8	—	6	14	14	0,22	63,71	
Hirschholm	—	—	7	7	5	0,22	22,73	Eine Quase.
Sönderbye	9	1	1	11	5	0,53	9,43	
Neuhof	1	1	2	4	3	0,59	5,08	
Skovbye	1	—	1	2	3	0,79	3,80	
Lysabbel	4	—	—	4	2	0,30	6,67	
Mummark	—	—	2	2	1	0,05	20	
Erteberg	—	—	7	7	4	0,13	30,77	
Kettingholz	1	—	4	5	2	0,14	14,23	
Adzerballigholz	2	—	5	7	5	0,38	13,16	
Nottmarkholz	—	—	2	2	2	0,21	9,52	
Tarup. Hinmark Klingenberg	} 5	—	—	5	5	1,40	3,57	
Lauensby	2	—	—	2	2	0,71	2,82	
Pöhl	—	—	2	2	2	0,57	3,51	
Holm	2	1	1	4	3	0,91	3,30	
Norburg	2	—	3	5	4	0,12	33,33	
Mels	3	3	16	22	16	0,20	75,14	
Broballig	—	—	1	1	1	0,03	33,3	
Kjär	1	—	2	3	2	0,22	9,09	
Sebbelev	2	—	3	5	4	0,15	26,7	
Ketting	—	—	1	1	1	0,15	60,06	
Ulkebüll	2	—	6	8	8	0,15	53,4	
Staugaard	4	1	—	5	4	0,81	4,94	
Satrup	9	—	—	9	5	0,78	6,41	
Blans	1	—	7	8	7	0,26	26,9	
Ortschaften 35. Summa	96	11	139	246	190	9,95	19,1	

8) Kreis Apenrade. (II.)

Ortschaften	Eigentliche Fischer	Gehülfen derselben	Gelegenheitsfischer	Sa. der Fischer	Zahl der Fahrzeuge	Grösse des Fischereibezirks □ Meilen	Zahl der Böte pr. □ Meil.	Bemerkungen
Transport Ortsch. 6 von (I.)	30	13	30	73	48	3,40	—	Hauptfang: Aal, Hering, Dorsch, Wittling, Schollen,
Warnitz	4	1	5	10	9	0,50	18,0	Flunder, Goldbutte, Lachs, Makrelen, Meerforelle
Schöbüllgaard	2	—	3	5	3	1,22	2,46	(Lachsforelle).
Feldstedtholz	1	—	5	6	5	0,40	12,5	
Süderhostrup	12	—	4	16	16	0,62	25,8	4 Quasen.
Stübbeck	6	1	—	7	4	0,62	6,45	
Landgemeinde Apenrade	1	—	2	3	3	0,52	5,77	
Stadt Apenrade	6	8	6	20	8	0,52	15,22	
Stollig	—	—	2	2	2	0,52	3,85	
Barsmark	5	—	3	8	8	2,14	3,74	
Loit (Kirchdorf)	4	8	5	17	4	3,25	1,23	
Gjenner	1	—	4	5	5	2,50	1,70	
Ortschaften 17. Summa	72	31	69	172	115	5,50	20,9	

9) Kr. Hadersleben.

								Hauptfang: Aal, Hering, Dorsch, Goldbutte, Lachs,
Süderballig	6	—	6	12	8	1,94	4,12	Makrelen.
Süderwilstrup	2	—	2	4	3	0,26	11,15	
Kjelstrup	4	1	2	7	4	0,58	6,90	
Heisager	6	—	—	6	4	0,58	6,90	
Halk	2	—	—	2	1	0,43	2,33	
Raad	—	—	6	6	3	0,03	100	
Aaroe (Insel)	7	—	20	27	11	1,55	7,1	
Stevelt	—	—	7	7	4	0,14	28,5	
Aastrup	1	1	—	2	1	0,13	7,7	
Wonsbeck	1	—	—	1	1	0,05	20	
Orby	7	7	10	24	8	2,72	2,94	
Knud	4	—	—	4	4	1,69	2,37	
Anslet	6	—	2	8	8	0,45	17,8	
Meng	—	—	3	3	3	0,58	5,17	
Ortschaften 14. Summa	46	9	58	113	63	5,50	11,5	

Provinz (Reg.-Bez.) Schleswig-Holstein. (Westküste.)

1) Kreis Tondern.

								Hauptfang: Aal, Hering, Hornhecht, Schellfisch,
Westerende	—	—	3	3	—	0,08	—	Schollen.
Baadsbyll	—	—	3	3	—	0,05	—	Flache und offene Böte.
Reisby mit Bünty	—	—	8	8	—	0,11	—	
Emmerlev	—	—	11	11	—	0,11	—	Ohne die Austernfischer.
Westerland auf Sylt	—	—	10	10	2	1,49	1,34	
Rutebüll u. Friedrichskoog	—	—	18	18	2	1,60	1,25	
Ortschaften 6. Summa	—	—	53	53	4	3,28	1,3	

2) Kreis Husum.

								Hauptfang: Aal, Hering, Anchovis, Hornfisch, Stint,
Amt Bredstedt	3	—	8	11	11	0,95	11,7	Schollen, Rochen, Seehunde, Taschenkrebse,
Halebüll	—	—	9	9	—	0,16	—	Hummer.
Schobüll	—	—	1	1	—	0,16	—	
Nordstrand (Insel)	—	—	3	3	1	1,01	1,00	
Simonsberg	1	—	6	7	—	0,42	—	
Ortschaften 5. Summa	4	—	27	31	12	2,16	5,5	

3) Kreis Eiderstedt.

								Hauptfang: Stör, Aal, Stint, Schollen, Schnepel.
Westerhever	1	—	1	2	2	3,30	0,66	
Oldenswort	1	1	—	2	1	0,12	8,33	
Kirchspiel Tönning	2	3	5	10	4	0,65	6,15	
Ortschaften 3. Summa	4	4	6	14	7	3,30	2,1	

4) Kreis Schleswig.

								Offene Plattböte.
Friedrichstadt	—	—	8	8	4	0,1	40	do.
Süderstapel	—	—	12	12	6	0,1	60	
Ortschaften 2. Summa	—	—	20	20	10	0,1	100	Hauptfang: Stör.

5) Kreis Norderditmarschen.

								Hauptfang: Stör, Aal, Hering, Sprott, Goldbutte,
Wrohn mit Lexfähr	—	—	1	1	1	0,2	5	Granat.
Tielenhemme	—	—	4	4	4	0,2	20	
Pahlen	—	—	2	2	2	0,2	10	4 Frauen. Fahrzeuge mit Vordeck. 3 Ewer, 1 Jolle.
Horst	—	—	6	6	6	0,2	30	
Büsum	—	—	20	20	4	2,14	1,87	
Ortschaften 6. Summa	—	—	33	33	17	2,34	7,3	

6) Kreis Süderditmarschen.

Ortschaften	Eigentliche Fischer	Gehülfen derselben	Gelegenheitsfischer	Sa. der Fischer	Zahl der Fahrzeuge	Grösse des Fischereibezirks ☐Meilen pr. ☐Meil.	Zahl der Böte	Bemerkungen.
Wöhrden	—	—	8	8	—	1,0	—	Hauptfang: Stör, Aal, Goldbutte, Granat.
Tatingburen								
Barsfleth	—	—	5	5	—	1,0	—	
Wiedersbüttel								
Elpersbüttel								
Ammerswurth	—	—	13	13	—	1,0	—	
Eesch								
Barlt	—	—	2	2	—	1,0	—	
St. Michaelisdonn	—	—	1	1	—	—	—	
Friedrichskoog	1	—	3	4	1	1,22	0,82	1 Frau.
Dahrenwurth	1	—	—	1	—	0,17	—	
Marne	1	—	3	4	—	0,17	—	3 Frauen.
Fahrstedt	—	—	2	2	—	0,17	—	
Kronprinzenkoog	—	—	5	5	1	1,22	0,82	Frauen.
Menghusen	—	—	1	1	—	0,17	—	Frau.
Schmedenswurth	—	—	1	1	—	0,17	—	Frau.
Kattrepeldeich	—	—	4	4	—	0,13	—	Frauen.
Ortschaften 17. Summa	3	—	48	51	2	2,23	0,9	
7) Kreis Steinburg.								
Borsfleth	—	—	5	5	6	0,40	15	Hauptfang: Stör.
Blome'sche Wildniss	—	—	14	14	7	0,90	7,77	
Glückstadt	—	—	8	8	4	0,83	4,82	Wallfischböte.
Gr. Colmar	—	—	15	15	15	0,63	23,8	
Kl. Colmar	—	—	16	16	7	0,63	11,1	
Neuendorf	—	—	6	6	3	0,31	9,68	
Ortschaften 6. Summa	—	—	64	64	42	1,14	37	
8) Kreis Pinneberg.								
Seestermühe	—	—	4	4	2	0,21	9,5	Hauptfang: Aal, Schellfisch, Schollen, Steinbutte, Zunge.
Blankenese	67	144	8	219	75	500	—	62 Fischerewer, 5 Fischerkutter.
Ortschaften 2. Summa	67	144	12	223	77	500	0,15	
9) Kreis Altona.								
Stadt Altona	1	1	—	2	1	1,83	0,55	Hauptfang: Aal, Stint, Plötz, Kaulbarsch. 1 Frau.

Staat Hamburg.

Finkenwerder	143	127	—	270	90	500	0,18	Hauptfang: Tobiasfisch, Schellfisch, Schollen, Zunge, Granat.
Dose	—	—	7	7	—	0,01	—	
Duhnen	2	—	4	6	—	0,04	—	
Sahlenberg	—	—	3	3	—	0,20	—	
Ortschaften 4. Summa	145	127	14	286	90	500	0,18	

Helgoland (Englische Insel).

Helgoland	205	270	—	475	85	52	1,6	Frühjahrsbezirk 20 ☐Meilen. Hauptfang: Dorsch, Schellfisch, Rochen, Haifisch.

Provinz Hannover.
Reg.-Bez. Stade.

1) Kreis Stadermarsch.								Hauptfang: Aal, Stint, Goldbutte.
Drochtersen	1	1	2	4	4	0,57	7,02	Fischerewer und Kähne.
Freiburg	—	—	4	4	4	0,60	6,67	Kähne.
2) Kreis Neuhaus.								Hauptfang: Aal, Stint, Goldbutte, Schnepel, Granat
Neuhaus a/Oste	8	—	—	8	4	0,88	4,55	2 kleine Ewer.
3) Kreis Otterndorf.								Hauptfang: Goldbutte, Schnepel, Granat.
Osterende Otterndorf	—	—	5	5	3	0,88	3,41	Kähne.
Westerende Otterndorf	—	—	4	4	1	0,43	2,33	Kähne.

4) Kreis Lehe. (I.)

Ortschaften	Eigent- liche Fischer	Ge- hülfen der- selben	Ge- legen- heits- fischer	Sa. der Fischer	Zahl der Fahr- zeuge	Grösse des Fischerei- bezirks ☐Meilen pr.	Zahl der Böte ☐Meil.	Bemerkungen.
Spikaer Neufeld	—	—	1	1	—	0,13	—	Hauptfang: Stint, Goldbutte, Granat.
Cappeler Neufeld	5	5	2	12	5	2,75	1,81	
Dorum	2	2	1	5	2	1,38	1,45	
Wremen	9	10	8	27	11	5,86	1,87	Wattenfahrzeuge.
Imsum	2	2	2	6	2	1,91	1,04	
Ortschaften 5. Latus	18	19	14	51	20			

Grossherzogthum Oldenburg.

1) Amt Elsfleth.								Hauptfang: Stint, Maifisch, Aal, Stör, Qualbe, Neun- augen, Schnäpel, Brassen, Weissfische, Lachs. Die Fischer unter 1 haben je 2 Dielenschiffe zu 3 und zu ²⁄₄ Last.
Elsfleth und Lienen	4	3	7	14	15	1,30	—	
2) Amt Brake.								Hauptfang: Wie bei Amt Elsfleth.
Käseburg	8	3	—	11	7	0,4	—	Dielenschiffe.
Oberhammelwarden	22	30	4	56	44	1,1	—	do.
Brake	2	—	1	3	4	0,9	—	2 Dielenschiffe.
3) Amt Oevelgönne.								Hauptfang: Granat, Butten.
Blexen	1	2	—	3	1	0,1	—	Jollen.
Waddens	4	—	—	4	2	1,09	—	do.
Burhave	13	—	—	13	5	1,09	—	do.
4) Amt Varel.								Hauptfang: Granat.
Varelerhafen	1	3	—	4	1	0,26	—	
Dangast	20	—	—	20	—	0,07	—	
5) Amt Jever.								
Bardterdeich	—	—	4	4	—	0,06	—	Darunter 1 Frau.
Vosslapp	1	—	2	3	—	0,06	—	
Hooksiel	—	—	4	4	1	0,1	—	Kleine Jolle.
St. Joost	3	—	4	7	—	0,1	—	Darunter 3 Frauen.
Minsen	—	—	5	5	—	0,1	—	
Ortschaften 14. Summa	79	41	31	151	80	2,15	37	Hauptfang: Granat, Taschenkrebse, Bütt, Aal, Stint.

Staat Bremen.

Bremerhafen	4	4	—	8	4	1,0	4	

Provinz Hannover.

Kreis Lehe. (II.)								Hauptfang: Aal, Stint, Goldbutte.
Transport Ortsch. 5 von (I.)	18	19	14	51	20	0,20	30	1 Fahrzeug mit Deck.
Rechtenfleth	3	—	—	3	6			
Ortschaften 6. Summa	21	19	14	54	26	10,02		

Reg.-Bez. Aurich.

1) Kreis Emden.								Hauptfang: Stör, Aal, Maifisch, Sardellen, Hering, Schellfisch, Stint, Schollen, Zunge, Schnepel, Granat.
Spiekeroog	—	—	2	2	2	62,0	0,03	
Neu-Harlingersiel	3	19	2	24	3	10,2	0,29	Fahrzeuge mit hohem Deck, 3 bis 9 Lasten Tragfähigkeit.
Norderney	199	186	—	385	64	60	1,1	Fischerschalupen.
Hintelermarsch	—	—	8	8	8	0,09	8,89	
Vestermarsch II	—	—	12	12	13	0,1	130	Schlickböte oder Aggeböte.
Pilsum	—	—	1	1	—	0,45	—	
Larrelt	8	8	—	16	8	0,02	400	1 Pünten, 8 Frauen.
Vorssum	6	1	6	13	2	0,24	8,3	
Larsum	—	—	1	1	1	0,03	33,3	
Oldersum	2	—	3	5	4	0,11	3,04	2 Bollschiffe von 4 Roggentragfähigkeit.
Borkum	8	16	—	24	8	14,7	0,54	Jachten.
Emden	180	—	—	180	9	200 ca.	0,045	Loggerschiffe der Emdener Actiengesellschaft.
Ortschaften 12. Summa	406	230	35	671	122	300 ca.	0,4	
2) Kreis Leer.								Hauptfang: Stör, Aal, Sardellen, Hering, Goldbutt, Granat, Karpfen, Hecht.
Leerorth	13	2	17	32	30	0,1	300	
Bemgum	1	2	2	5	2	0,05	40	
Critzum	—	—	1	1	1	0,05	20	1 Schalappe und 2 Bollschiffe.
Ditzum	5	6	2	13	11	10	1,06	Es wird mit Schlickschlitten s. g. Kreiern gefischt.
Jgum	—	—	10	10	—	1,85	—	Es wird mit Schlickschlitten s. g. Kreiern gefischt.
Ditzumer Hammerich	10	1	—	11	—	1,15	—	
Ortschaften 6. Summa	29	11	32	72	44	10	4,4	

Die Fischerei im Haff.
A. Curisches Haff.

Ortschaften	Eigentliche Fischer	Gehülfen derselben	Gelegenheitsfischer	Sa. der Fischer	Zahl der Fahrzeuge	Grösse des Fischereibezirks ☐Meilen	Zahl der Böte pr. ☐Meil.	Bemerkungen.
a) Kreis Memel.								
Schmelz	15	7	13	35	28			Die Fahrzeuge meistens kleine Kähne mit flachen Boden.
Starischken	—	—	30	30	15			
Schaeferei	—	—	40	40	20			
Piaulen	—	—	3	3	3			
Klischen	—	—	6	6	6			
Drawochnen	14	33	18	65	30			
Schwenzeln	5	13	15	33	20			
Ortschaften 7. Summa	34	53	125	212	122			
b) Kr. Heydekrug.								
Gaitzen	—	—	5	5	4			
Kischken	—	—	6	6	3			
Prätzmen	—	—	16	16	5			
Bliematzen	—	—	10	10	4			
Ogeln	—	—	16	16	8			
Kinten	—	—	14	14	9			
Szauken	—	—	9	9	7			
Paweln	—	—	7	7	7			
Suwöhnen	—	—	34	34	15			
Blaszen	—	—	5	5	3			
Stankischken	—	—	12	12	5			
Feilenhof	4	—	7	11	11			
Sturmen	—	—	12	12	12			
Windenburg	1	—	52	53	49			
Minge	55	—	10	65	65			
Pockallna	14	—	43	57	67			
Warruss	27	—	21	48	64			
Skirwith	21	—	23	44	50			
Ackminge	9	—	18	27	27			
Karkeln	4	4	31	39	35			
Ortschaften 20. Summa	135	4	352	491	448			
c) Kreis Heinrichswalde.								
Löckerort	—	—	5	5	5			
Loje	—	—	14	14	14			
Inse	58	116	28	202	86			
Tawe	30	60	78	168	108			
Ortschaften 4. Summa	88	176	125	389	213			
d) Kreis Labiau.								
Gilge	32*)	14	60	106	82			*) Darunter 4 Pächter der Stromfischerei.
Nemonien	20	35	40	95	40			
Agilla	2	2	13	17	12			
Pelszen	2	2	3	7	5			
Labaginen	4	4	44	52	26			
Neu-Rinderort	3	3	14	20	10			
Alt-Rinderort	8	8	8	24	12			
Kampken	8	3	7	18	9			
Ortschaften 8. Summa	79	71	189	339	196			
e) Kr. Königsberg.								
Willmanns	5	15	—	20	20			
Postnicken	14	14	8	36	18			
Steinort	10	34	—	44	17			
Conradsvitte	9	28	—	37	15			
Sand	4	5	3	12	7			
Schaaksvitte	11	11	7	29	18			
Stombeck	14	28	—	42	20			
Neufitt	2	5	3	10	8			
Rodahn	4	5	4	13	8			
Ortschaften 9. Summa	73	145	25	243	131			
f) Kr. Fischhausen.								
Sarkau (ex parte)	24	48	—	72	24			
Kunzen	—	—	2	2	2			
Rossitten (e. p.)	4	8	4	16	8			
Pilkoppen (e. p.)	14	28	—	42	20			
Ortschaften 4. Summa	42	84	6	132	54			

Ortschaften	Eigentliche Fischer	Gehülfen derselben	Gelegenheitsfischer	Sa. der Fischer	Zahl der Fahrzeuge	Grösse des Fischereibezirks ☐ Meilen	Zahl der Böte pr. ☐ Meil.	Bemerkungen
Kr. Memel Transp. Ortsch. 7	34	53	125	212	122			
Nidden (e. p.)	38	76	—	114	38			
Ortschaften 8. Summa	72	129	125	326	160			

Recapitulation.
Curisches Haff.
Regierungs-Bezirk Königsberg und Gumbinnen.

Kreis Memel Ortschaften 8	72	129	125	326	160			
,, Heydekrug ,, 20	135	4	352	491	448			
,, Heinrichswalde ,, 4	88	176	125	389	213			
,, Labiau ,, 8	79	71	189	339	196			
,, Königsberg ,, 9	73	145	25	243	131			
,, Fischhausen ,, 4	42	84	6	132	54			
Ortschaften 53. Summa	489	609	822	1920	1202	29,4		

B. Frisches Haff.

a) Kr. Fischhausen.								
Fischhausen	26	30	—	56	26			Fischt auch ⅛ Meile in See.
Peise	24	72	3	99	48			Fischt bei stillem Wetter auch ¼ Meile in der Ostsee.
Zimmerbude	30	74	5	115	41			
Gr. Heidekrug	39	107	18	164	35			
Kaporn	9	8	—	17	20			
Pokaiten	2	3	1	6	4			
Nautzwinkel	16	4	—	20	20			
Ortschaften 7. Summa	152	298	27	477	194			
b) Kr. Königsberg.								
Holstein	2	4	7	13	9			
Hafstrom	—	—	4	4	8			
Heide-Wundlaiken	—	—	7	7	14			Es sind diese Fahrzeuge theils Handkähne theils grössere Böte. (Jollen und Sieken.)
Heide-Waldburg	—	—	18	18	36			
Heide-Maulen	1	2	16	19	35			
Wangitt	1	1	3	5	10			
Ortschaften 6. Summa	4	7	55	66	112			
c) Kr. Heiligenbeil.								
Dümpelkrug	3	2	—	5	6			
Kl. Brandenburg	11	11	2	24	26			
Patersort	3	—	3	6	2			
Schölen	—	—	4	4	2			
Schölen zu Rippen	—	—	15	15	5			
Heide-Fedderau zu Pohren	1	—	3	4	8			
Fedderau	—	—	5	5	2			
Wolitta	7	—	8	15	15			
Kahlholz	12	—	38	50	10			
Balga	8	—	23	31	9			
Follendorf	—	—	12	12	4			
Rosenberg	55	—	—	55	20			
Poln. Bahnau	44	—	—	44	12			
Leysuhnen	—	—	32	32	8			
Alt-Passarge	31	78	—	109	41			¼ Meile in See.
Ortschaften 15. Summa	175	91	145	411	170			
d) Kr. Braunsberg.								
Neu-Passarge	—	—	40	40	35			
Frauenburg	39	5	5	49	24			
Ortschaften 2. Summa	39	5	45	89	59			
e) Kreis Elbing.								
Tolkemit	17	49	59	125	57			
Kadinen	1	—	—	1	1			
Succase	3	—	6	9	9			
Gr. Steinort	—	—	4	4	4			
Bollwerk	11	2	10	23	21			
Terranova	21	2	23	46	44			
Zeiersniederkampen	—	—	4	4	4			
Jungfer	9	15	45	69	57			
Neustädterwald	3	—	—	3	3			
Stobbendorf	20	12	19	51	51			

e) Kreis Elbing. (Fortsetzung.)

Ortschaften	Eigentliche Fischer	Gehülfen derselben	Gelegenheitsfischer	Sa. der Fischer	Zahl der Fahrzeuge	Grösse des Fischereibezirks ☐Meilen	Zahl der Böte pr. ☐Meil.	Bemerkungen.
Lakenwalde	—	—	1	1	1			
Tiegenhof	—	—	1	1	1			
Neuendorf	—	—	7	7	7			
Holm	1	1	—	2	1			
Elbing	21	2	4	27	23			
Kl. Hornkampe	4	—	17	21	21			
Grenzdorf	—	—	14	14	14			
Ortschaften 17. Summa	111	83	214	408	319			
f) Kreis Danzig.								
Stutthof (e. p.)	16	4	16	36	30			Die Böte sind im Seebezirk eingetragen.
Liep (e. p.)	—	—	26	26	—			
Vöglers (e. p.)	11	12	—	23	12			
Narmeln (e. p.)	13	12	—	25	14			
Ortschaften 4. Summa	50	28	42	110	56			

Recapitulation.
Frisches Haff.
Regierungs-Bezirk Königsberg und Danzig.

Kreis Fischhausen Ortsch. 7	152	298	27	477	194		
,, Königsberg ,, 6	4	7	55	66	112		
,, Heiligenbeil ,, 15	175	91	145	411	170		
,, Braunsberg ,, 2	39	5	45	89	59		
,, Elbing ,, 17	111	83	214	408	319		
,, Danzig ,, 4	50	28	42	110	56		
Summa Ortschaften 51	531	512	528	1561	910	15,6	

C. Pommersches Haff.

1) Kreis Usedom-Wollin.
a) Insel Wollin.

Zünz	8	—	2	10	8
Zirzlaff	18	—	3	21	28
Lüskow	—	—	2	2	2
Körtenthin	1	—	—	1	2
Jarmbow	2	—	8	10	10
Darsewitz	3	—	5	8	8
Wollin	82	76	28	186	82
Soldemin	1	—	4	5	5
Karzig	7	—	3	10	4
Kalkofen	11	—	3	14	14
Vietzig	5	—	19	24	19
Pritter	10	10	70	90	30
Werder	—	—	8	8	4
Klüss	—	—	20	20	10
Ortschaften 14. Summa	148	86	175	409	226

b) Insel Usedom.

Caseburg	37	—	21	58	29
Caminke	39	4	28	71	24
Garz	—	—	11	11	4
Newerow	—	—	5	5	1
Dargen	—	—	1	1	1
Cachlin	1	—	—	1	1
Pretenow	—	—	9	9	2
Gemmelin	—	—	23	23	5
Stolpe	—	—	12	12	4
Welzin	—	—	10	10	5
Paske	6	—	—	6	6
Usedom	9	—	12	21	13
Amtswick	1	—	1	2	1
Mönchow	2	—	8	10	4
Carnin	1	—	6	7	4
Gellenthin	—	—	12	12	2
Zecherin	—	—	12	12	6
Gnewentin	—	—	6	6	3
Suckow	—	—	2	2	1
Krienke	1	6	—	7	8
Ranckwitz	22	—	10	32	17
Quilitz	6	—	—	6	3

b) Insel Usedom. (Fortsetzung.)

Ortschaften	Ortschaften	Eigentliche Fischer	Gehülfen derselben	Gelegenheitsfischer	Sa. der Fischer	Zahl der Fahrzeuge	Grösse des Fischereibezirks □ Meilen	Zahl der Böte pr. □ Meil.	Bemerkungen.
Warthe		6	—	35	41	17			
Restow		—	—	10	10	4			
Grüssow		—	—	13	13	7			
Dewichow		—	—	3	3	3			
Balm		—	—	13	13	6			
Neppermin		—	—	15	15	7			
Neuendorf		2	2	1	5	1			
Lütow		1	—	6	7	7			
Neeberg		8	—	4	12	4			
Sanzin		—	—	4	4	2			
Mahlzow		—	—	11	11	5			
Zeckerin		—	—	17	17	10			
Ins. Used. Ortsch. 34. Sa.		142	12	321	475	217			
„ Woll. „ 14. „		148	86	175	409	226			
Kr.Us.-Woll. Ortsch. 48. Sa.		290	98	496	884	443			
2) Kreis Kammin.									
Kammin		18	1	6	25	8			
Hager		12	7	2	21	9			
Gaubitz		7	11	1	19	8			
Sager (e. p.)		22	37	—	59	22			Vor Schwinemünde, ausserdem v. S. 365.
Paulsdorf		10	2	31	43	12			
Schwantewitz		22	6	—	28	22			
Gr.- und Kl.-Stepenitz		11	11	1	23	12			
Ortschaften 8. Summa		102	75	41	218	93			
3) Kr.Uckermünde.									
Ziegenort		10	12	—	22	9			
Wahrlang		3	5	2	10	2			
Albrechtsdorf		1	6	1	8	3			
Neuwarp		35	23	115	173	68			
Altwarp		50	12	20	82	25			
Bellin		2	1	—	3	2			
Uckermünde (c. p.)		30	30	—	60	17			2 Fischer in der Ostsee s. o.
Grammbin		10	—	9	19	6			
Ortschaften 8. Summa		141	89	147	377	132			
4) Kreis Anclam.									
Anclam		11	16	7	34	37			11 Lieger-, 17 Polte-Böte zum Fischen und 9 Polte zum Transport.
Camp		5	5	12	22	27			
Fähre (e. p.)		7	7	2	16	16			
Ortschaften 3. Summa		23	28	21	72	80			

Recapitulation.
Pommersches Haff.

Kr. Used.-Woll. Ortsch. 48		290	98	496	884	443		
„ Kammin „ 8		102	75	41	218	93		
„ Uckermünde „ 8		141	89	147	377	132		
„ Anclam „ 3		23	28	21	72	80		
Summa Ortschaften 67		556	290	705	1551	748	11,8	
A. Curisch. Haff Ortsch. 53		489	609	822	1920	1202	29,4	40,9
B. Frisches Haff „ 51		531	512	528	1561	910	15,6	58,3
C. Pomm. Haff „ 67		556	290	705	1551	748	11,8	63,4
Summa Ortschaften 171		1576	1411	2055	5032	2860	56,8	50,3

Recapitulation.
1. Deutsche Ostseeküste.
Staat Preussen.
A. Provinz Preussen.
I. Regierungs-Bezirk Königsberg.

Memel	10	184	344	67	595	213	34,8	6,12
Fischhausen	26	138	200	121	459	139	31,3	4,44
Summa	36	322	544	188	1054	352	55,8	6,31

II. Regierungs-Bezirk Danzig.

Kreis	Ortschaften	Eigentliche Fischer	Gehülfen derselben	Gelegenheitsfischer	Sa. der Fischer	Zahl der Fahrzeuge	Grösse des Fischereibezirks ☐Meilen	Zahl der Böte pr. ☐Meil.	Bemerkungen
Danzig	23	355	237	259	851	452	28,9	15,9	
Neustadt	27	436	354	280	1070	418	23,0	18,3	
Summa	50	791	591	539	1921	870	42,0	20,3	
Totale	86	1113	1135	727	2975	1222	90,8	13,5	

B. Provinz Pommern.

I. Regierungs-Bezirk Köslin.

Lauenburg	5	40	29	97	166	45	18	2,5	
Stolp	12	134	74	191	399	91	12,8	7	
Schlawe	6	91	44	34	169	50	10,9	4,6	
Cammin (Fürstenthum)	12	153	184	92	429	184	34	5,4	
Summa	35	418	331	414	1163	370	72,1	5,14	

II. Regierungs-Bezirk Stettin.

Greiffenberg	5	76	17	24	117	39	19	2,1	
Kammin	4	66	75	—	141	51	13	3,9	
Usedom-Wollin	27	556	553	161	1270	442	63	7	
Uckermünde	2	3	1	—	4	3	0,55	5,3	
Anclam	1	1	1	—	2	1	0,55	1,8	
Summa	39	702	647	185	1534	536	67,2	7,98	

III. Regierungs-Bezirk Stralsund.

Greifswald	14	259	69	41	369	168	5,35	31,4	
Grimmen	7	42	9	11	62	26	7,93	3,3	
Rügen	97	584	192	507	1283	660	37,4	15	
Franzburg	25	274	146	75	495	228	13,3	17,2	
Summa	143	1159	416	634	2209	1082	46,1	23,5	
Preussen und Pommern	217	2279	1394	1233	4906	1988	171,0	11,6	

Staat Mecklenburg-Schwerin.

Totale	35	347	77	111	535	298	20,4	14,6

Staat Lübeck.

Totale	4	94	58	42	194	114	11,0	10,4

Staat Oldenburg.

	3	34	55	2	91	38	2,6	14,6

Staat Preussen.
C. Provinz Schleswig-Holstein.

Oldenburg	21	111	20	75	206	161	18,9	8,5	
Plön und Kiel	12	124	54	38	216	104	21,95	7,5	
Eckernförde	9	131	48	172	351	183	16,5	11,1	
Schleswig O.	2	98	96	4	198	67	1,55	43,2	
Flensburg	22	76	30	86	192	123	6,47	19	
Apenrade	17	72	31	69	172	115	5,5	20,9	
Sonderburg	35	96	11	139	246	190	9,95	19,1	Der Fischereibezirk von Holstein beträgt 30,9, von Schleswig 45,2 ☐Meilen.
Hadersleben	14	46	9	58	113	63	5,5	11,5	
Summa	132	754	299	641	1694	1066	67,0	15,9	

Fischerei der Ostsee deutschen Antheils.

Staat Preussen.									Auf ein Boot kommen Fischer
A. Preussen	86	1113	1135	727	2975	1222	90,8	13,5	2,45
B. Pommern	217	2279	1394	1233	4906	1988	171,0	11,6	2,47
C. Schleswig-Holstein	132	754	299	641	1694	1066	67,0	15,9	1,59
Seefischerei	435	4146	2828	2501	9575	4276	328	13	2,24
A. Curisches Haff	49	489	609	822	1920	1202	29,4	40,9	1,60
B. Frisches Haff	47	531	512	528	1571	910	15,6	58,3	1,73
C. Pommersches Haff	63	556	290	705	1551	748	11,8	63,4	2,07
Hafffischerei	159	1576	1411	2055	5042	2860	56,8	50,4	1,76
Preussische Ostseefischerei	594	5722	4239	4556	14617	7136	384,8	18,5	2,05

Fischerei der Ostsee deutschen Antheils. (Fortsetzung.)

Kreis	Ortschaften	Eigentliche Fischer	Gehülfen der selben	Gelegenheitsfischer	Sa. der Fischer	Zahl der Fahrzeuge	Grösse des Fischereibezirks ☐Meilen	Zahl der Böte pr.☐Meil.	Auf ein Boot kommen Fischer	Bemerkungen
Mecklenburg	35	347	77	111	535	298	20,4	14,6	2,40	
Lübeck	4	94	58	42	194	114	11,0	10,4	1,70	
Oldenburg	3	34	55	2	91	38	2,6	14,6	1,80	
Summe ohne Haffs	477	4621	3018	2656	10295	4726	345	13,7	2,20	
Summe mit Haffs	636	6197	4429	4711	15437	7586	401,8	18,9	2,04	

2. Deutsche Nordseeküste.
Staat Preussen.
A. Provinz Schleswig-Holstein.

Tondern	6	—	—	53	53	4	3,28		1,3	
Husum	5	4	—	27	31	12	2,16		5,5	
Eiderstedt	3	4	4	6	14	7	3,3		2,1	
Schleswig W.	2	—	—	20	20	10	0,1		100	
Norderditmarschen	6	—	—	33	33	17	2,34		7,3	
Süderditmarschen	17	3	—	48	51	2	2,23		0,9	
Steinburg	6	—	—	64	64	42	1,14		37	
Pinneberg	2	67	144	12	223	77	500		0,3	
Altona	1	1	1	—	2	1	1,83		0,55	
Summa	48	79	149	263	491	172				

B. Provinz Hannover.
I. Landrostei Stade.

Stadermarsch	2	1	1	6	8	8	—	—	
Neuhaus	1	8	—	—	8	4	—	—	
Otterndorf	2	—	—	9	9	4	—	—	
Lehe	6	21	19	14	54	26	10	2,6	
Summa	11	30	20	29	79	42			

II. Landrostei Aurich.

Emden	12	406	230	35	671	122	300	0,4
Leer	6	29	11	32	72	44	10	4,4
Summa	18	435	241	67	743	166	310	

| Summa Hannover | 29 | 465 | 261 | 96 | 822 | 208 | | |

Staat Hamburg.

	4	145	127	14	286	90	500	0,14

Staat Oldenburg.

Elsfleth und Brake	4	36	36	12	88	70	—	—
Oevelgönne und Varel	5	39	5	—	44	9	—	—
Jever	5	4	—	19	23	1	—	—
Summa	14	79	41	31	151	80	2,15	37

Staat Bremen.

	1	4	4	—	8	4	1	4

Englische Insel Helgoland.

	1	205	270	—	475	85	52	1,6

Fischerei der Nordsee.

Staat Preussen.									
A. Schleswig-Holstein	48	79	149	263	491	172	—	—	2,85
B. Hannover	29	465	261	96	822	208	—	—	3,95
Summa Preussen	77	544	450	359	1313	380	—	—	3,46
Staat Hamburg	4	145	127	14	286	90	—	—	3,18
Staat Oldenburg	14	79	41	31	151	80	—	—	1,76
Staat Bremen	1	4	4	—	8	4	—	—	2,0
Summa d. deutsch. Reichs	96	772	622	404	1758	554	500	1,1	3,37
Helgoland	1	205	270	—	475	85	—	—	5,59
Totale	97	977	892	404	2233	639	500	1,3	3,49

Fischerei der deutschen Küste

Provinz und Staat	Zahl der Orte	Fischer	Ge-hülfen	Ge-legen-heits-fischer	Summa der Fischer	Zahl der Fahr-zeuge	Zahl der Orte	Fischer	Ge-hülfen	Ge-legen-heits-fischer	Summa der Fischer	Zahl der Fahr-zeuge
	ohne die Haffe.						mit den Haffen.					
Staat Preussen.												
A. Preussen	86	1113	1135	727	2975	1222	182	2133	2256	2077	6466	3334
B. Pommern	217	2279	1394	1233	4906	1988	280	2835	1684	1938	6457	2736
C. Schleswig-Holst.	180	833	448	904	2185	1238	—	—	—	—	—	—
D. Hannover	29	465	261	96	822	208	—	—	—	—	—	—
Summe Preussen	512	4690	3238	2960	10888	4656	671	6266	4649	5015	15930	7506
Mecklenburg	35	347	77	111	535	298	—	—	—	—	—	—
Lübeck	4	94	58	42	194	114	—	—	—	—	—	—
Oldenburg	17	113	96	33	242	118	—	—	—	—	—	—
Hamburg	4	145	127	14	286	90	—	—	—	—	—	—
Bremen	1	4	4	—	8	4	—	—	—	—	—	—
Summe Deutschland	573	5393	3600[1]	3160	12153	5280	732	6969	5011	5215	17195	8130
Helgoland	1	205	270	—	475	85	—	—	—	—	—	—
Totale	574	5598	3870	3160	12628	5365	733	7174	5281	5215	17670	8215

[1]) Unter den Gehülfen und Gelegenheitsfischern 302 Frauen.

Schluss-Tabelle.

Ostsee.

Regierungsbezirk Provinz und Staat	Küsten-länge roh gemessen Meilen	\multicolumn{8}{c}{Es kommen auf die Meile Küste ohne Berücksichtigung der Haffs}
		Fischer-orte	gewerbs-mässige Fischer	Gehülfen derselben	Gelegenheits-fischer	Summa der Fischer	Böte	an befischter Fläche ☐ Meilen	befischte Fläche mal Böte. (Bedeutsamk.)
Bezirk Königsberg	27,5	1,31	11,7	19,8	6,8	38,3	12,8	2,03	26,0
Bezirk Danzig	19,5	2,56	40,6	30,3	27,6	98,5	44,6	2,19	97,7
Provinz Preussen	47	1,83	23,7	24,2	15,5	63,3	26,0	1,93	50,2
Bezirk Köslin	25	1,40	16,7	13,2	16,6	46,5	14,8	2,91	43,1
Bezirk Stettin	16,25	2,40	43,4	39,8	11,3	94,4	33,0	4,14	136,6
Bezirk Stralsund	29,5	4,85	39,3	14,1	21,5	74,9	36,7	1,56	57,3
Provinz Pommern	70,75	3,06	32,2	19,7	17,4	69,3	28,1	2,42	68,0
Staat Mecklenburg	17,75	1,97	19,5	4,3	6,3	30,1	16,8	1,15	19,3
Lübeck, Oldenburg, Prov. Schlesw.-Holst.	44,75	3,11	19,7	9,2	15,3	44,2	27,2	1,49	40,5
Obige vereint	62,5	2,79	19,7	7,8	12,7	40,2	24,3	1,38	33,5
Deutsches Reich	180,25	2,65	25,6	16,7	14,7	57,0	26,2	1,91	50

Nordsee.

Deutsches Reich, Helgoland	72	1,35	13,6	12,4	5,7	31,7	8,9	7	62,3

Vergleichung der Tabelle mit anderweiten Mittheilungen und Ergänzung des Details.

In dem Bericht*) des Herrn Ministerialdirektors MARCARD an Se. Excellenz den Herrn Minister für die landwirthschaftlichen Angelegenheiten finden sich eine Reihe statistischer Angaben, welche zu einer Kritik der vorliegenden Tabelle benutzt werden können. Jener Bericht war das Resultat einer Reise durch den grösseren Theil des preussischen Fischereigebietes und da dieselbe im officiellen Auftrage geschah, konnte von allen Ermittelungen, welche die Regierungen veranlasst hatten oder kannten, Gebrauch gemacht werden. Es war nicht die Aufgabe des Berichts eine erschöpfende Statistik zu geben, jedoch ist das Material, welches er bringt, für uns höchst werthvoll, denn es sichert einestheils die Angaben der Tabelle, anderentheils zeigt es, wo Unsicherheiten und Zweifel zu hegen sind.

Der Bericht giebt Kreis Memel 10 Orte, die Tabelle ebenso, jedoch wird von dem Bericht ein Ort Amtsvitte am Ausfluss des Haffs auf Neunaugen fischend angeführt, den die Tabelle nicht angiebt, derselbe scheint mit in die Fischerei der Stadt Memel aufgenommen zu sein.

Hier muss noch bemerkt werden, dass die Angaben darüber, ob es wirklich Immersatt oder ob es Nimmersatt sei, welches dicht an der russischen Grenze fischt, verschieden lauteten.

Für das frische Haff verzeichnet

der Bericht	die Tabelle
Ostpreussen 792 Fischer mit 360 Gehülfen,	370 Fischer, 401 Gehülfen, 272 Gelegenheitsfischer,
Westpreussen 583 „ „ 149 „	161 „ 111 „ 256 „
1375 Fischer mit 509 Gehülfen,	531 Fischer, 512 Gehülfen, 528 Gelegenheitsfischer,

es hat dennoch d. B. die Gelegenheitsfischer zu den eigentlichen Fischern gezählt und wahrscheinlich deshalb eine höhere Zahl erhalten, nämlich 1884 gegen 1561, weil die Gelegenheitsfischer nicht alle stark genug fischen, um den Verdienst eines Monats zu gewinnen. Auf eine erhebliche Abnahme der Fischer seit 1870 zu schliessen hat man keinen Grund und die Zahl der Gehülfen stimmt wenigstens in der Endsumme so genau, dass es nicht gerechtfertigt sein dürfte, der Aufnahme zu misstrauen.

Für den Danziger Regierungsbezirk giebt d. B. 917 Fischer, d. T. 791 Fischer mit 591 Gehülfen, im Ganzen 1921 fischende Personen. Für das Domanialamt Zoppot sind die Fischerorte namhaft gemacht und 110 Fischer gezählt, die T. enthält dieselben Orte und einen mehr, nemlich Koliebke mit 5 Fischern. Der B. giebt 110 Fischer, wir zählen zwar nur 90 eigentliche Fischer, jedoch einer der von dem B. genannten Orte, Rewa, hat gar keine ordentlichen Fischer, aber 32. Gelegenheitsfischer angegeben; da nun die Summe der fischenden Personen 278 Fischer mit 109 Böten beträgt, so scheint auch hier die Zählung eine grössere Anzahl Fischer, als man bisher kannte, zu Tage gefördert zu haben.

Für die Fischerei dieses Kreises ist hier die folgende Bemerkung von Hrn. MARCARD zu citiren:

»Anlangend die räumliche Ausdehnung der Seefischerei, so beschränkt sie sich im Wesentlichen auf die Danziger Bucht, und ungern entfernt sich der Fischer weiter als auf eine halbe Meile von der Küste. Man benutzt offene, flachgehende und mangelhaft gebaute Boote von 15—20 Fuss Länge; nicht alle führen Segel, und wenn dieselben vorhanden sind, so sind sie auf unvollkommene Art getakelt, und die Schiffer sind selten im Stande, gegen den Wind aufzukreuzen.«

Die für den Kreis Lauenburg angegebenen Orte des B. stimmen mit der T.

Im Kreis Stolp sind Genossenschaften gebildet, die der B. wie folgt aufzählt:

	Zahl der Genossenschaften.	Theilnehmer.	Böte.
Rowe	4	40	4
Gr. Garde	6	60	6
Holzkatten	2	20	2
Fuchsberg	2	20	2
Klucken	3	30	3
Rumbke	1	10	1
Dazu Schmolsin	—	1	1
	18	181	19.

Die Klucken stehen in der T. unter Zemminer Klucken und Selesen, sie giebt 4 neue Orte, aber Rumbke fehlt! Meine Karte giebt ein Rumke dicht bei Leba, welches wohl mit jenem identisch ist, es dürfte also dieser Ort in der Aufnahme vergessen sein. Die Zahlen der T. sind weit höher wie die des B.

Vom Schlawer Kreis giebt der B. 5 Orte an, die T. 6, der B. giebt jedoch einen Ort Seeshöft, den die T. nicht hat und den ich auf der Karte nicht finde.

---

*) Darstellung der preussischen Seefischerei und ihre jetzige Lage. Berlin 1870.

Die Orte des Fürstenthumer Kreises sind in B. und T. identisch.

Für Greifenberg nennt der B. einen Ort Fischerkaten, (ein Name, der sonst nicht für Ortsbezeichnungen üblich ist, da er überall an der Küste für einzeln stehende Fischerhäuschen gebraucht wird) ich finde den Ort nicht auf der Karte, die T. giebt neu einen Ort Pustchow.

Für Kammin kommt in T. Sager hinzu.

Für Usedom-Wollin finden sich übereinstimmend 27 Orte angegeben, aber der B. giebt 512 eigentliche und 119 Gelegenheitsfischer gegen 556 u. 161 im Ganzen 1070 Fischer d. T.

Für das Pommersche Haff hatte man, nach dem B., veranschlagt 1500 Fischer mit 400 Gehülfen und 1400 Gelegenheitsfischer in Summa 3300 und 2860 Böte, die T. ergiebt 1551 fischende Personen und 748 Fahrzeuge, also nahe $1/2$ resp. $1/4$ der Zahlen des Berichts. Die Fischer aus Usedom-Wollin waren in die Fragebögen ohne Unterschied ob sie im Haff oder an der Küste fischten aufgenommen, wie ich weiss, weil eine Reihe von Dörfern zugleich auf den von dort kommenden Fragebögen sich eingetragen hatte. Unter diesen Umständen muss angenommen werden, dass dort die Eintragungen genau geschehen sind, denn die Orte an der Küste stimmen wie schon gesagt genau mit der im Bericht angegebenen Zahl. Der ganze Fehler, wenn ein solcher vorläge, fiele folglich auf den noch übrigen Theil der Küste des Haffs. Nach den Angaben des B. würden auf die Quadratmeile 242 Böte kommen, sie würden also etwa so dicht liegen, wie im Breitling bei Rostock Karte VIII. Demnach ständc solche Dichte der Befischung nicht ohne Beispiel da, wäre aber doch für ein so grosses Gewässer auffallend.

Die Frage, welche der beiden Angaben die richtigere sei, muss offen bleiben, aber ich möchte doch die Vermuthung aussprechen, dass die höhere Zahl auf Kosten der Raubfischer entstanden sei, über welche unter anderem im Bericht der Hrn. Dr. WITTMACK[*]) sehr geklagt wird. Diese Leute gehören wohl nicht in die Tabelle, da der Reiz einer so verbreiteten Raubfischerei an einem so grossen Gewässer doch wohl mehr im Sport als im Gewinn zu suchen sein dürfte.

Ueber den Stralsunder Bezirk sagt der B. Folgendes:

Die Fischerei wird in Neuvorpommern vielfach als Nebenbeschäftigung von Ackerwirthen, Schiffern und Tagelöhnern betrieben, eine ungefähre Ueberschlagung derjenigen Einwohner aber, welche der Seefischerei als Haupterwerbszweig obliegen, ergiebt für den

    Kreis Rügen etwa . . . 650
    » Franzburg » 170
    » Greifswald » . . . 250
    » Grimmen » . . . 34
      Summa 1104

Die T. ergiebt:

    Fischer und Gehülfen in:
    Kreis Rügen 776
    » Franzburg 412
    » Greifswald 328
    » Grimmen 51
      Summa 1567

Hier ist jedoch die Fischerei in den Bodden mitgerechnet, was im Bericht nicht der Fall gewesen zu sein scheint.

In Schleswig-Holstein wurden incl. der Landseefischer gezählt 1035 selbstständige Fischer mit 1044 Fahrzeugen die T. giebt 1281 Fischer und Gehülfen mit 904 Gelegenheitsfischern sowie 1238 Fahrzeuge. Es ist zu bemerken, dass die Austernfischerei in der Tabelle nicht mit enthalten ist.

     Für die Nordsee giebt:

	der B.		die T.
Blankenese	60 Ewer.		75 Ewer.
Finkenwerder	79 »		90 »
Norderney	67 »		64 »
Borkum	8 »		7 »
Neu Harlingersiel	3 »		3 »
Spikeroog	1 »		2 »
Carolinensiel	1 »		nicht genannt.

Bei letzteren Angaben wird es sich um die wirkliche Zu- oder Abnahme in den letzten Jahren handeln.

---

[*]) l. c. S. 149.

Ueber die Fischerei in Ostfriesland haben wir genauere Nachrichten von Hrn. Dr. METZGER[1]). Es werden am Dollart genannt:

Ditzummer Verlaat mit 8 Haushaltungen
Pogum u. Dykterhusen » 7 »
Ditzum » 10 »
Kl. u. Gr. Borsum » 8 »
Larrelt » 7 »

Es fehlt mir Dykterhusen, aber bei Pogum allein sind 10 Fischer angegeben. Ditz. Verlaat möchte wohl mit D. Hammerich der T. identisch sein. Im Uebrigen finden sich in d. Tabelle mehr Orte angeführt[2]). Leider reichen meine Karten zu genauerem Vergleich nicht aus. Hr. METZGER giebt an, dass zwischen Finken-polder und Utlandshörn von 22 Familien gefischt wurde, beide Punkte finde ich nicht, zwischen Accummer- und Carolinensiel fischen 10 Haushaltungen, von der Westküste der Krummhörn, die ich nicht finde, 5 Familien, ausserdem werden 3 in Norden ansässige Fischer genannt, welche die Tabelle nicht hat. Im Ganzen ist die Tabelle über die Küstenfischerei der Landrostei Aurich schwerlich weniger vollständig, wie die Angaben von Hrn. Dr. METZGER.

Aus obigen Vergleichungen resultirt, dass zwar in der vorgelegten Tabelle noch manche Lücken sich finden werden, dass aber im Ganzen die Aufnahme eine genügend umfassende gewesen ist, um sie als gute Basis für theoretische Schlüsse und praktischen Gebrauch zu benutzen. Ausserdem dürfte nunmehr in jedem einzelnen Kreis, wo es in späteren Jahren von Wichtigkeit sein sollte, verhältnissmässig leicht mit Hülfe der vorliegenden Karten neue Zählungen ausgeführt werden können. Correctionen und Erweiterungen der gemachten Angaben werden übrigens von den Lesern erbeten und mit Dank angenommen.

Einige Ergebnisse der Aufnahme.

Schon oft ist der ungünstige Zustand unserer Fischerei gegenüber derjenigen anderer Länder hervorgehoben worden, daher kann hier über dies Verhältniss kurz hinweggegangen werden.

Während die deutsche Küste 17195 Fischer mit 8130 überwiegend sehr unbedeutenden Fahrzeugen zählt, schlägt der B. nach SCHMARDA's Ermittelung die englischen Fischer auf 134000 Mann mit 36000 Fahrzeugen an. Nach dem Reisebericht des Hrn. TOLLE[3]) sind in Frankreich 73757 Mann mit 16819 Fahrzeugen gezählt, davon an der atlantischen Küste 63381 Mann mit 12927 Fahrzeugen, in Italien[4]) schätzt man 60000 Fischer und 18000 Fahrzeuge, in Oesterreich[5]) 7196 Mann mit 1852 Fahrzeugen.

Es kommen durchschn. 3,72 Mann auf ein Fahrzeug, in Frankreich atlantische Küste 4,90 in Italien 3,3, Oesterreich 3,7, in Deutschland, Ostsee 2,04, daher können die Fahrzeuge nicht mehr als Maassstab der Vergleichung dienen, sondern wir müssen uns an die Zahl der Menschen halten. Es findet sich, dass England 7,8, Frankreich 4,2, Italien 3,5, Oesterreich 0,5 mal so viel Fischer hat wie Deutschland, die Haffe mitgerechnet.

Die Küstenlängen mit einem Zirkel, dessen Spitzen 5 Meilen fassen, gemessen, giebt für Deutschland 165, für Grossbritannien 550, für Frankreich atlantische Küste 190, für Italien ohne Sardinien 440, Oesterreich 90 geographische Meilen. Bringt man diese Länge in Rechnung, so kommen auf gleiche Küstenerstreckung auf einen deutschen Fischer 2,3 englische, 3,2 französische, (atlant. Küste) 1,3 italienische, 0,7 österreichische, oder ohne[6]) die Hafffischer respective 3,2, 4,4, 1,8 und 1 Fischer der genannten Länder.

Weit ungünstiger noch wird sich diese Lage jenen Staaten gegenüber gestalten, wenn die Kapitalien, welche in der Fischerei angelegt sind, veranschlagt werden könnten, hier liegt bekanntlich die hauptsächlichste Schwäche der deutschen Fischerei.

Betrachten wir den Fischereibetrieb in der Ostsee genauer. Die Zahl der Fischer, welche auf ein Boot kommen beträgt im Mittel 2,04, in den Haffen 1,76, an der Küste 2,20. Am westlichen Theile der Ostseeküste ist sie geringer, Schleswig-Holstein 1,59, Mecklenburg 1,8 im östlichen höher, Pommern 2,47, Preussen 2,45.

---

[1]) Die maritime Production der Ostfriesischen Wattküste. Circular d. d. Fischereivereins 1872 Nr. 1 S. 29.

[2]) Im Dollart wird für die Fischzäune Schilfrohr verwandt. v. BAER: Materialien zu einer Geschichte des Fischfangs Bulletin de l'Academie des sciences de St. Pétersburg. Tom XI S. 258 sagt: dass Plinius behauptet, ihre (der Chauken, von der Rhein-Mündung weiter nach Norden angesiedelt) Netze wären aus ulva et palustri junco geflochten, muss auf einem Missverständniss beruhen . . . eher sei an eine Art Wehra zu denken. Die Angaben von Hrn. METZGER bilden also ein interessantes Commentar zu jener Stelle des Plinius (Hist. naturalis XVI.)

[3]) Bericht an Se. Excellenz d. Herrn Minister f. d. landwirthschaftlichen Angelegenheiten. Die Austernzucht und Seefischerei in Frankreich und England 1871, den Circularen der deutschen Fischereigesellschaft beigelegt.

[4]) Gesetz-Entwurf, vorgelegt vom Minister für Ackerbau, Gewerbe und Handel (Castagnola). Circular d. d. Fischereigesellschaft 1871, No. 6.

[5]) KRAFFT, die neuesten Erhebungen über die Zustände der Fischerei. Wien 1874.

[6]) WIDEGREN, die Fischerei an den Küsten von Gothland, Kalmar und Ostgothland. Circular d. d. Fischereigesellschaft 1871, No. 5, S. 28, zählt an der Ostküste Schwedens von Bergwara bis Slätbacken 915 Fischer, die Strecke beträgt 30 Meilen; es kommt hier also auf einen deutschen Fischer 0,3 respect. 0,4 schwedische, jedoch dürfte eine Zählung nach Art der unseren wohl mehr Fischer ergeben.

Da in England und Frankreich die Mannschaften der Fahrzeuge zahlreich sind und in den östlichen Provinzen die Mannschaften zahlreicher sich finden, könnte man schliessen wollen, dass in den östlichen Provinzen seetüchtigere Fahrzeuge seien, wie in den westlichen; jedoch dies ist im Allgemeinen nicht der Fall, sondern die Erhöhung der Verhältnisszahl resultirt theils aus der etwas grösseren Anzahl von Gelegenheitsfischern im östlichen, theils aus einzelnen Fällen, wo der Fischer zwei Böte besitzt im westlichen Theil. Immerhin sind im östlichen Theil die Gehülfen relativ zahlreicher wie im westlichen Theil.

Vergleicht man die Küstenlänge der einzelnen Theile mit dem befischten Gebiet, so kommt man zu dem Resultat, vergl. d. Schlusstabelle, dass Mecklenburg mit nur 1,15 Quadratmeile und Schleswig-Holstein, inclusive Lübeck und Ostseeküste von Oldenburg, mit 1,49 Quadratmeilen pr. Meile am wenigsten ausgedehnt fischen. Das Mittel für die ganze Küste ist nämlich 1,91 Quadratmeile, und Pommern hat 2,42 (mit Köslin 2,91, Stettin 4,14, Stralsund 1,56 Quadratmeilen), Preussen 1,93 (mit Königsberg 2,03, Danzig 2,19 Quadratmeilen), pr. Meile. Es liegen diesen Angaben genügend grosse Zahlenreihen zu Grunde, um wenigstens noch die erste Decimale für ziemlich zutreffend halten zu können. Wenn man die Karten ansieht, wird man bemerken, dass in der That die Böte am westlichen Theil sich sehr in den Häfen und Rheden zurückhalten. Die grosse Zahl für Stettin entsteht namentlich durch die Ausdehnung der Aalbecker Fischerei; ich bemerke daher, dass Erkundigungen, welche ich persönlich in Lohme auf Rügen einzog, ergaben, dass die dortigen Fischer wohl wussten, dass die Aalbecker bis zu ihnen hinauf fischen.

Wenn wir die Anzahl der Böte per Meile mit der pro Meile befischten Fläche multipliciren, erhalten wir einen numerischen Ausdruck für die Bedeutsamkeit der Fischerei pro Meile Küste. Vergl. oben S. 351. Es ergiebt sich die mittlere Bedeutsamkeit = 1 gesetzt: Königsberg 0,5, Danzig 1,95, Köslin 0,86, Stettin 2,73, Stralsund 1,15, Mecklenburg 0,39, Schleswig-Holstein mit Lübeck und Oldenburg 0,81. Provinz Preussen 1, Pommern 1,36, Mecklenburg, Schleswig-Holstein u. s. w. 0,67. Der Fischereibetrieb in Mecklenburg ist also am unbedeutendsten, dann folgt Königsberg, Schleswig-Holstein, Köslin, noch dazu gestaltet sich für die westlichen Theile das Verhältniss in Folge einer grösseren Anzahl der Fahrzeuge günstiger. Auf die Summe der Fischer berechnet, haben wir Mecklenburg 0,23, Schleswig-Holstein 0,61, Königsberg 0,72, Stralsund 1,07, Köslin 1,24, Danzig 1,99, Stettin 3,58, als Ausdruck der relativen Bedeutung des Betriebes, wobei als mittlere Grösse die Zahl 1,00 zu nehmen ist.

Die Dichte der Böte in den Haffen ist fast viermal so gross wie die an der Küste, sie steigt von Osten nach Westen. Da keine Bezirke gemacht sind, ist eine Vergleichung dieser Dichte mit der Menge der Böte an einzelnen Küstenabschnitten nicht statthaft.

An der Küste ist das Mittel der Dichte 13,7. Dies Mittel wird an dem westlichen Theil ein wenig überschritten, östlich bleibt die Dichte etwas zurück, namentlich in Pommern mit 11,6 Böten per Quadratmeile. Dass Preussen günstiger steht, beruht namentlich auf der Dichte der Befischung im Danziger Bezirk (20,3).

Eine Betrachtung der Karten (welche zusammengefügt werden können) ergiebt, dass in den grossen Buchten der Fischereibetrieb intensiv und entwickelt, an den in die Ostsee vorspringenden Theilen dagegen schwach und wenig extensiv ist. Hier wird er auch, wie die Tabelle ergiebt, vorzugsweise von Gelegenheitsfischern betrieben.

Als Buchten nenne ich die der Kurischen Neerung vorliegende Bucht, die Danziger Bucht, die Einbuchtung zwischen Colberg und Rügen, zwischen Fehmarn und Alsen und die lübische Bucht. Die Dichte der Befischung nimmt zu in den kleineren Abzweigungen dieser Buchten und den Bodden, ist jedoch offenbar modificirt durch locale (politisch-geographische) Ursachen. Als ein mehr natürliches Verhalten ist die Beobachtung zu betrachten, dass die Fischerei um isolirt ins Meer vorgeschobene kleine Landzungen oder diesen ähnliche kleine Inseln herum, stark zunimmt. So bei Poel, bei Maasholm an der Schlei, Eckensund in der Flensburger Bucht, Kekenis auf Alsen, vielleicht gehört hierher auch Hiddensöe bei Rügen.

Dagegen findet man den Fischereibetrieb an der Küste von Samland schwach, ebenso, wenn gleich weniger ausgesprochen, im östlichen Theil des Bezirks Köslin, schwach im Gebiet an der Spitze von Rügen, am Dars, an der Insel Fehmarn und den anliegenden Theilen und an der Ostküste und Südspitze von Alsen.

Der Einfluss der grossen Städte und Eisenbahnen erscheint nicht so lokalisirt und scharf begrenzt, um auf den Karten deutlich zu werden.

Wie erklären sich die Verschiedenheiten des Fischereibetriebes?

Wenn in dem Nachfolgenden versucht wird, den Gründen für das Verhalten des Fischereibetriebes näher zu treten, muss vorausgeschickt werden, dass selbstverständlich die gegebenen Erklärungen nirgends als erschöpfende betrachtet werden dürfen.

Dass die deutsche Fischerei in der Nordsee recht unbedeutend ist, war schon so oft Gegenstand der Erörterung, dass hier nicht darauf zurückzukommen ist.

Wenn neuerdings in Emden eine zeitgemässe Hochseefischerei in's Leben getreten ist, so mag für die Entwicklung an diesem Orte neben der Tradition die holländische Nachbarschaft entscheidend gewesen sein. Es knüpfen sich an dies Unternehmen für Deutschland sehr ernste Interessen.

Dass von Blankenese und Finkenwerder aus Hochseefischerei betrieben wird, ist theilweise auch durch traditionelles Herkommen erklärlich, Orte wie Glückstadt, Brunsbüttel, Cuxhafen geben garkeine Ewerfischer an, und doch ist die Lage dieser Orte mindestens nicht ungünstiger, wie die der obengenannten. Brauchbare Häfen und Tradition sind also, wie es scheint, für die Nordseefischerei maassgebend.

Die Fischerei auf Helgoland und Norderney ist nicht so bedeutend, wie sie nach der Zahl der Fischer erscheint, im Sommer nehmen die Badegäste die Fischer für ihre Zwecke in Anspruch, und gewähren ihnen einen reichlicheren und leichteren Verdienst.

Die Fischerei an der ostfriesischen Küste, ist von Hrn. Dr. METZGER so eingehend geschildert, dass ich auf dessen Bericht verweisen kann.

Leider gilt von der Schleswig-Holsteinischen Westküste noch heute Dasjenige, was der B. des Hrn. MARCARD darüber sagt: »Die Fischerei ist unbedeutend. Die Gründe dieser Erscheinung liegen theils in der Küstenbildung, in dem Mangel an gut gelegenen und gesicherten Häfen und Landungsplätzen, theils in der wenig betriebsamen Eigenart der Bevölkerung; denn auch da, wo gute Häfen vorhanden sind, wie in Tönning und Husum, fehlt es an unternehmungslustigen Fischern. Nicht minder ist zu beachten, dass die dünne Bevölkerung der Westküste und die Höhe des Arbeitslohnes in diesen Marschdistrikten wenig günstig ist für ein so mühsames und gefährliches Gewerbe.«

»Ungünstiger noch (wie in Büsum wo energische Versuche einer ausgedehnteren Küstenfischerei gemacht werden) liegen die Aussichten für die Seefischerei auf den Westseeinseln, deren ganze männliche Bevölkerung dem Seemannsberufe obliegt und Jahre lang auf Reisen in fremden Welttheilen von der Heimath entfernt ist. Nach der Rückkehr von solchen Reisen, auf denen der Insulaner seine besten Kräfte und Jahre verbraucht und sich in der Regel so viel erwirbt, dass er ein Recht zu haben glaubt, sich einem behaglichen Müssiggange hingeben zu dürfen, denkt derselbe nicht mehr daran, sich zum Fischer zu degradiren.«

Es kommt hinzu, dass bei kleinem Betrieb der Absatz nicht genügend gesichert werden kann, um die Chancen der grossen Fänge zu verwerthen und dadurch die Fischerei genügend lohnend zu machen, zum grösseren Betriebe fehlen aber sowohl die Kapitalien als auch die Anhäufung der Fischer an bestimmten Orten.

Wenden wir uns zur Ostsee.

Als Gründe dafür, dass die Haffe durchgehends besser befischt werden als die Küste, kann angeführt werden: 1) Die Fischerei ist dort weniger gefahrvoll, denn es wird weder der Wellenschlag und die Brandung so erheblich wie in der Ostsee, noch haben die Fischer zu befürchten, auf das hohe Meer hinausgetrieben zu werden. 2) Die Küstenerstreckung ist sehr günstig, weil die Haffe rings vom Lande umschlossen werden. 3) Die Fische sind im Allgemeinen werthvoller. 4) Die Communication in's Innere des Landes per Boot ist erleichtert. 5) Durch die regelmässigen Wanderungen gewisser Meeresfische (Lachs, Neunaugen u. s. w.) in's Haff, wird die Fischbevölkerung dichter. In dieser Beziehung spielen die Einmündungen des Niemen, der Weichsel und der Oder eine wichtige Rolle. Ob dagegen eine bessere Ernährung der Fische durch Zufuhr aus diesen Strömen zur Erklärung zu Hülfe zu nehmen ist, scheint mir zweifelhaft, namentlich im Hinblick darauf, dass auf Rügen und dem Dars die Bodden eine ähnliche Dichte der Befischung zeigen, und dass die Weichsel doch nur theilweise in das frische Haff mündet.

Die grössere Dichte der Befischung im Pommerschen Haff findet eine Erklärung darin, dass nicht nur Neerungen, sondern gut bevölkerte Inseln ihm vorliegen und die Eisenbahn schon seit langer Zeit für die Hafffischerei ein gutes Transportmittel abgab. Die Fischerei des kurischen Haffs dürfte wohl schon unter der langen Winterkälte leiden.

Für die Küstenfischerei wird es nöthig, die Frage zur Erörterung zu bringen, ob etwa ein Unterschied in der Tüchtigkeit der Küstenbevölkerung für die Ausübung des Fischereibetriebes in Betracht komme. Für Mecklenburg und Schleswig-Holstein ist die Gelegenheit zur Seefahrt von alten Zeiten her mindestens ebenso ausreichend geboten worden, wie für die östlichen Küstentheile, in ungenügender Seebefahrenheit darf also eine Erklärung der geringeren Fischerei im Westen nicht gesucht werden. Sowohl die Häfen wie die Fahrzeuge sind im Westen besser.

Für Schleswig-Holstein kommt hinzu, dass es seiner Zeit mit den Lasten des Dienstes in der dänischen Kriegsmarine doch auch die Vortheile für seine Küste gewann, welche dem Gewinn grösserer Tüchtigkeit und grösseren Selbstgefühls der kriegsgeschulten Bevölkerung liegen, weiter, dass bis Kiel schon über 25 Jahre die Eisenbahnverbindung bestand. Unzweifelhaft wird selbst noch an der ostpreussischen Küste brav in's Meer hinaus gefischt, aber es scheint mir nach dem Gesagten doch nicht, dass man die stärkere Extensität der Fischerei in den östlichen Theilen auf Rechnung eines kühneren und energischeren Sinnes der betreffenden Küstenbevölkerung setzen könne.

Man könnte geltend machen, dass in den westlichen Theilen das Land an der Küste besser, und daher die Landwirthschaft relativ lohnender sei. Diesen Einwurf halte ich jedoch nicht für richtig, denn die eigentliche Fischerbevölkerung, und dahin gehören neben den Gehülfen gewiss noch viele Gelegenheitsfischer, mischt sich recht wenig mit der anderen Bevölkerung, sondern bleibt ganz auf den Fischereibetrieb (eventuell Fährdienst,

Flottmachen von Fahrzeugen u. s. w.) angewiesen; so ist es meines Wissens in Schleswig, Eckernförde, Neustadt, auch in Warnemünde, und war es bis vor Kurzem noch bei Kiel. In Rügen scheint die Landwirthschaft häufiger Nebenerwerb zu sein, dagegen wohl nicht auf Hiddensöe, sicher nicht auf den Neerungen und auf Hela. Wenn demnach, könnte man sagen, die Fischerei traditionell in den Familien bleibt, warum ist im Westen die Fischerei weniger bedeutend? ist hier das Meer ärmer, vielleicht erschöpft? Die Bedeutsamkeit des Fischereibetriebs in der Ostsee hängt jedoch nicht allein ab von der Fruchtbarkeit des Meeres, der Güte der Fahrzeuge, der Tüchtigkeit der Bevölkerung, sondern auch von deren Dichte, von den Kosten ihres Unterhalts und ihrer damit verbundenen Lebensweise, sowie von der Nachfrage nach Fischen.

Die Dichte der Bevölkerung ist im Westen ohne Zweifel ausreichend, im Osten scheint im Kreis Fischhausen die Fischerei durch zu geringe Bevölkerung der Küste zu leiden.

Dagegen sind die Kosten des Unterhalts wohl sehr maassgebend für die Bedeutsamkeit der Fischerei im Westen der Ostsee. Der Bericht des Hrn. MARCARD ergiebt S. 44 für die Halbinsel Hela einen Verdienst von 252 ℳ., für das Amt Zoppot 672 ℳ. pr. Mann; da es sich um Bruttoerträge handelt sind dies Summen, mit denen die Fischer im Westen kaum auskommen werden.

Nach den von Hrn. WITTMACK[1]) gegebenen Daten ist ein durchschnittlicher Unterschied von Erheblichkeit in den Preisen der Fische zwischen Westen und Osten der Ostsee nicht zu erkennen; es wird also der Fischer im Westen nicht nur mehr fangen müssen und weniger Zeit auf der Hin- und Rückfahrt von den Fischplätzen verlieren dürfen, um zu leben, sondern er wird auch eine Concurrenz weniger leicht ertragen können. Der Auszug aus dem Bericht d. k. Regierung zu Schleswig für 1871 enthält für den Mangel an Nachfrage[2]) mehrfache Belege.

Dass der westliche Theil der Ostsee weniger fischreich sein sollte, wie der östliche, ist kaum anzunehmen, denn im Osten gedeihen die Salzwasserfische weniger gut und soweit sich übersehen lässt, nimmt die Fischerei nach der Nordsee hin continuirlich zu. Aber auch unsere Karten deuten davon nichts an, denn wenn dem Fischereibetrieb durch Mangel an Fischen ein Halt geboten würde, müsste die Fischerei an den Küsten weit gleichmässiger betrieben werden, als dies der Fall ist. Die Dichte der Befischung im Breitling bei Rostock, vor der Mündung der Schlei, in der Flensburger Föhrde, gegenüber den gar nicht oder sehr schwach befischten Stellen an der mecklenburgischen und schleswigschen Küste sprechen deutlich gegen obige Vermuthung.

Dass Mecklenburg besonders schwach fischt, dürfte hauptsächlich an mangelhafter Nachfrage nach Fischen und ungenügender Entwicklung des Handelsverkehrs liegen, denn das Bahnnetz ist dort erst spät fertig geworden und liegt der Fischerei wenig günstig, ich kann im Uebrigen die Verhältnisse nicht beurtheilen.

Für die dichtere Befischung der grossen Einbuchtungen, gegenüber den Vorbuchtungen in die Ostsee, lassen sich einige Gründe anführen.

Die Küste der Buchten bildet im Allgemeinen immer die Peripherie eines Kreisabschnittes, von der aus nach dem Centrum zu die Fischerei convergirend vordringt, die Vorbuchtungen bilden dagegen ein Centrum oder den Durchmesser eines Kreisabschnittes, vom dem aus divergirend die Fischerei ausstrahlt.

An die Buchten knüpft sich die stärkere Nachfrage nach Fischen; in ihrer Nähe, gewöhnlich an den letzten Ausläufern der Bucht, liegen die grösseren Städte; Königsberg, Danzig, Stettin, Wismar, Lübeck, Kiel, Flensburg, hier münden die Verkehrsadern.

Ob eine grössere Dichte der Fische in der Bucht sich findet, ist zweifelhaft. Man kann allerdings sagen: wenn ein Zug von Fischen gegen eine Bucht trifft, so wird er nach dem Sack derselben zu sich dichter zusammenschaaren, aber ob das dabei angenommene mechanische Vorwärtsdringen der Fische in Wirklichkeit stattfindet, ist zweifelhaft. Man kann ferner erkennen, dass in der Danziger Bucht durch die Halbinsel Hela eine Art Reuse gebildet wird, die den Fischen das Verlassen der Bucht erschwert, dass in der pommerschen Bucht der Greifswalder Bodden die Fischzüge zu concentriren sehr geeignet ist, aber für die anderen Buchten finden sich gleiche geographische Verhältnisse nicht.

In die drei östlichen Buchten ergiessen sich grosse Ströme und dienen als Anziehungspunkt und Brutstätte für Lachs, Stör und Neunaugen, auch vermehren sie vielleicht (die Frage ist der sichern Entscheidung noch fern) das verwendbare Nahrungsmaterial, aber in den westlichen Buchten haben wir nur ganz unbedeutende Ströme und dennoch eine stärkere Befischung.

Ich glaube wohl, dass die Ufer der Buchten, namentlich wo sie einen verwickelten Verlauf haben, einen Lieblingsaufenthalt der Fische bilden, aber die vorliegende Statistik bringt für den Nachweis kaum ein genügendes Material.

Dass die Vorsprünge in die Ostsee so schwach befischt werden, darf nicht als ein solcher Nachweis angesehen werden, denn es kommen hier andere Momente zu sehr in Betracht. An der Küste von Memel ein Zins, der von dem Fiscus erhoben wird[3]), am Samland absorbirt die Bernsteinfischerei manche Kräfte und die

---

[1]) l. c. Besprechung d. Fischspecies.

[2]) Circular d. deutschen Fischereivereins 1872, S. 189, 191.

[3]) Vergl. MARCARD Bericht S. 47 u. Schreiben des Oberfischmeisters DOEPNER, Circular d. deutsch. Fischereivereins 1870 IV, pag. 11.

Communication ist schwierig. An der Küste des Bezirks Coeslin finden sich gleichfalls mancherlei die Strandfischerei betreffende Gerechtsame[1]), welche dem Fischereibetrieb nicht günstig sein können. Uebrigens ist die Fischerei bei Leba sehr gut entwickelt, dazu mag theils der Lebasee mitwirken, ich finde jedoch, dass Herr E. FRIEDRICH[2]) berichtet, es seien auf seine Veranlassung Fischer von Dievenow nach Leba und Stolpmünde übergesiedelt und hätten dort sehr guten Verdienst gehabt. Ob an der Nordküste von Rügen besondere Verhältnisse der Fischerei hinderlich sind, vermag ich nicht zu sagen, die Küste ist allerdings grösstentheils steinig und hoch, aber ähnlich verhält sie sich u. a. bei Lohme, ohne die Fischerei zu hindern. Auf dem Dars, bei Fehmarn und Alsen sind die Absatzwege ungenügend; dies erklärt die schwache Befischung.

Es wird demnach durch die genannten Verhältnisse eine Abnahme der Fische nach der Mitte der Ostsee zu nicht erwiesen, aber auch nicht eine Zunahme derselben angedeutet.

Bei der Betrachtung der einzelnen Bezirke fällt die Dichte der Fischerei bei Neufähr an der Mündung der Weichsel auf. Die Angaben von dort waren zuerst evident unrichtig, die volle Fischerei ergab sich erst bei weiterer Nachfrage. Es scheint dort eine Hauptpassage der Fische zu sein.

Ob die Fischerei bei Heisternest und Zeynowa nur ins Meer hinausgeht, ist mir zweifelhaft, die Angaben lauteten nur auf das Meer und ich habe keine Recherchen veranlasst.

In den Bodden nordwestlich Stralsund (Grabow, Saaler Bodden) werden wahrscheinlich gesetzliche Vorschriften die Fischerei modificiren, dieselben sind mir jedoch nicht bekannt. Aehnliches dürfte für den Breitling gelten, wo wahrscheinlich mehr die berechtigten, als die wirklichen Fischer aufgeführt sind.

Die Schonbezirke habe ich nach den gesetzlichen Bestimmungen construirt, doch da mir die Lokalkunde fehlte, werden die Grenzen schwerlich völlig genau sein.

Ein Ueberblick der Karten ergiebt, dass überall, wo es angeht, die Fischer den Schutz der Buchten ausnutzen. Dies würde sich wahrscheinlich viel ausgesprochener zeigen, wenn die Böte auf die Lieblingsplätze der Fischer hätten vertheilt werden können. Diese Aufgabe konnte nicht gelöst werden, aber da nun die Karten vorliegen, wird es vielleicht den Kennern der Fischerei gelingen, die einzelnen Karten so umzuarbeiten, dass sie ein besseres und mehr ausgeführtes Bild der wirklichen Fischerei geben. Gewiss würden solche Arbeiten für den Fortschritt des Fischereibetriebes werthvoll werden.

Bemerkungen über die Möglichkeit, den Fischfang zu heben.

Nachdem durch vorstehende Arbeit eine übersichtliche Kenntniss der Lage des deutschen Fischereibetriebes gewonnen worden ist, darf wohl die Frage erörtert werden, was der Anforderung gegenüber, welche die Volkswirthschaft an die deutsche Fischerei stellt, geschehen könne.

Wenn ich mir erlaube, meine persönliche Ansicht hinter dem Büchertisch heraus hier auszusprechen, so geschieht dies, weil ich dafür halte, dass es der Sache förderlicher ist, jetzt zu discutiren, als nach Jahren mit zwar besser begründeten, aber doch der Natur der Sache nach nicht sicheren Ansichten hervorzutreten. Ich muss dabei mir und dem Leser sagen, dass weitere Studien und Beobachtungen mich sehr wohl zwingen können, meine Ansicht zu ändern.

Nach der offiziellen Statistik des deutschen Reichs über Ein- und Ausfuhr der Fische in 1873[3]) ergiebt sich, dass die Einfuhr die Ausfuhr um folgende Summe übersteigt:

Frische Fische und Flusskrebse	342000 ℳ
Diverse Fische (namentlich aus der See)	3150000 »
Heringe	27798000 »
Schaalthiere	387000 »
Kaviar	973000 »
Summa	32650000 ℳ

Da der Fischfang der Blankeneser und Finkenwerder Fischer z. Th. als Einfuhr berechnet sein mag, kann man die Summe zu 10 Millionen Thalern ansetzen. Diese Summe wird von uns alljährlich als Busse dafür ans Ausland gezahlt, dass wir, statt selbst die uns offenen, wenn gleich nicht sehr bequem liegenden Fischgründe zu befischen, sie für uns von Anderen befischen lassen. Die Frage, ob das von der Natur sonst nicht begünstigte Deutschland diese Ausgabe als eine unvermeidliche hinzunehmen hat, ob es einen volkswirthschaftlichen Fehler begeht, die Fischerei so sehr den Nachbarn zu überlassen, oder ob es seine Arbeitskraft in anderer Weise besser und vollständig verwerthet, wird unter diesen Umständen schwer wiegend.

Der Bericht[4]) über den holländischen Herings- und Frischfischfang der Hrn. v. FREDEN, DANTZIGER und v. RENSEN legt seiner Berechnung über die Rentabilität des Emdener Fischereiunternehmens eine Einfuhr von 647000 Tonnen Hering zu Grunde (1873 war der Ueberschuss der Einfuhr über die Ausfuhr 771000 Tonnen).

---

[1]) MARCARD Bericht S. 41.
[2]) Mittheilungen. Circular d. deutschen Fischereivereins No. 2, 1871, S. 38.
[3]) Mitgetheilt von Dr. WITTMACK, Beiträge zur Fischereistatistik S. 216.

Es wird angenommen, dass, da kein Logger über 1000 Tonnen im Jahr fischt, 650 deutsche Logger dabei Beschäftigung finden könnten.

Von unserem differenten Standpunkte aus können wir sagen, dass das deutsche Reich mindestens 1000 Logger mit 15000 Mann Besatzung haben muss, um sich die Ausgabe obiger 30 Millionen M. an das Ausland zu sparen. Meiner Ansicht nach würden 9000 von den 17000 Fischern der Tabelle genügen können, um bei gehöriger Leitung, entwickeltem Betrieb, Nachweisungssystem über die Fischzüge u. s. w., die Küstenfischerei so auszubeuten, wie es zur Zeit der Fall ist.

Der oft citirte Bericht des Herrn MARCARD giebt uns eine Reihe vortrefflicher Vorschläge zur Entwicklung der Fischerei. Das königl. preussische Ministerium für die Landwirthschaft hat sich nachhaltig mit der Hebung der Fischerei beschäftigt und eine Reihe jener Vorschläge zu wirksamer Ausführung gebracht. Hierher rechne ich alle die Schritte, welche bestimmt sind, die öffentliche Aufmerksamkeit der Fischerei zuzuwenden. Die Entstehung eines deutschen Fischereivereins, dessen Bestrebungen das königl. Ministerium energisch unterstützt, ist ein wichtiger Erfolg. Mit ihm ist durch die Hebung der Brutanstalten, durch die Berliner Fischereiausstellung, durch die Unterstützung des Emdener Fischereiunternehmens, Herbeiführung von Erleichterungen des Fischtransport, durch die grosse Reihe interessanter Mittheilungen in den Circularen, unter denen die Statistik des Herrn Dr. WITTMACK, an vielen Stellen der Keim des Fortschritts gelegt. Hierher ist hoffentlich auch die Thätigkeit unserer Commission zu rechnen, die das königl. Ministerium ins Leben gerufen hat. Dazu kommt dann in letzter Zeit die Schaffung eines neuen Fischereigesetzes.

Ehe diese Werke vollbracht waren, konnte eine intensivere Entfaltung der Thätigkeit nicht nützen. Wenn nun eine solche allmahlg ins Leben treten soll, so muss, wie mir scheint, die Frage zu einem gewissen Abschluss kommen, wo unsere Hauptfischereigründe zu suchen sind, in der Ostsee oder in der Nordsee; denn ich glaube nicht, dass wir wagen dürfen, an verschiedenen Stellen zugleich Kapitalien einzusetzen. Wenn es vorwärts gehen soll, muss den Fischern stark nachgeholfen werden, denn unsere natürliche Lage ist wenig günstig, der Staat wird noch lange Zeit die Stütze abgeben müssen, an welche sich die kaum werdende Hochsee-Fischerei in Zeiten des Sturms anklammert; aber bekanntlich ist die richtige Weise, den Unternehmungen die Hülfe zu sichern, noch nicht klar zu bezeichnen.

Die Ostsee scheint das richtige Gebiet zur Entfaltung des Fischereibetriebes nicht zu sein. Ich zweifle gar nicht, dass auch sie einen besseren Fischereibetrieb lohnend erwiedern würde, aber es liegen dafür genügende Beweise nicht vor und wenn man unbefangen zu urtheilen versucht, muss man sich eingestehen, dass viel leichter der Versuch ausgedehnterer Fischerei in der Ostsee scheitern kann, weil die Wege, welche eingeschlagen werden sollen, noch zu suchen sind und weil ein sehr grosser Verdienst wegen des beschränkten Areals, aus dem die Fische zusammenströmen könnten, nicht zu erwarten ist, als in der Nordsee, wo die Ueberzeugung, dass es gehen kann und die Wege, wie es Andern geglückt ist, schon festgestellt sind. Fischen wir gut in der Nordsee, so wird auch die Ostseefischerei nicht zurückbleiben, denn die Nordseefischer werden der Ostsee entnommen werden; das Umgekehrte lässt sich nicht behaupten.

Als Haupthinderniss für die Nordseefischerei wird der Mangel an Mannschaft angegeben und Herr MARCARD weist mit guten Gründen nach, dass auf die Bewohner der Nordseeküste und Inseln für die Bemannung der Fischereifahrzeuge wenig zu rechnen sei. Ich möchte zwar an den interessanten Bericht[1]) des Herrn HANSEN zu Keitum erinnern, der nachweist, dass die Inselfriesen im 17. Jahrhundert gegen 4000 Mann für den Walfischfang stellten, doch dies sind lange vergangene Zeiten und die Tradition wird kaum noch lebendig sein.

In der That wird man für einen so schweren und mässig lohnenden Dienst anderswo zu werben haben. Die Bewohner des Dars, von Hiddensoe, der Halbinsel Thissow auf Rügen, von Hela, den Neerungen und den Haffküsten sind, wie ich glauben möchte, zur Zeit mehr geeignet die Mannschaft zu rekrutiren. Angewiesen ausschliesslich auf die Fischerei, die ihnen karg lohnt, sind sie an harte Arbeit und Entbehrung gewöhnt und leiden häufig, wenn die Fischerei in einem Jahre schlecht ausschlägt, geradezu Noth. So schrieb im Mai 1873 unser Berichterstatter aus Hela: Familienväter, die vor 10 Jahren die Schifffahrt aufgegeben hatten, müssen wieder zur See gehen, bei manchen Familien herrscht bereits grosse Noth an Lebensmitteln. Weder Flunder noch Heringe giebt es, wie man allgemein verzagt.

Vielleicht legen die anliegenden Karten die Möglichkeit solcher Werbungen, die zum Vortheil der einzelnen Fischer und der deutschen Finanzen scheinen ausfallen zu müssen, näher. Der Staat dürfte, meine ich, auf Ansuchen seine Unterstützung leihen, um die Schwierigkeiten der Anwerbung und der weiten Reisen, welche aus den deutschen geographischen Verhältnissen resultiren, compensirend abzuhalten.

Dass übrigens auch das Binnenland Kräfte abzugeben vermag, ergiebt sich aus dem oben citirten Bericht über den holländischen Heringsfang, wo mitgetheilt wird, dass 400 Mann aus Oldenburg, dem Bückeburg'schen und Westphalen in Holland an der Fischerei theilnehmen.

---

1) Circular 1871 Nr. 3, S. 39.

Immer dürfen wir nicht vergessen, dass uns keine Fischereigründe, wie bei den Lofoden zu Gebote stehen, wo der Fischfang erst dann beginnt, wenn das Loth wegen der sich drängenden Fische nur langsam zu Boden sinkt[1]).

Mit wenigen Worten erlaube ich mir noch auf das einzugehen, was ich betreffs der Ostseefischerei anzugeben wüsste.

»Es ist, sagt der Bericht des Hrn. MARCARD,[2]) irrationell von einem örtlich eng begrenzten Fischereirevier verlangen zu wollen, dass es alljährlich gleiche und reiche Ausbeute liefert, der Zug der Fische ist bekanntlich von den herrschenden Windströmungen und von unzähligen anderen Zufälligkeiten abhängig, und wenn der Fischer nicht im Stande ist, dem Zuge der Fische zu folgen, die Schwärme aufzusuchen, wo er sie findet, sei es in der hohen See oder nächst der Küste, so unterliegt sein Geschäft denselben Zufälligkeiten, wie sie der Zug der Fische bedingt.

Wenn die Ostseefischer sich aus dieser Abhängigkeit lösen und eine seetüchtige Ausrüstung beschaffen könnten, so würde es um den Wohlstand dieses Theils der Küstenbevölkerung bald anders stehen.

Die Ostsee ist wohl im Stande, den Fischer auskömmlich zu ernähren, nur muss das Geschäft, wie jedes andere auch, mit den nöthigen Mitteln und mit verständiger Umsicht angegriffen werden.«

Diese Worte drücken m. E. die Sachlage prägnant und vollständig richtig aus, nur in einem Punkt vermag ich ihnen nicht beizustimmen.

Wenn ich annehme, dass rationelle Staatshülfe gerne gewährt werde, so halte ich dennoch für höchst unwahrscheinlich, ja für unmöglich, dass die Fischer »sich« lösen werden von ihrer Abhängigkeit, ich glaube sie werden daraus gelöst werden müssen.

Es wird an den Fischer die Anforderung gestellt, den Zügen der Fische zu folgen, die Schwärme aufzusuchen, sei es in der hohen See oder nächst der Küste, wo er sie findet. Dies und was Alles dazu erforderlich ist, (wenn es sich um mehr wie das auch jetzt schon geübte Verfolgen des Revierwechsels der Plattfische u. s. w. handelt) kann der Fischer aus sich selbst heraus nicht leisten. Mancher Fischer mag alle dazu nöthigen Eigenschaften besitzen, aber der Vergeudung von Zeit, den Misserfolgen, Verlusten und Enttäuschungen bei verdoppelter Mühe, vielleicht noch den Verspottungen, welche der Versuch die Fische auf hoher See und in weiteren Touren aufzusuchen und ihnen nachzugehen, nothwendig mit sich bringt, diesen die Stirne zu bieten, ist nicht die Sache des eigentlichen Fischers. Wer aus eigener Kraft dies durchführt, d. h. neue nicht etwa von Anderen befischte Gründe in vervollkommter Methode ausbeutet, würde sich weit über das Niveau eines gewöhnlichen Fischers erheben, wird sich aber, wie sehr zu fürchten steht unter ihnen nicht finden. Man wolle dagegen nicht einwenden, dass doch in der Nordsee die Fische aufgesucht würden, hier liegen die Dinge ganz anders. Die Küsten und Inseln sind günstiger gelegen und geformt, die Fischzüge so viel bedeutender und breiter, der Gewinn so viel grösser, dass die Kapitalien sich zum Theil herbeidrängen und die Verfolgung zunächst von Schottland aus nicht schwer wurde. In der Ostsee, so viel kann mit Bestimmtheit gesagt werden, giebt es derartig reiche Fischgründe überhaupt nicht.

Ich halte nach dem Gesagten dafür, dass der Staat selbst die Führung übernehmen sollte.

Wenn ich versuche meinen Gedanken darüber conkreten Ausdruck zu geben, so weiss ich wohl, dass ein solcher Plan in Ausführung gebracht, wesentliche Modifikationen erleiden würde, aber es ist nicht die Aufgabe hier wohl ausgearbeitete Pläne vorzulegen, sondern die Absicht besteht nur darin, die in Folge der gemachten Studien und Ueberlegungen entstandene Ansicht in einer discutablen Form vorzuführen.[3])

Es sollte, meine ich, eine passende Persönlichkeit aus der Reihe unserer Kriegs- und Zoll-Marine gewonnen werden, welche die Aufgabe hätte, sich durch stete Ueberwachung der Fischerei an den verschiedenen Stellen der Küste die Fähigkeit und Kenntniss anzueignen, leitend und dirigirend auf den Fischereibetrieb einzuwirken. Diesem, sagen wir Fischereidirigenten, würde ein passend gebautes und eingerichtetes kleines Dampfboot, ich denke mit 6 Mann Besatzung, zur Verfügung stehen müssen. Er würde die Aufgabe haben, die Hauptfischereien zu verfolgen, und würde dabei die Gelegenheit finden, die Fischzüge aufzusuchen und ihnen nachzugehen. Wenn überhaupt Jemand, würde er bald in die Lage kommen, die Fischer, so weit es die Fahrzeuge gestatten, auf die zur Zeit besten Fangplätze zu dirigiren, im Anfang würde er vielleicht Einzelne direct dort hin befördern können. Er würde aber auch in den Stand gesetzt sein, mit viel grösserer Vollkommenheit und Sicherheit als dies bisher möglich war, die Punkte zu zeigen, wo der Hebel zur Hebung der Fischerei einzusetzen sind.

Ich nehme an, dass es einem solchen Mann, dem die polizeiliche Ueberwachung nicht aufgebürdet werden dürfte, leicht werden würde, durch kleine Freundlichkeiten und Hülfeleistungen das volle Vertrauen und den Beistand vieler Fischer zu gewinnen. Er würde im Stande sein, den Fischerinnungen die fehlende

---

[1]) Vergl. v. BAER l. c. S. 276.
[2]) S. 53.
[3]) In Skandinavien sowie in Frankreich finden sich Fischereibeamte, deren Stellung manche Aehnlichkeit mit der hier angeregten Einrichtung hat, ich habe jedoch darüber keine Literatur zur Hand.

Initiative zu Gesuchen zur Gewährung rationeller Staatshülfe zu geben und würde der Mann sein, Neuerungen die, wie z. B. die Geschichte des Hartlepooler Boots*) lehrt, trotz unzweifelhafter Berechtigung leicht scheitern, durchzubringen.

Ich denke mir diesen Mann in enger Verbindung mit unserer Commission zur Untersuchung der deutschen Meere, also auch in steter Communication mit unseren Beobachtungsstationen, deren limitirte Vermehrung wir anstreben. Diese Verbindung kann, wie mir scheint, beiden Seiten nur zum Vortheil gereichen, denn ein Theil der Aufgaben welche zu lösen sind, ich meine das Auffinden der Laichplätze, die Beobachtung des Nutzens der Laichschonreviere, die Auffindung der Fischplätze auf hoher See, treffen wissenschaftliche Probleme und andererseits wird die gemeinschaftliche Verfolgung solcher Aufgaben das Gewicht des supponirten Beamten nur erhöhen können.

Bei dieser nur skizzirten Einrichtung werden die aufzuwendenden Geldmittel weder in Anbetracht der Sache, noch anderen Resorts gegenüber, unverhältnissmässig sein.

Der Erfolg, welcher natürlich sehr von der glücklichen Wahl der Persönlichkeit abhängen wird, kann dreierlei Art sein.

Entweder es stellt sich etwa nach Jahresfrist die Unzweckmässigkeit der Sache heraus, dann ist die Ausgabe umsonst gemacht; allein wenn nachweisslich alle Umsicht angewandt worden ist, wird selbst der Staat sich gefallen lassen dürfen, ein gewisses Lehrgeld zu zahlen.

Oder, es wird eine Reihe von Kenntnissen gewonnen, die Fischerei hat einen Aufschwung bekommen und macht sich selbstständig. Dann wäre der Zweck erreicht, die Kosten würden sich lohnen.

Oder, es zeigt sich, dass die Fischerei dauernd einer Leitung bedarf, ja diese vielleicht noch vermehrt und verstärkt werden müsste. Dann würde zur Frage stehen, ob der Staat die Kosten dauernd tragen will, das Verfahren an sich würde aber bewährt sein.

Mit diesem Vorschlag schliesse ich die Arbeit. Ich bin der Ueberzeugung, dass das deutsche Reich die Fischerei nachdrücklicher wie bisher wird heben müssen, und dass es dies thun wird, sobald die Vorarbeiten genügen. Ich erlaube mir kein Urtheil darüber, wie weit Letzteres schon der Fall ist, indessen hoffentlich werden die wissenschaftlichen Arbeiten auf diesem Gebiet, welche sich zwar noch in bescheidenen Anfängen finden, bald diese Vorarbeiten ausreichend beschaffen.

*) Circular 1873 No. 2 S. 139.

METHODS OF PLANKTON RESEARCH

W. J. Dakin

METHODS OF PLANKTON RESEARCH.

By W. J. Dakin, M.Sc.
1851 Exhibition Scholar, University of Liverpool

Introduction.

During the last few years there has been a great development in the study of those organisms which are found floating in the waters of seas and lakes, and which, though having in many cases the power of swimming, are practically as much at the mercy of the winds and currents as inanimate floating objects. Such organisms are denoted by the term "Plankton," and, as their full importance in the metabolism of the ocean has become appreciated, so has this branch of zoology and botany, which may be termed Planktonology, advanced from qualitative to quantitative investigation. The work has been carried on by both botanists and zoologists, but owes its growth mainly to the latter, and in great part to the German School, from the time of Johannes Müller onwards to the present quantitative scientific study of the Plankton which originated through the outstanding and fundamental work of Victor Hensen, Professor of Physiology in the University of Kiel. Hensen's work (1) appeared in 1887, but as far back as 1867 he was interested in the investigation of the sea in the interests of Fishery questions, and held the view that an attempt to estimate the productiveness of the sea would be an important step, both scientifically and economically. Finally, whilst working at the question of coast fisheries he attempted to determine the number of fish in a defined region by counting and estimating the number of floating fish eggs. This led to the idea that it was both possible and necessary to investigate quantitatively the planktonic fauna and flora, the source of food for the larger sea animals.

With these facts in view, Hensen invented the nets and methods by which the German investigators have, since that time, diligently worked. These methods were described by Dr. J. T. Jenkins in the Trans. Liverp. Biol. Soc. for 1901, and the purpose of the present paper is to bring the description of the German Plankton methods up to date and to briefly discuss some results of the work. I have been in a fortunate position in this respect, that I have been able, myself, to handle the apparatus whilst participating in the actual expeditions in the North Sea and Baltic. I must thank Professor Brandt for his kindness in securing permission for me to travel on the German Investigation Steamer "Poseidon," and Professors Lohmann and Apstein for their ever willing help and explanations.

Essentially from first to last, Hensen and his co-workers have had one aim in view—the better determination of the "see-saw" of life in the water and the laws governing this. It devolved, therefore, into a determination of what plankton was to be found in the sea at a given time and place, and how this mass changed with the change of time or of place, or of both, in quantity and quality. How far this aim has been realised will be discussed after an account of the nets and apparatus used.*

The Plankton Nets.

As described by Jenkins, the quantitative net was the apparatus which Hensen invented as the most satisfactory means by which the organisms in a known

* The blocks for figs. 1, 3, 4 and 7 have been kindly lent by the Commiss. f. wissensch. Meeresuntersuchung. The apparatus described in this paper for the quantitative work is manufactured by Zwickert, Optician, Kiel.

quantity of water could be estimated These quantitative nets are lowered perpendicularly in the water to a certain depth and then raised to the surface, so that any towing in a horizontal direction is avoided. Thus a vertical column of water passes through the net, and the volume of this has to be calculated before the net is used for quantitative work, since it is obvious that not the same quantity of water passes through the net as would pass through the open mouth if no net was attached to it. In short, the problem has been to devise an apparatus which should be handy and workable from a small ship, which would take up a definite quantity of water from the sea, and abstract as thoroughly as possible the organisms contained therein.

Silk nets still form the apparatus most used for this purpose, but as Lohmann (**11**, **13** and **18**) has shown, the results are only accurate to a certain point and must be supplemented by other methods, according to the aim of the research. The large Hensen vertical net was described in Dr. Jenkins' paper, and therefore need not be mentioned further. Apstein has shown that a much smaller net can be conveniently used with much saving of time and labour, and under conditions where the larger net is impossible, in the absence of a steamer.

At the present time the net used in the German investigations above all others, and thus the chief implement for plankton research, is the Middle Plankton Net of Apstein (**7**). The large Hensen net is used only for special purposes, one of which is the quantitative estimation of fish eggs, where a large catch of plankton is desirable, or for other large organisms that do not occur frequently enough for accurate measurements to be made with the smaller nets. In form the Apstein net is almost the same as the Hensen. It is shown in fig. 1 to consist of three parts:—(1) The filtering net itself; (2) the metal filtering bucket; and (3) the conical mouth piece. The mouth piece is constructed of thick material, which does not allow water to filter through it. It is supported by two brass rings, one of which serves to keep the mouth of the net open, and to this ring are attached the cords, three in number, which support the net. This upper ring, and therefore the mouth of the net, has a diameter of 14 centimetres. The lower brass ring is thicker and is supported by the mouth-piece cloth itself and by three cords which are attached to the upper ring and take the place of the three iron rods that separate the two rings in the Hensen net.

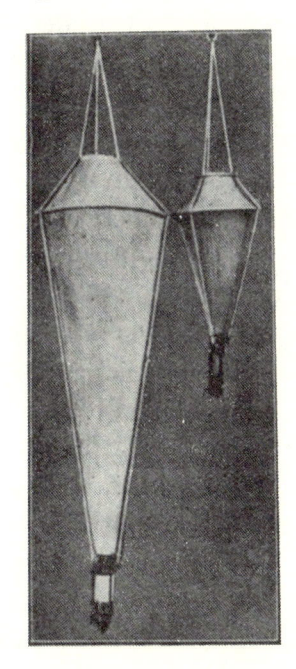

The length of the mouth-piece is 20 cm. down the side. This part of the net serves three purposes:— (1) It prevents mud passing into the net if it be lowered on to very soft ground; (2) it prevents the catch from being upset in a rough sea; (3) it performs the function of keeping the net mouth small in comparison with the filtering area of the net.

Fig. 1.—Apstein's middle and small plankton nets.

It is obvious that the water will be most completely filtered if as little as possible is allowed to enter and the greatest facility is given for it to leave the net. If the mouth-piece was not present, a greater quantity of water would attempt to stream through the large ring than could be filtered by the area of the net, and, therefore, as the net was hauled up, water would remain in the entrance and the water of the vertical column would be simply pushed aside without passing through the net. Thus a reliable sample would not be obtained.

The Net, which is the actual filtering agency, must be made of some material that will stand the work well—filter the organisms as thoroughly as possible, and most important of all, be so manufactured that the size and shape of the meshes will not easily alter, so that the formula calculated for determining the filtration co-efficient for the net may remain always applicable. The best material has been proved to be "Müllergaze" or "bolting silk," and the grade used for quantitative nets is denoted by the number 20. For this particular tissue the filtration capacity has been calculated. The net must be carefully made, because in order to determine the quantity of water from which the plankton has been abstracted, an extremely complicated mathematical and practical research is necessary, and this when once made for a particular silk and shape of net should be applicable to all others of the same size, shape and material. Furthermore, in order to compare the results of workers in different countries, it is of course advisable that a standard net such as the Middle Plankton Net should be used everywhere.

The silk net is attached above to a strip of strong linen cloth or canvas about an inch wide, which is fastened to the lower brass ring of the conical mouthpiece. It is conical in shape with a truncated end, which forms the bottom of the net and is fastened to a metal bucket by means of a clamp ring consisting of a strip of thin brass, which is bent into a ring with the two ends bearing projections perforated by holes so that they can be screwed together.

The brass bucket (fig. 1), being of considerable

weight, is not allowed to hang on the silk net, but is supported separately by three cords, which are fastened to the stout lower ring of the conical mouth-piece.

The size of the net and the method of cutting it can be seen from fig. 2, where the right-hand sketch is a representation of the net when sewn up, and the other figure is the same unrolled, showing the pattern as it should be cut out. For the middle plankton net the radius R of the mouth of the net is 20 cm. The radius r

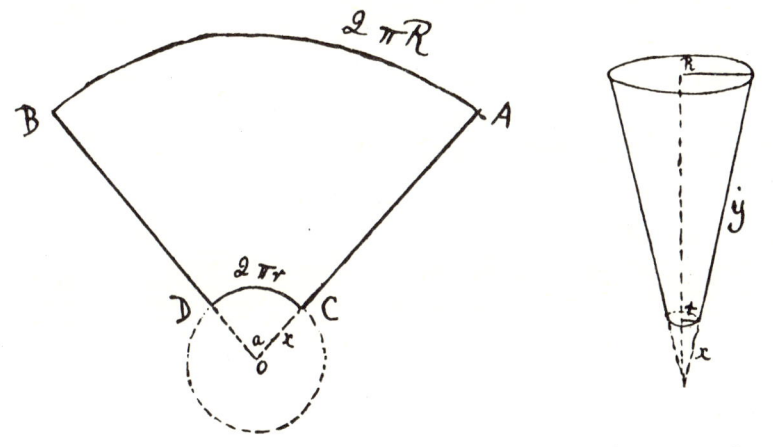

FIG. 2.—$R = 20$ cm.; $r = 3$ cm.; $x = 17{\cdot}65$ cm.; $y = 100$ cm.; and the angle $a = 61{\cdot}2°$.

of the bottom of the net, where it is attached to the metal bucket, is 3 cm. The portion of the cone cut off (fig. 2, x.) is 17·65 cm., and the length of side remaining 100 cm. Care must be taken to allow of a margin when cutting out the pattern, so that the edges AC and BD can overlap and be sewn together, the upper edge of the net be sewn to the linen strip, and the bottom of the net fixed by the clamp ring to the bucket.

The metal bucket serves as a receiver for the plankton and furthermore does away with the use of the filtrator invented for use with the great Hensen net. It can be described therefore as the filtering bucket. It consists of a 14 cm. long brass cylinder, whose sides are cut out, with the exception of three narrow pieces, so that there remains 3 cm. of cylinder above and 4 cm. below. A piece of No. 20 silk is then placed outside the three brass pieces and fixed above and below to the cylinder by the clamp rings. In addition, three brass plates are screwed against the pieces so that the silk filtering tissue lies between them and is quite taut over the three windows of the bucket. The upper part of the cylinder is supplied with a screw thread so that it can be screwed on to the brass ring, to which the net is attached. From the middle of the floor of the cylinder a tube descends, which is provided with a tap.

This completes the description of the ordinary Apstein net, but in order to make comparisons of the various layers, and the plankton at different depths (a most important factor, for it will be shown that the plankton differs both quantitatively and qualitatively considerably according to the depth and probably as a result of changes in the light conditions, salt contents or temperature), it is necessary to have a method by which the net can be hauled through a certain distance and then closed. In fact, when working with the vertical net, the column from the bottom to the surface should be divided into regions, and the temperature, and salt contents, together with the plankton, determined for each region separately. Various arrangements have been invented for closing the net. The following is about the most satisfactory now in use (fig. 3). The net differs from the one just described only in the possession of closing apparatus. The upper brass ring of the conical mouth-piece is replaced here by a much broader and heavier brass ring. Across the middle of this runs a stout bar, and directly above this two semi-circular brass lids are hinged so that they can fall down, one on each side, and lying over the brass ring completely close the opening. In order to make the closure more water-tight, the net ring bears a rubber ring on its upper surface, and there is further a piece of stout waterproof canvas fixed by screws over the hinge itself, so that no water can pass through the crevices into the net when it is being hauled up to the surface. One of these trap-doors (both together resemble a "butterfly valve") is perforated to allow air to enter the net when it is raised out of the water. This is necessary because when lifting the hermetically sealed net out of the water, the water contained in it filters out through the silk, and unless air is allowed to enter the filtration is hindered, and, moreover, the whole net collapses. In order that no water shall pass in through the opening for the air, it is provided with a balanced valve, so that it remains closed as long as the net is being pulled through the water, and opens immediately it is above the surface. The essential apparatus for closing the net at any particular moment is shown in the photograph (fig. 3). The apparatus in the photograph differs slightly from that

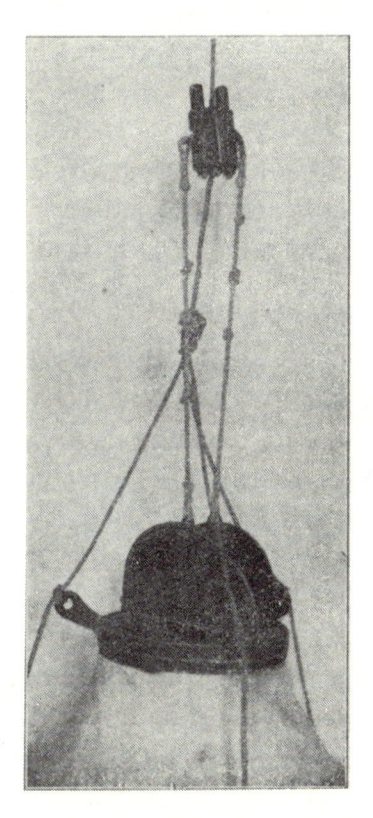

Fig. 3.—Closing apparatus at mouth of net.

here described, but only in one detail, viz., two ropes with knots are present in place of the two wires. It will be seen that the net can be supported in two ways—(*a*) by the lids; (*b*) by the three cords attached to the upper brass ring of the net. The first is the condition when the net is descending and whilst it is being pulled up through the region which is to be fished. The second is the condition when the net has been closed by a heavy weight run down the rope, which falls upon the closing apparatus and releases the two lids. In this condition it can be hauled up to the surface without any water entering. I have seen this closing apparatus used down to depths of 200 fathoms, and it has worked satisfactorily, though it should always be watched in case anything catches and the lids do not close at the proper time.

It will be seen that the net when lowered is open. This is the case for the ordinary net also, but since the water only enters the net through the filtering tissue, the various organisms will remain outside. It is only when the water enters the net through the mouth that it will make any catch.

Method of Using the Quantitative Net.

The wire or rope supporting the net should not run directly from the winch over fixed pulleys to the net, but should pass over a pulley which is supported by an "accumulator." This is particularly important when using the large net in a rough sea, or when the boat is rolling considerably, since, otherwise, the sudden pull as the side of the boat rises to a wave is liable to damage the net, besides rendering the results inexact, owing to the pressure constantly varying. It is essential that the net should be pulled up with an equal speed, and therefore pressure, and the accumulator aids this considerably.

When the boat is rolling very much, a clever winch man can so work as to wind the net up only as the ship's side descends, while the ascent serves to haul up the net. For systematic plankton work of this kind a steam winch is very desirable. On the German Investigation Steamer "Poseidon" there is a very complete set of winches, and the rope used for the net is of thin wire. This wire passes over a very essential, though small, piece of apparatus fixed to the deck, and consisting of a wheel about one foot in diameter, so geared that it records every metre of rope that passes over it, and therefore the exact depth of the net can be seen by simply looking at the figures on the meter face.

Furthermore, the gradual changing of the figures, as metre after metre passes out or in, can be easily timed, and from this the speed at which the net is being hauled in can be adjusted. In order to determine the volume of water filtered by the net, a formula has been calculated, depending on the size and shape of the net, the filtration capacity of the silk, and the speed at which the net is hauled through the water. The net must be hauled up at a speed of half a metre per second in order that the coefficient determined for this type of net can be used. One has then only to multiply the number of organisms found in the catch or the volume of the catch by 80 to give the number or volume present in a column of water of the length the net has been hauled through and of area equal to one square metre.

In any case, in quantitative work, where the results of different catches are to be compared, and whether the number of organisms in a particular quantity of water is needed or not, it is necessary that the net should always be hauled up at the same speed, otherwise the pressure in the net will not be the same and a greater or less quantity of water will be filtered, and different volumes of catch will be obtained. Thus, if the quantity and constitution of the plankton were exactly alike at two stations, but the net was hauled up with a greater speed at one than at the other, the volume, and moreover the constitution, of the two catches would be different, even though the distance through which the net was hauled was the same at both places. Hence the necessity of some recording machine.

It is further important, in quantitative work, that the net should descend and ascend vertically. If this is not the case, two errors may enter into the work :—(1) The net will be towed more or less in a horizontal direction and not give a true picture of the plankton in a column of water. (2) The depth recorded by the amount of rope paid out will not give the true depth of the net.

In order to obviate this oblique descent, which occurs when there is any considerable current in the water, the net should be weighted with a heavy leaden weight slung under the filtering bucket. The net provided with the closing apparatus does not need such a heavy weight as the ordinary Apstein open net, since it is naturally somewhat heavy. Both, however, should be weighted, otherwise even in a region with no current their descent would be too slow. It is always safer to sink the nets as rapidly as possible, particularly if there be much current.

Herdman (**17**) refers to the " Nansen " net and its easy method of working, due in part to its lightness. If, however, the Nansen net is to be used as a satisfactory vertical net, it must be weighted until it is relatively as heavy as the other plankton nets. Furthermore, it may be slightly inaccurate for quantitative work, since it lacks the conical mouth-piece, and, in addition, no formula has been calculated to determine the volume of water it filters. In order to determine the true depth of the net, when the rope shows that it has not descended vertically, the following simple method used by Apstein is of great service. It can also be used to determine the depth at which a dredge or any other piece of apparatus is trailing. The apparatus consists of a piece of plate glass about a foot square and ruled at regular intervals of one centimetre with horizontal lines. A flat strip or ruler of aluminium is fixed at one corner by a pivot passing through the plate glass, so that the ruler can be made to describe an arc over the glass plate. The rule is divided into centimetres and perforated along the middle line at each centimetre mark, so that when it is standing vertically, the perforations for each centimetre mark lie exactly over the corresponding horizontal lines on the glass. Both the lines on the glass and on the rule are numbered 1-2, 3..., similarly, but each space can be used to equal 10, 20 or 100 metres, as the case requires. If the net has been drawn somewhat out of the vertical, it is only necessary to hold the glass plate up so that its upper and lower edges AB, CD are perfectly horizontal and so that the net rope from the pulley to the water lies between the observer and the light. Supposing now that 80 metres of rope have been paid out, let each division on rule and glass equal 10 metres, then the rule is moved until it is parallel with the net rope as seen through the glass plate, and the line on the glass plate intersected by the 80 metre mark on the rule will be the true depth of the net. Let us suppose a certain station has been reached where it is desired to make a quantitative catch. The ship should be anchored so that it remains in the same position during the work, and all fear of towing the net in a horizontal position will be done away with, unless a strong current is present. This is most important where it is most difficult to perform, namely, where the depth is considerable and some time elapses during the lowering and raising of the net from the bottom. The net is swung over the ship's side, supported by its wire rope, which passes over the "accumulator pulley" and then over the recording apparatus on deck to the winch. When all is in readiness, word is given to lower the net. An assistant should note the moment that the *mouth* of the net reaches the surface of the water, and shout a word of warning. This is the zero, and at this moment another assistant who is observing the recording meter takes down the figures exposed. To these figures the required depth should be added, and then the net can be lowered until the meter gives the necessary numbers.

When the net has been hauled again to the surface, it is held over the ship's side and well washed down with a strong stream of water. This is most important, as a great quantity of the catch is often lodged under the mouth-piece lower ring. A strong stream of salt water from the hose is by far the best method of washing the catch down into the filtering bucket, and if no steam hose is available, a small hand pump worked on the deck is better than using buckets. A separate filtrator is unnecessary. After the net has been well washed and the water allowed to run out until only a little remains in the bucket, this is unscrewed and the catch can now be removed and fixed.

Preservation of the Catch.

When preserving the catch it is advisable to remove as much sea water as possible, and to use a reagent that will be simple in application and render the organisms easy of identification. For this purpose 90 per cent. of alcohol is used directly, it having proved the most convenient for ordinary purposes in quantitative work. It is applied as follows:—The filtering bucket which has been unscrewed from the net is inclined so that what little water remains in it lies over the silk. By carefully tapping or rubbing the latter, this water can be got rid of; but in doing this great care must be taken that the water, and consequently part of the catch, does not run over the edge of the bucket.

The filtering bucket is now held over a glass tube or bottle, the tap opened, and the whole catch washed out by a strong stream of alcohol directed from a wash bottle directly on to the organisms on the *inside* of the bucket. By this means the catch is easily removed by the fixing fluid itself, the sea water is reduced to a minimum, and the catch is fixed and put in its preserving fluid as soon as possible after leaving the water, by means of one operation. The bottles can be stored away and taken to land for further investigation. To run the contents of the filtering bucket into salt water and carry to land is, even when only an hour intervenes, not at all advisable and, of course, impossible on a long cruise.

The Estimation of the Catch.

There are two methods at present in use by which the plankton tables are constructed. One is a simple method of estimation by examining the catch under the microscope, noting down the forms that occur and denoting their frequency by letters such as *c. c.* (very common), *c.* (common), + (neither common nor rare), *r.* (rare), *r. r.* (very rare). This method is still the most general one in use. The other method is that carried out by the Hensen School, and forms as essential a part of the quantitative work as the nets themselves. By this latter, the actual organisms present in a known fraction of the catch are counted. Since the first method still is the most common, it will be necessary here to emphasise its great defects and almost worthlessness for quantitative work when not supplemented by the other. Suppose that one has a certain plankton catch obtained by a vertical haul of the net through 40 fathoms, that the catch has been estimated, and according as the various forms are relatively frequent or rare, they have been designated with letters as above described in the tables. Now we will assume that a second catch taken in another place from the same depth has all the organisms present in the same relative proportions as in the first, but in double or treble the quantity. This would make no difference whatever in the tables, the relative frequencies still remain the same, even though a form which is represented by "rare" in both catches may be present two, three, or four times as many in one catch as in the other. Thus the tables could not be directly comparable for quantitative purposes. We have, however, assumed here that the constitution of both catches was identical— a thing of almost impossible occurrence. Let us assume now that the constitution varies, and that three catches are taken (an example given by Apstein), as one makes a voyage out from the coast, and that these are estimated by both methods. By counting, the first is found to contain 50,000 *Ceratium fusus* and great masses of the diatom *Sceletonema*. In the second catch, taken further out, there are still 50,000 *C. fusus*, but the diatoms have disappeared. At the third station *C. fusus* still remains at about the same number, but *Ceratium macroceros*, up till now rare in the catches, appears rather abundantly. Now, an investigator who simply estimated the relative frequency of these organisms would state that *C. fusus* was very rare in the first catch (since they were overshadowed by the great masses of diatoms), common in the second catch (where in proportion to the diatoms and other forms they *seemed* abundant), and again very rare in the third catch (where the *C. macroceros* has appeared so abundantly). In reality, however, the number of *C. fusus* has remained the same. An observer estimating by relative frequencies would have constructed a table and curve showing a great increase at Station 2, and then sought for an explanation of this increase, which in reality did not exist. If plankton tables are to be constructed for a large sea area, in order to compare the plankton at different places under different conditions of salt contents, temperature, currents, and other changing conditions in the sea, *quite false results* would be obtained from the method of estimation without counting.

Moreover, the reliability of such estimations is not good. In order to determine this, Apstein and one of his colleagues took four catches and first simply estimated them in the usual way, and then counted and estimated by the Hensen method (**14**). A section of the table will show the results. The first column gives Apstein's estimate, the second gives that of his colleague, and the third gives the true number present as found by counting the various forms present and then using letters derived from the frequencies determined by the counting, in order to compare with the other two columns.

	A., by estimation.	R., by estimation.	By counting method.
Rhizosolenia alata	r	+	rr
,, semispina	...	rr	rr
,, shrubsolei	c	...	c
,, stolterfothi	r	r	+
,, styliformis	+	+	c
Ceratium tripos	cc	cc	c
,, longipes	cc	+	cc
,, furca	+	c	cc
,, fusus	r	r	+
Cyphonautes	...	r	r
Limacina	r	+	c
Molluscan larvae	+	+	c
Oikopleura	c	+	c

KK

Eighty-one species were estimated in the catches, and in only one case did both estimations agree with the numbering. It was possible for three things to happen: (1) For both estimations to agree with the numbering; (2) for estimation and numbering to give parallel results, but not be alike; (3) for estimation and numbering to be contrary to one another. Only one species agreed in every respect, 13 species gave parallel figures, and in 67 species the estimations and numbering were contrary. Thus the personal error forms an additional source of failure in the simple method of estimating a catch by the frequencies, whereas by the counting method two observers will practically agree, if both count the same catch. Thus, for tables to be of any scientific worth in comparisons made to show the dependency on hydrographical or other conditions, or of the various forms upon each other, the catches must be made quantitatively and the unfortunately tedious method of counting followed.

A very detailed account of the apparatus and method of counting has been given by Jenkins (**12**), but the method as at present carried out for general work will be briefly described here, in order to give the complete procedure The first work consists in the estimation of the volume and the construction of curves to illustrate this. In most cases, unless the plankton is caught on an expedition lasting some months, the volume estimation will be made on shore. If it is required to estimate the volumes of the catches on board ship, the usual swinging table is required. The catch which has been fixed and preserved in alcohol is allowed to stand, and the alcohol decanted and its place taken by distilled water. The catch in distilled water is now brought into the measuring vessel. If distilled water is not used instead of the alcohol, the volume will be quite inaccurate, because a precipitate forms from the salt water that has been round the organisms when first fixed. This in appearance is like a diatom deposit, and the volume of a catch may be reduced to one-third by transferring from alcohol to water, owing to the removal of this precipitate. Ordinary measuring glasses are of no use for measuring accurately small catches. A special make of glass tube is used, the bottom of which is drawn out into a cone ending in a blunt point, so that a small volume of catch will occupy a considerable depth of this narrow termination. The plankton catches in distilled water are transferred to these tubes and allowed to settle for 24 hours. A mark is then made with ink on the outside of the tube at the level to which the sediment attains, and the catch is again removed. The quantity of water measured out by means of a burette, which takes up the same space as the sediment, will be the volume of the catch.

This volume estimation is necessarily very rough, since, especially if diatoms be present, a quantity of liquid remains between the organisms and causes the sediment to appear much greater in volume than it really is.

Having found the volume of the plankton in cubic centimetres, it is multiplied by 80 (for the Middle Apstein Net pulled up $\frac{1}{2}$ metre in 1 second), and this gives the volume present in a column whose area is one square metre and whose length is the distance through which the net has been hauled. For purposes of comparison and the making of curves, the average volume per cubic metre is generally reckoned from the above.

The next division of the work consists of counting the organisms. For general use with the Middle Net the catch is brought into 50 c.cm. of distilled water. If the catch is very large a further dilution may be necessary. This 50 c.cm. with the catch is placed in a shaking flask, and, by means of the plankton pipettes, 0·1 c.cm. of the fluid is withdrawn after carefully distributing the organisms by thorough shaking.

It is necessary here to emphasise the use of these special pipettes devised by Hensen (**1** and **5**), since no other apparatus will allow of the accurate abstraction of such small quantities.

First, 0·1 c.cm. is taken and removed to the counting plate under the microscope, and the organisms counted. A sheet of paper is used with the names of the species to be counted, and, as each form is passed over, a stroke is placed opposite the name on the paper. Since 0·1 c.cm. is $\frac{1}{500}$ of the 50 c.cm. to which the catch was diluted, the numbers must be multiplied by 500 to give the full number for the catch, and then from this the number per cubic metre is calculated. In general use only one plate is counted with 0·1 c.cm., and then a pipette abstracting 0·2 c.cm. is used in the same way, but only those organisms occurring in very small numbers, or doubtful in the first plate, are counted in the second, so that whereas 50 species may be counted in the 0·1 c.cm., this number may be reduced to 12 in the 0·2 c.cm. Following these two plates, 0·5 c.cm. and then 1 c.cm. are taken, and finally the rest of the catch for the larger forms and for rare forms is counted, making a total of five plates. When greater accuracy is required, more plates are counted for the same pipette until the difference between the number of organisms on the last counted and the average number for the previous plates is less than 5 per cent. If an organism is required for a preparation or for further observation, it can be removed from the numbering plates by very small capillary pipettes about two inches long.

During the last few years it has become obvious that the catch with the fine meshed bolting silk only gives an incomplete sample of the plankton present in the sea at any given place. Kofoid (**8**) and Lohmann (**11, 13** and **18**) have both emphasised this error, but it is to the latter that we are indebted for a complete investigation of it and of the means of overcoming any failings in this direction. In an important paper, published in 1902, an account of the comparisons between various methods for catching the smaller plankton organisms was given in detail. The subject has since that time been further investigated, and whilst writing this a detailed and very elaborate account, bringing the plankton work up to date, is going through the press (**18**). By the kindness of Prof. Lohmann, I have been able to see his tables and read through the proofs of this work. Hensen's method rests on two hypotheses:—(1) That the pelagic organisms in the sea inside a region of *like conditions of existence*, with regard to time and space, are so equally distributed that by the investigation of relatively small quantities of water, a sufficiently accurate picture of the quantity and quality of the plankton for the whole region can be obtained. (2) That the apparatus used for these investigations, namely, the Hensen net, even with its uncontrollable errors, gave essentially a true estimate of the plankton. The first hypothesis will be discussed later. With regard to the second, there is the possibility of the net failing to catch an important part of the plankton, through small organisms passing through the meshes. Hensen himself in 1887 stated (**1**) that if he allowed the water filtering through the silk net to pass through close silk, filter paper, &c., and investigated the residue, many diatoms, peridinians and silicoflagellates would be found to have passed through the net. He believed, however, that the influence of this loss, on the constitution of the catches and the results given by numbering, was of no essential importance, since the mass of the forms slipping through was only small in comparison to the quantities caught by the bolting silk; and, in any case, Hensen gave his numbers as a *minimal value*, recognising that a loss must occur. Kofoid in 1897 (**8**) through new investigations in the fresh waters of North America came to the conclusion that the loss which enters into the results, when nets of "Müllergaze" are used, was much greater than Hensen had supposed. By using filters of hardened paper, he demonstrated that only 2-50 per cent. of the organisms were caught by the net. Lohmann (**13**), when investigating the Appendicularia, also found that the quantity of small forms going through the net must be of far greater importance than was formerly supposed. Nothing shows the loss more distinctly that the investigation of the food of the plankton organisms themselves. One finds in their alimentary canal the remains of the smallest diatoms, Peridinians, Coccolithophoridae and Silicoflagellates, of which an ordinary net used in the same water in which the "devourers" (Pteropods and Appendicularia, &c.), lived, would contain none or few. As showing the importance of this loss, Lohmann mentions the fact that the Coccolithophoridae, which play a great part as food for the plankton animals of the North Sea, are almost unrecorded in the tables when bolting silk nets are used.

The most favourable organism on which to study the food of the plankton animal is the Appendicularian, which does not take food directly into the alimentary canal, but secretes a special structure, the "house," for the purpose of catching its food. This is a perfectly transparent structure, and under the microscope can be seen to contain the uninjured and living food before it passes into the alimentary canal. Here in this filtering apparatus of the Appendicularian are numerous naked Rhizopods and Gymnodineae, together with smaller skeleton-carrying Rhizopoda, completely absent from the net catches. It is obvious, therefore, that if a complete knowledge of the plankton is to be gained, other methods must be applied.

It was assumed that the loss of plankton by the use of No. 20 silk in the net was unimportant, and that the real masses of plankton in the sea might be only 2 to 3 times greater than the figures given by the net. This would certainly be of no great importance if only the volume or weight of the plankton present was required without any reference to its constitution. If one requires however, the chemical constitution, it is quite incorrect, and this applies further to the qualitative and quantitative counting method, because the animals and plants in the catch will occur in quite different relative proportions from the true conditions present in the sea. Since the mesh work of the net itself has a large area, many of the small forms which could pass easily through the meshes will be caught on the net tissue itself; and this will give a more deceptive appearance of reliability than if these forms had altogether escaped.

Again, the fractions of these small forms caught is not always the same, because if the sea contains a great number of diatoms (as *Chaetoceros*), the meshes of the net will be gradually filled up, and the spines interlocking will cause the net to act as a much finer filtering material and hold back many species which would otherwise slip through. This accounts very often for the large catches with the nets, when diatoms are very abundant. By comparison of the net catches with the other quantitative catches made by the apparatus to be described below, it has been found that the Metazoa, with few exceptions, are completely or sufficiently caught by the plankton net, whilst of the Protozoa only a few large forms, *Noctiluca*, *Ceratium tripos*, &c., or species with long spines, will be obtained. In fact, the number of individuals present in the sea is from 5 to 100 times greater than is demonstrated by the net, and the species which form this loss are not of " no importance," but are the chief forms of food for the larger species, and therefore of great significance in the metabolism or see-saw of life in the ocean. If, therefore, the Metazoa or larger Protozoa and Protophyta only are to be studied, the net can be used as the best instrument by far for the capture, whether for quantitative or qualitative purposes. If a complete investigation of the plankton is to be made, and the relation of larvae to adults and food to the eaters of it are to be considered, other kinds of apparatus must be used. Of these, the most important is the Pump and Tube, by which water is pumped up to the boat and later is filtered. In shallow water, and up to 100 metres deep, the net can be more or less supplanted by the pumping method, but, unfortunately, for greater depths, and for regions where there are strong currents, the pumping method is hardly applicable.

Pump, Tube, and Filter Method.

Essentially the method consists in the pumping up of a vertical column of water, which is filtered on the vessel or, later, on shore. An indiarubber tube of sufficient length for the deepest regions is required, and this is lowered vertically in the water by means of a rope attached to the lower end, and in the same way is slowly pulled up, whilst at the same time the water is pumped out of the upper end by a small brass pump.

By lowering the tube it fills gradually with water out of the various depths through which the lower end sinks, so that finally it contains a water column, consisting of water from all depths between the surface and the lowest point reached. When the tube is again slowly raised through this column to the surface, it is once more filled by water from each layer. Thus by repeatedly lowering and raising, whilst the pump is worked, any quantity of water may be obtained, representing a vertical column whose height is that from the lowest point reached by the tube up to the surface and whose other dimensions can be reckoned directly from the volume of water collected.

The water in the sea will naturally rise in the pump tube of its own accord until it attains the same level as the surface. It is only necessary, therefore, to use the pump to lift the water from the surface of the sea into the boat, and a small pump is accordingly quite sufficient. It is even possible in a boat with a deep bottom, where the upper end of the tube can be placed lower than the surface of the sea, to siphon up the water, but usually, owing to the motion of the boat, this method is not successful. The whole length of the tube should be fastened to a rope which will bear its weight. If currents are present, the rope and pipe must be kept vertical by means of a sinker.

A very simple and cheap arrangement was constructed and used by Lohmann in the Mediterranean (**13**), so that the simple turning of a windlass both worked the pump and pulled up the tube. Thus the rate of pumping and the pulling up of the tube were always in the same relation, however quickly the windlass was turned. Moreover, the direction in which the windlass was turned had no effect on the working of the pump, which continued to raise water whilst the tube was either lowered or raised. Since it is questionable whether this method of sucking up the water would have an effect on the catch, the entrance of the plankton and its passage up a tube was observed by Lohmann by using a glass tube. It was distinctly seen that the organisms in the centre of the tube ascended more rapidly than those against the walls. This difference in the current is, however, of no importance for the equal raising of the whole water column, because from each section the same quantity of central and peripheral water will be taken up respectively. It was noticed that some of the large animals were sensitive to the streaming, and Copepods, for example, moved energetically against the current. If, therefore, the current is slow, it is possible for the larger forms to move out of the tube, but, since the average speed of the current is 57 cm. per second, this is impossible; and any loss occurring when the pumping method is used applies only to the destruction of fragile forms in the filtration.

The water must be filtered, either on the ship or when conveyed back to the laboratory. The latter is probably the more simple. The water is pumped into large sulphuric acid "carboys" of about 28 litres contents, and a $\frac{1}{2}$ litre of commercial formol is added so that a 2 per cent. solution results, which suffices to kill the organisms and to fix them. The filter is simply hardened paper, which is folded into a cone and is held in a zinc funnel of about 50 centimetres diameter at the mouth. It is best to construct an arrangement so that the water can run from the carboys into the filter at the same speed as the latter filters. The whole can be then left to run of its own accord, with but an occasional glance to see that the filter does not become stopped.

METHODS OF PLANKTON RESEARCH. 525

The filter is carefully washed down towards the point of the cone when all the water has been filtered, and this is perforated by a pointed glass rod, whilst held over a bottle for the reception of the catch. The filtrate is then carefully washed through the perforation by means of a wash bottle provided with a strong indiarubber ball, in order to get a powerful current. The original volume of water collected being known, the catch as now obtained can be diluted and portions extracted for counting as explained above.

By fishing with the net and also with the pump and tube simultaneously, or directly after one another, and comparing the results, it is possible to determine the loss by the net due to its inability to retain the smaller forms.

If, however, exactly like methods are employed at the same time and place, and close to one another, the catch is different. This divergence, due to an irregularity in the distribution of the organisms in the sea, has been emphasised by Herdman and will be mentioned later. It must be borne in mind, therefore, when comparing the *unlike* methods, that a certain irregularity in distribution already exists. After reckoning the volume of water filtered by the net and reducing the number of organisms found in the whole catch to the number present in a volume of water equal to that collected by the pump, it was found by Lohmann that in the case of Copepod nauplii the net lost 74·5 per cent. This is a very important constituent of the plankton fauna. Other organisms were present in 1,000 litres of water in the following numbers:—

	Silk net No. 20	Pump, tube and filter.
Globigerina	250	2,125
Radiolarians	2,350	3,860
Cystoflagellates	20	20
Tintinnidae	475	19,900
Naked ciliates	some	35,300

The Radiolarians were caught almost equally well by the net owing to their shape and large spines. The Tintinnidae could easily slip through when meeting the net in the direction of their long axis.

For Diatoms and Peridiniae, &c., the following results were obtained:—

	Silk net No. 20	Pump, tube and filter.
Sceletonema	...	418
Coscinodiscidae	141	6,444
Rhizosolenia alata	1,143	4,013
Chaetoceros Ehrenbergi	44,317	149,793
Ceratium tripos	about equal in each.	
Peridinium divergens	72	317
,, globulus	15	1,190
,, pellucidum	5	1,089
Coccolithidae	28	11,267
Dictyocha	54	6,241

The comparisons for the Protophyta show, therefore, that the constitution of the plankton catch was completely altered by the use of the pump and filter. Out of more than 2,000,000 plants only 110,000 were caught by the net with No. 20 silk, and 9,000 animals out of the one-third million caught by the pump and filter.

The actual numbers have, however, not so much value, because the smallest organisms give the largest numbers for the loss. If the mass, however, be calculated it is found that the pump and filter gave 52·4 c.cm. as total catch out of 1,000 litres, whilst only 21 c.cm. were caught by the net.

Hence the constitution of the plankton, formerly determined solely by net catches, must contain greater errors for many forms than were supposed.

Method of Investigation for the Smallest Organisms.

It has been demonstrated by several observers that many of the small and fragile forms, without skeleton, are either destroyed completely during the filtration or pass through the filter, so that even the method with pump and filter fails to give the smallest forms and also the bacteria.

This loss is easily seen by an investigation of the filtrates from a hardened paper filter, which reveals the fact that as much as 26 per cent. of the Gymnodineae and the same per cent. of naked Chrysomonads can pass through; and a much greater percentage of bacteria would be found to have done so.

Of even greater importance than this loss is the fact that many softened fragile forms are killed by the filtration, and Monads, Amoebidae, and small Gymnodinae will be absent for that reason from the filter catches. Moreover, it is very difficult to recognise most of these forms when fixed and preserved, and, therefore, for these forms alone it is necessary to use other apparatus which does not require any filtering mechanism and which will allow of the organisms being studied in the living condition.

The apparatus consists of (1) a means for procuring samples of water from different depths, and (2) a centrifuge. By this method small water samples can be taken and examined from the various parts of a water column down to the greatest depths in the ocean, and from these, by interpolation, the average number or volume of organisms present in the complete vertical column can be calculated.

A "Krümmel" water bottle (fig. 4) is the most satisfactory for the purpose of obtaining water samples. It is already made sufficiently large to bring up three litres of water from any depth required, which is sufficient to allow of a portion being used for the determination of the hydrographical conditions—an absolute necessity in plankton work. The water bottle is lowered open and closed at the required depth by a falling weight, sent down the rope from the surface. The water required for the centrifugal investigation is taken from the water bottle, received in glass-stoppered bottles and placed in a cool and suitable place until the laboratory is reached: the examination should take place as soon as possible after the catch is made.

The employment of the centrifuge for this work was first suggested by Cori (**6**), but it has not been much used, as it was pointed out that the action on various organisms is selective, and the sediment is therefore not in its constitution a true sample of the plankton present. This selection does not come into play if the organisms are dead and in a preserving fluid that is lighter than water, consequently Kofoid has used the method for catches preserved in alcohol. Dolley (**16**), by using a very powerful centrifuge which he termed the "Plankton-okrit," and which gave 8,000 revolutions per minute, had complete success in the sedimentation of living plankton. Other American workers, also, have used this method, and have found that for accuracy of determination it far exceeds all other methods at present employed.

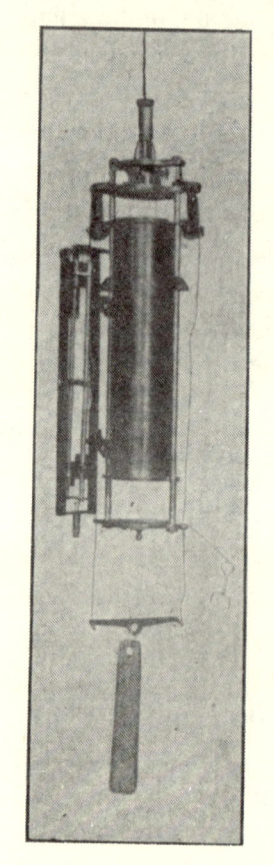

Fig. 4.—Krümmel water bottle.

Kofoid, however, raised the objection due to its selective influence, and found that many organisms would not form a sediment. One must remember that no method will be accurate for all forms, and none of the methods here described will alone give an accurate sample of *all* the forms present in the plankton.

Lohmann has used with success a centrifuge which carries four glass tubes and which gave easily 1,300 revolutions per minute when turned by hand. He found that 9,000 revolutions, in seven minutes, were usually sufficient.

For the investigation of the living forms, samples of only 5 to 15 c.cm. are taken. The tubes for containing the sample, on the centrifuge, are small cylindrical vessels, with the point drawn out slightly to form a cone-shaped end, in which the material will form a well-defined sediment.

After the completion of centrifugation, most of the water can be poured away, and the sediment remains undisturbed with the water that fills the conical end. By means of a pipette, the sediment, through repeated sucking up and forcing out, is finely distributed in the water remaining, and is finally completely sucked up and transferred to the glass numbering plate used with a specially constructed microscope stage. This is much smaller than the numbering stage for the large net catches, is comparatively cheap, and can be fitted to any microscope.*

The conical end of the tube is now washed out with a very little water (some of that originally poured off), and this is added to the main part of the catch on the glass plate. The whole catch should form only a single drop, such as can be covered with an ordinary 12 mm. cover glass.

If the water contains many flagellates and ciliates, the counting of such rapidly moving organisms is impossible. The cover glass should then be held over

* Zwickert, Optician, Kiel, is the maker of this stage.

osmium vapour for a short time before being placed on the drop of water and sediment, which is sufficient to cause narcotisation. When high powers are used, it is impossible to count the organisms in the complete area of the drop. In this case only a fraction is counted, in the following way. The glass counting plate on which the drop rests is crossed by a series of parallel lines running in one direction, from the observer. If the number of spaces between the lines, which are covered by the whole deposit when the cover glass has been applied, is divided by five, that gives the spaces in which the organisms should be counted in order to arrive at one-fifth of the total in the catch. These spaces counted should be equally distributed over the whole area, so that an average can be obtained. It is best first to count the organisms in one-fifth of the mass with a high power for the smallest and most frequent forms, and then, under lower magnification, the whole mass for the larger and less frequent.

It is well also to take another quarter of a litre of the same sample from the Krümmel bottle (fig. 4) and add formaline to make a 1 to 2 per cent. solution. After this liquid is partly removed from the catch by filtration through very fine filter paper, the residue can be centrifuged and compared with the centrifuged samples of living forms. The amount of water taken for the centrifugation of the latter must depend on the number of organisms present. If 15 c.cm. of water are first taken and centrifuged, and too many organisms are found for an easy count, then it should be discarded and a smaller quantity taken.

The extremely small quantity of water taken for these samples is astounding and might be considered insufficient for two reasons:—First, that not enough animal and plant life are present in such small volumes; and, secondly, that they are absurdly small for quantitative estimations of a column of water or for finding the true conditions in the area where the samples are taken. This does not, however, seem to be the case; and with regard to the first point, it is surprising what a mass of material 15 c.cm. of water gives with the centrifuge, so much that Lohmann had often to take less. The second difficulty is also only apparent, because when the Hensen plankton nets are used the sample of water taken for counting bears an equally extremely small relation to the quantity of water that has passed through the net. The only real difference between the two former methods, the net and pump and filter, and this method is, that in the former the plankton is collected from relatively large masses of water, and small quantities are taken out of this as samples for counting, whilst by the centrifugal method the small quantities are taken directly from the sea. In stating this, however, we are again confronted by the doubt as to the equal distribution of the plankton in the sea, which will be mentioned later. Lohmann thinks that the distribution is sufficiently like to allow of such small samples being reliable guides to the quantitative constitution of the plankton.

In order finally to calculate the average number of organisms present in a column of water, from which various samples have been taken at various depths, the following formula should be used. Assuming that the samples A, B, C, D are taken at increasing depths, separated by the distances a, b, c, then $\triangle$ is the average for the column.

$$\triangle = \frac{Aa + B(a+b) + C(b+c) + Dc}{2(a+b+c)}$$

If one wishes to make a complete investigation of

the plankton, either qualitatively or quantitatively, one must use all the three above methods side by side. If only a definite part of the plankton is to be studied, then the method must be chosen to suit the case. For large crustacea, fish eggs, and medusae and other large plankton forms occurring but seldom in the water, in comparison to the Copepoda and smaller forms generally, the large Hensen net described by Jenkins should be used, to work through much larger quantities of water. For the main constituents of the plankton, the Copepods, Ceratium, and, in fact, for general use, the Middle Plankton Net of Apstein is to be preferred. This has been the chief instrument used in German investigations, and holds its place because of the ease of working and the general applicability. It must be borne in mind, however, that when these nets are used there is a considerable loss, as has been shown above, and, therefore, when possible, the use of the net should be replaced by the pump, tube and filter. In fact, in water of moderate depths, and shallow water, the pump method is "the" method for plankton investigation, and the net and centrifuge should be used only to complete the results when the greatest possible accuracy is required and the complete constitution of the plankton is to be discovered. The points against the pump method are the difficulties encountered in deep water or when there is a strong current, together with the greater time that is required for pumping and filtering.

There is another possible method for arriving at these results, which could be applied to the greatest depths, and in comparatively stormy weather. It is to use a water bottle for collecting a volume sufficiently large to allow of its being filtered and examined in the same way as the water from the pump. In this way, however, the contents of a vertical water column would have to be calculated from the samples taken at various depths. For this purpose, too, it would be necessary to obtain more water than is brought up by the bottles now in use. The Krümmel bottle as now used by hydrographers has three litre contents, and there should be no difficulty in increasing the size to five litres, which would give a sufficiently large sample.

Other Plankton Apparatus Used for Qualitative Work.

The apparatus above described is intended for the quantitative estimation of the plankton in volume, chemical constitution, or by the counting method of Hensen. For mere purposes of qualitative investigation the procedure is naturally much more simple. The net described is, in any case, of great use as a vertical net, and would completely supplant the pump and filter, whilst the centrifuge would be used to catch organisms that pass through the net. Under special circumstances, however, other nets are used, which are coarser, and have their special use according as they are for surface or deep work, and for small, large, or very large organisms. Again, it is sometimes desirable to investigate the plankton of areas over which the ship is passing at a considerable speed, and for this purpose other devices are necessary.

For general use in qualitative work there is the ordinary small tow-net, well known at all biological stations. This is constructed out of bolting silk, and has the same conical shape as the vertical net, but does not have a mouth-piece as described for the quantitative nets. No calculations can be made as to the quantity of water it filters. These nets are generally used for horizontal fishing, and should be made with No. 20 silk for smaller organisms, and with No. 12 or No. 3 when larger forms are specially required, since with these latter silks more water will be filtered in a shorter time, and the catch will be free from the masses of small diatoms, which are not wanted.

All plankton nets should be fitted with a metal filtering bucket attached to a brass ring, which forms the base of the net, by means of a screw attachment, or by the simple device of a bayonet joint. This bucket consists simply of a brass cylinder, the size varying according to the size of the net; the lower end is closed by a piece of silk of the same mesh as that used for the net, and attached to the brass cylinder by means of a clamp ring.

When the net has been used, it is only necessary to wash it down with a few pails of water thrown on the outside, and to unscrew the bucket with the catch. If time is short, and the catch has to be preserved as soon as possible, the silk itself can be removed from the bottom of the bucket, rolled up, and dropped into a bottle of alcohol without removing the organisms; a new piece of silk is then placed on the bucket and it is ready for further use.

For fishing pelagic eggs, young larval stages of fishes, or when large catches are desired for histological or anatomical work of large plankton organisms, such as Medusae, very large Copepods, Sagittae, pelagic worms, &c., the German "Brutnetz" is a very successful instrument. This, as the name implies, was constructed for fish eggs and larvae. It is much cheaper than the silk nets, since it is constructed of "cheese cloth," or of good canvas. This is simply a conical net about three metres long, the mouth of which is kept open by a wooden ring of cane 80 to

90 cm. in diameter. About one metre from the mouth the net is attached to a second wooden ring, to one point of which is attached an additional rope from the ship, so that when the haul has been made the net may be rapidly pulled up edgewise without offering opposition to the water. The apex of the net where the usual bucket is attached has a diameter of 10 centimetres.

A modification of the "Brut" net, called the "Scherbrutnetz," has been constructed to allow of the application of such a net as the former to the collection of the plankton from deeper layers. The essential feature is a strong galvanised iron plate, hinged, as seen in fig. 5,

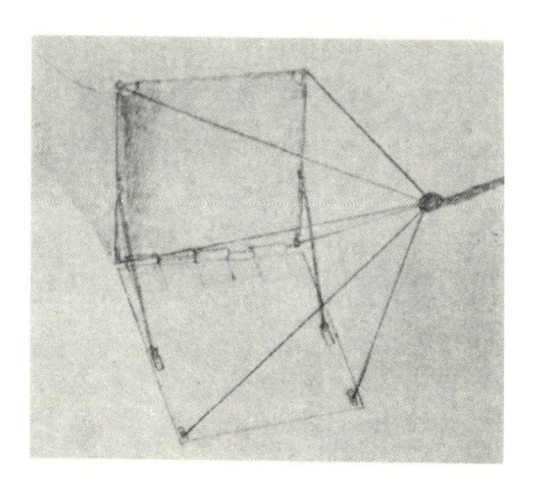

Fig. 5.—The "Scherbrutnetz."

to one side of the square mouth of the net. This "shear" board is, however, not allowed to move freely, but is fixed so that it makes an angle of 125° with the plane of the mouth of the net.

When this net is hauled the water presses against

the "shear" plate exactly as on the otter boards of the otter trawl, or like the wind on a "kite," causing, in this case, the net to sink in the water. Knowing the length of rope allowed to run out, the true depth of the net can be easily found by using the Apstein apparatus already described and measuring the angle the rope makes with the horizon.

A still larger net than the "Brut" net is sometimes desirable where very large catches are required of the larger plankton forms from deep water. For this purpose there is the so-called "Knüppel" net (fig. 6), which is

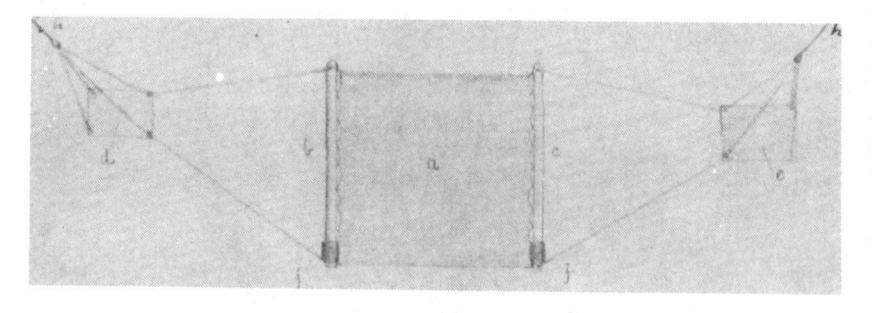

Fig. 6.—The "Knüppel" net.

worked on the principle of the otter trawl. The Knüppel net can only be worked satisfactorily when a fair sized vessel is available with a steam winch. The net which I have seen in use has the following dimensions. The net itself is made of strong canvas and is about 15 to 20 feet long, the mouth is square, each side of the square having a length of 8 feet, and each of the mouth edges is formed by a broad piece of sailcloth, to which the filtering canvas is sewn. The apex of the net is as usual fixed to a metal bucket, in this case about 9 inches in diameter. The two vertical sides of the mouth of the net are fixed at intervals to two stout poles (fig. 6, *b.* and *c.*) 9 feet long, and provided at the lower end with a heavy lead sinker (fig. 6, *f. f.*) in order to keep them vertical in the water. Each pole is attached by two strong ropes, fixed to the upper and lower ends respectively, to the "otter" boards (fig. 6, *d. e.*). The ropes are about 12 feet long, and represent the "foot rope" and "head line" of the otter trawl. The otter boards are strong wooden structures bound with iron, and measure 4 feet by 2 feet. When the net is lowered it sinks, owing to its weight, and the pressure of the water forces the two otter boards outwards, thus pulling the two vertical poles as far apart as possible, and in this way the mouth of the net is kept open. This net can be used satisfactorily at very considerable depths.

There remains to be described a very convenient and simple little instrument by which catches can be made whilst a vessel is travelling at a considerable speed, and, consequently, any changes in the nature of the plankton between two stations can be followed without interfering with the progress of the steamer. Several instruments have been invented for this purpose, but it will only be necessary to mention here the "Plankton Röhre," which was invented by Apstein and has not yet been described. It has the great advantage of being simple, and so small that it can be very easily carried about with one, so that plankton catches may be made on a sea voyage other than a scientific expedition. Fig. 7 shows the external appearance of the instrument. The Plankton Röhre consists simply of a brass tube 25 cm. long, one end of which, however, is not of the same diameter as the rest of the tube, but forms a truncated cone, making the mouth opening of the tube very narrow. The diameter of the cylindrical section of the tube is 3·5 cm., and the length 22·5 cm. The conical mouth-part is 2·6 cm. in length, and the opening is only 1 cm. in diameter. This narrow opening is for the entrance of the water, and is, therefore, the front end. The other end of the tube is closed by the filtering apparatus—simply a piece of No. 20 bolting silk, or coarser, if required, which is fixed in the usual manner by a clamp ring.

To one side of the tube is attached a heavy strip of lead (fig. 7) to keep the instrument from being pulled out of the water. This will consequently be the under side, and to the opposite and upper sides of the tube, at the front end of the cylindrical portion, two ring attachments are screwed, by which the whole apparatus is

Fig. 7.—The "Plankton-Röhre."

fastened to the hauling rope. The action of the instrument, when pulled at a considerable speed, depends on the small area of the opening, which allows but little water to enter, and therefore there is but little strain on the silk tissue, so that this is not torn nor are the organisms damaged. I have seen it used successfully at a speed of eight and a half knots. One disadvantage is that very small catches are obtained, even when towed rapidly for a quarter or half an hour, but since the apparatus was not intended for obtaining large quantities, this does not detract from its usefulness.

Results of the Plankton Work and Its Aims.

I propose now to discuss briefly some of the results obtained by the quantitative method, and the present position of the work. The first ideas came from Hensen's investigation into the distribution of the eggs of the plaice in Kiel Bay. These are planktonic eggs, which float as long as the salt contents of the water does not sink below 1·78 per cent., and this is seldom the case in the West Baltic. It became evident that these eggs extruded at many spawning grounds, must necessarily distribute themselves widely, and the longer they remain floating the more movement will take place and the more equal the distribution will become. On this equal distribution of the plankton particular stress must be laid, because it forms the foundation on which the value of the quantitative work depends. The investigation of these fish eggs led to notice being taken of the other planktonic organisms, and, finally, Hensen says—" The sea has its yearly production in animals and plants, just in the same way as a garden or field. For the land, it is an almost impossible problem to work out this production, because even if one, with extreme weariness, worked out the fauna and flora completely and quantitatively for a small area, at a short distance from this point the conditions and distribution would be altogether different, and we could never be certain that what was found in the small area would be a true sample for a large area. In the sea the conditions are quite different, the species and number remain to a certain extent everywhere constant."

Thus, with two fundamental hypotheses the quantitative method has been applied. These are, first, that the plankton organisms are equally distributed in the sea where like conditions of existence are found; and, second, that this equal distribution is sufficiently exact to allow of relatively small quantities of water being taken as samples of the total production of the area. Amongst applications of the plankton quantitative method, the following are perhaps the chief:—

1. To estimate the produce of the sea or ocean or any particular area per year, and to compare the productiveness of different regions.

2. To investigate the dependence of the plankton as a whole, and also of the different organisms, on the hydrographical conditions, such as light intensity, temperature of the water, salt contents, currents, and, at the surface, wind and waves.

3. To investigate the relations existing between the different plankton organisms themselves, their dependence on one another, and the relation between the "eaters" and the "eaten."

4. To investigate the reproduction of the various plankton organisms, and of others not planktonic, but whose eggs or larvae are pelagic; the relations existing between the number of eggs, the number of larvae and the adults, and the length of time occupied in the life history.

For the purpose of investigating the condition of the plankton on the high seas, the "Humboldt-Stiftung Expedition" already alluded to was fitted out, no doubt stimulated by the results of the English "Challenger" Expedition. Not all the German zoologists were in favour of the object, and Haeckel in particular wrote against it (2), arguing that the pelagic organisms were not equally distributed, but that they travelled in swarms, or at least were so irregular in their occurrence that samples taken at some distance from each other would be valueless for a quantitative estimation. He has been answered in detail by Hensen (3). The vessel chosen for the voyage, the "National," started from Kiel, July 6th, 1889, and proceeded northwards through the Kattegat and Skagerack, and then across the Atlantic Ocean to Greenland. Vertical plankton hauls were taken at intervals on the way. From Greenland the course was directed S.W. for the Bermudas, and consequently went over the banks of Newfoundland and across both the Arctic Labrador Current and the warm waters of the Gulf Stream. From the Bermudas the course ran almost parallel to 30° N. lat., and thence across the Sargasso Sea, until the meridian of 38° W. was crossed, and the Cape Verde Isles steered for. This direction was followed further to the S.E. up to the Island of Ascension, then west again over the ocean to the Amazon's mouth, practically along the South Equatorial Current. From the latter place the vessel returned direct to the English Channel.

The series of reports by specialists on the different groups of pelagic organisms are not yet all published, and the general conclusions have not yet been put together, but from some results given by Hensen (4), it was shown that the quantitative catches agreed, as far as volume is concerned, far better than was expected, and gave still further proof of the equal distribution of the plankton. Several interesting points were brought to light in connection with the distribution. An unexpected result was that, contrary to the conditions existing on the land for both animal and plant life, the plankton was decidedly more abundant in cold and temperate regions than in the tropics. The difference in volume in the catches between Greenland and the Hebrides and those taken from the Sargasso Sea is truly remarkable.

This result was quite unexpected by those who had worked at the material rich in species, taken by the

"Challenger" in the warmer seas, but since these were only qualitative catches, no comparison could be made, and we are faced with another problem—What condition is it in the sea which makes it more favourable for life in the colder regions? This has been referred to in an important paper by Brandt (**9**), which should lead to further investigation.

The catches, however, made in the cooler regions are in reality much smaller than they appear, because the bulk of the organisms present are diatoms. When the volume of plankton is estimated by allowing the catch to settle down for 24 to 48 hours, an easy, but not always reliable method, it will be seen that a diatom catch refuses to sedimate as one where Copepoda or Ceratium are present. Thus a diatom catch appears to have a much greater volume than is really the case, even after it has stood for weeks. Furthermore, as Lohmann has since pointed out, the presence of diatoms in considerable quantity increases enormously the catch because the diatoms entangle themselves over the meshes of the net, and render it a much finer filtering tissue.

In the tables given in the published results of the expedition there is in one case an increase in the catch between the stations of from 5 c.cm. to 156. This sudden increase was due to *Calanus finmarchicus*; and one must evidently consider this as a swarm. The nearest land was 500 miles distant, and, after the large catch, the catches at the following stations again showed quite a small volume.

It is a well-known fact that the Siphonophora, *Porpita* and *Velella*, are found travelling in great shoals together, and in the accounts of the expedition we find that south of the Cape Verde Isles shoals were very frequently met with, amongst which occurred swarms of

Physalia, Pyrosoma, Salpa, Schizopods, Janthina, Beroe, Pteropods. Thus, one of the nets with an opening of 1 13 square metres hauled up from a depth of 500 metres 520 Pyrosoma on one occasion. The question is—What has brought these together? Neither wind nor their own motion, unless governed in some way unknown to us, could do this. At another place 5,860 Doliolum were caught in one haul of the net, as against 1,500 in all the other catches together.

Darwin, and other observers, had previously recorded the fact that long stretches of the sea were frequently met with, deeply coloured by the abundance of some animal or plant species, as, for example, *Trichodesmium erythraeum*. It is this association of planktonic organisms in swarms that is now being investigated by Herdman (**17**), and it will be interesting to see how far it extends. (See also **19**.)

With the exceptions of some swarms, Hensen maintains that the equal distribution was never disturbed to such an extent, where the conditions remained the same, as to render the application of the quantitative method unsatisfactory. In the Sargasso Sea, for example, where there is no current practically speaking, the catches were astonishingly small, but the volume remained constant over a stretch of some thousand miles. It is possible, however, that the constitution of the catch was altered.

The results of this expedition tend to show that the ocean waters are very poor in plankton. There is a sharp distinction existing between oceanic and coastal forms; many of the oceanic species are never or only exceptionally seen near the coast, and one must visit an oceanic island in order to study them. What is the barrier to this distribution? The oceanic species are neither more frail nor more nor less active than many of the coastal forms. This brings up the whole subject of the different conditions to which the planktonic flora and fauna are subjected in oceanic and coastal regions.

It is an important point, because though the oceanic regions are of very great extent, the waters that are of practical importance for fisheries are our coastal seas, like the North Sea, the Irish Sea, &c., where the depth is nowhere very great, but where the plankton is very abundant, and where a thorough planktonic investigation should be of considerable economic value. The bottom of these seas and all coasts is inhabited by a large and varied animal and vegetable population; the laminarian zone, for example, is probably the richest area of the earth's surface in animal life. From the Echinoderms, the Crustacea, the fish, &c., found in these shallow seas arise myriads of larval forms, which, after a pelagic life, again migrate to the bottom and continue their existence as fixed or sedentary animals. Thus the plankton of the Irish Sea is made up to a very large extent of eggs and larvae of animals which are not pelagic when adult. In one group the Crustacea, for example, there are orders like the Copepoda, which are typical plankton forms and remain, with few exceptions, free-swimming and pelagic during their whole life; while the Cirripedia, on the contrary, have the pelagic larval forms, but their adults are fixed, and therefore not constituents of the plankton.

Then, again, the Hydrozoa contain forms which alternate between a fixed hydroid generation and the free medusoid of the plankton. Certain Nereis species are to be found creeping about the bottom or swimming sluggishly, but when sexually mature undergo a considerable change in structure, the parapodia become modified for swimming, and the so-called Heteronereis stage may then be caught in considerable numbers in the plankton nets on the surface itself. Thus the coastal plankton is made up of very diverse forms, partly always plankton, partly plankton during only certain periods in the life history. Out on the high seas, where the ocean floor is a waste as far as fixed living plants are concerned, and the water is 2,000 or more fathoms deep, the plankton contains no forms arising from the bottom. Thus the oceanic plankton is subjected to different conditions of existence, and the absence of these forms in general from our coasts is probably due to their failure to compete with the abundant pelagic life of the shallower waters. The ocean, according to the figures provided by the oceanic quantitative plankton expeditions, may be considered as a desert, receiving its life from all sides, and from this producing forms that are peculiar to it, and have in the struggle for existence been driven further out.

It is now obvious that the most important regions for the employment of quantitative methods are areas like the North Sea and the Irish Sea, or coastal water generally, where, since the plankton is of great importance as the food of fishes and contains the eggs and larvae of the latter, the results may be applied to the elucidation of problems in fishery work. It is necessary, also, to determine to what extent the plankton is dependent upon the various hydrographical conditions, and also what variations occur during the year. Since the year 1901, Great Britain, Germany, Norway, Sweden, and other countries in Europe bounding the North Sea and Baltic, have together investigated the hydrographical and biological conditions of these two areas. Grants have been given by the Governments concerned and suitable steamers provided, and an International Committee has drawn up a programme in accordance with which various stations are visited four times a year, and scientific observations are carried out simultaneously, with the object of making a complete investigation of the whole area.

One of the most important questions is, naturally, the condition of the plankton at different places in these areas at the same time, and the variations during the year. Since at the time that these plankton investigations are carried out, the hydrographical conditions are also very thoroughly observed, there is an excellent opportunity of comparing both. Unfortunately, so far as the plankton research is concerned, the only result of these voyages four times a year has been the publishing of a great series of tables, which, for purposes of comparison, are practically worthless, since only one country, Germany, has used the counting method of Hensen. The total failure of the ordinary methods of estimation has already been discussed above, and it was then pointed out that, if the problems are to be solved, the more scientific method of counting the organisms must be adopted.

The distribution of plankton and its relation to the hydrographical conditions has, to a certain extent, been worked out by Apstein and others for the Baltic and North Sea, from the catches made on the quarterly expeditions, for the stations belonging to the German section (**15**). It has been found of very great importance to use the closing net, and, in addition to a vertical haul from the bottom to the surface, to divide this column up into sections, as, for example, where the depth is 210 metres, a haul is taken from 210 to 65 metres, another from 65 to 25 metres, from 25 to 5, and, lastly, from 5 metres deep to the surface. This last catch is particularly important and very often differs markedly in its constitution and volume from the others. In all probability the surface layer of water to a depth of only one metre is the layer concerned, but, owing to wave motion, it is better to make the haul from a depth of five metres. This shallow surface layer of the sea appears to be particularly rich in plankton, and it is therefore conceivable from this how two tow-nets pulled along the surface may differ in contents if one of them is accidentally a little heavier than the other, or, for some reason, has been towed a little deeper.

The following figures from the German North Sea catches will illustrate the differences in the volume from different depths at the same stations:—

Depth at which catch was made.	c.cm. under 1 sq. metre area.	c.cm. in 1 cub. metre.
35—5 metres	112	3·7
5—0 ,,	56	11·3
44—5 ,,	72	1·8
5—0 ,,	72	14·4
63—47 ,,	56	3·5
47—5 ,,	144	3·4
5—0 ,,	144	28·8

In the Baltic, volume estimations have been made and the catch also quantitatively examined. On one expedition, for example, the volumes from Stations 1, 2 and 3 in the West Baltic, where the salt contents was 17 to 20 °/₀₀, were very large and above the average. At Station 8, a point further east, there was also a large catch, but the salt contents was only 8 to 10 °/₀₀. The constitution of the catch varies in the Baltic, probably with the salt contents, which, unlike the North Sea, varies within wide limits. Thus, *Aphanizomenon flos-aquae* increases as one travels east. *Chaetoceros decipiens* and *C. didymum* decrease and eventually drop out altogether. *Ceratium* also decreases in the same way. At Station 8, however, where there was a low salt contents, this decrease for some reason was not present. In the North Sea the simple hydrographical conditions of the Baltic do not prevail, and the whole matter is rendered far more difficult. The total volumes and the constitution of the catches made varies considerably at the different stations. Thus, in one of the expeditions at Station 9, North Sea, there was a greater quantity of *Ceratium macroceros* and an abnormal number of *Oithona* and *Pseudocalanus*. At Station 11 a great number of Actinotrocha larvae formed an important constituent in the catch. In May, 1903, there was in the North Sea a remarkable preponderance of plankton in the upper five-metre layer, far exceeding that of the deeper layers. This was quite independent of the salt contents, for it occurred where there was no difference in the constitution of the sea water between the surface and the bottom. Thus, at one Station the numbers for plant cells were in the proportion 0 to 5 metres deep, 400; 5 to 40 metres deep, 55; 40 to 75 metres deep, 8; 75 to 150 metres deep, 2; 150 to 450 metres deep, 1. In another example, however, there was a decrease as above from the surface down until the 25-metre depth was reached, but between 25 metres and 75 metres deep the average number of organisms present was twice as great as at the surface, that is, about twenty times what it should have been. This was due to an abundance of *Phaeocystis*. What determines these variations? Salt contents seem to have nothing to do with the diminution which occurs as one passes from the surface into deep water, though in the Baltic, as will be mentioned below, the salt contents seem to cause an opposite result. Light intensity might be connected with it, but very good catches are often obtained at depths of 75 to 100 metres. In the Baltic, in February, 1903, the figures gave different results for the vertical distribution, for the plankton was always more abundant in the deeper layers than at the surface.

The organisms which caused this increase were

Ceratium balticum, C. longipes, C. macroceros, C. fusus, Polychaete larvae, Copepod larvae, *Oithona similis, Centropages hamatus, Paracalanus* and *Pseudocalanus.* These are all forms which are characteristic of the North Sea and West Baltic water, where the salt contents are high. Owing to the peculiar conditions prevailing in the Baltic, a great variation occurs in the salt contents of the water, varying from 20 °/₀₀ in the West to fresh water in the North-east, and, moreover, at any station there is commonly a great difference between the salt contents at the surface and at the bottom. It is, therefore, natural to presume here that the greater abundance of the plankton in the deeper layers was due to the salt contents of the water, since that was greater in these layers than at the surface, and the organisms present were those characteristic of salt water. At the present time, however, a great deal still requires to be learnt with regard to the relations between the plankton and the hydrographical conditions, and in many cases the results obtained so far contradict each other.

Finally, it is necessary to examine some of the extremely interesting statistics given by the Hensen method quoted by Jenkins and others. I refer first to such estimations as the number of Copepods in the West Baltic or the number of Peridinians annually devoured by a Copepod. We have only to consider how little we know of the conditions under which these plankton forms live, and the admitted inaccuracies of the method, to see that such results must be so hypothetical as to be of very little practical importance.

To one of the calculations I must refer in greater detail. The number of floating eggs of the cod and flat fishes found in the Eckenförde waters, the area of which is 16 miles, was estimated at 30 per square metre of the surface for January, 45 to 50 for February, 60 for March, and 50 for April. The average depth of water is given as 20 metres, and the eggs take 15 days on the average, under the conditions prevailing in the Baltic, to develop, so that the above numbers must be doubled to give the number of eggs present per month under a square metre of surface water. This gives 370 eggs per square metre for the period January to April. From the returns of the Eckenförde fishermen, it was calculated that the cod and plaice annually caught would have produced 23,400 million cod and 73,895 million plaice eggs annually, if left in the sea.

These figures gave for every square metre of the 16 square miles over 26·6 cod and 84 plaice eggs, a total of 110·6 eggs, which represented the loss through the fish being caught. If this is added to the number 370 above calculated, the total 480·6 is the number of eggs produced by all the cod and plaice, captured and free, yearly for every square metre of surface water. The relation 110·6 : 480·6 = 1 : 4·4, and this is described as giving the ratio of the adult fish caught annually to the total number present in this area—a capture of a quarter of the total fish.

This argument is, however, incorrect for the following reasons. The number 110·6 represents the number of eggs under each square metre of the surface, assuming that all the eggs had survived which the fish caught annually were capable of producing in their ovaries. The numbers 23,400 million cod eggs and 73,895 million plaice eggs were arrived at from direct estimations of the number of eggs in the mature fishes. Now, it is well known that the cod and plaice produce a very large number of eggs, but that out of the enormous number only a certain proportion survive. Hence the need for such a large number, and hence, also, the attempts made by fish hatcheries to save a greater number of the embryos by rearing them through the early stages.

Then, again, since unfertilised plaice and cod eggs do not remain pelagic and other dead eggs fall to the bottom (when their death is not due to their being devoured), the floating eggs capable of being caught must be but a small proportion of the number actually produced. Hence the number 110·6 is much too high, as a calculation of the number of eggs per square metre lost by the capture of the adult fish, and cannot be compared directly with the number 370, which although the actual number of eggs fished, represents only a portion of those produced. The calculation has assumed that the relation between the number of eggs floating in the sea and the fishes that produced them is the same as that between the number of eggs in the ovaries and a mature fish.

In conclusion, it may be repeated that for a scientific quantitative study of the plankton, the complete apparatus and the Hensen method of counting must be employed. It is quite obvious that a certain amount of inaccuracy will occur with the use of each piece of apparatus, and the numbers must be considered approximate only; but since the errors will average the same for each catch, they do not invalidate the results for purposes of comparison. It is quite another matter, on the other hand, if the plankton is found to be not so equally distributed that the small samples taken will give reliable results for the whole areas. It is not to be expected that under varying hydrographical conditions the plankton will remain the same; but, at the present time, very little is known of the actual relations. Again, it has been pointed out several times in this paper that the results of recent plankton work have very often shown sudden and striking variations in the quantity and constitution of the catches at two stations where apparently the conditions prevailing were the same. Hensen and the later German workers regard these fluctuations that occur as of little importance; but it is clear that more knowledge upon this question of the unequal distribution is required, because if small samples (and they are only 15 c.cm. for the centrifuge) are to be taken, they will give no true picture of the plankton present in either quantity or quality, nor of the relations of larvae to adults, if swarms occur or if there is unequal distribution to any considerable extent. It is, therefore, very necessary to take a small region and to make sure that the hydrographical and other conditions are the same throughout, or to take catches in exactly the same way, side by side, or separated by the length of a vessel, in order, after a systematic research, to tabulate the fluctuations that have occurred. Herdman (**17**), who is working on these lines at Port Erin, has already given some surprising figures of the variations in the catch of two nets worked side by side, and the detailed account of his results, to be published in this volume (see **19**) should throw some light on this very important question.

Papers Referred to Above.

1. HENSEN. Über die Bestimmung des Planktons, 5 *Bericht der Kommission z. wissen. Unter. d. deutschen Meere*, 1887.

2. HAECKEL. Plankton Studien. Jena, 1896.

3. HENSEN. Die Plankton Exped. u. Haeckels Darwinismus. Kiel, 1891.

4. HENSEN. *Ergebnisse der Plankton Exped.* Band I, Kiel, 1892.

5. HENSEN. Methodik der Untersuchungen bei der Plankton Exped. *Ergebnisse der Plankton Exped.*, Band II, Kiel, 1895.

6. CORI. Über die Verwendung der Centrifuge in der zoologische Technik. *Zeitschrift. f. wissenschaft. Mikroskopie*, Band XII, 1895.

7. APSTEIN. Das Süsswasser plankton. Kiel, 1896.

8. KOFOID. On some important sources of error in the Plankton method. *Science*, N. S., Vol. VI.

9. BRANDT. Über den Stoffwechsel im Meere. *Wissensch. Meeresuntersuch.*, N. F., Bd. IV, Kiel, 1899.

10. VOLK. Hamburg Elbuntersuchungen. *Mitt. Nat. Hist. Mus.*, Hamburg, Bd. XVIII.

11. LOHMANN. Über das Fischen mit Netzen aus Müllergaze No. 20. *Wissensch. Meeresuntersuchungen*, N. F. Ab., Kiel, Bd. V., Heft 2.

12. JENKINS. Methods and Results of the German Plankton Investigations. *Trans. Liv. Biol. Soc.*, Vol. XV, 1901.

13. LOHMANN. Untersuchungen über den Reichtum des Meeres an Plankton. *Wissen. Meeresunter.*, N. F. Ab., Kiel, Bd. VII.

14. APSTEIN. Die Schätzungsmethode in der Planktonforschung. *Wissen. Meeresuntersuch.*, N. F. Ab., Kiel, Bd. VIII.

15. APSTEIN. Plankton in Nord-und Ost See auf den deutschen Terminfahrten. *Wissen. Meeresunt.*, Kiel, Bd. IX.

16. DOLLEY. The Planktonokrit. *Proceed. Acad. Natur. Sci.*, Philadelphia, 1896.

17. HERDMAN. Report of the Marine Biol. Stat. at Port Erin. *Trans. Liv. Biol. Soc.*, Vol. XXII, 1907.

18. LOHMANN. Untersuchungen zur Festellung des vollständigen Gehaltes des Meeres an Plankton. Kiel, 1908.

19. HERDMAN and SCOTT. Intensive Study of Marine Plankton, etc., in Lancashire Sea-Fisheries Laboratory Report for 1907, p. 94. *Trans. Liv. Biol. Soc.*, Vol. XXII, 1908.

THE OYSTER AND OYSTER-CULTURE

Karl Möbius

XXVII.—THE OYSTER AND OYSTER-CULTURE.*

By Karl Möbius,
Professor of Zoology at Kiel.

INTRODUCTION.

Since the first attempt in France, in 1858, to raise oysters by artificial means, very much has been written concerning the oyster and its culture. Authors, themselves astonished, have endeavored, by displaying long rows of figures indicating the great number of oysters that could be produced, to awaken like astonishment among their readers and arouse the inhabitants of the coast to propagate extensively in all their countries this most valuable of all sea invertebrates. These accounts of the immense production resulting from the artificial culture of the oyster went from paper to paper and book to book, and carried with them such an appearance of credibility that even practical oyster-breeders and acute biologists believed that, with little labor, great sums might be realized by raising oysters for the table. This is comprehensible, for the reason that the official reports of England, France, and America concerning oyster-culture, from which the large figures were taken, present either no information as to their true significance, or that only of a scattered nature, intelligible to those alone who are already acquainted with the subject. In order to gain this acquaintance and comprehend the true significance of such figures it is necessary to become informed as to the nature and the condition of life of the oyster; and in regard to both of these subjects biologists, as well as breeders and consumers of oysters, will find in the present work all that is necessary to enable them to form an opinion upon the questions which will arise in regard to the breeding and rearing of oysters. I believe I have clearly demonstrated that true oyster-culture must be conducted according to the same principles that are employed in the extensive cultivation of any other living commodity. If I have done so, then I have accomplished what should not have been necessary; for what is more natural than that both oysters and oyster-culture should be subject to the same universal, controlling, biological laws. And yet an explanation was

---

* Die Auster und die Austernwirthschaft; von Karl Möbius, Professor der Zoologie in Kiel. Mit einer Karte und neun Holzschnitten. Berlin, Verlag von Wiegandt, Hempel & Parey. 1877. Small octavo; pp. 126. Translated by H. J. Rice, B. Sc., by permission of the author, by whom electrotypes of the original cuts have been furnished.

necessary, for not only the ignorant in natural science, but men also who teach and write upon biological subjects have, even in our day, expected the most impossible results from the artificial breeding of oysters. The investigator has seldom to seek for new facts and ideas alone; generally, in the first place, he must be able to recognize and expunge from the system any errors which may exist in the knowledge previously acquired, and in their places establish those facts and ideas which he has found to be true. And while I am well aware that the little book hereby presented to the world contains but a very modest share of what we wish to know with certainty concerning oyster biology and oyster culture; still I have allowed it to appear because, incomplete as it is, it will give welcome information to many biologists and oyster-breeders, and will serve as a safe basis for the operations of those governments which have within the limits of their territories natural oyster-banks which they desire to have managed in the best interests of the general public. Those, of course, who delude themselves with the belief that, by means of artificial cultivation, oysters can be bred in great quantities wherever there may be sea-water, will scarcely agree with my book, and it is indeed quite certain that it will not convince them of their error. But the most dazzling error does not become transformed into truth, however long and firmly one may believe in it.

KARL MÖBIUS.

KIEL, *July* 8, 1877.

CONTENTS.

	Page.
1.—The sea-flats	3
2.—Oyster-banks and oystering	7
3.—The reproduction of the oyster	10
4.—Why are oysters not found over all portions of the sea-flats	14
5.—Artificial oyster-breeding in France	16
6.—Attempts to introduce the French system of artificial oyster-breeding into Great Britain	20
7.—Can the French system of artificial oyster-breeding be carried on in the waters of the German coast	21
8.—Can natural oyster-beds be enlarged, and can new beds be formed, especially along the German coast	25
9.—Growth and fecundity of the oyster	31
10.—An oyster-bed is a Biocönose or a social community	39
11.—Concerning the increase in the price of oysters and in the number of consumers, and the decrease in the number of oysters	47
12.—The chemical constituents and flavor of oysters	50
13.—The objects and results of oyster-culture	56

1.—THE SEA-FLATS.

Among those oysters which are produced in the waters of the west coast of Europe the Holstein oyster has, for more than a hundred years, maintained a well-merited celebrity. The beds which furnish them lie along the west coast of Schleswig-Holstein, in a territory only 74 kilometers long by 22 broad. The most and the best oysters are found on the east side of the island of Sylt and in the neighborhood of the islands of Amrum and Föhr.

Along the northern boundary of the German oyster-territory, near the island of Röm, and along the southern boundary, near the islands of Pellworm and Nordstrand, opposite the city of Husum, there are only a few insignificant beds. And since the flavor of the oyster is entirely dependent upon the quality and quantity of food in the water in which it grows, it becomes necessary, first of all, to examine into the character of the soil and water of the Schleswig-Holstein Archipelago. In comparison with the open North Sea this portion of our coast is a very shallow division of the ocean. Along the entire southern portion of the open North Sea, between Germany, Holland, England and Scotland, the general depth is from 35 to 45 meters. In no place in the Schleswig-Holstein Archipelago is the water as deep as this, the greatest depth being 15 to 20 meters, and this only in the channels which connect it with the open sea. The floor of this archipelago is raised above the deep

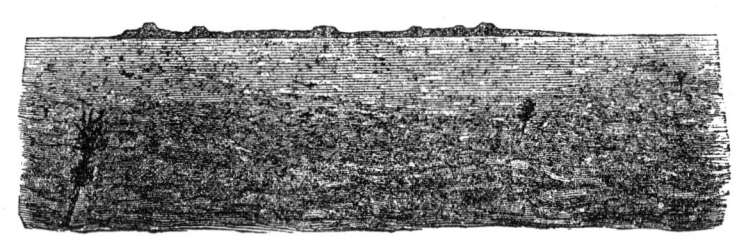

FIG. 1.

The *sea-flats*, with three buoys indicating navigable water. In the background the Hallig Langeness is seen above the surface of the water.

bottom of the open North Sea, very much like a high table-land. In this table-land valleys, varying in depth and width, have been cut out between the islands and the mainland. At high water, the entire floor is covered, but at the end of the ebb-tide, very much of this table-land lies dry above the surface of the sea. These stretches of sea-bottom which thus become dry are termed "Watten," (plains or flats,) and from these "Watten" this archipelago has received the name of "Wattenmeer," (sea-flats.) The water, which during the ebb-tide runs off from the flats, flows in both shallow and deep channels, called by the sailors "Leien" and "Tiefen," partly in a northerly, partly in a southerly direction, into the open sea, until the incoming flood-tide, which flows in from both sides twice daily, stops the ebbing water and turns it back. The water now

Map of the sea-flats of Schleswick-Holstein, showing the oyster-banks, the currents, and the depths in meters.

rises once more. The Leien and Tiefen can no longer hold it, and it pours over their banks and over the flats, finally flooding them to such a depth that small vessels can pass over places where only a few hours before men and wagons might travel with perfect safety. In an investigation of the oyster-beds our little steamer got into too shallow water between the island of Föhr and the mainland, and ran fast aground about nine o'clock in the morning. The water was falling, and in a few hours it was entirely out. We descended and went on foot to Hallig Oland,* which lay like a green plate, upon the level, grayish sea-bottom, about one kilometer to the eastward of our vessel. While upon this Hallig we visited a hill which had been formed by artificial means. Upon the hill was a fresh-water pond surrounded by a small group of dwellings, among which was a church encompassed by graves. We then returned to our vessel before the water had again flooded the flats. At about seven o'clock in the evening the water had risen so high that our vessel began to rock; it soon floated, and we steamed to Föhr, to anchor for the night in the harbor of Wyk.

Along the entire German coast, from Röm in the north, upon the Danish border, to Borkum in the west, near the islands of Holland, the sea is of a similar character. Thus, before the mouth of the Elbe, from Cuxhafen to the island of Neuwerk, the sea-bottom is laid bare with every ebbing of the tide, for a breadth of 7 to 8 kilometers. At such times one can reach the island on foot, on horseback, or with a wagon. In passing over this flat one finds himself at such times on a level with the sails of vessels which are passing by upon the sea, and along the border of the retreating waters and the emerging sea-bottom one sees scattered flocks of sea-birds hunting the uncovered worms, mussels, and crabs before they withdraw into the earth. When the flats, at the time of the lowest ebb, are lying, dry and silent, above the water, one can already hear in the distant depths the roar of the incoming flood. First it comes in slowly, then faster and faster, and finally more slowly again, until at the full flood the water stands over the northern por-

---

* *Halligen* is a name given to small, low islands in the Schleswig-Holstein Archipelago, composed of marsh land, and not protected by dikes from high tides. They are green plains, enlivened by pasturing cattle and sheep, and lie only a foot above ordinary high-water level. They are overflowed by the water during storms. The word *Hallig* is perhaps derived from *Haf-lik*. That portion of the coast which is dry during the ebb and covered during the flood tide is called *Haf; lik* means like, similar. No other land is so similar to the *Haf* land as the land of the Halligen.

The islands of the Schleswig-Holstein sea-flats consist either of low marsh land protected by dikes, or of higher sand tracts and downs.

Nordstrand and Pellworm are marsh islands; Föhr is marsh and sand together, and Sylt and Amrum have high sand tracts and downs.

The marsh soil is a gray, uniform, fine mass without any stones; when wet it becomes tough and sticky. It originated from muddy material brought down by rivers and streams and deposited in quiet places along the sea-coast. The high sand tracts are composed of old raised sea-bottoms. They are uneven, consist principally of coarse sand, and are much less fertile than the rich marsh soil, which, without manuring, yields abundant harvests.

tion of the flats nearly 2 meters higher, and over the southern portion, out from the mouth of the Elbe, nearly 3 meters higher than at the ebb. The tide generally attains three-fourths of its entire height about three hours after turning. In this short time immense masses of water move towards the coast, and in many places currents are formed as swift as the current of the Rhine between Coblenz and Bonn, the rate of which is from 1.5 to 2 meters per second. Yet the ebb-currents are nearly everywhere stronger than the flood-currents, since they not only carry off the sea-water which has been brought in, but also the fresh water from the land, which was checked in its flow during the flood. Hence the ebb-currents bring about much greater changes in the soil of the sea-flats than the flood-currents, and they displace and transport the constituents of the flats in the most powerful manner, wherever great fresh-water streams enter the sea, as at the mouths of the Eider, Elbe, Weser, and Ems. Here the floating buoys and the implanted buoy-stakes (Baken), which indicate navigable water for vessels, are changed nearly every year because of the changes in the channels.

The principal ingredient of the bottom of this changeful sea is quartz sand. In many places there are accumulations of mud, which is very slimy and sticky, and contains much organic matter. This mud is found along the shores of the mainland and on the east side of the island of Sylt, principally at those points where, after the changing of the currents, the water does not flow fast enough to carry away all of the muddy material which was deposited upon the bottom as the tide ran out. Along the slopes which lie between those portions of the flats, which the tide leaves dry, and the deep channels are long dry stretches of soil where the ground is covered with coarse sand, small and large stones, and shells. At such places colonies of oysters, so-called oyster-beds, are found, along with many other sea-animals.

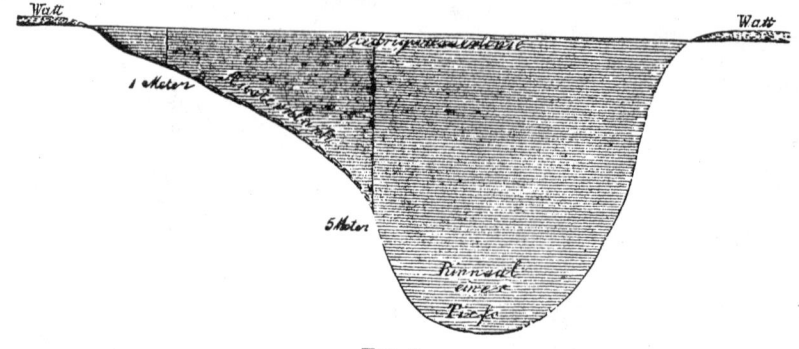

FIG. 2.

Diagram of a cross-section of a deep channel in the sea-flats, upon the left bank of which lies an oyster-bed. Upon both sides are flats which are left dry by the ebb-tide. (The breadth of the channel is drawn upon a much smaller scale than the depth.)

2.—OYSTER-BANKS AND OYSTERING.

By far the greater number of our oyster-beds are never exposed to view on account of the muddiness of the water of the sea-flats, from the continual stirring up of the sediment upon the bottom. Only when, during the lowest ebb of the spring-tides, easterly winds drive off a great deal of water from the land, does the sea along the border of many beds become so shallow that the oysters can be seen, and even taken up with the hand. This state of affairs occurs upon the oyster-beds which are numerous along the east coast of the island of Föhr, and in one autumn as many as 20,000 oysters could be gathered from these beds by hand and transplanted into deeper waters.

Generally one is obliged to use measuring-sticks or dredge-nets in order to tell when he is over a desired oyster-bed. The measuring-sticks are poles, five to six meters long, with the lower half divided off, by different colors, into feet. They are used from vessels, in shallow portions of the flats, in order to ascertain during the journey whether the depth increases or diminishes, so that the vessel may not run aground. The measuring-rod is pushed down to the bottom, and one can thus easily tell whether the bottom is composed of soft mud or pure sand, or whether it is covered with shells.

The dredge used by the oyster-men (Fig. 3) consists of an iron frame upon either side of which there is a shank. These shanks, or side-pieces, are brought together and united, at a short distance from the frame, so as to form a ring in which the dredge-rope is fastened. Fastened to the frame upon the opposite side from the shanks is a net whose upper half consists of coarse yarn or cord, and the lower half, that which drags along the sea-bottom, is, for greater durability, made of iron rings united together, each of which has a diameter of from six to seven centimeters. The entire net weighs from 50 to 60 pounds.

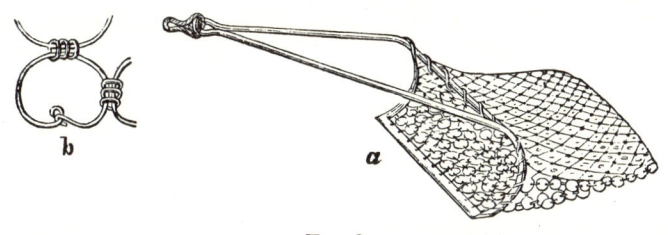

FIG. 3.

Oyster-dredge. The frame and handles are made of iron. The upper portion of the bag is made of coarse net-yarn, the under portion of iron rings from six to seven centimeters in diameter. The form of the rings and the method of uniting them is represented with greater exactness at *b*.

The older oyster-dredgers know the position of all the oyster-beds with great precision, and they guide their vessels to the desired places by reckonings from high-lying points of the coast and islands, from lighthouses, churches, windmills, and houses. Their vessels are yacht-like, with a capacity of from three to six tons. Each one generally carries two sailors in addition to the owner.

Upon the Schleswig-Holstein banks there are fourteen vessels engaged in the oyster business. When the wind is favorable and brisk, four dredges can be used at the same time; but with a light wind, two, or one only can be dragged. They are fastened by means of strong ropes to the windward side of the vessel. One hand is kept upon the dredge-rope, in order to tell by the feel whether it is passing over smooth ground or over oyster-beds, for the rope is given an irregular, jerking motion upon rough bottom. Generally the net is allowed to drag from five to ten minutes; then it is drawn up by two or three men, and the entire contents of the bag emptied upon the deck. This mass consists of old oyster-shells, mussels of various kinds, living oysters, snails, crabs, worms, star-fish, sea-urchins, polyps, sponges, and sea-weeds, which are generally mixed up with sand and mud. From this heterogeneous heap all the matured oysters are now picked out. As they pass singly through the hands of the fishermen, the coarsest of the foreign material is cut and scraped from the shells with a knife, and then the oysters are thrown into baskets. In these they are shaken about, in order to get off any material which has escaped the knife. Ropes are then fastened around the baskets, which are put overboard, and raised and lowered in the sea until all dirt is completely washed from the oysters. They are now for the first time in the condition in which they appear in commerce. Despite these manifold cleansings, many oysters when they are exposed for sale are covered with dead and living animals, and the peculiar odor which oysters have when carried into the interior arises from the death and decay of the organic material upon the outside of the shells, and does not pertain to the living oyster itself. In no place upon the sea-flats do oysters grow upon rocky bottom. They grow best where there is a substratum of old oyster and other shells. The most of them lie singly, and they are seldom found growing together in clumps or masses. The wide-spread notion that they are found growing firmly attached to the sea-bottom, and piled upon one another, layer upon layer, is accordingly false. Upon the best of the Schleswig-Holstein beds the dredge must drag over a surface of from 1 to 3 square meters, and often over a still greater distance, in order to secure a single full-grown oyster. Over the Schleswig-Holstein sea-flats there exist 50 oyster-beds of very different sizes. The largest is not far from 2 kilometers long, but the greater number are shorter than this. Their breadth is much less than their length, which is in the same direction as the channels along the slopes of which they lie. The greater number of the beds have a depth of water of at least 2 meters above them when the ebb-tide has left the neighboring flats dry.

There are no beds upon our sea-flats which have a greater depth of water over them than from 6 to 9 meters. Although all the beds lie within an area 74 kilometers long by 22 broad, yet the nature of the oysters, and especially the form and solidity of the shell and the flavor of the animal, differ very greatly. Upon two beds inside of the south point of the island of Sylt are found oysters which in fullness and delicacy of flavor are not inferior to the best English "natives."

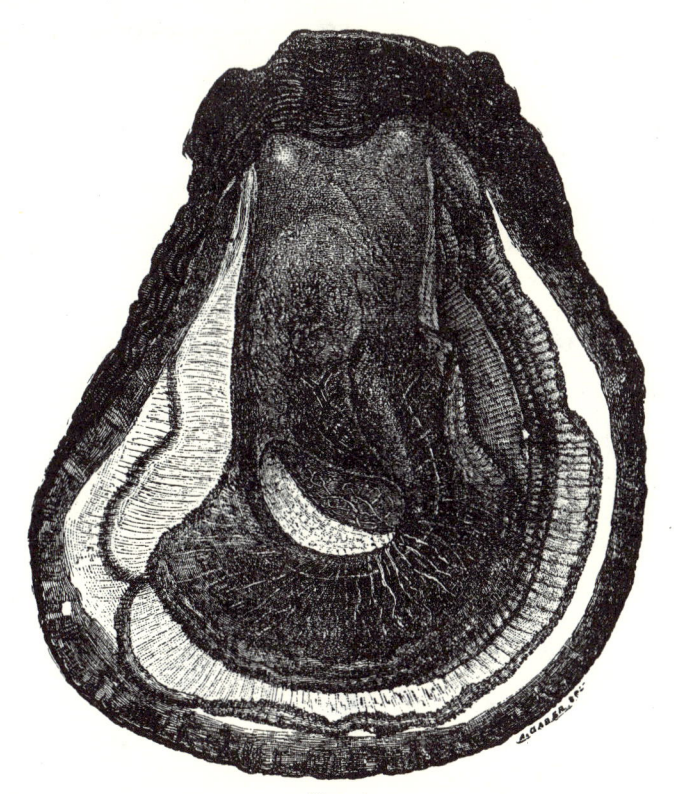

FIG. 4.

A full-grown Schleswig-Holstein oyster, about ten years of age. It is a female with eggs, and was drawn from life, on the 14th of June, 1871, by Mr. J. Wittmaack. The right, or upper, valve of the shell has been removed. The oyster lies in the hollow of the left valve, in its natural position. On the upper side the thickened layers of the shell can be seen. Each year new shell-layers are formed. The inner surface of the shell is white to near the edge, where it becomes of a brownish color. Above, close to the back of the animal, which is somewhat curved, is a crescent-shaped brown mass, the shell band or ligament. In separating the valves this band is broken across in the middle. The right side of the animal is exposed to view; the left rests upon the inner hollow surface of the left valve. The upper layer, with its edge turned back, is the mantle-lobe or fold of the right side. The white lines seen in it are muscular fibers. The left mantle-lobe lies close upon the shell, and is more expanded than the right. The gills are to be seen just below the inverted edge of the right mantle-lobe. In the oyster they are four in number. The outer gill of the right side is the most exposed; a narrow border of two others can be seen. All four have furrows running from the inside to the edges. Upon these furrows are situated cilia, by the motion of which water is driven over the gills for the purposes of respiration. Along the upper portion of the gills hang two pairs of furrowed folds, or lobes (the labial palps), between which is the mouth opening. The swollen upper portion of the body contains the generative organs, the liver, stomach, intestine, and heart. The bean-shaped organ near the center of the body is the adductor muscle, the so-called *stool*. This muscle which closes the valves consists of an upper grayish and a lower whitish portion.

3. THE REPRODUCTION OF THE OYSTER.

If the surfaces of all the Schleswig-Holstein oyster-beds should be united together they would not cover a space equal to the one-hundredth part of that portion of the sea-flats which remain under water. Why is this? Is it because from a lack of oyster-broods all the places between the banks are yet to be peopled? I cannot accept this view for the following reasons: The entire number of full-grown oysters existing upon the Schleswig-Holstein beds I estimate to be not far from five millions. According to my observations, 44 per cent. at least of these oysters will bring forth broods of young oysters in the course of a summer.*

---

*The data from which I arrived at the conclusion that at least 44 per cent. of full-grown oysters spawn during each spawning season were derived from the following observations:

I opened on—	Oysters.	White germs.	Bluish germs.	In all.	Per cent.
June 16, 1873	112	5	4	9	8
July 6, 1873	63	7	6	13	20.6
August 12–17, 1869	480			72	15.8
Total					44.4

I do not know the length of time of development, from the beginning of segmentation of the egg until all the embryos have passed out from the mother animal, but it is probably less than four weeks; for while, in the last weeks of May in the years 1871 and 1874, from June 4 to 6 of the year 1873, and June 6 to 9 of the year 1876, in hundreds of oysters which I opened, I found no embryos in the beard, yet of 112 oysters dredged on the 16th of June, 1873, five contained germs of a white color, and four contained germs already bluish, and possessed of shells and vela.

If by the end of the first week in June no eggs have been laid, but by the beginning of the third week germs are found of a bluish color, then the transformation of the white germ into the blue cannot consume more than a week, and these germs will hardly remain in the beard for an additional period of more than two weeks. Those oysters, then, which are found with eggs during each of the following months must be different individuals from those which spawned during the earlier periods; hence, it is right to add together the percentage of egg-bearing oysters found separately in June, July, and August in order to arrive at the percentage of egg-bearers for the entire summer. And since many oysters are found upon the Schleswig-Holstein beds with germs of a bluish color in the mantle even in the beginning of September, then the percentage of 44.4 per cent. surely cannot be too high. Oysters are hermaphrodite. In a large number of oysters which I examined I found ova in the generative organs, but no

Longitudinal cross-section of a seven or eight years old oyster.

The curved left valve (*lK*) of the shell is, as usual, somewhat thicker than the right (*rK*). They are bound together by the ligament *B*, which keeps them separated a certain distance from each other so long as the adductor muscle (*gS* and *wS*) does not by its contraction stretch the ligament and close the valves. The shell-muscle consists of fine gray (*gS*) and coarse white muscular fibers (*wS*). In the neighborhood of the ligament, in the left valve, are four small perforations made by a sponge (*Clione celata*) which lives housed in the oyster-shells of the sea-flats. Other larger holes are often found in the shells of old oysters. They are filled with water, which has a rank, gaseous smell; hence they are called "gas-holes." The thickest part of the shell is near where the shell-muscle is attached. The shell consists principally of carbonate of lime, arranged in firm, glistening layers. Close above the shell-muscle lies the heart (*H*). The mass over the heart, which in life is of a bright color, consists principally of the organs of generation (*G*). These surround the stomach (*Ma*) and the liver (*L*), which during life is brown in color. In the space above the intestines is seen the sections of the four mouth-plates (*Mp*). Under the shell-muscle and close to the shell are the two mantle-lobes (*Mt*), whose edges are thickened and beset with fringes. They contain muscle and nerve fibers. During life the mantle-lobes can be thrust out over the edge of the shell.

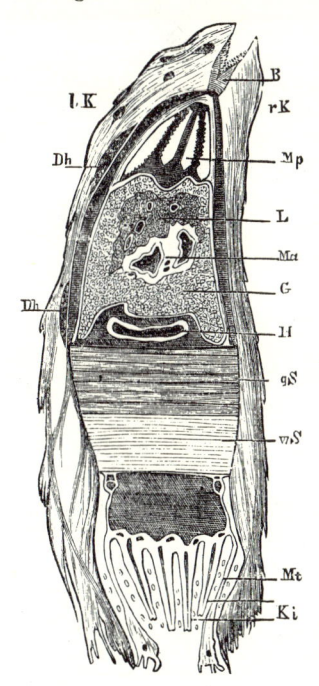

FIG. 5.

From the edges of the mantle-lobes all the shell material is secreted. Between the two mantle-lobes there is a wide space, in which hang the four gills (*Ki*). Each gill

spermatozoa; in many others I found spermatozoa, but no ova; and in seven oysters which bore embryos of a blue color upon the beard I found spermatozoa in the generative organs.

Three oysters with embryos of a white color attached to the beard had no spermatozoa in the generative organs. Most mature oysters produce either ova or spermatozoa, and not both at the same time. Of 309 oysters which were dredged on the 25th of May from four different beds along the east side of the island of Sylt, and which were examined from the 26th of May to the 1st of June, the sex of 18 per cent. could not be determined; of the remaining 82 per cent., one-half were males and one-half females In none of them were the generative products completely matured. From these results I conclude that the ova and spermatozoa do not arise in the generative organs of the oyster contemporaneously, but that one follows the other. The spermatozoa can arise ery soon after the expulsion of the ova, and probably one-half of the oysters of a territory during any spawning period produce eggs only, the other half spermatozoa only.

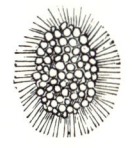

a. A mass of spermatozoa, still clustered together just as they arise in the generative organs, enlarged 275 times.

b. A single spermatozoan, enlarged 1,000 times. By the motion of the tail the body is driven forward.

The ripe spermatozoa pass from the generative organs into the water, with which they pass into the brood-chamber of the female oyster, where they impregnate the freshly-laid eggs by penetrating the yolk and uniting with it.

FIG. 6.

lobe consists of two plates, which grow together above and below. The mantle-folds and the gills taken together form the so-called "beard" of the oyster. In the spaces between the layers of the beard the development of the eggs takes place. In the figure a few germs are represented in this brood-cavity.

Now, a mature egg-bearing oyster (Fig. 4) lays about one million of eggs, so that during the breeding season there are upon our oyster-beds at least 2,200,000,000,000 young oysters, which surely would suffice to transform the entire extent of the sea-flats into an unbroken oyster-bed; for if such a number of young oysters should be distributed over a surface 74 kilometers long by 22 broad, 1,351 oysters would be allotted to every square meter. But this sum of 2,200,000,000,000 young oysters is undoubtedly less than that in reality hatched out, for not only do those full-grown oysters which are over six years of age spawn, but they begin to propagate during their second or third year, although it is true that the young ones have fewer eggs than those which are fully developed. At a very moderate estimation, the total number of three to six years old oysters which lie upon our beds will produce three hundred billions of eggs. This number added to that produced by the five millions of full-grown oysters would give for every square meter of surface not merely 1,351 young oysters, but at least 1,535. In order to determine how many eggs oysters produce, they must be examined during their spawning season. This begins upon the Schleswig-Holstein beds in the middle of June, and lasts until the end of August or beginning of September. The spawning oyster does not allow its ripe eggs to fall into the water, as do many other mollusks, but retains them in the so-called beard, the mantle, and gill-plates (Fig. 5) until they become little swimming animals (Fig. 7). The eggs are white, and cover the mantle and gill-plates as a semi-fluid, cream-like mass. As soon as they leave the generative organs the development of the germ begins. The entire yolk-mass of the egg divides into cells, and these cells form a hollow, sphere-like body, in which an intestinal canal arises by the invagination of one side (Fig. 7). Very soon the beginnings of the shell appear along the right and left sides of the back of the embryo, and not long afterwards a ciliated pad, the velum, is formed along the under side. This velum can be thrust out from between the valves of the shell at the will of the young animal, and used, by the motion of its cilia, as an organ for driving food to the mouth, or, in swimming, as a rudder. During these transformations the original cream-white color of the germ changes into pale gray, and finally into a deep bluish-gray color. At this time they have a long oval outline, and are from 0.15 to 0.18 of a millimeter in breadth. Over 300,000 can find room upon a square centimeter of surface. If an oyster in which the embryos are in this condition is opened, there will be found upon its beard a slimy coating thickly loaded with grayish-blue granules. These granules are the embryo oysters, and if a drop of the granular slime be placed in a dish with pure sea-water the young animals will soon separate from the mass, and spread swimming through the entire water. When the embryos are at this stage their number may be estimated in the following manner: The whole mass of embryos is carefully scraped from the beard of the mother oyster by means of a small hairbrush. The whole mass is then weighed, and afterwards a small portion of the mass. This small portion is then diluted with water or spirits of wine, and the embryos portioned out into a number of small glass dishes,

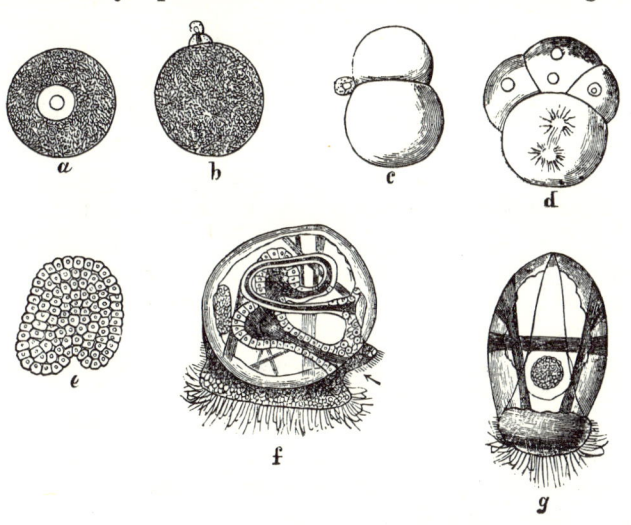

FIG. 7.

A few stages of development of the embryo oyster; *a* to *e* enlarged 125 times, *f* and *g*, 150 times.

a. The freshly-produced egg. In the yolk-mass is seen the germinative vesicle, with its nucleus.

b. Commencement of development. A part of the vesicle has passed out.

c. Division of the egg into two unequal portions.

d. A later stage.

e. The germ now consists of a layer of cells, which have arisen by repeated divisions of previous cells. They form a hollow vesicle, with a depression upon one side, which is the beginning of the digestive system.

f. The embryo is now represented at about the stage at which it leaves the broodcavity. It has a transparent two-valved shell, and inside of the body the course of the digestive tract can be made out. An arrow shows the position of the mouth, and those within the body indicate the course which the food takes. Behind the œsophagus is the stomach, with two enlargements. The end of the intestine is shown over the mouth. To the left of the first enlargement of the stomach is the shell-muscle. On the under side is the velum, which is the locomotive organ of the young oyster. The young oyster can, by means of muscles, draw the velum entirely within the shell.

g. An embryo, seen from behind. Upon the sides are seen the valves of the shell, and across the body, from one valve to another, passes the shell-muscle. Below this muscle is the velum, with the muscles, one on each side, which serve to withdraw it into the shell.

so that they can be placed under the microscope and counted. Thus, knowing the weight of the small portion and the number of embryos in it by count, we can estimate the total number of embryos from the weight of the entire mass, which is also known. In this manner I estimated the number of embryos in each of five full-grown Schleswig-Holstein oysters, caught in August, 1869, and found that the average number was 1,012,955.

4.—WHY ARE OYSTERS NOT FOUND OVER ALL PORTIONS OF THE SEA-FLATS?

It is now clear that the fruitfulness of the oyster is extraordinarily great, and that the extension of oyster-beds over the entire surface of the sea-flats does not fail of being accomplished from a lack of young oysters, but from other causes. It then becomes our duty to investigate into the characteristics of our sea-flats; in order to determine whether some portions are more suitable for the growth of oysters than others; and whether the saltness, temperature and movement of the water, the amount of food which it contains, and the nature of the ground composing the oyster-banks, differ in any respect from these same features as observed in other places over the bottom.

The saltness of the upper layers of the water of the open North Sea is from 3.47 to 3.50 per cent.* The water of the sea-flats is slightly less salt, being only from 3 to 3.3 per cent.† Here upon our sea-flats, and in other European coast-seas, where the water is less salt, the oysters acquire a much finer flavor than upon the ground of the open North Sea,‡ where they live in water 35 meters or more in depth, with a percentage of salt of about 3.5.

That coast-water is, then, the most desirable for oyster-culture which

---

*Dr. H. A. Meyer has published a paper concerning the *saltness, temperature,* and *currents* of the North Sea in the "Bericht der Commission zur Untersuchung der deutschen Meere über die Expedition zur chemisch.-phys. und biologischen Untersuchung der Nordsee, 1872. Berlin, 1875." (Report of the commission for the investigation of the German Ocean upon the expedition for the chemico-physiological and biological investigation of the North Sea.) (Specific weight and saltness, page 18.)

† I have myself repeatedly determined the temperature and saltness of the water during investigations of the oyster-beds of the sea-flats; and since 1872 the commission for the investigation of the German Ocean have caused regular stated observations to be made, which, since 1874, have appeared under the title "Ergebnisse der Beobachtungs-Stationen an den deutschen Küsten über die physik. Eigenschaften der Ostsee und Nordsee. Berlin, 1874, 1875, 1876." (Results of investigations into the physical characteristics of the North and East Seas made at observation stations along the German coasts.)

‡ Many oysters are taken north of Germany and Holland, east of England, and in the channel between England and France. The German fishermen of Blankenese and Finkenwärder, near Hamburg, who fish with great dredge-nets for flounders, turbots, and soles out from the mouth of the Elbe, often dredge oysters along with their fish. The oyster-grounds of the open North Sea lie mostly from 33 to 34 meters beneath the surface of the water. They begin with a small stretch to the southeast of the island of Heligoland, extend from this island in a west-northwest direction, and form a territory 15 to 22 kilometers broad, which spreads out far to the west. Fishermen from Holland and Germany dredge for oysters here, especially during the months of August,

contains about 3 per cent. of salt; and since not only over our oyster-beds, but over our entire sea-flats, the water possesses this degree of saltness, neither a lack nor an excess of salt can hinder the extension of the beds over the whole area. Even less can the temperature of the water hinder their extension, for the variation is the same over the oyster-beds as at other points, and it fluctuates, during the course of the year, from 20° C. above zero to 2° C. below. Nor can a lack of motion of the water or of nutriment be the cause why the oyster-beds have not during the past hundreds of years extended themselves beyond certain definite limits, for floating everywhere, in the ebbing and flooding water, are microscopic plants and animals, and much dead organic matter, which would nourish large numbers of oysters, just as they do multitudes of soft clams (*Mya arenaria*), edible mussels (*Mytilus edulis*), and cockles (*Cardium edule*). There remains, then, as the single natural hinderance to a further extension of the oyster-beds, the unfavorable condition of the ground over the greater portion of the sea-flats. Oysters cannot thrive where the ground is composed of moving sand, or where mud is being deposited, and one of these conditions or the other is found over the greater part of the sea-flats. The number and size of those places where, notwithstanding the daily ebb and flood currents, the ground remains unchanged and free from mud are very limited. Only along the slopes of certain channels to the north of the mouth of the Eider do we find united all the conditions favorable for such places, and only within these limited districts can young oysters grow to complete maturity.

When the young oysters attached to the beard of the mother have reached a diameter of 0.15 to 0.18 of a millimeter, when their digestive organs have reached such a stage that the young animal can receive nourishment through them, and when the velum, by means of its cilia, is in a condition to enable them to move about, they leave the brood-cavity, swarm at the surface, and after swimming about for a short time finally sink once more to the bottom. If the swarm of young oysters settles upon a spot covered with clean stones or mussel shells to which they can become attached, they have a prospect of growing to maturity; but if, on the contrary, they settle upon a changing sand-bank or upon a muddy bottom, they will surely be lost; for at the close of their swarming period their velum, which is their swimming organ, is absorbed, and

---

September, and October, and often catch, at a single drag of the dredge, as many as 5,000 oysters. Sometimes great bunches of oysters growing attached to one another are gathered into the net.

The deep-sea oysters grow much larger than those found along the coasts. Specimens are taken with shells 13 centimeters broad. Their flesh is tough, yet large numbers are consumed in England, France, and Germany; in England and France chiefly in pastries and sauces, but in Germany many are eaten fresh, especially in Hanover and Bremen. For general winter use they are kept under water in certain places adapted to them, especially near the island of Wangeroog. (S. Metzger's Beiträge zu dem Jahresbericht d. Commiss. zur Unt. d. deutschen Meere, 1873, page 171, s. 1875, page 252.)

no muscular foot, as an organ of locomotion, is formed in its place, as with most other bivalves. The oyster must thus remain upon that spot where it settles at the close of its swimming career. If currents and waves cover it with sand, if, during tidal changes, the quiet water allows mud to sink down upon it, if plants luxuriate over it, then, being unable to work its way out into free water, and wander to a better place, it must remain as it is, and, from lack of air and nourishment, soon perish.

5.—ARTIFICIAL OYSTER-BREEDING IN FRANCE.

The yield of many once rich oyster-beds along the west coast of France had fallen off to such an extent from 1850 to 1860 that Prof. P. Coste* of Paris, the originator of the celebrated fish-breeding establishment near Hüningen, in Alsace, presented, in 1859, to the Emperor Napoleon III, a plan for the artificial breeding of oysters, by which means he would prevent the destruction of a large number of young oysters at the beginning of their lives as independent animals. The first attempt to render the impoverished oyster-beds once more fruitful was made in the Bay of Saint-Brieuc, upon the north coast of Brittany. Here, where 1,400 men were formerly engaged yearly in fishing for oysters, and where the yield was of the annual value of from 300,000 to 400,000 francs the oyster-fishery, during the ten years from 1850 to 1860, had become almost entirely valueless.

In the months of April and May, 1858, under the direction of Professor Coste, vast numbers of the old shells of oysters and other mollusks were scattered over the ground, and great numbers of fascines were sunk and anchored with stones so as to float in the water just free above the bottom. After 1,000 hectares of sea-bottom had thus been excellently prepared for the reception of the young oyster-broods, three millions of mature oysters were planted upon it.

In the autumn all the shells and the twigs of the fascines were found so thickly covered with young oysters that even the wildest expectations were more than realized.

This abundance of young oysters was something indeed entirely natural. Professor Coste, in his report to the Emperor, January 12 1859, says, when speaking of this experiment in the Bay of Saint Brieuc, that every mature oyster produces from two to three millio embryos, but he does not inform us as to his authority for this state ment. If we allow that those oysters which were planted for breed ing purposes in the Bay of Saint-Brieuc produced each only the sam number of embryos as are produced by a Schleswig-Holstein oyster, th entire progeny would amount to the enormous sum of 1,320,000,000,00 young oysters. Such a number would allow 132,000 to fall upon ever square meter of sea-bottom, and for the reception of this numbe

---

* *Professor Coste* died in 1873. His chief work, upon the artificial breeding of oyster mussels, and fish, appeared under the title: " Voyage d'Exploration sur le Littoral de la France et de l'Italie, 2. éd., Paris, 1861."

there was enough suitable material already spread out about the mother oysters.

This experiment at Saint-Brieuc was considered to completely demonstrate the possibility of artificial oyster-breeding. It was believed by many that the whole coast of France might be bordered with oyster-beds, and they began already to reckon, according to the market-price of oysters at that time, which was 20 francs per thousand, how many millions of francs would be the result of this sea-harvest. Capitalists hastened to form companies for the purpose of engaging in the business, and obtain from the government the right to lay down oyster-beds upon certain definite portions of sea-bottom along the coast. But in not a single case were the rich earnings which had been reckoned upon beforehand as resulting from the sale of marketable oysters ever realized; and not only this, but the money which had been paid for the preparation of the ground and the purchase and transportation of breeding oysters from natural beds gave no returns, and for the most part proved an entire loss. The young oysters were nearly all covered up by sand or mud, or eaten by other sea-animals. This explains why, in the year 1869, I found in the Bay of Saint-Brieuc nothing remaining of the beds which had been thus artificially formed in 1858. The bottom of the bay had become unsuitable for the growth of oysters because of the wide-spread distribution of sand and the changes which it was constantly undergoing.

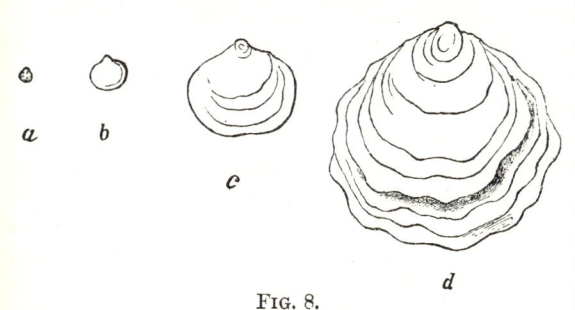

FIG. 8.

Outline figures of young oysters, natural size.—*a*. About one month old. *b*. About two months old. *c*. About four months old. *d*. From twelve to fifteen months old.

At the present time the extensive propagation of oysters by this method of breeding is carried on with success only in certain places along the French coast where the natural conditions are especially favorable. The Bay of Arcachon, south of Bordeaux, is one of the favored places. There, as I myself have observed, the soil and the saltness of the water are very similar to that of our sea-flats. We find there wide-spread shallow sand and mud banks which are covered with vast numbers of bivalve-shells. These banks are left dry by the ebb-tide, and between them are deep channels through which the water at ebb and flood tide flow out and in. In places which remain always under water natural oyster-beds are found, and at other places longer stretches along the soft, bare slopes of the water-courses are made use of as breeding-beds. Here mother oysters from natural beds are planted, and among them, towards the end of May, are placed old shells and tiles with a covering easily detached, as objects of attachment for the young broods.

In October the young oysters, which have become firmly attached, are freed by means of chisels from the larger of these objects of attachment, and are then placed in flat boxes, 2 meters long, 1 meter broad, and 15 to 30 centimeters high. In detaching the young oysters about one-third are destroyed.

The boxes into which the young oysters are placed are made of thick plank, with wire-sieve bottoms, through which the water can pass in and out. At the corners of each box are fastened stakes which serve to raise the box above the sea-bottom, so that there will be a depth of water of about 10 centimeters between the ground and the wire netting. The object of this protection is to guard the young oysters against small crabs (*Carcinus mœnas*), "drills" (*Murex erinaceus*), and other enemies, which formerly destroyed great numbers of young, as the breeders of Arcachon found out by bitter experience. At this period the shells of the young oysters are too thin to protect their soft bodies from their enemies. While in these breeding-boxes the young oysters must be kept continually under water, so as not to be destroyed, either by being left dry, by the heat of summer, or the cold of winter. In order to accomplish this square trenches, from 30 to 40 meters long and 4 to 5 meters broad, are dug in those portions of the oyster-territory which are left dry by the ebb-tide. The side walls of these excavations are made firm by means of posts and planks, and the spaces between the planking and the banks are packed with clay. The bottom is covered with sand and gravel to serve as a bed for the oysters. At one of the four sides a canal, with a gate, is formed, through which, at the pleasure of the breeder, water can be allowed to flow in during the flood or to pass out during the ebb-tide. In these artificial ponds, called *claires*, the boxes containing the young oysters are placed, nourishment being brought to them by the water which forces its way in through the sieve bottoms. As often as the condition of the water renders it possible, the breeder opens the tops of the boxes, in order to permit the free entrance of air and light and to remove any accumulations of dirt which may have lodged upon or around the oysters. Two months later he takes the oysters from the boxes and strews them about upon the bottom of the breeding-pond. In these ponds they must not be placed too close together if their best growth is desired. Even into these ponds their numerous enemies will make their way, and in order to protect the oysters from these hordes of spoliators a small-meshed net is drawn over them. I is very desirable to change the oysters, once or twice during the cours of the year, into neighboring ponds which have been purified by lyin entirely dry for several months. During the warmest and also during th coldest months, especially during ebb-tide, a depth of water of at leas 20 centimeters must be retained over the oysters. This troublesome an tedious handling is demanded for at least two years before the oyster can be brought to market. At least this is the case in the Bay of Arca chon. In the year 1874–'75 there were produced in this bay 112,000,00 artificially grown oysters, and in 1875-'76 about 196,000,000. This important yield of the last year, as compared with the poor returns of former years, may be accounted for principally through two causes:

First. The natural oyster-beds in the Bay of Arcachon had had complete rest for the entire two years immediately preceding these rich harvests. During the years 1870-'71 they had produced only 4,897,000 oysters, but after this period of rest, in November, 1874, 8,500 persons assembled, and in the space of three hours, during which time the gathering was in progress, 40,360,000 oysters were taken from the sea. A great number of these were transplanted, as breeding oysters, to the prepared beds, which covered altogether an actual area of sea-bottom of 2,669 hectares (about 5,338 acres).

Second. The former imperfect method of caring for the oysters had been improved to the extent that the young oysters were protected from their enemies, and care was exercised that during hot and cold weather they should always be kept under water.

With the earlier methods very many of the young oysters were destroyed by their enemies, and often, during a few unfavorable summer or winter days, when a low tide left the beds dry, all the young oysters died. The extraordinary yield of oysters in the Bay of Arcachon and at other points along the coast of Brittany, as a result of the improved method of artificial oyster-breeding, has very materially lessened the price of oysters in France, despite the greater consumption occasioned by this abundance. In 1873 oysters sold for 43 francs per thousand, while in 1876 the price was 25 francs per thousand. On this account only those oyster-breeders who attend personally to the work and are assisted in it by their families make anything over and above all expenses. Those who undertake the breeding of oysters, relying upon outside help to do the great amount of work necessary, can gain returns scarcely worthy of the name; at least this is the case in the Bay of Arcachon, as I know from trustworthy sources.

The cost of transforming a hectare of sea-bottom along this coast into an oyster-bed, together with the necessary apparatus for oyster-culture, and a guard-vessel as required by the government, is not less than seven to eight thousand francs.*

---

* Besides the works of Coste, which have already been mentioned, the following also treat of oyster-breeding in the Bay of Arcachon:

K. *Möbius.* Ueber Austern- und Miesmuschelzucht und die Hebung derselben an den norddeutschen Küsten. (A report to the hon. minister of agriculture.) Berlin, 1870. pp. 8.

A. *Tolle.* Die Austernzucht und Seefischerei in Frankreich und England. (A report to the hon. minister of agriculture.) Berlin, 1871. pp. 8.

De Bon. Notice sur la situation de l'ostréiculture en 1875. Paris, 1875. (Extract from the Maritime and Colonial Review.)

6.—ATTEMPTS TO INTRODUCE THE FRENCH SYSTEM OF ARTIFICIAL OYSTER-BREEDING INTO GREAT BRITAIN.

In Great Britain a large number of men are employed in oyster-dredging and in the oyster-trade, and, according to the published official estimate for the year 1870, the yearly value of oysters sold in the kingdom is not far from £4,000,00) sterling. If we take the average price of oysters as one penny (two cents) apiece, which is rather too much than too little, this amount would account for 960,000,(00 of oysters.

In the year 1864 there were brought to the London market alone more than 495,000,000 of oysters, which were worth over £2,000,000 sterling. The culture of oysters being thus of so much importance to Great Britain, it was very natural that attempts at artificial oyster-breeding in France should be watched with intense interest, and imitated at various points along the British coast. It was carried on most extensively upon the coast of the small island of Hayling, east from Portsmouth, by the South of England Oyster Company, organized in 1865 with a capital of £50,000. Inside of a dike upon the west side of the island five oyster-beds were prepared, having an extent of sea-bottom of about 32 hectares (about 80 acres). May 11 and 12, 1869, when I visited these beds, several of them had not been overflowed. The natural bottom, which was a sticky mud, had been covered with gravel and mussel-shells, and upon the largest bed hurdles, each 2.4 meters long by 75 centimeters broad, and composed of birch twigs, had been placed so as to rest horizontally at about one-half a meter above the ground. Besides these hurdles, laths, with oyster-shells and bundles of small rods nailed to them, were stuck about over the ground, so that there should be plenty of objects of attachment for the young oysters. The inward and outward flow of the water was regulated by means of a sluice and gate. The mother oysters are generally placed in the beds just before the breeding season.

In 1869 they expected to place upon the beds 50,000 breeding oysters. The water is generally changed every day, except during the winter months, when there would be danger of freezing the oysters, and also except during the swarming period, when the young would be liable to escape into the sea with the changing water. In 1867, 600,000 mature deep-sea oysters were placed on an oyster-bed which covered a surface of 7.3 hectares, and over which 10,000 hurdles were placed as objects of attachment. Upon an average over 12,000 young oysters were found attached to each hurdle, making for all the hurdles a total of more than 120,000,000. In these and other experiments at artificial oyster-breeding in England all the experiences of French oyster-breeders were made use of as far as possible, but, notwithstanding this, at no single breeding station were the expectations of a great yield of marketable oysters ever realized. In London, on the 4th of May, 1876, Mr. Blake, the inspector of fisheries, made, before the commission for the investigation of the oyster-fisheries, the astounding statement that every oyster grown by means of artificial culture near Reculvers, at the mouth of the Thames, cost £50 sterling, that every one grown in Herne Bay cost £100, and in a third place about £500, and that he was prepared to furnish several other examples of a like character. Mr. Blake, who is very well acquainted, from personal observation, with French and English oyster-culture, considers artificial oyster-breeding according to the French method impossible along the British coast, on account of the unfavorable character of the climate.

The most important source whence I have drawn my information in regard to the culture of oysters in England is the report of the select committee on oyster-fisheries, together with the proceedings of the committee, minutes of evidence, appendix, and index, ordered by the House of Commons to be printed July, 1876. This report contains 3,941 questions and answers concerning oyster-culture.

What I have been able to learn, through my own observations, of the English oyster industry I have described in a work referred to in chapter 4. To this, Mr. A. Tolle, who accompanied me as hydraulic engineer of the commission of the Prussian minister of agriculture, has, in a report to the honorable minister, issued a supplement, which is also referred to in the same chapter.

7.—CAN THE FRENCH SYSTEM OF ARTIFICIAL OYSTER-BREEDING BE CARRIED ON IN THE WATERS OF THE GERMAN COAST?

What German who loves oysters has not wished that the whole German coast might be bordered by fruitful oyster-beds? For this reason we wish to investigate as to whether the necessary conditions for artificial oyster-breeding are to be found in the coast-waters of Germany. As regards the saltness of the water, the currents, the food, and even the composition of the soil, our sea-flats will compare as favorably for the artificial gathering of the young broods and for the raising of the same as the Bay of Arcachon, but not as regards temperature and the depth of water.

In the Bay of Arcachon the difference between ordinary high and low tide is 4.5 meters, and during a storm a meter more. But along our North Sea coast during a storm the water rises with the tide even more than twice as high as during ordinary flood-tide. The power of the water during a storm, as compared with the power of the water during an ordinary flood-tide, is much greater along our coast than in the Bay of Arcachon. Hence, we would be obliged to give to our oyster-beds a much greater firmness than the French breeders have to give to theirs. We would also be obliged to place them so far out in the sea that they would be entirely covered with water, even at the lowest ordinary tide, and also give them sufficient stability to withstand, during a storm, a rise of water of from 2 to 2.5 meters, as well as the great and powerful force of this water-mass. Beds thus laid down would cost much more than the ditched and planked ones of Arcachon. But even if they were so placed as to bid defiance to the most severe flood-storm, they would indeed hardly suffice to protect the breeding oysters from being covered with mud and sand; and thus one flood-storm, or storm in connection with a flood-tide, might destroy the accumulated oysters of many generations. A visit made to the island of Norderney showed us how destructive nature can be to the oyster-beds of our sea-flats. Upon the inner side of this island, early in the year 1869, a surface of 825 square meters was dug out, and made firm by double-planked walls to about the height of half tide. The space between the walls and banks was filled in with sand and mud, and the inclosure itself was divided into two compartments, one of which was longer than the other.

In the smaller division the water was detained to enable it to deposit its coarser materials before it was allowed to pass into the larger one. In the beginning of June, 20,000 mature oysters were placed in these artificial beds, with the expectation of reaping a rich harvest of young oysters; but the harvest never came. Star-fish and crabs attacked the oysters, in the beginning of August flood-storms broke down the walls, and the storms of autumn completed the work of destruction, so that very soon nothing was left of the entire enterprise. If the situation of the free sea-flats is not suitable for the formation of oyster-beds, perhaps there is still a possibility of artificial oyster-breeding being carried on inside of the dikes which protect the fertile marsh-land along the German coast from the encroachments of the waters of the North Sea. For this purpose basins would have to be dug out inside of the dike and placed in connection with the sea by means of canals. Where these canals cut through the dike it would be necessary to build a gate, in order to prevent the sea-water from passing in during high-tide. Then, oyster-beds could not be laid down in the neighborhood of this gate, because it would serve not only as an inlet for salt water, but as an outlet for the fresh water from the marsh-land, and so fresh water instead of sea-water would cover the oyster-beds. But even if it is admitted that oyster-beds might be laid down inside of a dike without danger to the diked lands, and with sluices and gates to permit the inflow of sea-water, there are yet several questions to be answered. How will oysters thrive in such beds? Will they receive enough nourishment to become fat? How will they exist during continued cold weather? And will they produce young in such a place? It is certain that they will not receive as much food as in the open sea, since they cannot have nearly as much water as will pass over them upon the natural beds; and the quantity of nourishment varies in proportion to the amount of water which passes over the beds. In these beds the oysters would also be in danger of being buried in the deep mud, and in order to prevent this they must either be changed very often into clean beds, or else a cleaning-pond must be formed beside the breeding-pond. But while the water is rendered clear by being allowed to stand quiet, yet by this means a large amount of organic matter which serves as food for the oysters is taken from it. Especially dangerous, however, to oyster beds within the dikes would be the cold during winter weather, for along our North Sea coasts the water is lowest during an east wind, and at the same time such a wind is accompanied by the lowest degree of temperature. Hence, at such times, when a great depth and a constant change of water over the beds would be the best protection from freezing, we cannot have high water, nor can the water then standing over the oysters be constantly changed; thus during every cold winter, a large number of oysters would be sure to perish in their beds. Even now, upon the shallow oyster-beds of our sea-flats, oysters are frozen exactly in proportion to the depth of the water over them during these cold spells; the shallower the water the greater the destruction upon the beds during a severe winter. During the severe winter of 1863-'64, when, on account of ice, no oysters could be taken from December 21 to February 17, and during the winter of 1864-'65, when the fishing was interrupted from January 24 to March 26, dead oysters were found upon a large number of the banks. The greatest destruction of oysters within the memory of the oldest fishermen took place during the severe winter of 1829-'30, when Schleswig-Holstein was visited by an unusually low temperature, which continued from the middle of November until the beginning of the next February. Most of the beds suffered greatly, and it was many years before they again recovered their former fruitfulness. In cold weather slime collects upon the gills and mantle-lobes of the oyster, the power of the muscles and cilia being weakened by the cold. Accordingly, the oyster is no longer in a condition by means of its rapidly-moving cilia, and the quick closing of the valves of its shell, to drive out the particles of slime brought in with the water. But the power of the cilia and the elasticity of the muscles are again restored as the water becomes warmer, providing the cold has not lasted too long. The gills become clean once more, and respiration and nourishment, which have been disturbed by the sliming, proceed again as before. If the cold spell is prolonged, then, in addition to the sliming of the gills and mantle, there are yet other pernicious results. The shell-muscle becomes so soft that it can no longer close the valves. The cilia move slower and slower, and finally, when the shell-muscle has allowed the valves of the shell to gape wide open, cease moving altogether. The mantle and gills become pale in color, infusoria nest in them and hasten their destruction, and soon their ciliated layer separates and disappears. The softest portions of the body, the generative organs, the liver, and the stomach quickly vanish, probably consumed by snails, crabs, worms, and starfish as soon as they can make their way unhindered into the open shell. The last part of the mollusk which is to be found in the shell is the shell-muscle. It remains free between the two valves, or attached to only one of them, until finally but a trace of its fibers is to be seen at the points of attachment, the so-called muscular-impressions. During the latter part of March, 1870, I was able to follow out for myself, at the Schleswig-Holstein beds, the entire course of the changes produced in the oyster by freezing. Long-continued east winds had kept the water extraordinarily low, and for more than a month thick ice had covered the flats, so that from the 4th of February to the 7th of March no oysters could be taken. On the 14th of February the water in the neighborhood of an oyster-bed at the north end of the island of Sylt was found to be of a temperature of 2° C. below zero. At this point the depth of water was 3.5 meters. Of those oysters which were taken in my presence from the shallower beds 7 to 8 per cent. were frozen. Upon beds which lay in deeper water, nearer the open North Sea, the cold had killed only from 2 to 3 per cent. Evidently, then, these latter beds had suffered less damage because at every flood-tide they received water of a somewhat higher temperature from the open sea. I have frozen the mantle and gill lobes of oysters in North Sea water and allowed them to remain inclosed in ice for an hour at a time, with the temperature of the water varying in degree from 4° C. to 9° C. below zero. When the ice had melted, the cilia began to move feebly, and four hours later, when the temperature of the water had risen to 5° C. above zero, their movements were once more fully established. Other gill and mantle lobes which had been three hours in water of a temperature of 1° C. to 2° C. below zero moved quite lively on the following day. This recalls to me a very weighty difference between fresh and salt water, which is often overlooked. It is generally known that fresh water is densest and heaviest at a temperature of 4° C. above zero. When any portion has arrived at this temperature during freezing weather, it sinks to the bottom of the body of fresh water, where it remains until the entire mass above it is of the same density. That portion which first becomes lower in temperature than 4° C. then expands, rises to the surface, and stiffens into ice as it reaches the temperature of 0°.

The fact is less known that with sea-water the lower the temperature the greater the density and weight of the water. Therefore, it also sinks to the bottom until it has reached the temperature at which it forms ice, which, when it holds 3 per cent. of salt in solution, is 2.28° C. below zero. It is evident, then, that water may be found at the bottom over the sea-flats of a temperature of 2° C. below zero, while, during the most severe cold, water at the bottom of the lakes and deeper rivers of North Germany is found to be constantly several degrees warmer than this. When, finally, the sea-water, from the surface to the bottom, has reached its freezing point, it does not become solid ice for the whole thickness, but thin layers of ice, at greater or less distances apart, are formed in it. These layers, which are crystallized from the salt water, are free from salt, are hence lighter than the surrounding water, and accordingly ascend to the surface; consequently, those animals which live upon the ground of the deeper portions of the sea-flats remain surrounded by water whose temperature is 2° C. lower than the freezing point of fresh water. In shallow places which the ebb-tide leaves dry, the frost kills all animals which have not the ability to dig their way to such a depth in the sand and mud that they will be beyond the influence of frost, and where the water remains liquid.

Here only a few kinds of mussels, worms, and crabs possess this ability; hence all of those sand and mud banks of the sea-flats which are left dry by the ebb-tide are comparative "barrens," occupied only by very few animals and plants.

Our investigation, then, has led to the grievous conclusion that profitable artificial oyster-breeding, according to the French system, is not possible along our North Sea coasts. Whoever should attempt to carry out this system, despite the unfavorable conditions of our waters and climate, would be certain to find that his breeding oysters were more costly than many English oyster-breeders have found theirs to be; for upon the English coasts the difference between ordinary high tide and the tide increased by a storm is much less than upon our sea-flats, the lowest water does not occur simultaneously with the coldest winds, as along the southeast shore of the North Sea, and the climate there is milder than upon our coasts.

8.—CAN NATURAL OYSTER-BEDS BE ENLARGED, AND CAN NEW BEDS BE LAID DOWN, ESPECIALLY ALONG THE GERMAN COAST?

It will thus be seen that the German oyster industry remains dependent now, as ever, upon the natural oyster-banks of our coast-seas, where oysters have lived for thousands of years, and where they exist to-day fruitful and well-flavored. And in regard to these beds we have now to consider the important questions:

First. Is it possible to increase their size?

Second. Can we still farther increase the surface of our oyster-territory by laying down new beds?

FIG. 9.

In the foreground are the *sea-flats*, with two can-buoys which indicate the course of the channel for vessels. In the background are seen the dunes or sand-hills of Hörnum, the southern point of the island of Sylt.

The water in the neighborhood of the banks, and over all the stretches between them, has the same character as over the banks themselves. All that is necessary, then, in order to increase the size of these beds is to render the sea-bottom between them habitable for oysters. Old beds increase naturally in size whenever the shifting and slimy sea-bottom which borders them becomes changed into stable and clear ground. This can take place if changes occur in the force and direction of the ebb and flood currents. In such cases the extension can be hastened artificially by placing upon the newly forming ground shells of oysters and other mollusks, in order to furnish just outside the borders of the old bed the most judicious objects of attachment for the young broods as they swarm out from the mother oysters. For the establishment of new beds, within the limits of the German sea-flats, in places where no oysters are found at present, it will be necessary to find stretches of sea-bottom which are free from mud, where the soil is not being constantly shifted about by currents, and where the ebb-tide will leave at least one to two meters in depth of water over the beds. But nearly all such places are at present occupied by oyster-beds. In the year 1876 the buoy-tenders, who are best acquainted with the bottom over the entire Schleswig-Holstein sea-flats, and who have to mark out the channels for vessels, by means of cask and stake buoys (Figs. 1 and 9), sought to find some places upon the flats suitable for oyster-beds, where no oysters yet existed. They found within their whole territory only eight such places where it might be possible for oysters to thrive; and it would be very hazardous to immediately distribute over all these places a great number of breeding oysters, since it is yet doubtful whether new beds would be able to flourish there or not. It would be much wiser to experiment with one only of these places at first. Upon this let oyster and other mollusk shells be scattered in May and again shortly before the breeding period; then, upon the ground thus prepared place several thousand mature oysters. If, by next fall, a deposit of young oysters is found to have taken place, it will not be certain even then that the experiment will prove successful, but only after three or four years, when a large number of half-grown oysters are found lying beside the old mother oysters, and when these young are found in turn to have produced other broods which locate upon the new bed. Over the entire German sea-flats lying south and southwest of Schleswig there can hardly be found a single place which is suitable for the formation of a profitable oyster-bed; for in front of the mouths of the Eider, Elbe, Weser, Jahde, and Ems the sea-bottom is so covered with mud, or so subject to change, that oysters could not live and multiply there. In the fall of 1868, when I investigated with a dredge-net the sea along the German coast from the Eider to Borkum, I found over this entire territory but one single locality upon the coast of Hanover, between the mainland and the island of Juist, west of Norderney, which in any manner would be suitable for such an experiment. Here, in the spring of 1869, a large number of breeding oysters from the Schleswig-Holstein beds were distributed. But no permanent bed has been established there, for in June, 1875, during an investigation of the bottom near Juist and Borkum, only seven oysters were taken, notwithstanding the dredge was used for three whole days. The sea-bottom in the neighborhood of Juist is, therefore, not suited to the growth of oysters. It is too muddy, and already in possession of the edible mussel (*Mytilus edulis*). During the last century, and the first half of the present one, the Hanoverian Government was accustomed to lease the oyster-fisheries along its coast. These fisheries were principally in the neighborhood of Juist and Borkum, and from 1841 to 1846, inclusive, 193,684 oysters were taken there, making an average yield of 38,727. In 1851, in a survey of the beds, very few oysters were found, and in 1855 the beds were so impoverished that no one would rent them.

The exhaustion of the beds resulted from excessive fishing and from the increase of mud upon the ground occupied by the oysters. Whoever, therefore, would establish new oyster-beds along the German portion of the coast of the North Sea, between the Eider and the mouth of the Ems, must begin his difficult work by changing the ebb and flood currents in the southern portion of the North Sea, in order to prepare a surface upon which oysters can thrive; for to attempt to adapt oysters to a bottom of shifting sand or mud is not natural, nor is it conducive to an industry which is to last for a hundred years. For thousands of years innumerable young oysters have been scattered from the oyster-beds over changing mud and sand banks, and yet not one has so altered its organization as to become adapted to such a bottom and transmit its new nature to its progeny; they have all been destroyed.

Since the sixteenth century, along the west coast of France, on both sides of the mouth of the Seudre, near Marennes and La Tremblade, the oyster-breeders have been in the habit of transplanting oysters, one year old, from natural oyster-beds to prepared ponds in order to fatten them and improve their flavor. These ponds, called *claires*, are shallow excavations of various shapes and sizes. The greater number are square or rectangular, and cover from two to three thousand square meters of surface. They lie near together, but irregularly, and are divided off into sections by deep trenches or canals, by means of which the sea-water flows in and out during spring-tides. The bottom of the ponds is somewhat higher near the center than around the edges. The walls surrounding the ponds are formed of the earth dug from within, and are about one meter in height. The neighboring ponds are placed in communication with one another by means of ditches or wooden pipes in the walls. Flood-gates are placed in the larger trenches, by means of which the water can be retained in the ponds from one spring-tide to another. In the fall, when fishing upon the sea-beds is permitted, young oysters are taken and transplanted to these ponds. From August until the breeding season next year, these transplanted oysters acquire a cloudy, dark-green color in the tissues of the mantle, gills, liver, and stomach. The delicate flavor, for which the green oysters of Marennes are especially famous in Paris, is only acquired after three or four years. During this time they must often be cleansed from the mud which has accumulated upon them, and transferred to fresh ponds, if they would be kept healthy. In these feeding-ponds the oysters spawn well, and at times, when there are any objects of attachment free from mud, such as stones, shells, and pieces of wood, the young oysters become attached, but they do not mature into marketable oysters. Oyster breeders, after three hundred years of practice in rearing young bank-oysters in the mud of feeding ponds, have not as yet been able to transform the oyster into a mussel which can live and propagate in the mud. The breeders of Marennes and La Tremblade have been able to change the color and flavor of mollusks, but they have not been able to give the oyster a foot for the purpose of locomotion. Along the German coast, in the East Sea, the sea-bottom, over many extensive tracts, is firm, and also free from mud. These places possess, then, in this respect, one of the most important conditions for the successful formation of oyster-beds. Yet several attempts to plant oysters in the Baltic have proved entire failures. In 1753, 1830, and 1843, oysters were planted along the coast of Pomerania. The last of these attempts was made by a company, of which the Kings of Prussia and Hanover and the Prince of Putbus were members. Fifty thousand oysters, taken in the northern portion of the Cattegat, near Frederikshavn, were placed, on the 6th and 13th of April, 1843, in the waters southeast of the island of Rügen, near Greifswalder Oie. Two years later, investigations showed that they were all dead, since not a single living one could be found. The much talked-of attempt at oyster-breeding by Coste gave a new impulse to the question of planting oysters in the Baltic. In the Bay of Kiel, on the south coast of the island of Laaland, in the neighborhood of Korsör, and in the Isefjord, on the coast of the island of Seeland, mature oysters were planted, upon apparently suitable ground, but the desired result was not attained in either place. The water of the Baltic is not salt enough for the propagation of the oyster. East of the island of Rügen the water at the bottom contains only 1 per cent. of salt, and near the surface still less, since the rivers bring in much fresh water. West of Rügen, south from the Great Belt, to near the coast of Mecklenburg, the water at the bottom contains indeed as much as 3 per cent. of salt, but here also the surface-water everywhere contains a less degree. The young oysters, as soon as they had left the mother oysters, would then ascend to the surface, and thus come into water which throughout the entire southern portion of the Cattegat contains less than 2 per cent. of salt, while they need water with at least 3 per cent. of salt. This I infer from the fact, that such a degree of saltness is to be found at all places along the European coast where natural oyster-beds exist. There are two other conditions of the Baltic besides the low percentage of salt, which certainly hinder the growth of the oyster—the long-continued low temperature of winter, and the lack of regular tidal-currents; for in the North Sea, where there are strong and regular tidal-currents, the oyster, which is a stationary animal, will receive daily a greater quantity of oxygen and food in the water brought to it than it will in an interior sea, where the water is in less regular motion. These chemical and physical differences between the North and East Seas render it not only impossible for the oyster to live in the latter, but also for many other North Sea animals, of which I will mention only the lobster, the larger punger (*Platycarcinus pagurus*), and the edible sea-urchin (*Echinus esculentus*).

If nothing further were necessary in order to establish a permanent settlement of oysters in the Baltic than to plant there several thousand fresh and healthy mature oysters, why then cannot lobsters, crabs, sea-urchins, and all the other animals which are found associated with the oysters upon the banks, and indeed the entire fauna of the North Sea oyster-banks, flourish in the Baltic? If this could have been accomplished, I should long ago have had a large number of the animals of the North Sea naturalized in the Bay of Kiel, in order to facilitate my own investigations, and for the purpose of instruction to students. Nature has already made frequent efforts to introduce not only oysters, but other North Sea animals, into the Baltic. Nearly every year fish and other animals from the North Sea appear in the Baltic, but they are not permanent, and soon disappear again from our fauna.

The great storm-flood of the 13th and 14th of November, 1872, brought *Noctiluca scintillans* from the North Sea into the harbor of Kiel in such numbers that for weeks they made the waters of the harbor brilliant with their phosphorescent flashes, but very soon they had entirely disappeared. Under the present geognostic and physical conditions the oyster can advance no farther towards the Baltic than into the southwestern part of the Cattegat. Here a line drawn from Samsöe over the island of Anholt to Gothenburg represents the limits of those conditions suited to their welfare. Along this extreme border of their existence one could not expect such productiveness and size among the oysters as a costly artificial system of breeding would demand in order to be profitable.

Every change in the saltness of the water below the general mean, or in the temperature of the sea-water, would incur heavy loss to any artificially conducted system of oyster-breeding which might be carried on here. That oysters of their own accord spread out from their great breeding home in the North Sea into all places where they find the external conditions favorable, is proven by their substantial immigration into Lim Fiord, in the north of Jutland. This fiord, up to the year 1825, consisted of a number of connected brackish-water lakes, with an eastern out-flow into the Cattegat. During the last century futile efforts were made to establish oyster-beds in these seas; but on the 3d of February, 1825, a fearful storm-flood broke through the dam which separated the western portion of the Lim Fiord from the North Sea, and after this the water of the fiord became more salt every year, the brackish-water animals and plants which had lived there vanished, and in their place came North Sea animals, and among them, in 1851, the oyster was first noticed. From year to year they spread over more surface. In 1860 only 150,000 were dredged; presently 98 places were known where oysters had become established, and in 1871-'72 the oystermen were able to take for foreign consumption seven millions of mature oysters from the beds of Lim Fiord. Their distribution was very rapid. In 1851 the first were found; had there been many there before this time, they would certainly have been noticed by the fishermen. The water must first contain a percentage of salt of 3 per cent before they can enter a new territory. If we admit the first appearance of oysters here in 1840, then in an interval of thirty years they had spread over an extent of surface 15 miles (German) in length, which shows a yearly advance, in territory covered of about one-half mile in length, or rather more, about 3,700 meters. The beds of the Lim Fiord are from 1 to 8 kilometers from one another. Their length is from 1 to 7 kilometers and their breadth somewhat less. These facts show that the young swarming oysters are capable of moving over a stretch of bottom 8 kilometers in length. In the same manner as it has thus immigrated into the Lim Fiord the oyster would have established itself in the Baltic had the water been similar in its characteristics to that of the North Sea, and this would be the condition of affairs if the connection between the North Sea and the Baltic were broader and deeper than it is at present. At one time it was broader and deeper, and, for this reason, oysters once lived four miles east of the point where the city of Kiel now stands. This is proven by the fossil oyster-beds found near Waterneversdorf, in the eastern part of Holstein, which, together with the entire bottom of the western portion of the Baltic, have been raised more than 30 meters. By this elevation the Cattegat, the Belt, and the Sound were made shallower and smaller pathways for the water coming in from the North Sea than they were in olden times, when the oyster-beds of Waterneversdorf still produced oysters. Yet, by this elevation of the sea-bottom, which took place thousands of years ago, the percentage of salt in the water has been lessened but very little. Thousands of years later, when the oyster-beds of Waterneversdorf had been dry land for a long time, oysters were found in such abundance along the coast of the Danish Islands that they served as food for the people of the Stone Period who lived in this vicinity, since great masses of oyster-shells are found in the heaps of kitchen refuse of that time.

And since the oyster-shells of Waterneversdorf and of the kitchen-heaps of the Stone Age fully agree with those of to-day, since they are also bored like ours by the boring sponge (*Clione celata*), and since the whelk (*Buccinum undatum*) and other animals at present found upon the sea-flats lived with them, conditions favorable to their growth must have existed at that time in the meridian of the present Cimbrian Peninsula, the same as now to the west of Schleswig-Holstein. The oyster has thus not changed during the course of at least ten thousand years. It has not accommodated itself to the changes which have taken place in the territory occupied by it, but has yielded to those changes, although they were brought about very slowly. Hence it is impossible for any human power to change their nature in a short time and accustom them to the water of the Baltic as it is to-day.

The following Danish works treat of the oyster-beds of the Lim Fiord, the extension of the oyster into the southern portion of the Cattegat, and of the unsuccessful attempts to plant oysters in the Baltic:

Jonas Collin. Om Östersfiskeriet i Limfjorden. (With a chart of the oyster-beds.) Copenhagen, 1872.

G. Winther. Om vore Haves Naturforhold med Hensyn til konstig Östersavl og om de i den Henseende anstillede Forsög. Copenhagen, 1876.

F. Krogh. Den konstige Östersavl og dens Indförelse i Danmark. Hadersleben, 1870.

In the royal archives at Stettin and Stralsund are to be found the acts under which the attempts to locate oysters along the coast of Pomerania in 1830 and 1843 were made.

9.—SIZES AND PRODUCTIVENESS OF THE OYSTER.

The delightful hopes of bordering the entire German sea-coast with fruitful oyster-beds, and of seeing German oysters as food upon every table, must, therefore, be given up. The nature of our waters, as well as the nature of the oyster itself, forces us to do so. Yet it is especially difficult for those to understand this who share the widespread opinion that all eggs which are spawned by oysters are destined to become transformed into young mollusks. Most animals, however, whose ova and young are exposed to attacks and liable to be destroyed, produce a large number of eggs, while those animals, on the contrary, which guard their broods until they can take care of themselves, as is the case with mammals, birds, and some invertebrates, generally produce but few eggs; but in those cases where care for the brood is entirely lacking, or lasts for a very short time only, eggs are produced in such great numbers that the numerous enemies which regularly attack them are not able to destroy them all. A certain number escape destruction and arrive at maturity. The tape-worm of man (*Tænia solium*) produces from its eight hundred segments not far from forty million germs, and the parasite *Ascaris lumbricoides* forms in its ovary about sixty million eggs. Under the normal condition of affairs for the development of these worms, only a very few of the great number of eggs laid ever go so far in growth that they in turn produce eggs. This is satisfactory to everybody, since none desire that all of the forty million eggs of the tape-worm or the sixty million eggs of the "itch-insect" should ever become mature parasites. It would be a horrible state of affairs if such a thing should happen. On the contrary, every one very much desires that all the young broods which the oyster sends forth into the water should become mature table-oysters, since, when fully grown, they become one of the most delicious of delicacies. But in observations and investigations which have for their object the discovery of the methods and means by which nature brings these things about, such desires as these must not be allowed to have an influence upon our opinion; for whoever would have nature especially attractive, beautiful or useful, whenever he is in immediate contact with her becomes easily led away from the pathway of strict scientific investigation and lost in the dark and boundless territory of speculation.

Nature accomplishes at every place just what she is obliged to accomplish there with her united forces, according to the conditions upon which the development of the world has proceeded. Throughout her entire limits there are no such distinctions as useful or injurious. The terms agreeable or disagreeable, beautiful or frightful, useful or harmful, as applied to the workings of nature, exist only in the thoughts and comprehension of intelligent and sensitive beings. Yet very frequently we hear it said, when speaking of the fossil oyster-bed which now lies near Blankenese, below Hamburg, 80 meters above the level of the Elbe, that it did not make any difference that oysters should once have lived there and produced young, of which only a small proportion should ever come to maturity, since no human beings were there at that time who could have fed upon them.

Oysters belong to that class of animals which secures the continuance of the species, not by guarding the young for a long time, but by producing a vast number of embryos every season. They are able to produce so large a brood that enough of the number will be certain to arrive at maturity to maintain the status of the bed, and supply the places of those old oysters which die or are destroyed; and this result takes place notwithstanding many of the young are destroyed by sand, mud, or unfavorable temperature, and many others are eaten before their shells are thick and large enough to protect them from the numerous enemies which live upon the same banks with them. The number of descendants from any one oyster which thus arrive at maturity is so small even upon the best beds, where for more than a hundred years the finest and most productive Holstein oysters have been caught, that I am persuaded no one would give credit to my words if I was not able to substantiate them by means of figures. In 1587, Frederick II, King of Denmark and Duke of Holstein-Gottorp, appropriated the oyster-beds of Schleswig-Holstein as royal prerogatives.* They

---

* The public order by which the Ducal-King Frederick II took possession of the oyster-beds of the sea-flats along the coast of Schleswig-Holstein and Jutland, is printed by H. Kröyer in his work "De danske Östersbanker," Kjöbenhavn, 1837, page 110. Translated into English, through the German, it reads as follows: "We, Frederick, &c., make known to all by these presents, that since it has been brought to our knowledge that in the waters of the West Sea, in the fief of Ribe, a kind of fish called an oyster can be found and caught, therefore we have commanded our liege Albert Friis, superintendent and guardian at our castle at Ribe, that he permit this kind of fish to be caught in our name and sent to us; and in order that a future lack of them may not occur, we forbid one and all, whoever he may be, from taking oysters or allowing

were then leased, generally for a long term of years. From time to time the government caused the banks to be officially examined, in order to find out their condition and prevent their depletion by overfishing. The examination was conducted by commissioners appointed by the government, and the dredging carried on in their presence was performed by fishermen specially sworn for the purpose. The smaller beds were dredged in three, the larger in six, different places, and all the oysters taken were divided, according to size and age, into three classes, known as—
1. Zahlbar Gut, or marketable.
2. Junggut, or medium (half grown).
3. Junger Anwachs, or young growth.

The *marketable* oysters are those which are large and full grown. Their shell is at least 7 to 9 centimeters in length and breadth, and when closed the greatest thickness must be more than 18 millimeters. The left valve, or the one which is most curved, is from 6 to 9 millimeters thick at the point of attachment of the shell-muscle, and also under the ligament.

The greater number of full-grown oysters are from seven to ten years of age, yet many older ones are found, which can be distinguished from the younger ones by the greater thickness of their shell. Oysters more than twenty years old are seldom seen. The oldest which I have personally examined I estimated to be from twenty-five to thirty years old. The left valve, at the muscular impression and below the ligament, was 20 to 25 millimeters in thickness.

The shells of the half-grown oysters, when closed, show a thickness of from 16 to 18 millimeters. The valves, where thickest, are, at the most, only 5 millimeters thick, and their breadth is less than 9 centimeters. They are cleaner than the old oysters, upon whose shells are generally to be found many animals and plants. The *young growth* are those small and thin oysters which are not older than from one to two years (Fig. 8 *d*).

In the record of each inspection we find indicated the number of marketable and the number of medium oysters caught in each haul of the dredge, but the number of the young growth is not given, mention only being made as to whether there were many or few.*

---

them to be taken in that place. We except, however, those who take them in our name by the authority of our liege at Ribe. Whoever shall dare to act contrary to this command, and he can be justly convicted of so doing, shall be punished according to his deserts. Each one is then to govern himself accordingly, and guard against transgressing. Given at Skanderborg, the 4th day of February, 1587."

* The work of Kröyer contains also a tabular review of the numbers of mature and medium oysters of the official investigations which took place from 1709 to 1830. This table, and also tables for which I have to thank the royal government at Schleswig, have furnished me the numbers from which I have estimated the proportions between half and full grown oysters. I have not considered the investigations previous to the year 1730, partly because in the beginning of the eighteenth century a number of beds were unknown, and partly because the numbers of the first five inspections (1709 to 1728) give no positive results. On six official investigations made between the years 1839 to 1876 I have participated myself. The results of these I will give later.

During the period from 1730 to 1852 ten records were made of all the oyster-beds of the Schleswig-Holstein sea-flats. If from these records the numbers of all the marketable and all the medium-sized oysters are taken and added together, we will have a series of very different totals, showing no particular general law. But if for each of these reports the proportion of marketable to medium oysters is taken, then we will arrive at the surprising result that this proportion fluctuates but very slightly during all the records.

The following table gives a summary of the marketable and medium oysters recorded as caught during each of the ten investigations. From these numbers I have reckoned for each record the proportion of medium oysters to every thousand of those which were full grown.

Year of record.	Number of marketable oysters.	Number of medium oysters.	Proportion of marketable to medium.	
1730	5,394	2,602	1,000	486
1734	16,770	5,205	1,000	310
1740	7,185	3,007	1,000	418
1756	6,793	3,333	1,000	490
1795	2,078	1,006	1,000	484
1799	2,705	831	1,000	307
1819	2,828	1,087	1,000	388
1830	1,956	797	1,000	417
1839	3,272	1,552	1,000	440
1852	3,534	1,673	1,000	473
Total			10,000	4,213
Mean proportion			1,000	421.3

The following table gives the quantities of oysters which were taken during these investigations from two of the largest and most productive beds of the Schleswig-Holstein coast, the Huntje and the Steenack Banks:

Year of record.	Huntje. Marketable.	Huntje. Medium.	Steenack. Marketable.	Steenack. Medium.
1730	355	164	158	69
1734	1,353	874	465	90
1740	323	158	26	110
1756	736	99	149	35
1771	931	66	607	197
1795	87	461	261	106
1799	183	119	236	56
1819	363	173	53	79
1830	40	3	11	4
1852	128	64	116	58
Total	4,499	2,181	2,082	804
Proportion	1,000	484	1,000	385

The proportion of marketable to medium-sized oysters is thus seen to be almost the same upon single beds as in a mean of all the beds taken together.

In this similarity of proportions between the marketable and medium oysters in different years and upon different beds a natural law is very strikingly manifested. The medium-sized oysters of any bed consist of the descendants of the marketable ones. They are those members of the young broods which have escaped the numerous enemies living upon and around the beds, and which, despite the numerous attacks made upon their lives, have grown into very respectable-sized animals.

The medium oysters thus represent the total number of embryos from the bed which, in the struggle for existence, have continued to exist. A thousand mature oysters will produce during a breeding period, as I have already shown in chapter 2, at least 440,000,000 of young; but upon the beds alongside of these 1,000 mature oysters are to be found, on an average, not more than 421 half-grown ones; so that, as a rule, for every Holstein oyster which is placed upon the table more than 1,045,000 young are destroyed or die; and indeed even more than this, for not only do those oysters which are over six years of age produce eggs, but those which are two and three years old also reproduce their kind to a certain extent. The younger oysters, however, produce much less spawn than those which are mature, so I estimate that those half-grown oysters lying beside the mature ones on the same banks, and which are their offspring, will produce 60,000,000 young oysters.

We thus have, upon a surface of oyster-bed occupied by 1,000 full-grown and 421 half-grown oysters, at least 500,000,000 of young produced during the course of the summer, and of this immense number only 421 arrive at maturity. *The immolation of a vast number of young germs is the means by which nature secures to a few germs the certainty of arriving at maturity.* In order to render the ideas of germ-fecundity and productiveness more easily understood, I will make a comparison between the oyster and man.

According to Wappäus,* for every 1,000 men there are 34.7 births. According to Böckh,† out of every 1,000 men born 554 arrive at maturity, that is, live to be twenty years or more of age; thus, on an average 34.7 children are produced from 554 mature men, or 62.6 children from 1,000 mature men. Since 1,000 full-grown oysters produce 440,000,000 of germs, then the germ-fecundity of the oyster is to the germ-fecundity of man as 440,000,000 to 62.6, or as 7,028,754 to 1. On the other hand, the number which arrive at maturity, is 579,002 times as great with mankind as with the oyster; for of 1,000 human embryos brought into the world 554 arrive at maturity, or of 440,000,000 newly born 243,760,000 would live to grow up, while of 440,000,000 young oysters only 421 ever become capable of propagating their species. The proportion is then 421 to 243,760,000, or as 1 to 579,002. I am fully per-

---

* Wappäus, Handbuch der Geographie und Statistik. Band I, 1855, Abth. I, p. 197.
† R. Böckh, Sterblichkeitstafel für den Preussischen Staat im Umfange von 1865. Jena, 1875.

suaded that these figures represent the number of oysters which arrive at maturity more favorably than is really the case, since from every thousand of full-grown oysters it is certain that, on an average, more than 440,000,000 young are produced. The correctness of my argument that the number of oysters which arrive at maturity is very small indeed as compared with the exceedingly large number of germs produced is corroborated by the experience of those who have engaged in oyster-culture in France and England. In the year 1870 a small oyster-bed was discovered at the mouth of the Thames, northeast from Whitstable.* It was about 18 meters long by 6 meters broad. Forty-eight hours later 75 boats were there, close alongside of one another, fishing up the oysters. Upon every old oyster which was taken were found only from nine to ten young ones of different ages. This bed had never been previously disturbed, and the oysters were accordingly found in their natural condition. Whoever is not informed in regard to the small number which arrive at maturity, but knows only of their immense fecundity, will, in thinking of the growth and production of oysters, consider the oyster-beds as inexhaustible. It has, indeed, really been thought that if millions and millions of oysters were taken from a bed no harm would be done to its prosperity, since it was the opinion that the dredges would leave everywhere as many breeding oysters as would be necessary to supply the place of those taken away, by means of the immense number of young which would be produced. In accordance with this view, the oyster-fisheries were made entirely free in England in 1866. But the consequence of the continuous fishing which followed was everywhere a quick impoverishment of the beds, concerning which result the official reports upon the oyster-fisheries in France and England contain a vast number of authentic proofs. According to the statement of Mr. Webber, mayor of Falmouth, 700 men, working 300 boats, were profitably employed in oyster-fishing in the neighborhood of Falmouth so long as the old laws of close-time were observed. But since the year 1866, when those old laws were set aside, the beds have become so impoverished that now, in 1876, only about 40 men, with less than 40 boats, can find employment, and even with this greatly diminished number of boats no single boat takes daily more than from 60 to 100 oysters, while formerly in the same time a boat could take from ten to twelve thousand. About the year 1830 an oyster-bed was discovered upon the English coast near Dudgeon Light, containing an immense number of oysters, among which were very many old ones.

---

* The statistics concerning English and French oyster-fishing were taken partially from my own notes, made during a visit to the English and French coasts, and partially from two official English reports: I. Report on the Oyster and Mussel Fisheries of France, made to the Board of Trade by Cholmondeley Pennel, Inspector of Oyster-fisheries. London, 1868. II. Report from the select Commission on Oyster-fisheries. 1876.

These reports are attached to summaries of the profits arising from oyster-fishing in France, which were delivered to the authorities at the French department of marine and fisheries.

During the next three or four years this bed was fished so perseveringly and disastrously that since then it has not produced enough oysters to be worth recording. Between the years 1840 and 1850 there were in the harbor of Emsworth so many oysters that one man in a single tide (five hours) could take from 15 to 20 casks, each containing 1,600 oysters. Later, 70 to 100 sailing vessels from Colchester came into the harbor and fished up so many young and old oysters during the two or three weeks they were there that, in the year 1858, scarcely ten vessels could load there, and in 1868 the beds were so impoverished by this fishing that a dredger in five hours could not gather more than 20 oysters. These figures are taken from the statement of Mr. Messum, oyster-dealer, and secretary of an oyster company at Emsworth, made before the commission for the investigation of the British oyster-fisheries, on the 1st of May, 1876. From the beds of the districts of Rochefort, Marennes, and the island of Oléron, on the west coast of France, there were taken, in the years 1853–'54, ten millions of oysters, and in 1854–'55 fifteen millions. By means of long-continued and exhaustive fishing they were rendered so poor that in 1863–'64 only 400,000 oysters were furnished for market.

The very celebrated rich oyster-beds of the Bay of Cancale, on the coast of Normandy, have produced, according to official reports, the following numbers of oysters:

Year.	Number of oysters.	Year.	Number of oysters.
1800	1,200,000	1835	43,000,000
1801	1,500,000	1836	40,000,000
1802	1,300,000	1837	36,000,000
1803	900,000	1838	44,000,000
1804	1,400,000	1839	42,000,000
1805	800,000	1840	52,000,000
1806	500,000	1841	56,000,000
1807	1,090,000	1842	63,000,000
1808	1,800,000	1843	70,000,000
1809	1,200,000	1844	68,000,000
1810	700,000	1845	67,000,000
1811	1,130,000	1846	65,000,000
1812	1,100,000	1847	71,000,000
1813	600,000	1848	60,000,000
1814	400,000	1849	52,000,000
1815	800,000	1850	50,000,000
1816	2,400,000	1851	47,000,000
1817	5,600,000	1852	20,000,000
1818	5,300,000	1853	49,000,000
1819	6,800,000	1854	20,000,000
1820	6,700,000	1855	20,000,000
1821	6,000,000	1856	18,000,000
1822	11,800,000	1857	19,000,000
1823	18,000,000	1858	24,000,000
1824	20,000,000	1859	16,000,000
1825	20,000,000	1860	8,000,000
1826	25,000,000	1861	9,000,000
1827	28,000,000	1862	3,000,000
1828	33,000,000	1863	2,090,000
1829	31,000,000	1864	2,200,000
1830	36,000,000	1865	1,100,000
1831	42,000,000	1866	1,960,000
1832	38,000,000	1867	2,000,000
1833	41,000,000	1868	1,079,000
1834	46,000,000		

The records of inspections of the Schleswig-Holstein oyster-beds have furnished the means by which such an impoverishment of these rich beds can be explained. From 1800 to 1815 there were taken yearly from the beds of Cancale less than two million oysters. The oysters, both marketable and spawning, had thus an opportunity to accumulate in greater quantity to form the increased production which occurred in 1822. If the French oysters live as long as those of the Schleswig-Holstein beds, some of this stock, which had accumulated during the period of comparative rest at the time of the Napoleonic wars, would have been lying upon the banks as late as about 1830. From this time on for nearly a score of years it is probable that the ever-increasing yield was the produce of only those oysters existing upon the beds from 1820 to 1830. From 1840 to 1847 the number of oysters taken was extraordinarily great—evidently too great for the productiveness of the beds, since from this time they produced fewer oysters each year.

The total number of oysters taken between the years 1840 and 1847 was about 512,000,000, being on an average about 64,000,000 per year. If this average represents the natural stock of marketable, full-grown oysters upon the beds of Cancale, then the number taken yearly should not have been over twenty-six to twenty-seven millions, if it was desired that this degree of productiveness should be maintained. This I assert upon the supposition that the productiveness of the oysters in the Bay of Cancale is no greater than upon the Schleswig-Holstein banks. If this productiveness was higher than upon our sea-flats, then we ought to have at Cancale, not 421 half-grown oysters for every 1,000 full-grown ones, but, for example, 500. Under these circumstances the presence of 64,000,000 matured oysters would permit the fishing of 32,000,000 yearly, but no more if the fruitfulness of the beds would be kept at that number, since such a stock would be absolutely necessary in order that a sufficient number of young should be produced to secure the maturing of 32,000,000 yearly.

After the impoverishment of the beds of Cancale the inspection officers enforced once more the laws protecting the oyster, since they did not believe that all the mature breeding oysters had been taken off the beds. Upon some of these beds there has already been a very significant increase of oysters through this action; for in 1872–'73, 7,300,000 oysters were taken, in 1873–'74, 9,056,000, and in 1874–'75, 9,342,000. To preserve oyster-beds a stock of full-grown oysters, for the purposes of propagation, must be left lying upon the banks. The number thus left must depend upon the fruitfulness of the oysters of each section, or, still better, of each single bed. According to the experience of oystermen, the most fruitful as well as the largest of the Schleswig-Holstein oyster-beds is the Huntje Bank. The proportion of medium oysters upon this bed is 484 per thousand, which is thus greater than the mean productiveness of the whole Schleswig-Holstein oyster-beds. The productiveness of smaller beds is below the average of 421 per thousand. As examples, we give the proportions of the following beds: Steenack, 385 per thousand; Hörnum, 319 per thousand; West Amrum, 165 per thousand.

These beds are only from 1,000 to 1,050 meters long by about 300 meters broad, while the Huntje Bank is more than 1,800 meters long by about 900 broad. The speedy extension of oysters in the Lim Fiord has taught us that young swarm-oysters can wander from 4 to 8 kilometers away from their home-bed before they become attached to any object. So if the oyster-bank is of small extent, the young oysters are in danger of swimming out beyond the limits of the bed and settling upon unsuitable ground, and thus of being destroyed in much larger numbers than upon a larger bed. Of such broods of swarm-oysters the large banks will retain many more than the small beds; and if upon both all other conditions are the same, a much larger number of young will grow to maturity upon those beds which have a large extent of surface than upon the smaller beds.

10.—AN OYSTER-BANK IS A BIOCÖNOSE, OR A SOCIAL COMMUNITY.

The history of the impoverishment of the French oyster-beds is very instructive. When the beds of Cancale had been nearly deprived of all their oysters, by reason of excessive fishing, with no protection, the cockle (*Cardium edule*) came in and occupied them in place of the oyster; and vast hordes of edible mussels (*Mytilus edulis*) under similar circumstances appeared upon the exhausted beds near Rochefort, Marennes, and the island of Oléron. The territory of an oyster-bed is not inhabited by oysters alone but also by other animals. Over the Schleswig-Holstein sea-flats, and also along the mouths of English rivers, I have observed that the oyster-beds are richer in all kinds of animal life than any other portion of the sea-bottom. As soon as the oystermen have emptied out a full dredge upon the deck of their vessel, one can see nimble pocket-crabs (*Carcinus mœnas*) and slow horn-crabs (*Hyas aranea*) begin to work their way out of the heap of shells and living oysters, and try to get to the water once more. Old abandoned snail-shells begin to move about, caused by the hermit-crabs (*Pagarus bernhardus*), which have taken up their residence in them, trying to creep out of the heap with their dwelling. Spiral-shelled snails (*Buccinum undatum*) stretch their bodies as far out of the shell as they can, and twist from side to side, trying, with all their power, to roll themselves once more into the water. Red star-fish (*Asteracanthion rubens*), with five broad arms, lie flat upon the deck, not moving from the place, although their hundreds of bottle-shaped feet are in constant motion. Sea-urchins (*Echinus miliaris*), of the size of a small apple, bristling with greenish spines, lie motionless in the heap. Here and there a ring-worm (*Nereis pelagica*), of a changeable bluish color, slips out of the mass of partially dead, partially living, animals. Black edible mussels (*Mytilus edulis*) and white cockles (*Cardium edule*) lie there with shells as firmly closed as are those of the oysters. Even the shells of the living oysters are inhabited. Barnacles (*Balanus crenatus*), with tent-shaped, calcareous shells and tendril-shaped feet, often cover the entire surface of one of the valves. Frequently the shells are bedecked with yellowish tassels a span or more in length, each of which is a community of thousands of small gelatinous bryozoa (*Alcyonidium gelatinosum*), or they are overgrown by a yellowish sponge (*Halichondria panicea*), whose soft tissue contains fine silicious spicules. Upon many beds the oysters are covered with thick clumps of sand which are composed of the tubes of small worms (*Sabellaria anglica*). These tubes, called "sand-rolls," resemble organ-pipes, and are formed from grains of sand cemented into shape by means of slime from the skin of the worm. The shell forms a firm support upon which the worms can thus live close together in a social community. Upon certain beds near the south point of the island of Sylt, where the finest-flavored oysters of our sea-flats are to be found, there lives upon the oyster-shells a species of tube-worm (*Pomatoceros triqueter*) whose white, calcareous, three-sided tube is very often twisted about like a great italic *S*. The shells of many oysters upon these beds also carry what are called "sea-hands" (*Alcyonium digitatum*), which are white or yellow communities of polyps of the size and shape of a clumsy glove. Often the oyster-shells are also covered over with a brownish, clod-like mass, which consists of branched polyps (*Eudendrium rameum* and *Sertularia pumila*), or they may be covered with tassels of yellow stems which are nearly a finger long and have at their distal ends reddish polyp-heads (*Tubularia indivisa*). Among these polyps, and extending out beyond them, are longer stems, which bear light yellow or brown polyp-cups (*Sertularia argentea*). Within the substance of the shell itself animals are also found. Very often the shells are penetrated from the outside to the innermost layer, upon which the mantle of the living oyster lies, by a boring sponge (*Clione celata*), and in the spaces between the layers of the shell in old oysters is found a greenish-brown worm (*Dodecaceræa concharum*), armed with bristles, and bearing twelve large tentacles upon its neck. I once took off and counted, one by one, all the animals living upon two oysters. Upon one I found 104 and upon the other 221 animals of three different species. The dredge also at times brings up fish, although it is not very well adapted for catching them. Soles (*Platessa vulgaris*), which seek by jumping to get out of the vessel and once more into the water, stone-picks (*Aspidophorus cataphractus*), and sting-rays (*Raja clavata*), which strike about with their tails, are abundant upon the oyster-banks. Besides those already mentioned, there are many other larger animals which are taken less frequently in the dredge. There are also a host of smaller animals covered up by the larger ones, and which can be seen only with a magnifying glass. Very few plants grow upon the banks. Upon only a single one of the oyster-beds of the sea-flats has eel-grass (*Zostera marina*) taken root. Upon other beds reddish-brown algæ (*Floridiæ*) are found, and, floating in the water which flows over the beds, occur microscopic algæ (*Desmidiæ* and *Diatomaceæ*), which serve as nourishment to the oysters. If the dredge is thrown out and dragged over the sea-flats between the oyster-beds, fewer and also different animals will be found upon this muddy bottom than upon the sand. Every oyster-bed is thus, to a certain degree, a community of living beings, a collection of species, and a massing of individuals, which find here everything necessary for their growth and continuance, such as suitable soil, sufficient food, the requisite percentage of salt, and a temperature favorable to their development. Each species which lives here is represented by the greatest number of individuals which can grow to maturity subject to the conditions which surround them, for among all species the number of individuals which arrive at maturity at each breeding period is much smaller than the number of germs produced at that time. The total number of mature individuals of all the species living together in any region is the sum of the survivors of all the germs which have been produced at all past breeding or brood periods; and this sum of matured germs represents a certain quantum of life which enters into a certain number of individuals, and which, as does all life, gains permanence by means of transmission. Science possesses, as yet, no word by which such a community of living beings may be designated; no word for a community where the sum of species and individuals, being mutually limited and selected under the average external conditions of life, have, by means of transmission, continued in possession of a certain definite territory. I propose the word *Biocœnosis** for such a community. Any change in any of the relative factors of a biocönose produces changes in other factors of the same. If, at any time, one of the external conditions of life should deviate for a long time from its ordinary mean, the entire biocönose, or community, would be transformed. It would also be transformed, if the number of individuals of a particular species increased or diminished through the instrumentality of man, or if one species entirely disappeared from, or a new species entered into, the community. When the rich beds of Cancale, Rochefort, Marennes, and Oléron were deprived of great masses of oysters, the young broods of the cockles and edible mussels which lived there had more space upon which to settle, and there was more food at their disposal than before, hence a greater number were enabled to arrive at maturity than in former times. The biocönose of those French oyster-banks was thus entirely changed by means of over fishing, and oysters cannot again cover the ground of these beds with such vast numbers as formerly until the cockles and edible mussels are again reduced in number to their former restricted limits, because the ground is already occupied and the food all appropriated. The biocönose allows itself to be transformed in favor of the oyster, by taking away the mussels mentioned above, and at the same time protecting the oysters so that the young may become securely established in the place thus made free for them. Space and food are necessary as the first requisites of every so-

---

* From βίος, life, and κοινόειν, to have something in common.

cial community, even in the great seas. Oyster-beds are formed only upon firm ground which is free from mud, and if upon such ground the young swarming oysters become attached in great numbers close together, as happened upon the artificial receptacles in the Bay of Saint Brieux, their growth is very much impeded, since the shell of one soon comes in contact with that of another, and they are thus unable to grow with perfect freedom. Not only are they impeded in growth in this manner, but each oyster can obtain less nourishment when placed close together than when lying far apart.

In an oyster-breeding trench upon the island of Hayling, in the south of England, I saw, in May, 1869, oysters three years old which had grown thus far towards maturity attached to hurdles. Nearly all had twisted shells, which were not larger in diameter than from 2 to 3 centimeters, while a Holstein oyster three years old is from 5 to 6.5 centimeters broad. Evidently, the reason for their small size is to be found in the fact that in the trench they receive less nourishment daily than they would in the open sea. In the Bay of Arcachon the breeders are obliged to loosen the oysters from their artificial points of attachment and place them in boxes and trenches where they can grow to maturity, and in these the oysters must not be placed too close together or they will not grow to the best advantage. Even upon the best beds the oysters will remain poor if they are allowed to lie too thick upon the bottom; but if a portion of these poorly nourished oysters are taken away, those which remain—as has been found out by experience upon the Huntje Bank, the largest and most fruitful of all the Schleswig-Holstein beds—will soon become fatter, that is, their generative organs will become larger because more eggs or spermatozoa are produced than with poorer oysters. Thus, if a bed is above its mean in productiveness, every single one of the excessive throng of full and half grown oysters will not receive sufficient nourishment to enable them o generate a full number of germs, so that the number of germs produced and the number of young which arrive at maturity being thus regulated the entire bank will very soon be brought back to its former or normal condition. Since this law is in operation upon even the most productive of the Schleswig-Holstein beds, where the number of young which arrive at maturity is 484 for every thousand of mature oysters, while the average for all the beds is only 421 to each thousand, then a productiveness of 484 to the thousand is the highest which can be reached and maintained among the oysters existing under the biocönotic conditions of our sea-flats. Near Auray, in Brittany, the oyster-breeders collect many more young than they can grow to maturity, since they possess comparatively little oyster-territory, and this territory is not supplied with sufficient suitable food to nourish large numbers of oysters; so that whenever the breeders fail to find a purchaser for their extra stock of young they lose all the profits of their labor. Thus, it is with oysters as with all other animals; their increase in size and numbers depends upon the quantity of food which they get and consume. The peasants of Jutland are great breeders of horned cattle, but they have not sufficient food to grow and fatten all their calves. Accordingly, many are sold to the peasants of the marsh-lands on the west coast of Schleswig-Holstein, and upon these extensive pasture-lands great numbers of cattle can be raised. Upon the estate of Hagen, near Kiel, there is a carp-pond of more than 80 hectares in size which is drawn off every three years, and while in this condition sown with oats and clover. It is afterwards refilled with water and 30,000 yearling carp placed in it. In three years from this time, as a rule, the production is 40,000 pounds of food-fish. In order to obtain a still greater profit, another lot of young carp was once placed in the pond, and this time more than 30,000. In three years the produce is indeed a larger number of fish than before, but they weigh, taken all together, only 40,000 pounds. The quantity of food which the pond supplies in three years is thus sufficient only for the growth of 40,000 pounds of carp.

I do not consider it practicable to fatten oysters by artificial means, although in North America and Europe an effort should be made to fatten planted oysters upon corn-meal. The food of oysters consists of very small organic particles which float in the water, and if one should attempt artificial feeding by carrying to the oysters of a bed water containing pulpy pulverized flesh, bone-meal, fish-guano, or corn-meal, it would be necessary to prevent the water from flowing off from the bed until all the organic matter had been eaten. But by so doing a large quantity of foul gas would certainly be generated upon the bed and remain there, so that the oysters, instead of fattening, would become sick and die. Among the external life conditions of a biocönose, temperature plays an important part.

In our seas, with their equitable temperature, a mild winter, followed by a spring and summer with the temperature much higher than usual during spawning time, is especially favorable to the production of a vast number of embryos. All living members of a social community hold the balance with their organization to the physical conditions of their biocönose, for they live and propagate notwithstanding the influence of all external attractions, and notwithstanding all assaults upon the continuance of their individuality. Although every species is differently organized, in each the different forces act together for the growth and maintenance of the individual, and although each species has from this fact its own organic equivalent, yet they all possess the same (balancing) power for the totality of the external conditions of life of their biocönose. Hence all species must respond to a deviation in the conditions of life from the ordinary mean by a corresponding action of their forces, so that their efficacy may increase or diminish uniformly. If favorable temperature makes one species more fruitful, it will, at the same time, increase the fertility of all the others. If more young oysters exist upon an oyster-bed because the old ones receive more warmth and food than during ordinary years, then the snails, crabs, sea-urchins, and star-fish, and all other species living together upon the bank, will also produce more young, as repeated observations have shown to be the case. But since there is neither room nor food enough in such a place for the maturing of all of the excessively large number of germs, the sum of individuals in the community soon returns to its former mean. The surplus which nature has produced by the augmentation of one of the biocönotic forces is thus destroyed by a combination of all the forces, and the biocönotic equilibrium is by this means soon restored again. Where it is possible for one to furnish suitable ground and food for an excessive number of young germs, a greater proportion of them can arrive at maturity than in an entirely natural biocönose. The oyster-breeders of Arcachon and Auray increase very much the mean number of oysters which arrive at maturity upon their beds by placing tiles in the water, upon which the young can attach themselves. These young are then provided with a suitably prepared ground over which water containing food is allowed to flow. If in a community of living beings the number of individuals of one species is lessened artificially, then the number of mature individuals of other species will increase. Thus, upon the west coast of France cockles and edible mussels took the place of the oysters which had been caught from the beds; and upon the fertile prairies of North America herds of tame horses and cattle are now pastured where immense throngs of wild buffaloes (*Bos americanus*) once ranged in full liberty.

If the germ-fecundity of a species is lessened by the artificial distinction of many mature breeding individuals, while all the other forces of the community are working with their accustomed vigor, so surely must there be a decrease in the number of individuals of this species which arrive at maturity. A large number of the most productive oyster-beds upon the west coast of Europe have been devastated by overfishing, and many fresh waters have, through the incessant catching of half-grown fish, been almost entirely depopulated. It is very natural that those years during which a large number of herring, salmon, or sturgeon are caught upon a certain stretch of territory should be followed by years when fewer fish appear, because in the years when large catches are made very many breeding individuals are destroyed.

If in a case of subtraction the minuend is lessened while the subtrahend remains the same, the remainder will be lessened also. By the continued artificial destruction of breeding individuals, the fecundity of any one species of a community may sink so low that it is no longer able to produce sufficient germs to insure in all cases the maturing of a sufficient proportion which shall escape the ordinary natural assaults to which they are subject in the community; the species therefore dies out. In this manner the dodo (*Didus ineptus*) became extinct upon the island of Mauritius in the seventeenth century, after the Portuguese, in 1507, had disturbed the biocönose of the island by the introduction of swine and other animals, and after the Dutch, still later, had ruthlessly killed many of these birds. Also, at present, there are no turtles at Mauritius, while up to the year 1740, according to written testimony, hundreds were caught there for the provisioning of ships. Certainly many young dodos and turtles must have been devoured by the pigs. The beaver (*Castor fiber*) will perhaps very soon have vanished from our biocönotic transformed portion of the earth. The Greenland whale (*Balæna mysticetus*) is now seldom seen in the neighborhood of Spitzbergen and Greenland, on account of the persecution to which it has been subjected since the seventeenth century. Every biocönotic territory has, during each period of generation, the highest measure of life which can be produced and maintained there. All the organic material which is there ready to be assimilated will be entirely used up by the beings which are procreated in each such territory. Hence at no place which is capable of maintaining life is there still left any organizable material for spontaneous generation. If, in a biocönose the number of individuals which arrive at maturity would be maintained at the highest point, even though the number of breeding individuals is being artificially lessened, the natural causes which act towards the destruction of the embryos must be diminished at the same time. In the Bay of Arcachon the breeders raise to maturity an unusually large number of young oysters by guarding them artificially from their enemies.

In an example in subtraction the remainder may be kept unchanged, or even increased, if the subtrahend is decreased at the same time as the minuend; and the mass of individuals of any species may be increased permanently if the biocönotic territory is extended. Thus, when the Lim Fiord became filled with water from the North Sea, the number of mature oysters over the territory of the North Sea coast of Denmark increased to more than seven millions (chapter 8, p. 30). The oyster-beds in the Bay of Arcachon and the *claires*, or fattening-ponds, at the mouth of the river Seudre (p. 27), are artificial extensions of oyster-territory.

The individual number of cultivated plants and animals has been immensely increased because man has artificially extended their biocönotic territory; and this artificial increase in the number of plants and animals by means of cultivation is the foundation for the increased fecundity of the human species and the greater number of individuals which arrive at maturity—that is, for the extension of the biocönotic territory of *Homo sapiens*. The average yield of our woods, fields, and gardens is the result of natural force and human labor, for in addition to the chemical and physical forces of earth and air, and the organic forces of wild and cultivated plants and animals, the bodily and mental forces of man play an important rôle in the culture of field and forest, and a very significant share of the large yields of harvests is due not only to the numerous workmen of the woods and fields, but also to the makers of implements of labor, to the mechanics and opticians who produce instruments for the investigation of natural phenomena, and to the careful studies of the many investigators of nature and of those interested in land and forest culture. And these manifold interdependent human powers must unceasingly oppose the average uniform workings of natural forces if a permanent mean profit would be derived from the artificial communities of cultivated lands, or if Nature would be prevented from introducing again into each such territory her own communities. This was entirely disregarded in the case of the banks along the west coast of Europe. Millions and millions of oysters were taken from these beds, and great astonishment arose when it was noticed that their productiveness had diminished. Notwithstanding that the number of breeding animals was extraordinarily diminished at each annual gathering, yet every succeeding year an equally large number of mature descendants would be harvested. Oystermen wish the oyster to be exempted from the workings of those communal laws, according to which field and forest culture must be conducted in order to achieve a certain measure of success; and there are oyster-breeders in England who desire, for the entire satisfaction of their great yearly demands upon both the oyster-banks and the oyster-beds, that every year, during the breeding period, the temperature of the water should remain at from 18° to 20° C., that no wind should blow, and that no storm should suddenly disturb the good weather, in order that none of the young swarming oysters may be destroyed.

In France they expected that all of the million germs produced by a full-grown oyster would grow into marketable oysters, if only suitable objects were provided to which they could attach themselves. It was thus believed that some miracle would be wrought by means of which oysters would reach maturity, for there existed in the water which passed over the beds where the millions of young oysters were laying not a particle more food than was brought there years before, and that was only enough to feed the much smaller number of oysters existing there at that time. In Germany they desire that oysters should live and thrive upon changing sand-banks and mud-bottoms, and accustom themselves to the brackish water of the Baltic, and that at the same time they should remain animals of the same tenderness and delicacy of flavor as the oysters of the good Schleswig-Holstein beds. Such desires could only be realized by means of miracles or by the exemption of certain single cases from the necessary workings of Nature's laws. There must be an entire change in the form of our coasts and in that of the islands lying along them, in the direction of the mouths of the rivers and in the flood and ebb currents, before oysters can be made to thrive over our entire sea-flats. It would thus be necessary to supersede the natural oyster biocönose by an artificial one, which would have to be cultivated as farmers and gardeners cultivate their fields and gardens; and in order that oysters should be able to live in the waters of the Baltic to-day, their physiological activities would have to be so changed that they could thrive in water in which the percentage of salt is much more vacillating than in the water of the North Sea; that is to say, it must become another animal, and yet, at the same time, retain the flavor of the oyster. People have experienced a thousand times that the best-flavored and most agreeable animals and plants are brought to perfection only under entirely definite external conditions of life, yet they wish an exception to this law of nature in favor of the oyster. They wish for miracles in order that oysters may be supplied to the many who are now oyster-eaters, as cheaply and plentifully as they formerly were to the few who at that time appreciated their value.

11.—CONCERNING THE INCREASE IN THE PRICE OF OYSTERS AND THE NUMBER OF CONSUMERS, AND THE DECREASE IN THE NUMBER OF OYSTERS.

In England there are breeders of oysters and others who are well versed in oyster economy who maintain that the oyster-banks have become impoverished because of a long series of seasons which have been unfavorable as breeding years, and not because of overfishing upon the beds. According to their observations, there have been no large broods of young oysters since 1857, 1858, and 1859. This may be the case in regard to a number of localities, but it has no significance in the management of a permanent, profitable oyster-culture, since such culture is not conducted according to an unusually favorable summer, but according to the average of climatic conditions. And that these conditions have not changed in the west of Europe in our century, and thus during the time of the impoverishment and exhaustion of many beds along the west coast of Europe, is proven by the temperature observations which have been made at the Observatory at Paris since the year 1806. According to these, the mean yearly temperature of Paris during this century has remained, up to present time, at 10.8° C., from which it follows that the climate is the same now as before any impoverishment took place. In 1859 there were many young oysters spawned upon the beds along the west coast of France. In 1860 there were many young broods upon the beds near the island of Ré and near Rocher d'Aire, and but few broods at Arcachon; 1861 was a good brood year for all three places; 1862 bad for the island of Ré and good for both the others; and in 1865 there were very many young in the Bay of Arcachon and but few near Rocher d'Aire and the island of Ré. These facts show that local conditions can either favor or prevent the production of broods of young oysters in one and the same year.

On the 6th of April, 1876, Mr. F. Pennell made a communication to the commission for the investigation of the British oyster-fisheries, and at the same time remarked that, according to his experience, the number of young oysters in each brood period was dependent upon the number of breeding oysters, but that, nevertheless, at times, extraordinarily large numbers of young were produced.* Whoever has followed thus far the detailed statements which I have made must be obliged to confess that Nature is not to blame for the impoverishment of the oyster-beds along the western coast of Europe during the last century, for neither have the external conditions of life for the oyster become less favorable nor has the fecundity of single animals become less.

Nothing else but excessive fishing, without protection, has depopulated the beds. Most of the oystermen and those thoroughly acquainted with oyster industry, who reported their experience and opinion, in London, in 1876, to the commission for the investigation of the British oyster-fisheries, were entirely of this opinion. But the question will be asked, Why were the beds of the west of Europe not overfished in olden times? Because, before the time of steamboats, locomotives, and railroads, there was a much smaller number of consumers than at present. Then genuine connoisseurs were rarely to be found except along the coast where the oyster lived.

In the autumn, when oyster-fishing began, those only were very costly which were first caught, but as more were brought in the price rapidly fell. On the 21st of September, 1740, the first hundred fresh Schleswig-Holstein oysters sold in Hamburg for 1.42 marks (about 35 cents) of present money. Later the same day 900 were sold at 1.20 marks (30 cents) per hundred; then 3,400 at 15 cents; and finally 10,800 at 7½ cents per hundred. On the 15th of October of the same year, and at the same place, the first hundred fresh, newly arrived oysters sold for 2.40 marks; the second hundred for 2.10 marks; then 1,025 were sold for 1.80 marks per hundred; then 1,000 at 1.50 marks; then 2,000 at 1.20 marks; and finally 12,500 at 60 pfennige (15 cents) per hundred. These numbers are taken from the report upon the Schleswig-Holstein oyster-banks,† and show that it was necessary to lower the price of oysters very soon after the arrival of a large importation into Hamburg harbor, if they were to be disposed of in an eatable condition and not entirely lost, because there was no adequate means of transporting them into the interior. Such a fall in price guarded the oyster-beds from too destructive fishing. Soon, by means of steamers and railroads, oysters fresh from the beds could be spread far and wide into the country; then oyster-eaters began to increase in number; and so, despite the rapid advance in price, the demand for oysters increased from year to year. This demand was very much in proportion to the spreading of the network of railroads in England, France, and Germany. It did not come into the heads of the oystermen that a more exhaustive fishing would tend to depopulate the beds. Year after year they had found an ever ready supply of oysters upon the same beds; why should they not, then take away whatsoever came into their dredge?

---

* Report on Oyster-fisheries, 1876, p. 116, Nos. 2386 and 2387.

† *H. Kröyer.* De danske Oystersbanker. In this work several examples are given at pages 92 and 93, taken from the reports of oyster-culture in Schleswig-Holstein, o the decrease in price of oysters upon one and the same day at Hamburg.

In former times, fishing was carried on only in those places where the oysters lay thickly together, for where only a few oysters could be caught it did not pay to fish, because of the low price; hence all of those banks which were covered only with scattered oysters were left to rest until a sufficient number of mature oysters had accumulated upon them to repay the labor of fishing. But when, however, the number of oyster-eaters increased, and likewise the price of oysters, it became profitable to fish upon less fertile stretches, and the dredges were used so persistently that finally very little more could be found upon the banks. Before the time of railroads the decline in price of oysters regulated the fishing in favor of a good condition of the beds; but since the time of railroads the ever-increasing price has acted as an incentive to the oystermen to depopulate their banks. The official reports upon French and English oyster-breeding contain abundant proofs of this, as evidenced by the facts there set forth. The oyster-fishers of Cancale were made happy by receiving, each succeeding year, for those oysters which they sent fresh to Paris, more money than they had received the year before, and the possibility of depopulating their rich banks was not thought of.

Learned authorities had said that every mature breeding oyster produced from two to three millions of young. They believed, then, that if they left upon the beds only a hundred breeding oysters they would be doing all that was necessary in order, in a short time, to find upon the overfished beds two hundred to three hundred million descendants of the same. Up to 1854, the oyster-beds of Rochefort, Marennes, and the island of Oléron were fished with some regard to their preservation, since their oysters found a market only in those places which were situated along the neighboring coast. But in 1854 Rochefort was placed in connection with the interior by means of the network of railroads, and the market for these oysters, and the profits from them, increased so much that they were taken until these beds were almost entirely depopulated. From 15,000,000 in 1854–'55, the catch fell off to 400,000 in 1863–'64. (See chapter 9, p. 37.)

The last report upon the English oyster-fisheries in the year 1876, contains many instructive instances of the great advance in price as the result of the decrease in number of oysters. At Whitstable, where the finest kinds of native oysters are produced, the price for a bushel, 1,400 to 1,600 oysters, was, during the period from 1852 to 1862, never higher than £2 2s. sterling. In 1863–'64 it had risen to £4 10s., in 1869 to £8, and in 1876 to £12 sterling. Thus, in 1876, a single oyster cost there about 16 pfennige (3 to 4 cents) in our currency.

At Colchester, another celebrated market for oysters on the east coast of England, a bushel of oysters cost, during the years 1856–'63, 66s.; 1864–'65, 80s.; 1865–'66, 95s.; 1866–'67, 100s.; 1867–'68, 130s., English money.*

---

* Report upon oyster-fisheries, 1876, p. 63.

At Falmouth, a tub of oysters (1,600) cost, in 1830, 1s.; 1860, 2 to 2½s.; 1863, 4 to 14s.; 1867, 9 to 37s.; and in 1869, 45s. A cask of Schleswick-Holstein oysters (700 to 800) was sold, in 1875-'76, to oyster-dealers, for 105 marks (about $26.25).

Fifteen years previous the price was only a third of that sum. By the incorporation, in 1864, of Schleswig-Holstein in the German tariff-union, the territory into whose markets the Schleswig-Holstein oysters could be brought free of duty was very significantly increased, and at this time English oysters were becoming very rare in German markets. The political changes in Germany, and at the same time the great increase in the consumption of oysters, evidently increased the incentive to a more complete fishing of our oyster-beds than in former times, and this accounts for the extraordinary decrease in the number which arrive at maturity to-day, for at the inspection of these beds in 1869 there were found only 282 half-grown oysters for every 1,000 full-grown, and in five inspections during the years 1872 to 1876 there were found, on an average, only 107 half grown. This is in striking contrast to former inspections, where the average was 421 half-grown oysters for every 1,000 full-grown, as has been shown in chapter 9, p. 35.

12.—THE CHEMICAL CONSTITUENTS AND THE FLAVOR OF OYSTERS.

The heaviest portion of an oyster is its shell, and this, on an average, constitutes about 84 per cent. of the total weight of an ordinary Holstein table-oyster. Internally, the oyster is a soft animal; externally, it is a stone animal. The dried shells of very old oysters weigh from 250 to 320 grams. In such heavy, thick-shelled oysters the soft portion is generally very poor, and the body-space is smaller than at the time when it first attained its complete maturity. From this it follows that the edges of the last-formed shell-layers do not pass over those which were formed earlier, but lie under them. The principal constituent of the shell is carbonate of lime, which forms about 96 to 97 per cent. of the whole weight. The shell also contains 1.2 to 1.3 per cent. of sulphate of lime, 0.09 per cent. of phosphoric acid, 0.03 per cent. of oxide of iron, and traces of magnesia and aluminum. If these inorganic constituents of the shell are dissolved in acid, there will remain undissolved brownish bits and flakes of an organic substance which has been named *conchyolin*. This contains the elements oxygen, hydrogen, nitrogen, and carbon. The left or arched valve of a Holstein oyster contains from 1.01 to 1.025 per cent. of conchyolin, the right somewhat more, from 1.10 to 1.15 per cent. This increase in the percentage of conchyolin makes the right valve less brittle than the left.

At times pearls are found in oysters. They generally lie in the mantle, but also in the shell-muscle. Pearls are isolated deposits of shell-material. Their chemical constituents are accordingly the same as those of the shell. In some pearls taken from Schleswig-Holstein oysters there has been found a proportionally greater amount of carbonate of lime than is found in the shells themselves. Brilliant pearls, suitable for ornaments, are seldom found in oysters, those generally taken being white and without brilliancy. However, an oyster-eater in Hamburg once discovered, by means of his tongue, a pearl which he sold to a jeweler for 66 marks (about $16.50). Nearly all mussels have more beautiful shells than that of the oyster, but in delicacy and fineness of flavor the oyster surpasses every other mollusk. Only those materials can be tasted which, dissolved in fluids, come in contact with the organs of taste. Hence the flavor of the oyster depends upon substances which are either in solution in the juices of the body of the oyster or which become dissolved in the mouth of the eater. Fresh, living oysters, as is the case with all sea-animals, contain very much water. In order to estimate the proportion of water the greatest care must be taken in removing the oyster from the shell, especially when the shell-muscle uniting the two valves is cut. The exterior of the body must first be dried with blotting-paper, the body then weighed and finally placed under the air-pump and all the water drawn out. Two Schleswig-Holstein oysters which were taken from the shell and dried, weighed together 14.70 grams, and after all the water had been drawn from them they weighed only 3.05 grams. They thus contained 79.25 per cent. of water and only 20.75 per cent. of solid material. Two other large oysters, which had been previously deprived of their gills and mantles, weighed together 20.55 grams; after being thoroughly dried they weighed 4.809. Thus, their edible portion contained 76.64 per cent. of water and 23.36 per cent. of solid material.

A large number of investigations upon Schleswig-Holstein oysters demonstrated that the entire animal contained from 21.5 to 23 per cent. of solid material, while the body, without the gills and mantle, contained from 23 to 24.5 per cent. of solid material.* Making due allowance for size, there is a somewhat greater difference in the proportion of solid material between oysters and fish than between oysters and birds and mammals, for—

	Per cent. of solid material
Trout-flesh contains	19.5
Carp-flesh contains	20.2
Pork contains	21.7
Veal contains	21.8
Beef contains	22.5
Fowl-flesh contains	†22.7

* I am indebted to Prof. O. Jacobsen, of Rostock, for all the information that I have given in chapter 12 in regard to the chemical constituents of the Schleswig-Holstein oysters. In June, 1871, when he lived in Kiel, he analyzed, at my request, a number of oysters which I had received fresh from the Schleswig-Holstein beds.

† The figures quoted in the comparison of the amounts of solid materials in different kinds of flesh are based upon the analyses of Schlossberger and Von Bibra. They were taken from the Elements of Physiological Chemistry of Gorup-Besanez, third edition, 1874, p. 682.

One hundred parts of dried oyster-meat contain 7.69 to 7.81 parts of nitrogen, and 100 parts of fresh oyster-meat contain 1.85 to 1.87 parts of nitrogen. By burning the dried oyster-meat we can obtain the amount of inorganic material which it contains. By this method it has been found that the meat of completely dried Schleswig-Holstein oysters, when deprived of the board, contains 7.45 per cent of inorganic substances, while in those which have not been dried the amount is only 1.79 per cent. According to these determinations, 100 parts of the bodies of fresh edible Schleswig-Holstein oysters, contain—

77.00 parts of water.
21.21 parts of organic material.
1.79 parts of inorganic material.

The principal inorganic substances are salt (*sodic chloride*) and phosphoric acid. The proportion of salt in fresh oysters is 0.58 per cent., and of phosphoric acid 0.38 per cent. In fresh beef the proportion of salt is 0.49 per cent., and of phosphoric acid 0.22 per cent. From these results it is evident that the oyster contains as much food substance as the better sorts of meat used for food, or even somewhat more. In addition to this, it is still farther distinguished from the greater number of animal foods by being more easily digested. But if we compare the price of oysters with the price of equal quantities of the best kinds of ordinary meats, we find that, with us, the oyster furnishes a much more costly means of nourishment than the others. If the edible portions of a dozen Holstein oysters weigh 125 grams, or one-fourth of a pound, and if that number cost 2 marks (50 cents), then the oyster, as a means of nourishment, is $6\frac{1}{4}$ times more expensive than beef-steak, at 1 mark 20 pfennige (30 cents) per pound. The value of an oyster does not depend principally upon the amount of nourishment which it contains, but chiefly upon its delicacy and uniformly fine flavor. Oysters form the finest article of food which our seas produce—food which can be eaten fresh from the water, and which requires no artistic cooking to develop its excellencies. They resemble the noble pearls, which attain their greatest perfection in the place of their growth. What particular constituent of the oyster it is which gives it its flavor is as little known as the origin of the flavor of various other kinds of food. The liver and the generative organs contain glycogen and grape-sugar. Pure glycogen has no taste, and it is composed, as is grape-sugar, of carbon, hydrogen, and oxygen. Probably the fatty matters aid greatly in giving flavor to the oyster. I have repeatedly found that in May and the first half of June, when the generative organs are very much developed, the females have a much finer, nut-like, and full flavor than the males.

I have repeatedly placed fresh oysters, whose sex I had previously ascertained by means of the microscope, before different people in order to get their opinions of the flavor. They also, without knowing anything about the difference in sex, found the female superior in flavor to the male. Those females which are well developed are generally somewhat thicker and more cream-like in color than the males, whose bodies are more transparent and watery. In the middle of winter these differences are not so apparent as shortly before the breeding season. Immediately after the emission of the generative products, oysters are poorest and they are more watery then than at any other time. After the breeding period their size increases from month to month, and, in case their nourishment is not interrupted by long-continued severe cold, their flavor becomes fuller and richer in proportion to the rapid development of the generative organs. From this it follows that winter, but more especially spring, are the periods of the year for the enjoyment of oysters. I have repeatedly heard people, who rated themselves as genuine oyster-eaters, say that "oysters ought never to be bitten, but should be swallowed whole." If this were so, then one might better use, in the place of the high-priced oyster, a succedaneum made of tasteless thin paste, and having merely the form of the oyster.

As with all other kinds of food, the flavor of the oyster is more effective and can be better appreciated the more intimately its constituent elements come in contact with the surface of the organs of taste. Therefore, if one would obtain the full flavor of the oyster, it must be bitten to pieces and chewed, in order that all the constituents may be free to produce their greatest effects. The Schleswig-Holstein oyster-banks produce oysters of very different flavors. Those having the finest flavor exist upon beds which lie not very far from the deeper channels, through which the water passes in and out during the flood and ebb tides. (See the chart of these beds on page 4.) Thus, very superior oysters are found upon the beds at the northern and southern ends of the island of Sylt, and upon a single bank north of the island of Röm; but the very finest oysters are found upon the beds near Hörnum. The oysters of these beds are especially distinguishable by the large growth of their organs of generation. Their flavor is very delicate, and never bitter and watery, as is the case with the oysters of many other beds. This superiority in form and flavor must be the direct result of the action upon the oysters of these banks of the external conditions of life under which they exist. The oysters upon the beds near Hörnum lie deeper and nearer the open sea than those farther in upon the flats. The water, also, which flows over them during the course of the day and year is less subject to fluctuations in temperature than that which flows over the beds lying nearer the mainland, and there is here a somewhat greater percentage of salt than in the water over the beds of the shallower portions of the sea-flats. To these external physico-chemical properties of the Hörnum banks are also united faunal peculiarities. Here are to be found the three-sided worms (*Pomatoceros triqueter*), and colonies of polyps which, from their form, are called "sea-hands" (*Alcyonium digitatum*). Both of these forms are found abundantly upon the bottom of the open North Sea. They are not to be found upon any other beds of the sea-flats, and it is very evident that they cannot live upon the inner beds, since these do not furnish the necessary conditions for their growth. Thus, just in those places where the extreme limiting line of the territory inhabited by the "sea-hands" and three-sided worms passes across our oyster-beds, the most favorable conditions exist for the growth of the finest-flavored oysters. A three-sided worm upon the shell of a Schleswig-Holstein oyster is, therefore, a sign of its arrival from one of our best beds. A pastor living upon the island of Sylt was fond of good oysters, and was also well acquainted with this external indication of them; so he was accustomed to say to the out-going oyster-fishers: "Bring me some fresh oysters when you return, but only such as the good Lord *has marked*." In Paris the green oysters of Marennes and Tremblade are especially prized on account of their delicious flavor. This cannot come from the green constituents of their body, for if old oysters are taken there during the winter months and placed in a fattening-pond, they will, indeed, become green, but by no means so well-flavored as those oysters which were placed there when young and have lived there several years. (See chapter 8, p. 27.)

The flavor of oysters is best at the banks themselves, if they are opened very carefully and all the sea-water which is inclosed in the shell when shut is allowed to escape. This can be done most judiciously if the oyster is placed upon the flat right valve, after the loosening of the shell-muscle. This valve is a superior natural plate for the oyster, since it has no cavities like those of the left valve, filled with disagreeably smelling water, which flows out when the shell is opened and contaminates the flavor of the oyster. (See Fig. 5, p. 11.) The oyster can live for days perfectly dry without dying, but it gradually loses its softness, and soon begins to smell, from the dying of the animals which inhabit the outside of its shell. It is very seldom that these can be entirely removed by the usual means of purification (p. 8), so that the flavor of the oyster inland is almost always affected by these contaminating odors.

In order that oysters may be furnished to those who want them, in the freshest and best-flavored condition possible, only such a number should be caught at any one time, and for any one market, as can be disposed of in a very few days. But since wind and weather are often so unfavorable, just in the height of the oyster-season, that vessels cannot go out to the beds and fish, the oystermen are obliged to dredge a large supply of oysters during good weather and plant them in some place where they will live and at the same time be available whenever they are wanted. For this purpose large reservoirs have been built near Husum. These consist of four four-sided ponds, with perpendicular walls, lined with plank. The length of the ponds is 14 meters, the breadth 12 meters, and the depth about 2 meters. The bottom is paved with tiles. The ponds can be divided into compartments by means of perpendicularly placed wooden partitions. In these compartments the oysters are planted as soon as they are brought from the banks. They are then covered with water which has stood for a while in a neighboring pond, the clearing-basin, in order to let the mud settle. During cold weather 500 tons, or 350,000 to 400,000 oysters, can be kept in these storage-ponds, but during warm weather only 200 tons. If it be necessary, oysters can be stored in the clearing-pond and in the trenches which lead to the ponds. In order to preserve the oysters in the storage-ponds in a healthy condition they must not be placed too close together, especially in warm weather. They must also be changed very often from the compartment which they have occupied to a clean one, and be subjected to a rapid flow of water in order to wash off all the dead material from their shells. Most of the English oysters which are eaten in Germany come from Ostend. They are kept there in basins similar to those at Husum, which have been built behind the walls of the old fortress. In 1869 I found there nine of these basins, which could be filled to a depth of about 2 meters with sea-water, supplied through sluices which connect the basins with the sea. In these ponds oysters are only stored and fattened. Those not sold by the close of the oyster-season are generally sent back to the English beds, because they are kept with difficulty during the summer, but principally because, after lying in the ponds for a long time, they become very poor from the lack of food. In the years 1875-'76 it is reported that a weekly supply of 500 bushels (750,000 oysters) were received at Ostend from England.

The most celebrated oyster-port in England is Whitstable, situated on the southern side of the mouth of the Thames. Here the best natives are found. Their shells are indeed not very large, but their bodies are thick and very full, on account of the great depth of the cavity of the left valve.

The oysters which are sent over to Germany, by way of Ostend, are smaller varieties than the celebrated Whitstable oysters. These last are seldom sent to the continent, nearly all being demanded for the London market, where they command a higher price than any other kind. Oysters for exportation are packed in casks. In these they are placed with the left valve always undermost, and are packed so close together that, when the cask is closed, no room is left for them to open the valves of their shells. Upon many oyster-beds along the west coast of France those oysters which have very nearly arrived at a marketable size are at frequent intervals left uncovered, and longer and longer each successive time. As long as they are deprived of water they will keep their shells closed, and thus they are trained to retain, while in the dry casks, and until the knife prepares them for the table, the water which they inclosed in their shells when taken up. If ice is used to keep oysters fresh, care must be exercised that the water from the melting ice does not come in contact with the mollusks, or their flavor will be injured. Care should also be taken, especially with shelled oysters, that the ice used to cool them off does not entirely cover them.* Preserved oysters, packed in tin cans, are brought into the markets from North America. In these the natural flavor, for which the fresh oyster is so highly prized, is as much destroyed as is that of the tropical fruits which come to us cooked in sugar. If they were not preserved oysters they would hardly find purchasers. They serve merely as suggestions of fresh oysters.

13.—THE OBJECT AND RESULTS OF OYSTER-CULTURE.

The object of a good oyster industry is to gain from the territory cultivated the greatest possible profit, and at the same time to render the industry permanent. From a bed of inanimate material one can take away as much of the mass as he can use with profit. Such a proceeding does no harm to the prosperity of the bed, since what is left has nothing to do with the production of a new supply. With living objects, on the contrary, it is different. They are not quiet, immovable masses, but combinations of materials and active forces, which are engaged among themselves in a continual renewal; and if one is to derive the greatest possible benefit from them, their mass must not be indiscriminately reduced, as with minerals in a mine, but care must be taken that their powers of renewal are not weakened by a lessening of their available forces and materials. A breeder of cattle who would maintain a certain definite degree of productiveness in his herd must keep a definite number of breeding animals. If it is desired to have a definite permanent production of wood from a given extent of forest, only such an amount must be cut yearly as will be offset by the yearly growth. If a permanently larger quantity is desired, the forest surface must be increased. A profitable permanent system of oyster-culture is also dependent upon these same laws. Hence its foundation is the preservation of a stock of mature breeding oysters. No artificial system has yet succeeded in bringing to maturity, in inclosed parks, generation after generation of oysters, and the most clever breeders are obliged to rely upon the natural banks in order to obtain breeding oysters or young for their fattening-ponds. Hence the foundation of all oyster production, whether artificial or natural, is *the preservation of a stock of full-grown breeding oysters upon the natural oyster-banks.*

In France, ever since the government undertook to retain upon the natural banks along its coast a sufficiently large number of breeding oysters to keep up the stock, artificial oyster-breeding has maintained a secure basis. By this means the French Government has been enabled, through its fishery commissioners, to determine the beds which are in a suitable condition to be fished and the time at which they can

---

*The Romans were in the habit of cooling their oysters with ice from the mountains: "Addiditque luxuria frigus (ostreis) obrutis nive, summa montium et maris ima miscens." C. Plinii Sec., naturalis historia, lib. xxxii, 6, 21.

be profitably worked. In the rich oyster-regions of Cancale and Saint-Vaast-de-la-Hogue, on the coast of Normandy, and in the Bay of Arcachon, there are great banks which, during spring-tide (at the time of full and new moon), run dry, or are covered with so little water that people can wade over them and pick up the oysters with their hands. Near Cancale crowds, resembling caravans, of from 500 to 1,000 persons, mostly women and children, fish for oysters upon these exposed banks. One of the best of these beds in the Bay of Arcachon, called Le Cés, has an extent of 11 hectares (about 26.73 acres). When oyster-fishing is permitted on this bank, it is generally performed by women, who are placed in rows of about ten each, and, headed by two men, proceed over the bed. The oysters are gathered into sacks which are carried by women following behind the others, and who empty the sacks, as they become filled, into larger baskets. The gathering cannot continue longer than from two to two and one-half hours in any one spring-tide, because the bank is not exposed for a longer time. Yet, in this time, 40 to 50 persons can gather about 60,000 oysters. Immediately after any place is fished in this manner it is marked by four cask-buoys, so that it may not be fished again the same year, and in order that it may be readily found later, when they scatter oyster and mussel shells over the ground for the attachment of the young oysters. About the year 1870, the beds in the Bay of Arcachon had become almost entirely exhausted, but by this strict method of protection, the fecundity of the 19 beds which are located there has once more become so great that the water of the bay, from June till into August, is filled with swarms of young oysters. Hence it is no wonder that at times, and in favorable places, single tiles can be found to which from 1,000 to 1,200 young oysters have become attached. According to an official report* upon French oyster-culture which appeared in January, 1877, there were, in 1876, in the fattening-ponds upon both sides of the mouth of the Seudre, 80,000,000 oysters; near Oléron, 7,000,000; near Sables-d'Olonne, 10,000,000; near Lorient the same number; and near Courseulles-sur-Mer, 20,000,000 to 30,000,000. This extraordinary fruitfulness of the oyster-beds along the west coast of France is the result of the careful preservation of a rich stock of mature breeding oysters upon the natural banks, especially in the Bay of Arcachon, on the coast of Brittany near Auray, and on the coast of Normandy near Saint-Vaast-de-la-Hogue, Cancale, and Granville. Thousands of persons are industriously employed, during the season, in taking, upon shells, tiles, &c., the im-

---

* Rapport adressé au ministre de la marine et des colonies sur l'ostréicultur, par M. G. Bouchon-Brandely, Published in the official journal of the French Republic, January 22, 1877. Under an act of Parliament of May 17, 1877, an English translation of this appeared, with the title: Copy of translation of a report made to the minister of marine in France, by M. G. Bouchon-Brandely, relative to oyster-culture on the shores of the channel and of the ocean.

From this report I have taken all the remarks which I have made in chapter 13, in regard to the latest condition of the French system of oyster-breeding.

mense swarms of young oysters which are produced upon these beds, in guarding them from their enemies, and then in transplanting them to the numerous fattening-ponds along the coast, where, at last, by careful rearing, they are brought to marketable size. The number of persons employed daily in oyster-culture near Saint-Vaast-de-la-Hogue is 300, and near Cancale 4,000. In the district of Auray, for the year 1876, the total number of days' employment for all the men, women, and children who were engaged in this industry was 89,678. In Arcachon an oyster-breeding company laid out 110,000 tiles as objects of attachment for the young oysters. In 1876, 300,000 tiles were used for the same purpose near Vannes, and as many at Oléron. At Auray, in 1874, as many as 2,580,000 tiles were employed as objects of attachment for the young broods. At Lorient 60,000 troughs of cement, each trough 50 centimeters long by 30 to 40 broad, are used in rearing and fattening young oysters. In these the water remains constantly, during the lowest tides, 10 to 12 centimeters deep. Near Saint-Vaast-de-la-Hogue there are 185 oyster-beds, which cover an area of 88 hectares (about 213.84 acres); near Cancale the amount of surface which has been artificially changed into oyster-beds is 172 hectares (about 411.96 acres).

At Auray the amount of oyster-ground is over 300 hectares. Here there are 277 storage-beds and 20 fattening-ponds. In 1876, in the neighborhood of Marennes and Tremblade, on the Seudre, there were 13,526 artificially prepared beds, covering an area of 4,000 hectares (9,720 acres), and at the same time the Bay of Arcachon contained 3,317 such places. The production of oysters from these beds is so great that machines have been invented to sort them. With the help of a machine, two women can sort in a day 30,000 to 40,000 oysters. Railroads connect the feeding-ponds with the cleansing-basins, packing-houses, and landing-places of the boats which bring the young oysters from the banks and brood-beds for fattening. From these accounts it will be seen how large a surface of sea-bottom, how much money, and how much human labor are requisite in order that the embryos which under natural conditions originate in the sea shall be transformed into the immense number of full-grown oysters which the French oyster-breeders now place upon the market. The original plan of Coste to line the entire coast of France with a network of oyster-beds has indeed not been carried out; but in consequence of his exertions and experiments many oyster-parks have been established in favorable places along the coast from Normandy to south of the mouth of the river Gironde. The French, favored with innumerable bays and with a mild sea temperature along their coast, have, by diligence, perseverance, and the invention of new methods, brought oyster-culture to such a high degree of perfection, and given it such wide range, that now, in that favored land, it is to be reckoned as one of those cultured industries in which man converts to his service vast numbers of plants and animals. The large number of oysters produced as a result of the French system of oyster-culture has been held up very often to the inhabitants of the German coasts, in order to incite them to establish in their seas similar places for the artificial harvesting of oysters. The writers who did so knew neither the nature of the oyster nor the character of our seas. They might just as well have said to the inhabitants of the lower portion of the Elbe: " Lay out vineyards, for in 1874 the department of the Lower Loire produced 1,914,427 hectoliters of wine, and the department of Gironde 5,123,643 hectoliters." In Egypt there is nothing lacking, except water, in order to produce dates and wine in abundance upon the desert which stretches from Cairo to Suez. So it is with us; all we lack in order to carry on successfully artificial oyster-breeding upon the mud-flats of the North Sea are mild winters, with no ice, and security against the force of storm-floods. There is food enough there to feed billions of oysters. The old English method of oyster-culture was much simpler than the new French method. The work consisted chiefly in transplanting young oysters from the natural banks along the coast to suitable beds in the mouths of rivers, where they became fat and well flavored. They also removed the mud and plants from these new beds, destroyed as many of the enemies of the oyster as possible, and improved the ground by scattering over it the shells of oysters and other mollusks. This industry is carried on in a much better manner at Whitstable, where there is an oyster company which, it is claimed, has been in existence for six or seven hundred years. It numbers over 400 members, who work 120 vessels. Only the sons of those who are, or have been, members are admitted into the guild. In 1793 an act of Parliament adjudged to this company, as their property, an extent of oyster-ground about two miles long and the same in breadth, situated in the mouth of the Thames, and which they had claimed up to that time only by right of possession. This territory consisted partially of natural oyster-banks, partially of beds upon which oysters from the open sea had been placed to spawn, and partially of beds upon which oysters from along the coast had been placed to fatten. In order to still further improve these beds, empty oyster-shells, sent back principally from London, were often scattered over them. The Whitstablers consider that a thick layer of oyster-shells forms the very best bed for oysters, and they pride themselves that they possess the "best oyster-grounds in the world," as I myself have heard them say. The fecundity of the oysters upon their fattening-beds is very small. The cultivation of the oyster is carried on at Colchester, Burnham, and other places along the coast of England very much as it is at Whitstable. From these places many oysters are taken to Ostend. The efforts which have been made to bring living oysters from North America to England and plant them there have not succeeded well enough to warrant imitation. But if they could succeed in transplanting large quantities of young oysters from the breeding-stations of Normandy and Brittany to the excellent feeding-grounds of England, English oyster-breeding would probably soon take a very significant upward tendency.

The oyster industry is conducted in North America very much as it is in England. In protected muddy bays and mouths of rivers near the coast, great quantities of young oysters, which have been taken from the natural beds, are planted for the purpose of fattening, the method thus resembling that in vogue at Whitstable and at other places along the west coast of Europe. In North America places are also chosen where the oysters will be protected from frost and heat. In localities rich in food they arrive at marketable size in from two to three years. The North American oyster is a different species from that of Europe. Its scientific name is *Ostrea virginiana*. It is longer from the hinge-ligament to the shell-muscle than is the European oyster, *Ostrea edulis*, and the left valve is generally more curved than with our oyster. Very few efforts have ever been made in North America to catch and grow oysters artificially according to the French system. The natural banks produce such an abundance of young oysters that all the beds artificially planted can be abundantly and cheaply supplied from them. During late years the North American beds have furnished an annual average of about thirty million bushels of oysters for market; this is about nine to twelve billions of oysters, since there are from three to four thousand oysters in a bushel. In 1859 the number of oysters sold amounted to from six to eight billions.

The principal markets for North American oysters are New York and Baltimore. In 1867 there were over 10,000 men employed in the oyster-trade in Baltimore. The yearly capital employed in this business in New York was, about 1870, over $8,000,000.* The North American oysters are so fine and so cheap that they are eaten daily by all classes; hence they are now, and have been for a long time, a real means of subsistence for the people. This enviable fact is, however, no argument against the injuriousness of a continuous and unprincipled fishing of the beds. The size of the territory over which oysters are found, and the number of inhabitants, must not be left out of account, however, if a right judgment would be formed in regard to those great sums which appear in the oyster statistics of North America. The territory of the North American oyster-beds is of very great extent, comprehending the greater portion of the east coast of the United States. Oysters occur from Cape Hatteras, in North Carolina, to the mouth of the river Saint Lawrence, and Chesapeake and Delaware Bays are especially rich oyster localities. In the United States there are now 52,000,000 of people; in Germany, France, and England, altogether, over 109,000,000. Hence, in North America, with a less number of inhabitants, there is a much greater supply of oysters per person than there is in Europe. But as the number of consumers

---

* In the following works will be found more detailed statements in regard to oyster-culture in North America:

P. de Broca. Études sur l'industrie huitrière des Etats-Unis. Paris, 1865.

Spencer F. Baird. Report on the condition of the sea-fisheries of the south coast of New England in 1871-'72. Washington, 1873. Oyster-beds, p. 472, by A. E. Verrill.

increases in America, the price will also certainly advance, and then the desire will arise to fish the banks more severely than hitherto; and if they do not heed in time the unfortunate experiences of the oyster-culturists of Europe, they will surely find their oyster-beds impoverished from having defied those biocönotic laws which have been given in chapter 10.

As man has uprooted the greatest forests, so can he also annihilate the richest oyster-beds. In England it is now understood to be absolutely necessary that the natural oyster-banks should be regularly and systematically protected if they are to remain uniformly and permanently productive. A commission for the investigation of the English oyster-fisheries, which met in London early in the year 1876, recommended to Parliament that fishing for oysters be forbidden by law from the 1st of May until the 1st of September each year, and that definite limits of time be designated, during which certain definite oyster-territories must be allowed entire rest. During the close-time all handling of oysters for the purposes of food should be prohibited under penalty of fines; yet it should be permitted, even during close-time, for the purposes of transplanting, with the design of preservation and improvement, oysters taken in a lawful manner upon public beds. Upon the banks in the open sea the close-time was to last only from the 15th of June until the end of August, since these banks can very seldom be fished during the stormy seasons of the year. The size of the sea-oysters brought to market was to be at least $2\frac{1}{2}$ to 3 inches in diameter.*

A close-time has been enforced upon the Schleswig-Holstein beds for a long period. This time extends from the 9th of May to the 1st of September, and, furthermore, no oysterman is permitted to take away any oysters which are less than $2\frac{1}{2}$ inches in diameter. All oysters which are not of this size must be thrown back into the water. Both of these laws have been carried out; yet, nevertheless, in the course of the last twenty years the fertility of the beds, in comparison with earlier rental periods, has very significantly fallen off. These laws in regard to close-time and a minimum size for marketable oysters, which were designed to preserve to the banks an undiminished power of renewal, did not, therefore, attain their object at the very time of the high price of oysters, and when oysters should have been plentiful It is, therefore, not enough to regulate the time of catching and the size of oysters, if, at the same time, care is not used to prevent too large a number of oysters from being taken from the beds during any one fishing season. *But what number is too great?* A foundation for an estimate of the number of oysters which may be taken away from the beds without injury to their productiveness can be obtained, for the Schleswig-Holstein beds, by means of the inquiry in regard to their productiveness. This productiveness is, upon an average, 421 per thousand; so for every 1,000 full-grown

---

* Report on oyster-fisheries, 1876, p. iii.

oysters which are now upon the beds not more than 421 ought to be taken away annually. Upon a number of banks where the productiveness is less than this the number taken should be less. Upon the Huntje Bed, where the production is more than 421 per thousand, as many as 484 for every thousand can be taken yearly without endangering or lessening the productiveness, since that number of medium oysters grow into marketable oysters every year. But although the productiveness is thus expressed by a proportional number, yet the absolute number of full-grown oysters which may be taken from a bed during any one season cannot be arrived at without further consideration. One must know how thick the full-grown oysters lie upon the beds; whether, in fact, there is a sufficient number to secure an average fecundity to the bank. Upon banks such as those in the Bay of Arcachon and near Cancale, which are left dry during spring-tides, it is not difficult to observe the number of oysters necessary per square meter, in order to maintain the fecundity of the bank at its highest point, for at such times they are so exposed that they can all be counted. But those beds along the German and English coasts and in the open North Sea, which, on the contrary, remain continually under water, are much less favorably situated for the purposes of these inquiries. I have often been told that "such beds could be best investigated by means of divers." The general impression is that the divers can see, through the glass in the front part of their helmets, everything which lies upon the sea-bottom. But this is erroneous, for in those shallow coast-seas which have ebb and flood tides the water is so clouded by the floating particles of mud that very little light can penetrate to the bottom. But even in clear water a diver would not be in a condition to ascertain by sight the number of oysters, for whenever he steps he renders the water cloudy, by stirring up the lighter particles lying upon the bottom; and so he would have to depend principally upon his hands, and ascertain, by feeling, those oysters which could just as well be taken up by means of the dredge, for the dredge brings from the bottom not only some of the soil, but also a portion of its inhabitants. And if the contents of the dredge be placed in large vessels or aquaria, with sea-water, the animals will very soon assume their customary positions and motions, so that we can see, in quiet and clear water, just how they live at the bottom of the sea. An aquarium with the living inhabitants of an oyster-bank is thus a segment of the bank itself.

When, in imagination, I have united many such segments together, I can picture to myself the sea-bottom, with its inhabitants, as a diver would never be able to see it. I can see the ground covered with oyster-shells, and here and there among them a living oyster with open shell, out of which protrude the fringed borders of its mantle. Upon the upper valve polyps are growing with expanded heads, looking like delicate, many-rayed stars. Hermit-crabs, bearing their snail-shell houses, are crawling hither and thither over the rough surface, and groping about with their claws for something to eat. Worms stick their heads out of holes and crevices; sea-urchins stretch out their sucking feet beyond the points of their spines and pull themselves slowly up on to a stone; a star-fish, with greatly arched back, has fastened itself about a mussel in order to suck it out of its shell; and a small fish has stationed itself under an open oyster and snaps up the embryos as they come from the shell.* Of this life of the oyster-bank the diver would see little or nothing, even if he happened to be a zoologist, for as soon as he had descended to the bottom the oysters would shut themselves up, the crabs and worms creep out of sight, and the fish swim away. I thus sketch this picture of a small portion of the really abundant life of an oyster-bank in order to show that one may become really very well acquainted with an oyster-bed by means of a dredge. It can also be used to estimate the thickness of oysters upon a bed, if the distance passed over by the dredge while it is taking oysters be measured. In the inspections of the Schleswick-Holstein banks, during the last few years, this has been accomplished in the following manner: At those points, where the dredge is dropped upon the bed, an empty cask, attached by means of a rope to a heavy weight, is cast overboard. The weight sinks to the bottom and holds the cask securely anchored, floating upon the surface of the water. Connected with the rope of the cask is a measuring-line, which is wound upon a roller, and which runs off as long as the vessel is going forward and the dredge drags over the bottom. The mouth of our larger dredges is one meter in width. Thus, if we let the dredge drag over the bottom until 100 meters of line have run off, and find that we then have 50 oysters in the bag, we can conclude that one oyster came from every two square meters of bed-surface; and if an oyster-bank, the length and breadth of which are known, is dredged over in this manner in different directions, a foundation is obtained from which to estimate the number of oysters upon the bed with certainly as much accuracy and with far greater speed and ease than a diver; and when the proportional productiveness of a bed thus examined is ascertained, we can estimate the number of oysters which can be taken from the bed without injury to its productiveness.

Practical persons will object to these methods as being too detailed, and yet not leading us to a sufficiently high estimate of the number of oysters; but they will be obliged to admit that there is no better means of finding out, with any degree of certainty, the number of oysters upon these banks. A skillful oysterman, one who has been acquainted with the beds for a number of years, will notice, without the use of a measuring-line, whether the oysters lie upon the banks in sufficient

---

* An oyster-breeder, Captain Johnston, saw, at some oyster-station for artificial culture, small fish of the genera *Gobius* and *Mullus* swallowing young oyster-swarmlings. He caught the fish, opened their stomachs, and found therein partially digested embryos. (Report on the oyster-fisheries, p. 87, Nr. 1711.)

numbers for the prosperity of the banks, or whether the beds have become impoverished. He will reach this conclusion from the number of oysters which he can catch with a certain speed of his vessel, and during a certain definite time which his dredge drags over the bottom.

Those authorities who have control of the inspections of the oyster-fisheries might, therefore, be able to avail themselves of the services of skillful dredgers to find out the condition of the banks before they decide, each season, the particular places which can be fished and the number of oysters which can be taken from each. The inspectors at Arcachon, after observations extending over many years, have arrived at a definite conclusion in regard to the number of breeding oysters which it is absolutely necessary to retain upon the banks, in order to maintain them at that stage of fruitfulness necessary for a permanent and profitable oyster-culture.

The report of January, 1877, upon oyster-culture in France says: "Although the natural oyster-beds in the Bay of Arcachon are regarded as breeding-beds, yet, nevertheless, the government allows them to be fished for some hours ever year, in order to remove the surplus of oysters." This is a fundamental proposition which a judicious oyster-breeder must carefully consider if the greatest amount of profit would be gained. In accordance with this proposition, oysters should never be allowed to remain upon a bank after they have passed the period of their greatest growth and fecundity, or until they die of old age; but we should anticipate nature, which demands the death of the old and weak as an indispensable condition for the production and bringing to maturity of the greatest number of young upon any bed. I do not consider it, then, as for the best interests of the beds to prevent dredging upon one or all of them for any long periods of time. The French Government has not, therefore, in my estimation, acted in the best interests of the beds, in entirely forbidding dredging upon a strip of territory which lies along the edge of the oyster-banks of Cancale and Granville. The object of this protection is to retain there an undisturbed stock of breeding oysters, from which to rejuvenate the impoverished beds of both these places. Upon such unfished beds the natural biocönotic balance, from which a certain definite average germ-fecundity results, will very soon become established. But this will become less if, with the same proportion of nourishment, more superannuated than mature oysters are to be found upon the beds. The productiveness of any territory will thus be much less, if it is left entirely undisturbed than if it is judiciously-fished, and, moreover, the profits which result from the food-oyster taken from such territory are lost. Upon the Schleswig-Holstein banks the oysters are best when from about seven to eight years old. In warmer regions they become fully matured in a shorter time.

The amount of increase in the length of oyster shells during a given time is very different upon different portions of the Schleswick-Holstein sea-flats, but their average growth in thickness is much more uniform;

hence it would be more correct to estimate the minimum size for marketable oysters according to the average thickness of the shell than according to its breadth. Estimated thus, a thickness of shell of 18 millimeters would be a judicious minimum size for the Schleswig-Holstein oysters. In conclusion, I hereby give, as a foundation for all oyster-culture, the most important rules for the preservation and improvement of natural oyster-banks.

An oyster-bank will give permanently the greatest profit if it possesses such a stock of full-grown oysters as will be sufficient to maintain the productiveness of the bank in accordance with its biocönotic conditions.

Whenever the natural conditions will admit of it, the yielding capacity of an oyster-bed may be increased by improving and enlarging the ground for the reception of the young oysters.

The natural banks should be improved by removing the mud and sea-weeds with dredges and properly constructed harrows, and by scattering the shells of oysters and other mussels over the bottom. When circumstances will permit, all animals which are taken in the dredge, and which kill the oysters or consume their food, should be destroyed.

It would be much more judicious, and much better for those who eat oysters, if the close-time could be extended until the 15th of September or the 1st of October, so as to allow the oysters sufficient time, after the expulsion of the contents of the generative organs, to become fat before being brought to the table.

If it is desired that the oyster banks should remain of general advantage to the public, and a permanent source of profit to the inhabitants of the coast, the number of oysters taken from the beds yearly must not depend upon the demands of the consumers, or be governed by a high price, but must be regulated solely and entirely by the amount of increase upon the beds.

The preservation of oyster-beds is as much a question of statesmanship as the preservation of forests.

INDEX.

	Page.
Alcyonium digitatum, in oyster beds	40, 53
Alcyonidium gelatinosum in oyster beds	40
American coast, oyster territory of	60
Animal life in oyster beds	39
Appropriation of oyster beds by the King	32
Arcachon, oyster beds in	17, 21, 22, 42, 45, 57, 62, 64
Artificial extension of oyster territory	45
oyster beds	16, 17, 42
on the British Coast	20
oyster breeding	42
Artificial oyster breeding:	
in France	1, 16, 57, 58
in Great Britain	20
on German coast	21
cost of	19
Artificial oyster ponds	18
Aspidophorus cataphractus, on oyster banks	40
Asteracanthion rubens in oyster beds	39
Auray, Brittany, oyster beds at	57, 58
breeding in	42
Baird, Spencer F., report on the condition of the sea fisheries of the south coast of New England, cited	60
Balanus crenatus, in oyster beds	40
Baltic Sea, experiments in oyster planting	28
freshness of, excludes oysters	28, 29
Bay of Arcachon, oyster beds	17, 21, 22, 42, 45, 27, 62, 64
Biocœnosis of oyster banks	39, 41
Blake, Mr., on cost of artificial oysters	20, 21
Böckh, R., Sterblichkeitstafel, &c., cited	35
Breeding boxes for young oysters	18
of oysters, artificial	16
British coast, artificial oyster breeding	21
Broca, P. de, Études sur l'industrie huitrière des États-Unis, cited	60
Buccinum undatum, in oyster beds	39
Burnham, England, oyster industry at	59
Cancale, oyster beds at	37, 38, 39, 41, 49, 57, 58, 64
Carcinus mænas injurious to young oyster	18
in oyster beds	39
Cardium edule, in oyster beds	39
Chemical constituents of the oyster	50 et seq
Clione celata in oyster beds	40
in oyster shells	11
Coast-water desirable for oyster culture	14
Colchester, England, oyster industry at	59
Cold, effect of, on oysters	23, 24
Collin, Jonas, Om Östersfiskeriet i Limfjorden, cited	31
Coste, Prof. M., artificial breeding of oysters	16
Cross-sectional view of oyster	11
De Bon, Notice sur la situation de l'ostréiculture en 1875, cited	19

	Page.
Deep-sea oysters	15
Desmidiæ, on oyster banks	40
Destruction of oyster beds	44
of young oysters	19
of oysters, causes of	23
Diatomaceæ, on oyster banks	40
Didus ineptus, extinction of	44
Difficulties in establishing artificial oyster beds along the German coast	21, 22, 26, 27, 50
Dodecaceræa concharum, in oyster beds	40
Dodo, extinction of	44
Dredging of oyster beds	40, 41
for oysters	7, 8
Dudgeon Light, England, oyster bed at	36
Eastern limits of the oyster in Germany	29
Echinus miliaris, in oyster beds	39
Eggs, number of, produced by oysters	12, 16
Exhaustion of oyster beds	27, 37, 38
Embryo oysters	12, 13
development of	13
Enemies of the young oyster	18
Enlargement of natural oyster beds	25
Examination by a royal commission	33, 34
Extension of oyster beds, hindrance to	15
Extension of oyster territory, artificial	45
Eudendrium rameum, in oyster beds	40
Fattening ponds in France:	
at mouth of the Seudre	57
at Oléron	57
near Sables-d'Olonne	57
near Lorient	57
near Courseulles-sur-Mer	57
Fattening oysters, artificially	43
Female oyster, superior to male in flavor	52
Flavor of the Marennes oysters	27
of the oyster	50
dependent on saltness of water	14
Floridiæ, on oyster banks	40
Food of the oyster	43
Formation of the young oyster	12
Fossil oyster bed near Blankenese	32
oyster beds near Waterneversdorf	30
France, artificial oyster breeding in	1, 16, 57, 58
fattening ponds in	57
oyster fisheries in	64
oyster production in	58
transplanting practiced in	27
Freezing oysters, results of	13, 24
French system of oyster breeding	20
Fruitfulness of the oyster	14
German coast, artificial oyster breeding	21
coast waters, inadaptiveness	21, 25, 26, 27
oyster territory	3
Granville, France, oyster beds at	57, 64

[67] 749

	Page.
Great Britain, artificial oyster breeding in	20
Halichondria panicea, in oyster beds	40
Hallig Oland	5
Hayling, island of, oyster breeding trench.	42
Holstein oyster	3
Hörnum, oyster bed of	38, 53
Huntje Bank oyster bed	34, 38, 39
oyster bank	34
Husum, oyster reservoir at	54, 55
Hyas aranea, in oyster beds	39
Ice used in cooling of oysters	5
Immigration of the oyster	29
oyster	39
Impoverishment of oyster banks	47, 48, 49
Jacobson, Prof. O., analysis of oyster	51
Johnston, Captain, cited	63
Krogh, F., Den konstige Östersavlog dens Indförelse i Danmark, cited	31
Kröyer, H., De danske Östersbanker, cited	32, 48
Lim Fiord, appearance of oyster in	29, 30
oyster in	39, 45
Limits of the oyster in German waters	29
Location of oysters along the coast of Pomerania, acts respecting	31
Marennes, France, oysters	27
oyster bed at	39, 41, 49
Marketable oyster, size of	33
Market value of oysters	48, 49, 50
Maturity of oyster	64
Möbius, Die Auster und die Austernwirthschaft, cited	1
oyster and oyster culture	1
Ueber Austern-und Miesmuschelzucht, etc., cited	19
Measuring sticks for oyster beds	7
Medium oyster, definition of	35
Messum, Mr., statement of oyster fishing.	37
Meyer, Dr. H. A., on saltness, temperature, and currents of the North Sea, cit.	14
Murex erinaceus, injurious to oysters	18
Mytilus edulis in oyster beds	39
Natural oyster beds, conditions favorable	25
investigation of	25
Nereis pelagica, in oyster beds	39
Noctiluca scintillans in the harbor of Keil	29
North America, oyster industry in	60
North Sea, bottom of	6
oyster grounds of the	14
saltness of	14
tides in	5, 6
Number of eggs produced by oyster	12, 16
Ostend, oyster reservoir at	55
Ostrea edulis	60
Oyster and oyster culture (Title)	1
bank is a Biocönose	39
banks, how exposed	7
causes of impoverishment	47, 48, 49
inspection of	63, 64
protection required	65
bed at mouth of Thames	36
near Dugeon Light	36
Oyster-beds	3, 4, 6, 15
appropriation of, by the King.	32
at Röm	53
at Arcachon	17, 21, 22, 42, 45, 57, 62, 64

	Page.
Oyster-beds Cancale	37, 38, 39, 41, 49, 57, 58, 64
conditions favorable	42
Dudgeon Light	36
exhaustion of	37
formation of	42
fossil	30, 32
Granville	57, 64
hinderance to extension of	15
Hörnum	38, 53
Huntje Bank	34, 38, 39
Island of Oléron	39, 41, 49
Island of Ré	47
Island of Sylt	53
Marennes	39, 41, 49
near Auray, Brittany	57, 58
near Rocher d'Aire	47
on the American coast	60
production of	37
productiveness of	61, 62, 64
Rochefort	39, 41, 49
Whitstable	36, 46
Saint-Vaast-de-la-Hogue	57, 58
Schleswig-Holstein	3, 8, 9, 38, 39, 64, 65
Steenack	34, 38
West Amrum	38
Oyster breeding, artificial, in France	1, 16, 57, 58
artificial	42
causes of failures in	46
in Brittany	42
temperature of water for	46
trench	42
Oyster, chemical constituents of the	50 et seq.
culture in Great Britain	20
Oyster culture, object and results of	56
dredge	7
fisheries, restriction in	61
in England	61
in France	64
in Schleswig-Holstein	61
of Hanover	27
Oyster fishing, extent of, in England	36
increase of	49
processes in use	57
statistics of	37, 38
flavor of the	50 et seq.
grounds in the open North Sea	14
in Lim Fiord	3, 9, 45
industry in England	59
in North America	60
Oléron, island of, oyster beds at	39, 41, 49
Oysters, life of explained	1
Oysters, market value of, in Hamburg	48
at Whitstable	40
at Colchester	49
at Falmouth	50
Oyster, maturity of	64
preservation of	55
production of	58
quality of, at different periods	53
reservoirs	54, 55
tenacious to life	54
territory, artifical extension of	45
cleaning of, for commerce	8
Oystering	7
Pagurus bernhardus, in oyster beds	39

	Page.
Pearls found in oysters	50, 51
Pennel, Cholmondeley, report on the oyster and mussel fisheries of France, cited	36, 47
Platessa vulgaris, on oyster banks	40
Pomatoceros triqueter, in oyster bedst	40, 53
Pomerania, coast of, location of oysters	31
Prehistoric oysters in German waters	30
Productiveness of oyster beds	61, 62, 64
sizes and, of the oyster	31, 35
Propagation of oysters, artificial	16, 17
Quantities of oysters taken from beds on the Schleswig-Holstein coast	34
Raja clavata, on oyster banks	40
Ré, island of, oyster bed in	47
Reports on artifical oyster culture	21
Reproduction of the oyster	10
Reservoirs for oysters at Husum	54, 55
at Ostend	55
Rochefort, France, oyster bed at	39, 41, 49
Rocher d'Aire, France, oyster bed near	47
Romans used ice in cooling of oysters	56
Röm, island of, oyster bed at	53
Sabellaria anglica in oyster beds	40
Saint-Brieuc, oyster breeding at	16, 17, 42
Saint-Vaast-de-la-Hogue, oyster beds at	57, 58
Saltness of North sea	14
Schleswig-Holstein, oyster, flavor of	9
oyster, size of	9
oyster beds	3, 4, 37, 38
oyster beds, 3, 8, 9, 38, 39, 64, 65	
beds, composition of	8
oyster beds, extent of	3, 8, 9
Wattenmeer	4
Schlossberger, analysis of oyster, cited	51
Sea-flats	2, 4

	Page.
Sertularia argentea, in oyster beds	40
pumila, in oyster beds	40
Sizes and productiveness of the oyster	31
South of England oyster company	20
Spawning season of oyster	12
Spermatozoa of the oyster	11
Steenack oyster bank	34
oyster beds of	34, 38
Structure of oyster	11
Sylt, island of, oyster beds near	53
Temperature an important factor in life, conditions of a biocönose	43
Temperature for oyster breeding	46
Tides in the North Sea	5, 6
Tolle, A., Die Austernzucht und Seefischerei, etc., cited	19
Tolle, A., report on oyster culture	21
Transplanting of oysters in France	27
Tubularia indivisa in oyster beds	40
Von Bibra, analysis of oyster, cited	51
Wappäus, Handbuch der Geographie und Statistik, cited	35
Wattenmeer, Schleswig-Holstein	3, 4
Waterneversdorf, fossil oyster beds	30
Webber, Mr., on extent of oyster fishing	36
West Amrum, oyster bed of	38
Whitstable, England, an oyster port	55
England, oyster bed near	36, 49
oyster industry at	59
Winther, G., Om vore Haves Naturforhold med Hensyn til konstig Östersavl, etc., cited	31
Young oysters, formation of bed by	15
outline figures of	17
Zostera marina, on oyster banks	40

THE SEA BOTTOM
AND ITS PRODUCTION OF FISH FOOD

C. G. J. Peterson

I. Apparatus for Investigation of the Sea Bottom.

In the work of marine surveying, which has as a rule been carried out by the Naval Departments of the various countries concerned, it has generally been considered desirable to ascertain, not only the depth at each particular spot, but also the nature of the bottom. The easiest means of procuring information on this head is by affixing a lump of tallow to the lower end of the sounding lead, so that on hauling in, a sample is brought up sufficient to show whether the bottom is of sand, gravel, clay, or the like. This does not, however, give any very clear idea as to the nature of the bottom in other respects; at any rate, not as regards the animal life there. In order to procure something in the shape of a sample from the deeper layers, the more recent deep-sea expeditions have employed an iron tube, which is driven down into the bottom by means of heavy weights, and thus takes up a sample of the matter with which it is filled. These tubes, however, are only of slight dimensions, and likewise give us no idea as to the fauna of the ground worked.

Sir John Ross appears to have been the first to construct a clutching apparatus designed to bring up a bottom sample of sufficient extent to include animal forms. A contrivance of this description was made and used on his voyage to Baffin's Bay in 1817—1818, and brought up mud and animal forms from a depth of 1000 fathoms.

Nordenskjöld, Sven Lovèn, and Otto Torell have since employed similar apparatus, some of which are preserved at the Swedish Zoological Station at Christineberg; in 1914, I had an opportunity of inspecting and testing them there. They do not appear to have been described in any printed work; the construction, with metal springs and cannon-ball weights, is extremely complicated.

Recently, about the same time as I resumed my own bottom-sampler investigations, another Swedish Zoologist, Sven Ekmann, constructed a sampler for fresh-water work, and used the same with very good results. Ekmann and I have worked entirely independently of one another, and in 1911, we each issued the first printed report of our respective researches.

Our types of apparatus also differ widely one from another, each being apparently best suited to its own particular purpose; Ekmann's for investigations in fresh water, where the bottom is generally soft, and mine for work from on board a steamer at sea. His principle is to render the apparatus as light as possible, whereas I have purposely endeavoured to make mine heavy. (See Sven Ekmann: Die Bodenfauna des Vättern, qualitativ und quantitativ untersucht. Internationale Revue 1915.)

Generally, however, in investigating the fauna of the sea bottom, scientists

Fig. 1. Dredges and hand nets.

have almost exclusively employed apparatus of the dredge type, as exemplified in the accompanying illustration Fig.. 1 (the first five implements from the left).

These dredges consist of an iron frame, triangular, foursided, or with two curved sides, on which a short bag is stretched; they are intended to be dragged over the bottom, whereby parts of the "soil", together with animals resting or moving thereon, or scraped up by the frame, are brought into the bag. — For the more swiftly moving creatures, such as fish, etc. various adaptations of ordinary fishing implements have been employed; these, however, I will not discuss here, as they do not properly come into the discussion of bottom organisms proper.

For work in quite shallow water, hand nets are used, such as the four shown on the right of the illustration above referred to.

The first to introduce the dredge into Scandinavia was the Dane, Otto Frederik Müller, (1773); it had been used, in a somewhat different form, in Italy some years earlier, by Marsigli and Donati. Since then, it has been practically speaking the universal implement for all investigations of the sea bottom. The idea was taken originally from the oyster dredge.

The dredge is excellently adapted to the purpose of collecting, in a short time, a quantity of animals from the bottom, especially those lying on the bottom, and as long as the sole object was to procure specimens for anatomical and systematical research, it fulfilled these requirements well enough. Anyone, however, who has seen how a dredge hops and bounds over the bottom, if hard sand, or fills at once on soft clay, will realise that the catch must be largely a matter of

Fig. 2. Six bottom samplers. Station and S/S „Sallingsund" in the background.

chance, and can in particular give no sort of idea as to the true character of the local bottom fauna. The numerical frequency of the various species in a given area cannot possibly be ascertained by this means; the dredge will often only take in the rarer forms, and pass over almost all the animals most commonly found in the sandy bottom. (See Report XX p. 47—50 and XXI p. 18—27.)

In order therefore, to procure more precise information as to the quantity of animal life on the sea bottom, the fish-food, I was obliged to seek for some other implements better adapted to the purpose. I was not acquainted either with the apparatus used by Ross or those of others, as little has been written concerning these, and they have not been extensively used; I was consequently obliged to rely on my own resources.

The accompanying illustration (Fig. 2) shows a series of apparatus invented by myself, the oldest types being those farthest to the left. The earliest implement

2

was fastened to a pole, and was so constructed as to be closed by the pulling of a string, and opened by another. It covered an area of one square foot, and could be thrust into the bottom, closing over the space covered, when on soft ground, and thus take up the bottom material from the area embraced, together with the organisms living in and on the same. By a system of sieves, and a continuous flow of water from a pump, the animals were sifted off from the bottom

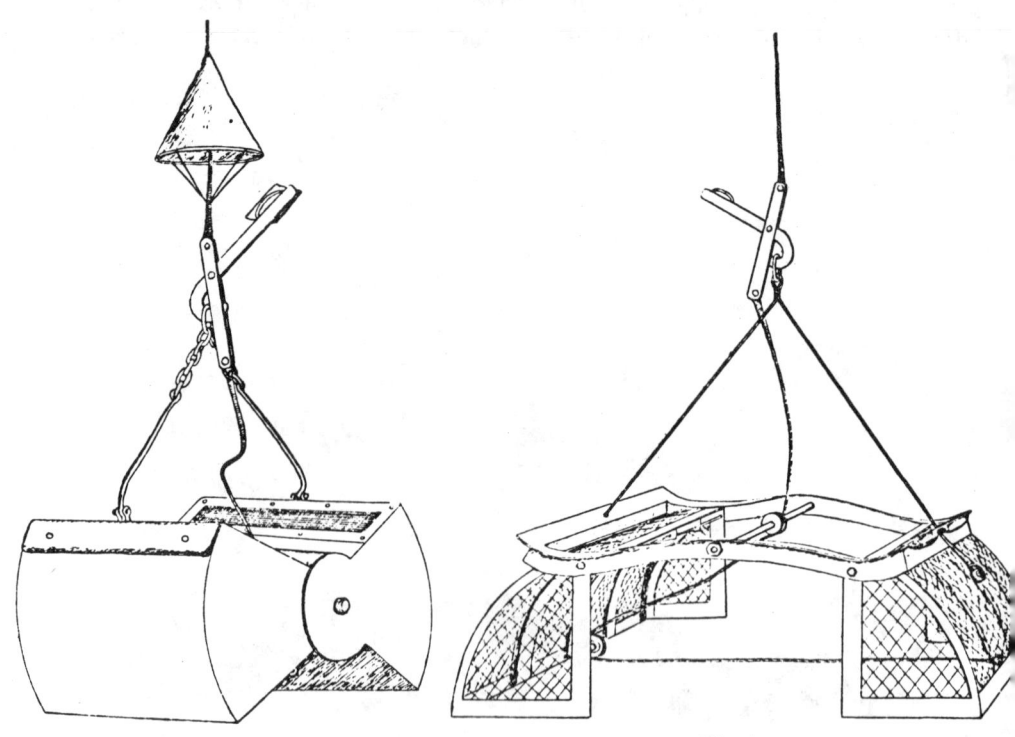

Fig. 3. $0_{,1}$ sq. m. bottom sampler, latest pattern, with parachute for deep-water work.

Fig. 4. 1 sq. m. bottom sampler for larger animals.

material, and we were thus able to determine what animals, and in what quantity, had lived within this small area.

This apparatus was made and used in the Limfjord in 1896, but could only be employed for depths to abt. 5 fathoms.

It was not until 1908 that I resumed these investigations, and then constructed, on board, my real bottom sampler, No. 2 from the left in the illustration, this being arranged to be lowered on a line, so that greater depths could be reached. The construction of this type has since been improved, and I have at last stopped at the small and heavy bottom sampler No. 2 from the right; this covers an area of $0_{,1}$ sq. m. Fig. 3 shows the very latest improvement of this bottom sampler, with a parachute arrangement above, especially intended for use in deeper water, as it is not required for depths of less than abt. 100 m. A heavy leaden cylinder is fastened round the great median axis, and the weight thus given renders it easier for the implement to dig down into the bottom. The large bottom sampler No. 1 on the extreme right is shown more clearly in Fig. 4; this is more particularly designed to take all the larger animals living on the bottom, such as starfish, etc., and is intended for investigations of a special character, so that it need not be dealt with further here; it covers an area of 1 sq. m., but does not take the smaller animals, owing to the size of the mesh, as if altogether filled with bottom material, it would be too heavy for the little vessel from which I have to work.

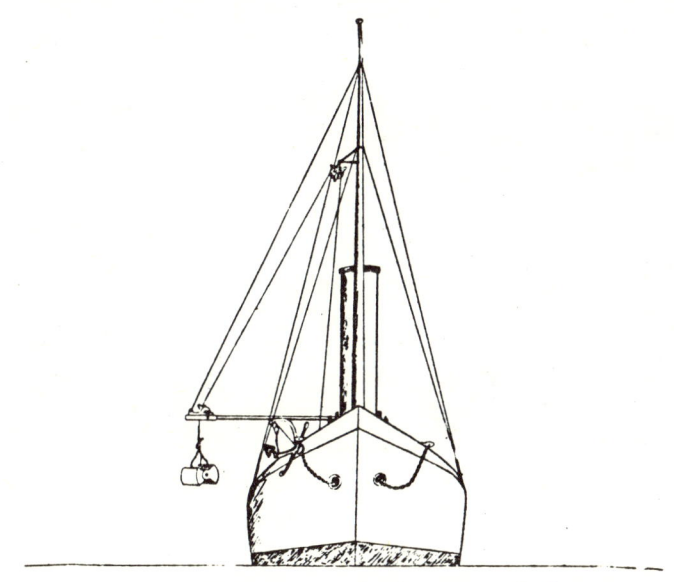

Fig. 5. The bottom sampler 0,₁ sq. m. ready for use.

Fig. 5 shows the small 0,₁ sq. m. bottom sampler hoisted up to the beam from which it is to be lowered down; on striking the bottom, a clutch is released, and by hauling on the line, the apparatus is made to close, by its own weight, about the section of bottom embraced, before it is lifted from the ground. On hard sandy floor it will not dig so far down as in softer bottom; for taking the organisms which live deep down in the sand, it is necessary to use continually larger and heavier implements. We have also apparatus of this type on board.

By taking a number of samples with the 0,₁ sq. m. apparatus, a good idea is obtained as to the common animals on and in the bottom; the number and weight per 1 sq. m. can for instance be calculated with a high degree of accuracy. On very stony bottom, of course, the sampler cannot be used, and the same applies to places near the coast where there is a particularly thick growth of *Zostera*. In such cases, other types of apparatus are called into play; these will be further described in Section VIII b.

In shallow water, it is an easy matter to take some hundreds, or even thousands, of samples in a single day; there is, however, a practical limit to the number that can be handled, owing to the work of sifting and preparing the specimens taken — for almost every single sample taken is found to contain some animals; I do not remember, in all my experience, having taken more than one or two samples containing none. At times there may be only a few, but generally they are numerous. About 50 samples can be taken and dealt with in a single day as a rule, but this means ten or twelve hours of uninterrupted work.

Several thousand samples have been taken in Danish waters with this bottom sampler, and the specimens counted, weighed, noted, and a considerable quantity of them preserved.

As already mentioned, apparatus of this type will not take all the animals living on the ground, and consequently, if we desire to procure information as to the entire fauna of a given water, it will be necessary to have recourse to other implements, and other sources, e. g. the fishery statistics. This will be further referred to in the following pages.

II. The Sea Bottom and its Vegetation, and Importance of the same for Animal Life.

The nature of the sea bottom in the Danish waters has been closely investigated, partly through the work of the Danish Naval Department in drawing up marine surveys; further also by the Danish fishermen with their fishing implements; and finally, by the scientists with their special apparatus, the dredge and the bottom sampler, so that it is an easy matter to give a rough idea of the general features.

On the coasts themselves, the sea floor consists of the same material as the land, viz. sand, gravel and stones, either mixed, or separately. The finer particles, on the other hand, of clay or mud, with quantities of minute organic remains, are not deposited on coasts where there is any current or wave action which could carry them away, but are precipitated for the most part in deeper water, where there is less disturbance. It is therefore only in quite small and sheltered bays, or landlocked waters, that mud is met with at slight depths, say of 3—4 metres, as for instance in the Bredning of the Roskilde fjord at Roskilde, and in many other fjords, as also in the more sheltered parts of the Great and Little Belts (see chart II, Report XXI). Out on the more open coasts, the finer particles are met with only in deeper water; first mixed with sand to a greater or lesser degree, and later, in the deepest part of the eastern Kattegat, in the form of almost pure bluish-grey clay, which is as soft as gruel in the upper strata. It is especially in the eastern Kattegat and in the Skagerak, that this blue clay is found, and from there it passes, through all transition stages, to sandy clay and pure sand, in the southern

Kattegat and the northern part of the Sound. In the Skagerak, there are extensive tracts of clay in the deep water, — and we find here, in the deepest parts, abt. 700 metres depth.

In the North Sea, on the exposed west coast of Jutland, on the other hand, and far out at sea, there is no clay deposit, as the depth is here but slight, 20—40 metres, and the action of the waves heavy, so that we here find sand, gravel and stones alone. The more distant parts of the North Sea need not be referred to in this connection.

There is thus but little variation in the nature of the bottom in our waters; we must imagine it as consisting of flat plains of sand, mud or clay, or transition stages between these, with or without stones. These last are met with almost everywhere, save where the clay or mud deposits are so thick as to cover even the largest stones; this state of things is mostly encountered in the deepest and calmest hollows.

The vegetation, however, does present a certain variation, and may be divided into two groups, 1) flowering plants with roots in the bottom itself, to which the *Zostera* belongs, and 2) the lower plants (algæ) without roots, which are often only attached to objects on the bottom, such as stones, shells and the like, or at times themselves lying loose on the bottom.

Of all plants in the Danish waters inside the Skaw, the *Zostera* is the most common, despite the fact that it does not extend out beyond a depth of abt. 14 metres, and even this only in the most open and transparent waters; in the fjords, it goes out as a rule only to 4—5 metres depth. It is evidently the amount of light which determines the vertical range of this plant. Of the algæ, which are often of the brown and red colours, several penetrate out into deeper water, but beyond 40—50 metres their occurrence is only slight, and even in shallower water it is almost a rare occurrence to find them in the mass brought up by the bottom sampler. In quite shallow water, on the other hand, of less than 1 metre's depth, there is often, in sheltered, waters, a surprisingly rich and rapidly growing algal vegetation in summer; but it is undoubtedly the *Zostera* which produces by far the greatest part of the vegetable matter in our waters. This plant has accordingly been made the subject of special study (see Mindeskrift for Japetus Steenstrup I. Del IX: Om Bændeltangens *(Zostera marina)* Aars-Produktion i de danske Farvande, by C. G. Joh. Petersen, p. 1—20, 1914) and not only the existing quantity, but also the annual production, has been determined by examination of the rhizomes, which bear joints showing the number of leaves annually produced. The winter joints of the rhizomes are fewer and shorter than those of the summer rhizomes, and it is therefore possible, from specimens in a good state of preservation, to count the number of joints — and leaves, these being one to each joint — produced each year. See accompanying figure 6.

It is shown on the chart, that the area covered by the *Zostera* inside the Skaw amounts to something like 2000 square nautical miles, and since the mean annual production may be reckoned as abt. 1200 gr. dry matter per sq. metre, the total

1912	1911	1910

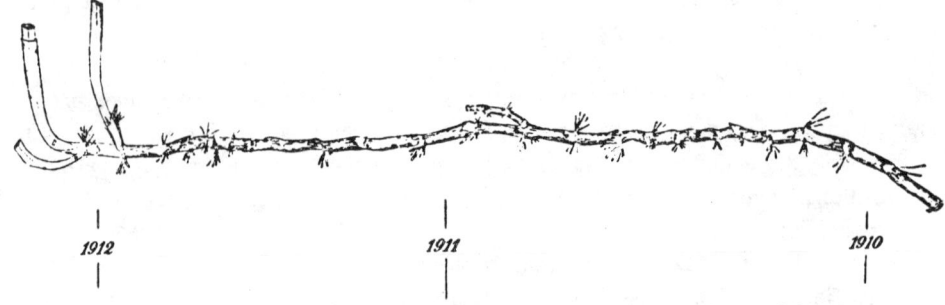

Fig. 6. Rhizome of *Zostera*, from Ebeltoftvig, 18 Maj 1912. Half natural length.

annual production will thus be over 8,000 million kg. dry matter, or about four times more than the quantity of hay annually produced by Danish fields and meadows.

The percentage of dry matter in the *Zostera* is abt. 16, and the plant, in a fresh state, contains a considerable amount of nutritive matter (see K. Rørdam, p. 61 in the present paper).

The *Zostera* grows only in about $1/_7$th of the water area from the Skaw to the Baltic, and the remaining vegetation, in deeper water, is, as mentioned, very scarce; nevertheless, the vegetation of the sea bottom must be regarded as the chief source of nourishment for the bottom fauna in all these waters. As to the importance of the plankton in this respect, very little is known at all (see Boysen Jensen, Studies concerning the organic matter of the Sea Bottom. Report XXII, p. 1—15); I have therefore disregarded it altogether in the present connection.

The bottom vegetation is, however, as a rule not consumed in a living state, or even on the spots where it grows, but is spread, after death of the plant — and to some extent also while still alive — throughout the entire area, though probably for the most part in the form of detritus, fine as dust, after the plants have been broken up by action of the waves or other means, and the fragments distributed throughout the water (see Boysen Jensen, Report XX, p. 7—34).

It was through examination of the stomach contents of common marine animals that we found, that the food consumed does not as a rule consist of plankton organisms to any great degree, but rather of a fine detritus, closely corresponding to the upper layer of the bottom itself. The existence and nature of this upper layer was determined by means of a glass tube, thrust down into the bottom, and bringing up a sample, so that the different layers could then be examined under the lens, and afterwards by the microscope. I had never imagined that the food of the animals in question could consist of the black malodorous mass of sulphureous mud often brought up by the dredge from the bottom, but the glass tube samples showed that above this lay a thin layer of different composition, brown or greyish in colour, containing, besides fine, inorganic particles, also minute vegetable remains, the structure of which often showed that they had originated from the vegetation of the sea bottom along the coasts.

Even before this, I had again and again found it surprising that oysters should never be found with anything in the stomach beyond a mass of generally indeterminable matter, the significance of which I could hardly appreciate; I was then of opinion that oysters lived on the plankton organisms which are so easily recognisable in the water, and that it was my misfortune to have been unable to discover them in the stomach. It was not till long after that I realised that the plankton organisms only comparatively rarely enter the stomach of oysters and other bivalves living on the bottom, the true food of these in fact consisting chiefly of the fine detritus dust which is found, either in suspension in the water, or deposited as the thin upper layer of the bottom itself, lifting and spreading at times, in stormy weather, but only to be precipitated anew later on. Then at last one day in the Limfjord, when Dr. Johan Hjort was on board, a happy thought led me to seek for the upper bottom layers with a glass tube fixed to the lead weight. Since then, I have frequently found these thin layers brought up in the bottom samplers, without being washed away by the water in hauling up.

The manner in which certain mussels and worms literally stuff themselves with this upper layer of fine detritus has been described by H. Blegvad (Report XXII, p. 54—61).

In the plant region proper, near the coasts, there are several animals, especially among the molluscs and crustaceans, which very largely feed upon living plants; here, however, it is as a rule chiefly the softer plants (algæ) and parts of the same, which are devoured, as also the tiny microscopical vegetation, found either on the larger plants themselves, or forming a slimy covering on all objects exposed to the rays of the sun.

That many of the marine animals are carnivorous, or predatory, is another matter; the great buk of the bottom animals are, and must necessarily be, herbivorous.

III. The Animal Communities of the Level Bottom.

It is an easy matter to form a correct idea as to the main features in the Danish vegetation on land, and characterise the different areas as meadow land, moorland, cultivated field, and the like; it is quite another matter, however, when we come to consider the marine fauna from a similar point of view, for the simple reason, first of all, that we can only see a very small part of the sea floor, in the immediate vicinity of land. It is only, therefore, by examination of numerous bottom samples, that we are able to realise that it is possible at all to distinguish between a small number of definite types of animal communities, as characterised by the principal and most commonly occurring species therein contained.[1]

---

[1] It might perhaps seem more natural to compare the fauna of the sea bottom with the fauna upon land; as a matter of fact, however, strange as it may seem, there does not exist any

This idea first arose in the course of the work with the bottom sampler. The original aim of these researches was merely to determine the amount of animal matter (fish-food) at different places, in a purely quantitative respect, but the experience gained with these samplers has gradually shown that certain common animal species are found to be distributed throughout areas so large and continuous that they can actually be charted in our waters, and quite empirically, the bottom sampler showed what species predominated in each area. Time after time the species in question appeared in the samples, with such regularity that we could always reckon on finding them, if not in one sample, then certainly in the next, as long as we kept to the area concerned. It must be borne in mind that the samples are quite small, only 0,1 sq. m., so that the animals must lie very close together on the bottom to be taken in every single sample; several, however, of these characteristic animals do lie so close, though the larger ones, such as *Brissopsis* and *Echinocardium*, do not always do so.

By means of some 10—12 different species, we learned in course of time to distinguish between 8 animal communities from the deep water of the Skagerak to the Baltic, each having its own separate area or areas, which, it was found, could easily be marked off on an ordinary chart. It was not always easy, however, at first, until the idea of epifauna had arisen and been understood; this will be further dealt with in the next section.

The character of these eight communities will be best made clear by actual illustration of their appearance (see Plates I—VIII appended to the present paper, and the Chart).

Each of these plates represents an area of $1/4$ sq. m. of bottom, or $2^1/_2$ samples with the 0,1 sq. m. sampler, with the animals thereto pertaining, sifted off; these have been drawn, in their actual numbers and actual size, on paper of $1/4$ sq. m., the drawing being then photographically reduced to abt. $1/6$th of the original area, so that both the area and the animal content appear in abt. $2/_5$ of the linear dimensions. Each plate is based on the animals taken from at least ten, or often more, samples taken with the bottom sampler close together on the bottom, and within the area of a given community. The illustrations are thus not real photographs of the animals naturally grouped upon $1/4$ sq. m. of bottom, since the great majority of the animals in question lie buried in the mud, only maintaining, by some means or other — it differs in the various forms — some slight connection with the surface of the sea bottom and the water above it. The plates are thus really a sort of composite average pictures of the normal stock, and they have been so often revised in the course of these investigations, that I dare maintain them to be not far from the actual conditions. A more serious difficulty is the fact that they can only represent so small an area of ground, 0,25 sq. m.; this means, that

---

survey of the animal communities on land based upon quantitative investigations of the commoner species. In this connection, reference may be made to F. Dahl: Über die Fauna des Plagefenngebietes, 1912. (Beiträge zur Naturdenkmalpflege. Conventz. Bd. III: Das Plagefenn bei Chorin.)

the somewhat rarer forms, which may also be characteristic for the different communities, are not included. They will be found noted: howewer on the lists in Report XXI. The larger predatory animals, then, in the different communities, are not shown in the illustrations; they are of course not to be found on cvery 0,₂₅ sq. m. of bottom, and should best be studied by other means than by the bottom sampler, which is the more easy, since they do not as a rule lie buried in the bottom, but move above it.

Take now the four communities from the deepest water; Nos. 4—7, Pl. IV—VII. All these live in the soft bottom; Nos. 5—7 in pure soft clay, 4 in clay mixed with sand, or sand with an admixture of clay. Immediately noticeable here is the great number of *Ophiuridæ*, though in No. 6, this is less marked than in the others. In every single sample with the 0,₁ sq. m. bottom sampler, one or more of these *Ophiuridæ* will be found, but in No. 7, the deepest community of the Skagerak, it is *Amphilepis norvegica* which takes the prominent place. In No. 6, the next-deepest community from the Skagerak, *Ophioglypha sarsii* is the most frequent, and we find here also two *Amphiura elegans*. No. 5, the clay community of the eastern Kattegat, has *Amphiura chiajei* in the principal place, and No. 4, the next deepest community of the Kattegat, shows *Amphiura filiformis* predominant over all others. In No. 7, we find the small transparent bivalve *Pecten vitreus* constantly and numerously represented, and when we designate this as the *Amphilepis—Pecten* community, the meaning will readily be understood; it is obviously necessary to have some name, not too long, when writing or speaking of these communities. No. 6 I have called the *Brissopsis-Sarsii* community, from the *Brissopsis lyrifera* and the *Ophioglypha sarsii*, which are always found together here; *Brissopsis* also occurs in No. 5 together with *Amphiura chiajei*, and I have termed this the *Brissopsis-Chiajei* community, while No. 4, with *Echinocardium cordatum*, and *Amphiura filiformis*, is noted as the *Echinocardium-Filiformis* community. Abbreviating these terms, we may refer to the four communities here mentioned as **Al. P.** No. 7, **B. S.** No. 6, **B. Ch.** No. 5 and **E. Fil.** No. 4.

Various other species are characteristic for each of these communities, as is evident from their occurrence in the samples, but in the first place they are less numerously represented, and in the second, it will not do to make the terms employed too long, and I therefore omit further reference to them here. Other species, again, are found in certain of these communities, but are not characteristic of the same, as they also occur in others. This is the case, for instance, with the small white bivalve *Abra nitida*, which occurs in great quantities in No. 6; it is also found right up in the Limfjord, in other communities. Some species, indeed, such as *Terebellides strømi*, appear to live everywhere, from the Baltic to the deepest parts of the Skagerak. Experience has shown us which species occur together and form the natural communities, and these are as a rule the species most predominant both in bulk and number; possibly also the species which are characterised by greater longevity of the individuals.

No. 3, Pl. III, is a community from the sandy part of the Kattegat, and as a characteristic species here, we often find *Echinocardium cordatum*, as also in the previous community, but there are likewise the small bivalves *Venus gallina* and *Tellina fabula* and other allied species; I have called it the *Venus* community, with or without *Echinocardium*, using the abbreviation **v. ± E.**; as the latter may be lacking at the shallowest stations. Instead of *Amphiura filiformis*, we here find *Ophioglypha albida* and other species of the same genus. — Up in the Belt Sea, this community and the preceding one are replaced by a related group, No. 2, Pl. II, likewise with or without *Echinocardium*, but with *Abra alba* as characteristic species; in addition, we here find at times other bivalves, such as *Macoma calcarea* and the three species of the genus *Astarte*, in great numbers; I have called this the *Abra* community, with the species mentioned as accompanying forms, using the abbreviation **b. ± E. a. c.** Near the coasts, and in the fjords, this community terminates as a pure *Abra* community, noted as **b.**, without the accompanying species before mentioned. — The community shown in No. 1, Pl. I, is found right up to all our coasts in the Belt Sea, and extends all round the Baltic; it also occurs, though less strongly, and only in some spots, along the coasts of the Kattegat and in all the fjords of Jutland, even on the coasts near Esbjerg, and, at any rate in 1914, in the greater part of Ringkjøbing Fjord. The characteristic species here is a small red or yellow bivalve, *Macoma baltica*, accompanied as a rule by *Cardium edule* and *Mya arenaria*. The sand worm, *Arenicola marina*, also, is very often found here, but may also go out into the *Abra* and *Venus* communities, e. g. in the Kattegat. In quite shallow water, the small gastropods *Hydrobia ulvæ* and *Hydrobia ventrosa* also occur in this community, which I have called the *Macoma* community, abbreviated **d.**, but which might perhaps better be termed the Baltic community, from *Macoma baltica*, its most characteristic species, and also because it is extremely widely distributed especially in the *Mare balticum*. This community appears in many forms, according as it is found on sandy or muddy bottom, or as the bottom is covered by stones or plant growth, and it is therefore not easy to recognise it as a distinct community; the point will be further referred to in the next section. It is found both in shallow and deep water in the Baltic, out to abt. 40—50 m., but in the Kattegat, it occurs only in water of a very few metres' depth, and only in certain spots; on the west coast of Jutland also, it is only met with in some particular places, and quite close to the shore; it is found in very brackish water, of far less than 1 % salinity, but also in water of over 3 % salinity, e. g. the Thyborøn Canal. Salinity, temperature and nature of bottom thus seem to be of no importance for its existence, and the fact that it does not occur in many other places, as for instance, the middle parts of the Kattegat, is altogether hard to explain. We shall have occasion to refer to this later on.

This community is generally noted as **d.** in the Reports.

Of echinoderms, there are in this community often none at all, as for instance in the Baltic and close to the coasts; in other parts, we find *Asterias rubens*, which may be met with in almost all communities except the very deepest.

A single local community. No. 8, Pl. VIII, is met with on the soft bottom of the south-eastern Kattegat, in a large continuous area, where the small *Haploops tubicola* builds its vertical tubes out of the soft clay bottom, in such numbers — up to abt. 3500 per 1 sq. m. — that the character of the bottom is altogether altered thereby, and the usual *Brissopsis* community therefore partly disappears. Instead, we here find *Pecten septemradiatus*, which lives a more or less natatory existence, *Lima loscombii*, and other forms. This community has been called the *Haploops* community, (abbreviated **Ha**), but possibly it may not be of the same importance as the others; the local nature of its occurrence might seem to suggest this.

It should be noted that in the Kattegat, we here and there find a sand community related to the above-mentioned *Venus* community, where *Echinocardium cordatum* is replaced by *Spatangus purpureus* and *Echinocardium flavescens*, and where bivalves such as *Psammobia færoensis*, *Abra prismatica*, *Mactra elliptica*, etc. occur. I have called it the deep *Venus* community in the Reports; it is of but slight importance inside the Skaw, but seems to be widely distributed in the North Sea. It will not be further referred to here. This, then, is the 9th community of the level bottom, noted as (**v.**).

We have the following communities on the level bottom:

I. The *Macoma* or Baltic community, **d.**, on all our southern coasts, and in the Baltic.
II. The *Abra* community, **b.** ± **E.**, especially in the Belt Sea and the fjords.
III. The *Venus* community, **v.** ± **E.**, on the open sandy coasts of the Kattegat and in the North Sea.
IV. The *Echinocardium-Filiformis* community, **E. Fil.**, at intermediate depths in the Kattegat.
V. The *Brissopsis-Chiajei* community, **B. Ch.**, in the deepest parts of the Kattegat.
VI. The *Brissopsis-Sarsii* community, **B. S.**, in deeper parts of the Skagerak.
VII. The *Amphilepis-Pecten* community, **Al. P.**, in the deepest water of the Skagerak.
VIII. The *Haploops* community, **Ha.**, locally in the south-eastern Kattegat.
[IX. The deep *Venus* community (**v.**), sporadically in the Kattegat. Not grafically shown here.]

Transition stages between the successive communities are doubtless found as a rule in nature, but the transition areas seem to occupy only a very small part of the whole, and could not therefore be indicated on the accompanying chart.

How far I have been able to find the correct delimitation for the animal communities above mentioned, it is not easy to say, until investigations have been carried out in numerous other waters on the coasts of Europe; a cruise with the „Sallingsund" however, to South Norway and the Christiania fjord showed that the same communities are also recognisable there, and my experience off the west coast of Jutland tends in the same direction. I am therefore inclined to believe that these communities are of wide geographical distribution, some more so than others. (See Appendix to Report XXI, 1913, with chart of their presumed distribution from the coast of France to the North Cape, etc. The Appendix is issued with Report XXII).

My original aim with these investigations was to ascertain roughly the weight and number of animals per sq. m. in our waters. This it was only possible to accomplish by dividing the whole into areas with uniform population of animals; i. e. according to the communities. And this I have, in the main, succeeded in doing, as will be seen from the chart appended to the present Report. As to whether I may have drawn too narrow limits for these communities, i. e. established more of them than necessary, this will depend upon the purpose to be served. I did not originally seek for these communities, for the simple reason that I was not aware of their existence; they have come to light as a result of the investigations. We could, of course, by taking related communities together under one head, reduce the number to be dealt with, but it would not be advisable to do this before it has been seen whether or no the communities as here established are recognisable as such elsewhere in large areas, and thus having geographical importance, which I believe will prove to be the case.

IV. The Epifauna.

We have seen in the foregoing section, that where the sea bottom is level, i. e. formed of fine sand, mud or clay without foreign bodies of any considerable size, the animal population is uinform throughout large tracts. But wherever any such objects are found, even a stone or shell which is not too frequently disturbed by the action of the water, we find as a rule animals of quite another type attaching themselves thereto and living thereon. It is much the same as in the case of a moorland tract with boulders lying about here and there; we do not expect to find heather growing on the stones, we encounter lichens instead. The animals of the level sea floor, with the exception of the predatory species, live as a rule buried in the bottom; the animals of the other category are found on the bottom. And where suitable conditions prevail, they may lie so closely packed as altogether to displace the ordinary fauna of the level bottom. This type of animal life I have called the Epifauna, as living upon or attached to other objects.

Animals belonging to the epifauna can attach themselves to stones, shells of other animals, living or dead, and to plants, as for instance the waving meadows of the *Zostera*, and differ in character accordingly; the nature of such an epifauna is, however, to a certain extent determined by the nature of the animal community of the level bottom upon which the epifauna itself is found.

In deeper and more saline waters around our coasts we find, for instance, one epifauna whose most important characteristic species are *Modiola modiolus*, *Ophiopholis aculeata*, *Trophonia plumosa*, *Balanus sp.*, accompanied by various other species, more or less mixed with the fauna of the level bottom. Plate IX. This *Modiola* Epifauna (**M.**) is found in the southern Kattegat almost exclusively on the *Echinocardium-Filiformis* community, and to some extent also on the *Venus* community, and extends, in the Sound, the Samsø Belt and the Great Belt, into the continuation of the first-mentioned community, the *Echinocardium-Abra*, but never into the *Macoma* community, nor have I ever found the *Modiola* epifauna in deeper commuties with *Brissopsis*. Other finds of this epifauna are noted on the chart appended.

The *Modiola* epifauna occurs especially where there is a strong current close to the bottom, clearing away the finer particles, so that stones and shells lie exposed, affording sites for attachment for the *Modiola*, which may here occur in such numbers as entirely to cover the bottom (see Pl. IX) for great tracts in the current channels. See chart, where *M*. indicates the rich *Modiola* stations.

The quantity of animal matter in this epifauna can be, and is as a rule, far greater per 1 sq. m. than the communities of the level bottom ever attain to[1]); this is doubtless connected with the fact that in these current channels, fresh detritus is washed to and from over the animals as they lie, thus affording them a constant and abundant supply of fresh food.

In deeper waters, the epifauna has as yet been so little studied that I will omit discussion of these for the present, merely mentioning that we shall here doubtless find *Crania anomala, Terebratulina, Gorgonocephalus* etc., *Paragorgia, Lophohelia*, and many other forms, as important constituents.

We know, then, but little as to the quantitative occurrence of the epifauna in deep water; in shallow waters, on the other hand, in the *Macoma* community, the question has been very well investigated. Several different types of epifauna are met with in this community. I would here call to mind that there is a special epifauna attaching to our harbour moles in very salt water, with laminaria vegetation, and another with hydroids etc. on the stones under the steamer quays, where the timbering entirely alters the light conditions; these are, in Denmark, of so purely local occurrence that they may be regarded as insignificant, but they should be studied om the rocky coasts of Norway and Sweden. We do find, however, commonly distributed along our coasts, especially where stones or shells abound, and where the action of the waves is not too violent, an epifauna whose characteristic species is the edible mussel, *Mytilus edulis*, accompanied by *Balanus, Littorina*, etc., and another epifauna associated with the rapidly fading leaves

---

[1]) Owing to the difficulty of preparation for photographing after counting, I have here merely photographed a single bottom sample of $0_{,1}$ sq. m. on Pl. IX, just as it came up from the water, save for a preliminary rinsing. It gives a good idea of the appearance of the epifauna, but does not correspond altogether either in respect of species or of numerical values to the accompanying example of valuation of this epifauna with *Modiola*, nor was it taken at the same place.

Pl. X, the *Mytilus* epifauna, on the other hand, corresponds more nearly to the valuation example; the animals are here less tangled together. The photograph is likewise based on the quantity of animals derived from $0_{,1}$ sq. m; the valuation example, on the other hand, gives the animals from $0_{,25}$ sq. m., i. e. $2^1/_2$ times as many as shown.

of the *Zostera* vegetation, the characteristic species here being especially the small *Rissoa membranacea* and *R. inconspicua*, *Idothea*, and many other small species well known to Danish zoologists. These two types of epifauna are very closely related, and are met with practically speaking only within the area of the *Macoma* community; they can, however, follow the *Zostera* out into the Kattegat where it enters the area of the *Venus* community.

The mussel, *Mytilus*, is found in both these types of epifauna, but can of course never attain any age on the short-lived *Zostera* leaves; it falls off while still quite small, or is carried away with the drifting leaves to places outside the *Macoma* area, where, as a rule, it is not able to continue its existance. On the other hand, wherever the small young mussels find some more permanent site of attachment than the *Zostera* leaves, as for instance a stone, a dead shell or another live *Mytilus*, they will grow to a good size, and may in suitable localities cover large patches of the bottom, just as the *Modiola* outside the *Macoma* area. (See Pl. X).

If a pole be thrust down into the bottom at some place where the *Mytilus* breed, especially in the months of May and June, it will often, after a short time, be found quite covered with small *Mytilus* from 1—2 mm. long. If the pole has been set out too long before the breeding time, it will become coated with slime, formed by the minute algæ, and the young *Mytilus* will then hardly be able to find a hold, so that the mussels fails to attach. Again, even when the young have attached themselves to the pole in thousands, they may afterwards disappear, owing to the attacks of the starfish, *Asterias rubens*, which can at times devour the whole brood. I have also, in the Limfjord, seen the small young of the *Mytilus* become overgrown by colonies of *Botryllus*, and suffocated by them; they have thus to fight for their position with other organisms, both plants and animals, and again to fight for their life against the predatory forms. There are, however, enormous quantities of young to take up the battle, and in some places they gain the victory, but it is almost exclusively in the area of the *Macoma* community that this is the case.

V. How the animal communities are maintained.

It is clear then, that the character of the bottom is of fundamental importance for the presence or absence of epifauna; nevertheless, the succession of the various types of epifauna and of the communities belonging to the level bottom cannot be explained by the character of the bottom alone. It is a factor of importance, of course, for the communities of the level bottom, whether the bottom itself is composed of sand, mud, clay, or mixtures of these, but on the other hand, we find areas of pure sand, pure clay, etc. with entirely different communities in different waters, or at different depths. The pure sand, for instance, will in our more enclosed waters, and in the Baltic, give a pure *Macoma* community, while in the Kattegat, we find on the same type of bottom the *Venus* community. Pure clay, again, may in the eastern Kattegat be found to be the home of the *Brissopsis-Chiajei* community, while in the Skagerak, we find the *Brissopsis-Sarsii* and *Amphilepis-Pecten* communities also on pure clay bottom. One community, the *Macoma*, seems to be altogether independent of the composition of the level bottom, living equally well on pure sand, as for instance at Esbjerg, and on pure mud, as in our enclosed fjords.

Among other factors, besides the nature of the bottom itself, which determine the distribution of the various species, it is natural to consider the temperature and salinity as likely to be of importance; the latter, varying as we know, very considerably between Gedser and the Skaw, is doubtless largely responsible for various features in the occurrence of many species. The temperature, also, varying as it does with the seasons in shallow water, while remaining constant throughout the year at greater depths, is doubtless also responsible in some degree. Or more correctly, we find in these factors certain lines of geographical distribution running parallel with those of the animal communities, just as the various zones of vegetation on the slopes of a high mountain run parallel with the lines for certain physical features, such as temperature, snow limit, etc. — Whether the outer conditions, however, really are the directly influential, regulating factors, is another matter; sometimes, perhaps, they may be, but there are undoubtedly cases where they are not, and in all probability, the position will as a rule be simply that the outer conditions favour a certain species, which will then be able to displace another species, though the latter might equally well have thriven there had it not been thrust out by its rival.

Most common marine animals living on the bottom commence their existence as minute larvæ in the water, and sink to the bottom at a very early stage, as for instance the bivalves. And it is remarkable to note how in those communities where the *Amphiura* spread their arms abroad, forming a network in the bottom, (see Pl. IV and V) extremely few bivalves are found at all. The young will here doubtless as a rule be devoured, while still quite small, by the *Amphiura*, and only a very few individuals of certain species manage to survive.

Both in the shallow water near the coasts, and farther out where it is deeper, where few or no *Amphiura* are found, there are quantities of small bivalves (see Pl. VI and I, II and III) of many different species, for instance *Mactra, Tellina*, as also in summer on grounds near land, where few or no *Echinoderms* at all are found. That the great majority of these young individuals never attain full growth is doubtless primarily due to the fact that the environment is here unfavourable in the long run; the action of the waves, for instance, will at times be too violent, very low water will kill off numbers of the young, as also severe cold in winter, etc.; presumably the same factors which account for the absence of *Echinoderms* in the same localities. It is in such places as these that the species of the *Macoma* communities can live and thrive continually; they are the only forms that are able to withstand the severe conditions prevailing, in a degree sufficient to ensure the maintenance of the species. It is remarkable, having in mind the hardiness of these *Macoma* species, that they should not be found deeper out in the Kattegat, throughout the whole of the *Venus* area, where we might imagine they would find the most favourable environment of all, and where *Mytilus* also make their appearance on any buoy set out, but hardly ever live on the bottom itself. It cannot be the depth which keeps the *Macoma* species away from these areas; we find for instance *Mya arenaria*, *Cardium edule*, *Macoma baltica* and *Hydrobia* out in at least 20 metres' depth in the Baltic, where their predominance is undisputed; in the Baltic, however, east of Gedser, there are, as we know, no *Echinoderms*, nor are such found in the low water on the shores of the Kattegat. I must therefore suppose that it is just certain *Echinoderms* which prevent the animals of the *Macoma* community from spreading over larger areas than they occupy in fact.[1])

---

[1]) In a work by R. Southern: Marine Ecology. Clare Island Survey. (Proc. Royal Irish Acad. Vol. XXXI, 1915) the various types of local fauna in an area of the West Coast of Ireland, out to 50 fathoms' depth, are subject to detailed consideration. The writer notes the fundamental division of the bottom fauna into two groups of organisms, those living on the level bottom consisting of fine particles *(Microlithic)* and those found in parts where the bottom is formed of larger objects, stones and rocks *(Macrolithic)*. These two terms coincide very nearly with mine; the fauna of the level bottom and the epifauna; the writer also quotes the Reports XX and XXI from the Danish Biological Station. As far as I can see, the coast in question presents an enormous number of epifauna types, but they are not always distinguished from the fauna of the level bottom, which it would perhaps be very difficult to do here, without knowing the communities beforehand. — In Southern's lists, which are based on dredgings, and therefore often do not include the commonest of the forms which live buried in the sand, I recognise many of our species from the Danish *Macoma* and *Venus* communities; also, indeed, something approaching the *Echinocardium-Filiformis* community, including *inter alia* many *Turritella communis* and some *Amphiura filiformis*; i. e. properly speaking, only two or three well-defined communities in all. The *Venus* community, however, appears extraordinarily dominant. It is therefore hardly surprising that he has not been able to carry the description farther, as so few communities are well represented, and these, moreover, are masked by the epifauna. His comparisons between this fauna and that of other seas are somewhat lame; they ought properly only to be compared with the *Macoma* and *Venus* communities of other waters.

Southern concludes with some observations on the geographical distribution, and remarks that the limits generally drawn for „faunistic regions" are of no real importance, as every species has its own particular laws governing its occurrence.

I can to a great extent concur in this view, but must make an exception in the case of animal communities based on characteristic species, and would in this connection refer to the Appendix to the Report XXI (issued with Report XXII). From the chart in the Appendix, it will be seen that I had already at that time considered it likely that on the west coast of Ireland, in water not too deep, there would only be *Macoma* and *Venus* communities. I had on that occasion, from lack of closer investigations in the larger waters, taken *Amphiura filiformis* and *Turritella communis* as belonging to the *Venus* community, though they can and should, in waters where their distribution is known, be kept apart as a distinct community in themselves.

When a whole area, or nearly the whole, as in the case investigated by Southern, is covered by *Venus* communites – albeit comprising both the shallower, with *Venus gallina* and *Tellina fabula*, and the deeper, with other species – it becomes difficult for the investigator to appreciate properly the importance of the communities; this can only be realised by comparison between different communities in practice.

VI. The Sound.

The particular occurrence of the animal communities in the Sound, below Elsinore, I have considered worthy of more detailed consideration here; the water in question has, from the days of A. S. Ørsted, (circa 1844) to the present time been a constant field of research among the Zoologists of northern Europe.

1) South of Amager and Saltholm, the *Macoma* community dominates exclusively the entire water from the coast out to the deepest waters; it extends up north of these islands along both coasts, and over the shallower grounds out to abt 9—10 m. depth. At Elsinore, its area on the coast narrows considerably, but the community extends up along the shores of Sealand towards Gilleleje, out to abt. 12 m. depth. *Tellina tenuis* here becomes common, as on all the more open coasts; on the shores of the Kattegat, in salter water, the *Venus* community as a rule takes its place.

Below Elsinore, on the *Macoma* community, we often find great quantities of *Mytilus edulis* as a special epifauna. For the contents of this community see Report XXI, Appendix p. 53—54, St. 5, 6 and 8.

2) Outside the *Macoma* community, we find in the Sound, as generally also in the Belt Sea, the *Abra* community (l. c. p. 52—53, St. 3 and 7) accompanied by *Macoma calcarea* and *Astarte sp.* at depths abt. 10—abt. 15 m.; in deeper water, abt. 15—20 m., we encounter *Echinocardium* (l. c. p. 52, St. 2 and 9). The *Abra* community occupies the principal area of the Sound from Lomma Bay to Elsinore. The *Abra* community with *Echinocardium* extends from Elsinore down both sides of Disken to west of Hveen, and east of Hveen down to west of Skabbrefvet, off the Swedish coast. This last community surrounds the deep channel in the Sound (20—over 40 m.) from Elsinore along the coast of Sweden to north and east of Hveen. In this deep channel we find:

3) The *Echinocardium-Filiformis* community, with traces of *Haploops* and often with a rich epifauna of *Modiola modiolus* and its usual accompanying forms, especially *Echinoderms* (l. c. p. 51—52, St. 1 and 2). — Thoroughly detailed investigations of the Sound with the bottom sampler have not been made up to the present, but the various communities there residing have been sought for, and charted, on the basis of numerous samples from various localities, taken for the purpose. (See chart accompanying the present paper.)

We find then, in the Sound south of Elsinore, the following communities: On the coast: 1) The *Macoma* community, from 0—abt. 10 m. depth; at abt. 10—abt. 20 m. 2) the *Abra* community without, or in deep water abt. 15—20 m. with *Echinocardium;* then, abt. 20—40 m. 3) the *Echinocardium-Filiformis* community — exactly the same communities, and in the same order of succession, as at the northern entrance to the Belt Sea, at Sjællands Odde. In both these waters, the *Echinocardium-Filiformis* community has its two southern limits in towards the Baltic, and does not therefore occur with the quantity of *Amphiura filiformis* noted at the typical stations in the Kattegat proper.

Between Elsinore and Helsingborg, the current runs so swiftly along the bottom, despite the depth (up to abt. 40 m.) that the bottom is found to consist of hard, comparatively barren sand; the *Echinocardium-Filiformis* community is not met with here, but any valuation of the ground is extremely difficult, owing to the hardness of the bottom. At Lappegrund, however, we at once find the mentioned community again; the bottom in this deep water is soft, though not so soft or with so pure clay as up in the Kattegat, where we have the *Brissopsis-Chiajei* community.

Ørsted's „*Regio Trochoideorum*" coincides in the main exactly with the area of the *Macoma* community, both as regards the species comprised and the limits marked on his chart. He was even then able to perceive that this community occurs in greatly varying types of composition under different conditions: i. e. epifauna.

His *Regio Gymnobranchiorum* comprises only quite narrow strips of the chart from north of Saltholm to Kullen, and should most properly be regarded as parts of epifauna types belonging either to the *Macoma* community or to other deeper ones.

His *Regio Buccinoideorum* embraces all communities on the level bottom with *Modiola* epifauna outside the area of the *Macoma* community, and is thus of less importance here; it shows how poor an idea of the true state of things can be gained by the use of the dredge alone. (See A. S. Ørsted: De regionibus marinis. 1844.)

E. Lønnberg (Undersökningar rörande Öresunds Djurlif. I Meddel. från Kongl. Landtbruksstyrelsen, Nr. I, 1898 Nr. 43) distinguishes between the *littoral* fauna (or brackish-water fauna) and the *marine* fauna proper, of which the former more or less corresponds to the *Macoma* community, the latter covering all the rest.

W. Björck (Biologisk-faunistiska Undersökningar av Öresund, II. Lunds Univ. Årsskrift, N. F. Afdl. 2, Bd. 11, Nr. 7, 1915) distinguishes between the same two types, but designates the latter as the *sublittoral* fauna.

Both these writers have thus only two divisions of the fauna in the Sound, as regards vertical distribution, whereas I have three, or, dividing the *Abra* community into two, according as it contains *Echinocardium* or not, four in all. If we were to disregard this intermediate *Abra* community in the Sound, with its accompanying species such as *Macoma calcarea, Astarte banksii,* and *A. borealis,* the fauna of the Sound would not at all correspond to that of the Belts in deeper water. In the Belts, the *Abra* community is the deepest, but in the Sound, it occurs at intermediate depths; the deepest community of the Sound on the other hand, *Echinocardium-Filiformis,* has its proper counterpart in the Kattegat. The Sound thus contains one community more than does the Great Belt.

The remarkable *Abra* community in the Sound, with *Macoma calcarea* and

Astarte, has also been met with inside Drøbak (Report XXIII, 1915), and there are analogous occurrences in the enclosed Koljefjord near Lysekil, and at several other places. This point is further dealt with in Report XXIII.

Once the communities have been found and charted in a water, it is obvious that we can no longer be content to speak of the distribution of the fauna in general, either vertically or horizontally, as this will vary for each particular community: the proper course will be to consider the distribution of the communities themselves. We may speak of the distribution of the single species, or of that of a single community, but generalities with regard to the "fauna" are no longer justifiable. I cannot therefore agree in any delimitation by a straight line dividing the Sound into a northern and a southern section of different faunistic character, but can only maintain that here a certain community ends and here another begins; any straight line drawn through the Sound will always have parts of the *Macoma* community on either side. We are forced to admit that the distribution of animal communities is not definable by straight lines.

VII. Quantity of animals in the Kattegat.
(Pl. XI.)

Having seen how it is possible to determine the quantities of vegetation and of animal life belonging to the various species living in or on the sea bottom, we are naturally led to enquire about the quantity of fish found in the waters concerned. And here, at the outset, we are met by a serious difficulty in the fact that fish cannot be counted in the same way as the slow-moving animals or stationary plants; we have only the fishery statistics to rely upon, and these do not give us the size of the stock. They do, however, in the case of many fish species, doubtless give a good idea as to the quantity annually produced by the stock (see Section IX), as in a water like the Kattegat, we may safely reckon that the annual yield of bottom fish is very near the limit of what the stock can bear; the entire stock of food fish is probably not many times greater in point of weight than the figure noted in the annual fishery-statistics. Moreover, the Kattegat has been fairly well valuated with the bottom sampler as regards the bottom animals, so that we can use this water to give a rough survey of the quantity of organisms found.

Pl. XI shows, that the plants here represent several millions of tons (24); this figure applies, as a matter of fact, to all the Danish waters inside the Skaw, but a great portion of it at any rate will fall to the Kattegat. In the case of the bottom fauna, also, we have also to reckon with millions of tons; the useless species, which do not serve as food for others, and are thus not even indirectly of use to man, make up 5 million tons, and the useful forms, which furnish or may furnish food for the fish, 1 million. Fish, such as plaice, cod, herring, on the other hand, are far less numerous; according to the statistics, only some few thousands of tons (5—7000) for each species. Starfish make up 25,000 tons, i. e. more than all the mentioned food-fish together, while crabs and gastropods amount to no less than 50,000 tons.

The food fish, then, make up only an insignificant part of the total stock of animal life in the Kattegat, even reckoning the former as perhaps from four to eight times greater than the annual catch noted in the fishery statistics. This is undoubtedly an undisputable fact; it is another matter, however, to arrive at a proper comprehension of why it is so. In order to realise the position, we must further consider the various animal forms found in this water, and endeavour to ascertain what parts they play in the general metabolism.

Obviously, the quantity (the stock) of plants, and especially the annual production, must be greater than that of the animals, since the latter all subsist, directly or indirectly, upon the former, and for every kilogramme of animal life produced, more than a kilogramme of vegetable matter is consumed. Somewhat the same applies to the carnivorous animals; they all require several kg. of food-animals to produce one kg. in their own weight. If 10 kg. of vegetable matter is required to produce 1 kg. of herbivorous animal species, and these again must furnish 10 kg. in order to produce 1 kg. carnivorous animal life, this would mean, that 1 kg. of carnivore costs 100 kg. of plant stuff to produce. Given yet another link in the succession, it would cost 1000 kg. Now all fishes are carnivorous, as for instance the plaice, which feeds on small bivalves and polychæta, these again subsisting on minute particles of vegetable matter. The cod, on the other hand, devours chiefly larger animals, such as the larger crustaceans, large gastropods, other fish, etc. — all predatory forms in themselves, and thus all in the first instance costing 100 kg. of vegetable matter per kg. of their own weight. The plaice then, ultimately costs 100 kg. of plant stuff per kg., whereas the cod will cost 1000. As, moreover, there are, apart from the useful animals, also quantities of others serving no purpose as food for fish, it is obvious that the bulk of the fish must be very small indeed when compared with the quantity of vegetation and that of the lower useful animals.

The food fishes are thus, being predatory animals, far from economical to produce, and the quantities of other animals, especially of the useless species, produced in the sea, are far greater. In other words, the conditions in the sea are not calculated to produce any great number of food fishes per unit of area.

Up to the present, we have considered only the rough weight of the animals, and it is this also which is shown in the table; it is obvious, however, that from the economical point of view, a jellyfish weighing 1 kg. is far less valuable as a source of nutriment, either for man or for other animals, than another creature of the same weight, but with a greater content of organic dry matter, as for instance a crab. The amount of water contained in the animals must be subtracted from their weight in calculating their value in the metabolism, and the same applies to

Plate XI.

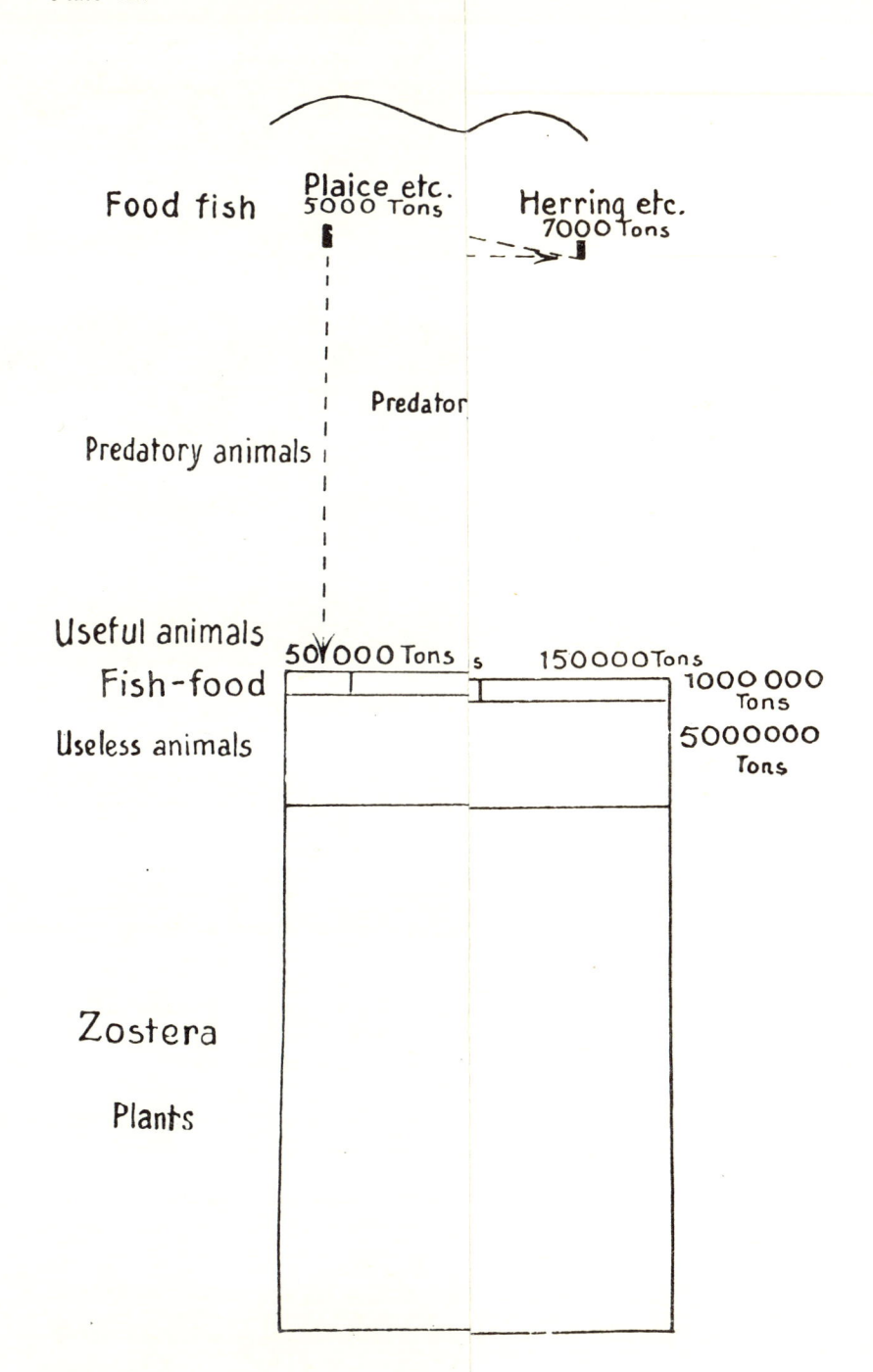

Food fish — Plaice etc. 5000 Tons — Herring etc. 7000 Tons

Predatory animals — Predator

Useful animals
Fish-food — 50000 Tons — 150000 Tons — 1000 000 Tons

Useless animals — 5000000 Tons

Zostera

Plants

the content of calcium carbonate. Water and lime are too easily accessible in the sea to be compared with the organic compounds of carbon and nitrogen. Analyses of the content of organic dry matter in the various species of animals have therefore been made. Using these results to recalculate the values in the table, the bulk of the fish will be found to increase somewhat in proportion to that of the other animals. Of the last, the useless animals in particular have a very poor content of organic dry matter, but the result; that the bulk of the fish is but slight in proportion to that of the lower animals in the sea, will not be essentially altered thereby.

By means of the various dry matter percentages for the different animal groups, we can often arrive at an estimate of the value of a certain investigated bottom area as a source of fish-food; animals such as the large bivalves: *(Cyprina, Astarte)*, and echinoderms *(Echinus, Echinocardium, Ophioglypha)* have a dry matter percentage of abt. 1—3 %, while small bivalves, *(Solen, Mya, Abra)* have abt. 5—8 %, and polychæta *(Annelidæ)* abt. 16—20 %. The echinoderms and the large bivalves are therefore of but little value, and are often entirely rejected by many fish. And it must also be borne in mind that though a given area of the sea bottom may have a great number of these large animals per hectare, giving an enormous dry matter content per unit of surface, this does not necessarily imply that the annual production is particularly large, as many of these animals are many years — possibly scores of years — old. The annual production of a grass meadow may perhaps be greater per hectare than that of an oak plantation. The polychæta and the small bivalves are doubtless as a rule the pasture grounds of many fish. In order to determine the animal production, it will be necessary to study each species separately; this, however, has not been carried out hitherto for the Kattegat; an example of production investigations will subsequently be given for the Limfjord.

Production investigations and percentages of dry matter give valuable information as to the quantity of fish food, but it is imperatively necessary also to ascertain what the fish actually have in their stomachs, and what they really digest, if we are to arrive at a proper appreciation of the part played by the different animals as sources of nourishment for the fish. Such investigations have also been carried out on a large scale (see Report XXII) and it has also been found necessary to investigate the manner of life among the various lower animals (see Report XX and XXII) in order to determine their respective importance in the metabolism of the sea, on the one hand as producers, and on the other hand as consumers of organic animal matter. It is only by means of all these detailed investigations that we have been able to arrive at the general view given in outline above.

VIII. a. Special Valuation of the Limfjord.

That the Limfjord was selected as the site of these closer valuation investigations is due to the fact that in 1908, extensive transplantation experiments were commenced there, at the expense of the Government, with young plaice, and it was necessary to ascertain what number of young plaice could properly be transplanted per hectare, without risk of their growth being impeded, or retarded by overstocking. Thisted Bredning, the most enclosed part of the Fjord, was chosen as the special field of investigation here, and also Nissum Bredning, from which latter water the young plaice were taken, and where the conditions are known to differ considerably from those in Thisted Bredning with regard to growth, and no. per hectare, of the plaice. From time to time, however, partial valuations have also been carried out at other places in the western part of the Limfjord.

Only those parts of the Brednings where the "Sallingsund" could be navigated were valuated, — the vessel draws 2 m. of water. That is to say, the shore grounds and the slope, with its often steep incline, were excluded, but the large areas of the Bredning outside — the parts where the plaice are mostly found — were examined.

The waters are well suited to the purpose of valuation investigations, as the bottom is level, and soft or sandy; only comparatively rarely were stones with epifauna encountered here. In Thisted Bredning, at first, 100 stations were taken, later only 50; in Nissum Bredning, on the other hand, generally 40, all of $0,_1$ sq. m. area, so that in all, we have only small areas of 10—5—4 sq. m. investigated each year, but as a rule both in spring and autumn.

Report XX contains calculations showing to what degree we may trust to the accuracy of such valuations; these calculations were kindly made by Prof. W. Johannsen. As was to be expected, the degree of accuracy is seen to differ for the different species, according to whether they are numerous, as generally the smaller animals, or more rare, as generally the larger animals, and also according as they are found to be distributed evenly over an area or collected in groups. In the case of the species affording nourishment for the eel and the plaice, the degree of accuracy is very high.

In order to give some idea as to the species more particularly concerned in these valuations, Plate XII has been drawn up, with names of the species in English and Latin. All the animals are shown in their natural size.

Plate XII shows in natural size all the commonest forms met with on the soft or sandy bottom in the deeper parts of Thisted Bredning. Most of these animals are mentioned in the present paper on several occasions, and are ordinarily found in the commonest communities; they will therefore be somewhat further described here.

No. 1: *Trochus cinerarius*, is a small gastropod, occurring in large quantities on the *Zostera* in the Limfjord.

Nos. 2 and 3: *Nassa pygmæa* and *N. reticulata*, are small rapacious gastropods, both of which attack the plaice and other fish caught in the nets, and eat away the flesh.

No. 4: *Buccinum undatum*, the common whelk, is also a predatory gastropod, which attacks the captured plaice and other fish: it is itself largely eaten by the large cod.

No. 5: *Littorina littorea*, the common periwinkle, is found on all coasts which are protected to any degree; it is harmless, but not easily eaten by the fish, owing to its thick shell.

Nos. 6 and 7: *Acera bullata* and *Philine aperta*, are a couple of small gastropods, which appear in certain years in great quantities, and serve as food for the eels and other fish. It is possibly the oily flesh of these animals which renders the flesh of the eels in some years yellowish, and less well-flavoured, for a season. No. 6 has large flippers with which it is able to swim.

No. 8: *Abra alba*, a small white bivalve, the favourite food of the plaice in these waters. It has a thin shell, and grows very rapidly. In the Limfjord, it never attains an age of more than abt. 2 years, presumably because it is devoured within that time. From spring to autumn, the Bredning can be filled with masses of these bivalves, so rapidly do they grow. It is from this species that the *Abra* community has been named.

No. 9: *Cardium fasciatum*, is a cockle. Its near relation, *Cardium edule*, is larger, and is common on all our coasts; the two species are much alike in appearance.

No. 10: *Macoma baltica*, is a reddish or whitish, at times yellowish, thin-shelled bivalve, which is characteristic for the *Macoma* community; it furnishes good food for many fish.

No. 11: *Nucula nitida*, is a small thick-shelled bivalve which is only eaten by the fish in times of scarcity; it is very common in the Limfjord and other waters.

No. 12: *Solen pellucidus*, the razorfish or spoutfish, is a reddish, thin-shelled bivalve, much sought after by plaice; this form is also able to grow up in a very short time, though it takes some years to attain its full size.

No. 13: *Corbula gibba*, called by the fishermen of the Isefjord „Hampefrø" (hempseed) is reddish as a rule, thick-shelled, and of slow growth, but is eaten by the plaice; it forms, as do the common mussels, byssus threads, which in certain years give rise to the formation of a byssus ring in the mouth of the plaice, passing through the mouth of the fish and out of the gills. Such fish are popularly known as ring-plaice.

No. 14: *Mytilus edulis*, the common mussel.

No. 15: *Cyprina islandica*, hardly eaten by any food fish, owing to its thick shell. It attains a great age. The specimen shown is perhaps twenty years old.

Nr. 16: *Mya truncata*, the truncated clam, found on our coasts in shallow water. Its young, 16 a and 16 b, are a favourite food of eels and plaice; the eel seems also to be able to devour the larger specimens, though they keep far down in the mud or sand. The eels attack particularly their siphons, which reach up to the surface of the bottom.

No. 17: *Aphrodite aculeata*, a polychæt, called the golden sea mouse on account of the glittering golden colour of its stiff bristles. It is much sought after by the cod.

No. 18: *Nephthys coeca*, a polychæt. Good fish food.

No. 19: *Pectinaria koreni*, a tubiferous polychæt, of very rapid growth, lives in tubes built from grains of sand plastered together. Good fish-food.

No. 20: *Gammaridæ*.

No. 21: *Diastylis rathkii*. A crustacea, *cumacea*.

No. 22: *Idothea*. No. 20—22, good fish food.

No. 23: *Ophioglypha texturata*, the brittle star. Contains much lime, and is little use as fish food, but is eaten at times. Many of these brittle stars, often 100 to 200, are found on every sq. m. of Thisted Bredning; it is possibly these animals which by their numbers and voracity keep down the stock of small useful animals, devouring them while they are still quite small.

No. 24: *Asterias rubens*, the starfish, is a voracious predatory creature, attacking all animal forms which it encounters, from bivalves to half-dead fish. It is not eaten by any of our Danish fish species.

No. 25: *Echinocyamus pusillus*. A small greenish-yellow sea-urchin, of little economical importance.

No. 26: *Echinus miliaris*. A sea urchin of the small type, which can occur in great numbers on the *Zostera* leaves. Of little importance, being not usually eaten by the fish.

In 1911 (Report XX, p. 58—73) when dealing with producers and consumers i the Limfjord for 1909—10, I worked out a preliminary calculation over the metabolism among the animals living *inter alia* in Thisted Bredning, on the basis of valuation experiments from $1^1/_4$ years. We now know more about this question, especially about the stock of *Mya*, which is of great importance there, and I will here, without attempting at this early stage to give any new calculation for the entire metabolism, give a survey of the results attained up to the present. For further convenience, a survey of the stock of the principal groups of animals is here given, calculated as the mean of 7 years from 1910 to 1916.

Pl. XII. Biolog. Stat. XXV.

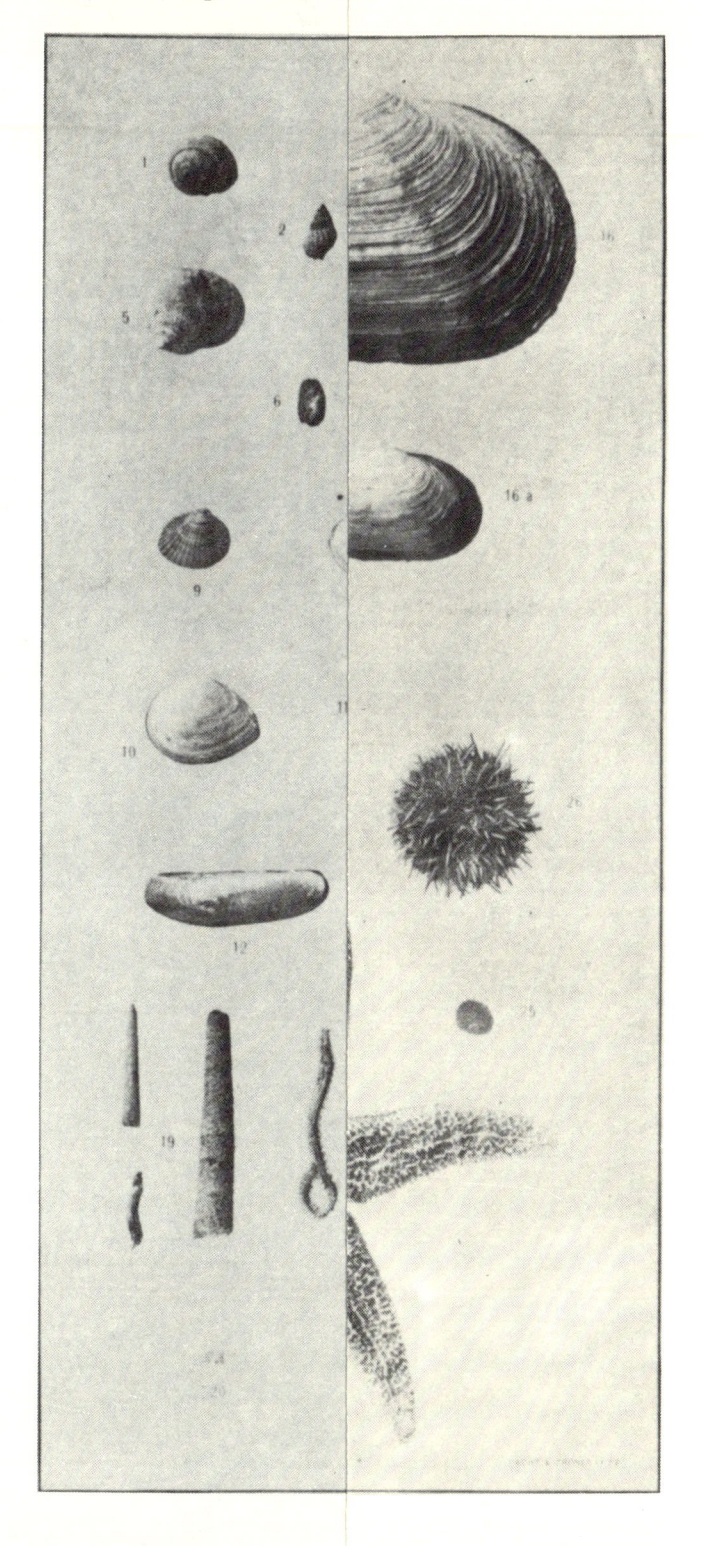

Principal groups of animals in the stock,

expressed in grammes rough weight per 1 sq. m., calculated as mean of the stocks from 1910—1916, excepting August 1915. The spring of 1910 in Thisted Bredning is likewise not here included.

The figures underlined denote quantities taken in Nissum Bredning, the others those from Thisted Bredning.	Thisted Bredning.	Nissum Bredning.	
Quantities in grammes.	gr. pr. 1 sq.	gr. pr. 1 sq.	
Large bivalves: Mytilus 38,35, 1,56, Modiola 0,02, 0,002. Cyprina 32,61, 0,83, Ostrea 3,46, 2,28.	74,44	4,672	Practically useless as fish food.
Mya truncata.	229,91	0	Little use as fish food.
Mya truncata juv. Mya sp. juv.	8,68	0,03	Fish food.
Small bivalves: Solen 9,22, 0,70, Corbula 1,78, 27,98, Nucula 4,21, 29,67, Abra alba 7,20, 3,42, Abra nitida 1,45. Macoma baltica 2,48. Cardium 0,32.	22,73	65,70	Fish food.
Small gastropods: Acera bullata 0,19, 0,03. Philine aperta 0,68. 0,83.	0,87	0,86	Fish food.
Polychæta.	10,62	9,77	Fish food.
Large gastropods: Buccinum 4,18, 4,06. Nassa reticulata and N. pygmæa 2,08, 2,70.	6,26	6,76	Predatory.
Starfish: Asterias rubens.	4,68	5,47	Predatory.
Brittle Star: Ophioglypha.	11,59	12,20	Predatory.
Spatangidæ: Echinocardium cordatum.	0	88,80	Useless.
Spatangidæ juv.: Echinocardium cordatum juv.	0,01	46,44	Useless.
Tunicates: Ascidiella.	0	30,53	Useless.
Average quantities taken in 1908 and 1909 in Thisted Bredning: Plaice 78,000 kg = 1,2 gr. pr. 1 sq m. Eel 35,600 „ = 0,5 „ „ Cod 50,000 „ = 0,8 „ „	2,5	?	

The area outside the 6 m. limit in Thisted and Nissum Brednings is respectively 65,000,000 sq. m. and 110,000,000 sq. m.

1 gr. pr. sq. m. in Thisted Bredning thus answers to a total quantity for the species concerned of 65 tons, in Nissum Bredning 110 tons.

A series of species belonging chiefly to the epifauna are not included in this list, **as they** represent, both in number and weight, but extremely small values.

Both Brednings belong, as far as regards their large middle area, to the *Abra alba* community, the one with *Echinocardium*, and the other almost altogether without this form; up in shallower water we may at times, in certain bays, encounter the *Macoma baltica* community, but this is of no importance in the present connection, save inasmuch as certain of the shallowest parts of the valuated area in Nissum Bredning belong to the *Macoma* community; in point of numbers, it is insignificant. In the western parts of Nissum Bredning there are also traces of the *Venus* community, but this is even more negligible here. In the main, then, both Brednings, as treated here, may be said to belong to the *Abra* community.

A transition area between the *Macoma* and *Abra* communities as met with *inter alia* also in the Belt Sea, is formed by the region with *Mya truncata;* it is found, far more frequently than would be imagined, along the slope on the hard bottom, where it keeps buried deep down, and is thus well hidden and difficult of capture. In the Limfjord also, it keeps to corresponding localities, but may at times spread out over the plains properly belonging to *Abra*; this is the case, for instance, in Thisted Bredning (see Tab. p. 27), where its bulk exceeds that of all other forms, with abt. 240 gr. pr. 1 sq. m., whereas in Nissum Bredning, only occasional traces of it are found. It is numerous as a rule, and very evenly distributed, in Thisted Bredning, and thus easily valuated.

It is otherwise, however, with some other of the large bivalves in Thisted Bredning, as *Cyprina*, on account of its rarity, oysters, which are relatively scarce and partly local in their occurence, *Modiola modiolus*, scarce, and finally the common mussel, *Mytilus edulis*, which occurs in groups. The average per sq. m. for these four species together is 74,₁ gr.; this figure is, however, as indicated, far from reliable. In Nissum Bredning these forms only amount to 4,₇ gr. pr. 1 sq. m.

Small bivalves of various species in Thisted Bredning amount to 22,₇ gr. pr. sq. m. and in Nissum Bredning 65,₇; these are the most reliable of all for valuation purposes.

Of the small gastropods *Acera* and *Philine*, we find but small quantities pr. sq. m., under 1 gr.; of polychæta, all species, but not including *Pomatoceros*, abt. 10 gr. in both places.

Of the predatory forms, *Asterias, Buccinum, Nassa*, there are 11–12 gr. pr. sq. m. in both Brednings; of *Ophioglypha* abt. 12 gr. and finally, in Nissum Bredning, we find the useless animals *Echinocardium* and *Ascidiella* with abt. 166 gr. pr. sq. m.

Common to these two waters, both rich in fish food, are the following:

Predatory forms, amounting to abt. 11–12 gr. pr. sq. m.

Ophioglypha,	"	"	12 "	"
Polychæta,	"	"	10 "	"
Small gastropods,	"	"	1 "	"
Small bivalves,	"	"	66 & 30 "	"

They differ in the quantities of large bivalves, of which Thisted Bredning has abt. 300 gr. pr. sq. m., while Nissum Breaning has hardly anything corresponding to this; on the other hand, the latter water has *Echinocardium* and *Ascidiella* in quantities answering to abt. 166 gr. pr. sq. m.

Investigations of stomach content in plaice and eels have shown beyond doubt that those very forms which occur in such great numbers, the large bivalves, *Echinocardium*, and *Ascidiella*, are of quite subordinate importance as food for these fish; the eels can certainly eat a large *Mya*, but it is impossible to say at present to what extent they ever do so. On the other hand, all the large bivalves are eagerly devoured by *Asterias*, and also to some degree by *Buccinum*.

The true food of the eels and plaice, then, consists of small bivalves, small gastropods, and polychæta, and these groups are also found to be well represented in both Brednings.

The importance of *Ophioglypha* in the general concourse of animals is difficult to determine; besides detritus, it also eats small young individuals of such forms as come down to the bottom, and is itself devoured, though to a slight degree, by plaice. Its food value is only small, but it occurs in numbers, up to abt. 200 pr. sq. m., and is doubtless very energetic; no small animal escapes its attention, though it does not devour them all.

It is very desirable to have reliable information as to the quantity of fish pr. sq. m. in these waters. In Thisted Bredning, the annual statistics often run up to abt. 35,000 kg. for eels, abt. 78,000 kg. for plaice, and in certain years abt. 50,000 kg. cod, besides herring, and other species with which we are not here concerned; altogether, abt. 2—3 gr. pr. sq. m.

As regards Nissum Bredning, no figures are available from the Fishery Statistics for corresponding information as to the stock; this consists, however, chiefly of plaice, which are many times, probably abt. 40—60 times, more numerous per hectare than i Thisted Bredning. To this I shall have occasion to refer later on.

The quantity of fish in Thisted Bredning seems thus but small in comparison with that of the bottom animals; in the case of the eels, however, the stock is probably several times as great as the quantity caught, whereas almost the entire stock of plaice is taken each year; there are hardly any plaice here beyond those purposely transplanted for restocking the water. If we now reckon that the number of silver eels annually escaping the fishermen's implements in Thisted Bredning is equal to the number of eels taken, this would give a total growth of abt. 70,000 kg. annually, and taking the remaining stock of younger year-classes as weighing abt. 3 times as much as this, then the average stock of eels will be abt. 280,000 kg., or abt. 4 gr. pr. sq. m. The average stock of plaice may probably be estimated at abt. 1—2 gr. pr. sq. m. and both together as abt. 5—6 gr. pr. sq. m.; a similar figure to that for the quantity of *Asterias*, but 4—5 times less than the bulk of the small bivalves.

These figures for the average stock of animals, however, give but a very vague idea as to what actually takes place every year in these waters; it would be better to know what each species produces each year, and what happens to the production in question; this applies more particularly to the species which serve as food for the fish. We must therefore have recourse to the more detailed tables, showing the size of the stock for the various species both in spring and autumn, in the different years since 1909, and see if we can find out anything from these.

Number and rough weight in gr. pr. sq. m. of the common bottom animals in Thisted Bredning.

A. = Number. V. = Rough weight in gr. pr. sq. m.		1909. Octbr.	1910. April.	1910. Octbr.	1911. May.	1911. Octbr.	1912. May.	1912. Octbr.	1913. April.	1913. Sept.	1914. April.	1914. Sept.	1915. June.	1915. Aug.	1916. May.	1916. July.	1916. Sept.
Oyster.	A.	0,1	0,2		0,1	0,1		0,4	0,2	0,2						0,2	
1. Ostrea edulis	V.	?	?		2,3	13,5		16,3	2,2	2,7						8,0	
Mussel.	A.	0,5	0,8	0,8	1,2	?	0,4	5,2	0,4	0,2	3,4	1,6	1,9		42,2	9,4	0,8
2. Mytilus edulis	V.	?	?	4,0	8,7	45,6	0,8	74,0	16,2	9,4	80,0	49,4	115,9		20,1	84,3	0,1
Great bivalve	A.	13,9	24,3	24,4	17,8	21,9	20,2	21,4	16,6	11,2	16,0	6,8	7,8	14,0	6,4	11,4	10,2
3. Mya truncata	V.	66,0?	199,6	250,0	225,0	294,0	304,0	306,0	270,0	190,0	312,0	93,9	192,8	268,0	140,0	246,0	165,1
4. „ „ juv.	A.	4,8	0,2	1,6	0,8	1,3	2,2	1,6	3,6	1,4	1,2	2,6	5,8	4,0	67,4	83,6	48,5
	V.	18,8	?	?	1,5	1,0	1,2	1,1	0,7	0,9	0,9	0,5	0,9	2,6	20,1	42,0	42,0
Small white bivalve.	A.	6,4		83,8	50,1	7,9	2,0	70,8	60,4	4,0	11,0	164,0	486,8	105,0	9,2	9,8	19,2
5. Abra alba	V.			11,2	9,3	0,8	×	6,3	6,0	0,9	2,0	10,3	44,2	27,8	1,2	0,4	1,0
Razorfish.	A.	29,2	38,1	9,8	15,6	226,1	226,0	79,2	103,0	62,0	52,8	61,2	15,6	14,0	12,4	4,2	5,0
6. Solen pellucidus	V.		4,6	1,7	4,0	19,4	32,8	14,6	20,0	3,6	11,6	5,0	3,9	3,4	2,1	0,9	0,3
Small bivalve.	A.	39,5	43,9	45,3	16,4	57,3	83,2	45,4	54,6	38,2	46,4	21,2	61,6	63,0	35,6	32,4	31,6
7. Nucula nitida	V.		3,6	4,7	1,6	5,9	7,1	5,2	5,3	4,2	4,8	2,1	4,7	5,0	3,1	3,0	3,2
	A.	0,3	0,6	0,2		50,4	161,8	11,2	90,4	22,6	57,2	10,8	43,0	15,0	21,0	26,0	19,4
8. Corbula gibba	V.			0,1		1,7	3,9	0,5	3,4	1,1	3,4	0,8	3,2	0,5	1,6	2,0	1,3
9. Cardium fasciatum	A.	2,6	7,1	2,5	0,9	4,8	7,2	1,8	1,8	1,2	9,4	5,6	2,8		4,6	1,2	21,2
	V.		1,2	0,3	0,3	0,2	0,6	0,1	0,3	0,1	1,2	0,2	0,4		0,2	×	0,3
Polychæta.	A.	92,4	66,0	10,5	11,7	97,3	58,0	28,4	44,6	215,2	118,4	24,2	3,4	61,0	99,4	5,0	4,0
10. Pectinaria koreni	V.		7,0	2,2	3,2	14,3	12,2	6,2	7,0	13,3	15,5	3,9	0,4	5,4	5,6	0,9	0,5
	A.		4,1			1,1	1,8	3,8	4,8	1,8	2,2	0,8	1,0			0,4	1,2
11. Terebellidæ	V.		0,7			0,2	0,9	0,7	2,4	0,5	1,0	0,1	×		×	×	×
Large polychæta.	A.	0,8	1,1	0,8	0,7		0,4	0,4	0,4		0,2			1,0			0,2
12. Aphrodite aculeata	V.		1,8	?	1,4		0,3	0,6		1,0			3,3				1,0
	A.	34,7	40,3	30,1	20,0	56,7	49,2	48,2	c. 34,0	41,2	43,8	27,4	41,8	21,0	30,4	37,2	39,8
13. Vermes var.	V.		3,3	2,9	2,1	3,8	4,6	3,1	3,4	2,9	5,4	1,8	2,9	2,5	4,5	2,9	2,1
Small gastropod.	A.	11,3	16,0	6,3	0,9	17,8	15,8	12,6	4,0	54,6	66,0	5,0			14,4	7,0	3,1
14. Philine aperta	V.		0,4	0,2	0,1	0,7	0,8	0,6	×	2,6	3,1	0,2			0,5	0,1	0,1
	A.		3,2	2,3	1,6	3,5	1,6	0,8	1,8	2,6	1,2	0,4					1,1
15. Acera bullata	V.		0,4	0,3	0,3	0,3	0,6	0,1	0,4	0,3	0,3	×					
Whelk.	A.	1,3	0,5	0,8	0,7	0,5	0,4	1,0	0,2	0,6	0,6	0,4	0,4	1,0	0,2	0,4	
16. Buccinum undatum	V.	c. 4,1	2,7	3,3	4,7	2,8	1,0	13,0	1,4	6,2	12,0	5,0	0,2	0,4	1,0	3,7	
Large gastropod.	A.	4,5	5,3	3,0	2,2	2,9	2,4	1,4	2,6	1,6	1,0	2,2	1,0	1,0	1,0	1,6	0,6
17. Nassa reticulata	V.		4,9	2,5	3,4	3,2	2,4	2,0	1,7	2,4	1,2	2,6	0,7	1,5	1,2	2,2	0,6
Small gastropod.	A.	?				0,2				1,8	0,2	9,8	1,8	3,0	0,2	0,4	0,4
18. Nassa pygmæa	V.					×			0,1	×	0,5	0,1	0,2	×	×	×	
Brittle star.	A.	102,6	86,0	45,7	53,5	155,2	101,4	211,2	145,6	140,8	157,2	121,0	86,6	78,0	52,6	71,8	88,2
19. Ophioglypha texturata	V.		10,2	7,2	9,5	16,4	9,8	20,4	13,8	13,5	17,0	11,1	8,4	7,9	5,1	8,0	10,
Starfish.	A.	0,9	1,1	0,6	0,8	2,3	3,6	1,0	1,6	0,8	2,2	0,4	0,4		0,2	0,6	1,1
20. Asterias rubens	V.		4,1	0,5	2,5	0,4	2,2	3,6	0,5	30,5	2,3	1,1	1,7		2,7	10,0	3,4
Sea urchin.	A.	1,1	1,8	1,2	1,8	1,7	1,6	1,6	0,2	1,8	2,0	0,8	4,4	1,0	1,2	0,4	1,2
21. Echinus miliaris	V.		4,0	1,6	2,0	4,1	1,3	0,9	0,3	1,2	1,2	0,8	1,3	0,3	0,9	0,3	0,8

In 1909, 1910, and 1911, 100 samples of 0,1 sq. m. were taken, later only 50; and in August 1915, only 10 samples, distributed throughout the whole of the Bredning.

We have omitted from these lists some animals of the epifauna from the Zostera and from a few stones in the *Abra* community; they represent only a small quantity. *Modiola* and *Cyprina* are also not included here; for these, reference may be made to the list p. 27.

× indicates that the weight is under 0,05 gr.

Number and rough weight in gr. pr. sq. m. of the common bottom animals in Nissum Bredning.

			1910.		1911.		1912.		1913.		1914.		1915.		1916.	
A. = Number. V. = Rough weight in gr.			April.	Octbr.	May.	Octbr.	May.	Sept.	April.	Sept.	April.	Sept.	June.	Aug.	May.	Sept.
	Small white bivalve.	A.	30,3	1,0		27,8	4,5	112,0	73,0	112,8	62,5	4,5	11,0		1,5	17,3
1	Abra alba	V.	3,0	0,2		3,4	0,3	17,6	7,3	6,1	2,9	0,3	0,4		0,5	2,7
	Small white bivalve.	A.	6,8	16,5	11,3	69,5	21,3	69,5	36,0	10,8	8,8	31,5	31,3	11,0		46,5
2.	Abra nitida	V.	?	0,9	0,8	3,9	1,2	3,3	3,6	0,5	0,3	0,8	0,9	0,5		2,6
	Razorfish.	A.	8,8	16,5	3,0	20,0	2,8	10,8	4,5	18,5	23,5	3,5	1,0		2,5	7,8
3.	Solen pellucidus	V.	1,2	1,4	0,4	1,7	0,6	0,6	0,5	0,8	1,0	0,2	0,1		0,1	0,8
	Small bivalve.	A.	266,0	456,8	499,3	607,3	591,0	457,8	536,0	455,0	396,0	290,3	190,5	31,0	117,3	150,8
4.	Nucula nitida	V.	14,1	26,4	28,8	48,9	48,3	46,4	44,7	41,8	34,3	17,8	16,9	2,2	9,0	8,7
	Small bivalve.	A.	14,3	32,3	18,5	66,8	135,3	54,3	73,5	164,8	231,5	1067,0	1520,8	2365,0	898,8	984,0
5.	Corbula gibba	V.	1,6	2,8	1,8	4,9	10,5	3,9	3,0	9,7	8,3	57,0	83,9	191,0	67,5	108,9
	Small bivalve.	A.	5,8	3,8	7,0	6,8	7,3	13,5	8,5	34,5	46,3	6,0	3,5		4,5	6,3
6.	Macoma baltica	V.	2,8	1,5	5,6	2,1	3,9	5,4	2,5	1,9	2,5	0,8	1,2		1,6	0,5
	Small bivalve.	A.	10,3	0,5	1,3	0,5	0,8	2,5	1,5	7,0	6,5	1,3	0,8		1,0	1,0
7.	Tellina fabula	V.	0,2	0,1	0,1	×	0,1	0,2	0,1	0,4	0,3	0,1	0,1		0,5	0,1
	Polychæta.	A.	21,5	39,3	5,5	53,3	2,8	105,3	19,5	28,8	12,5	5,3		19,0	2,8	24,8
8.	Pectinaria koreni	V.	4,8	8,6	1,7	12,1	0,9	19,7	5,1	4,6	2,4	0,4		1,4	0,4	2,3
	Polychæta.	A.	5,0	1,3		26,0	1,0						0,3			
9.	Terebellidæ	V.	1,2	0,3		5,0	0,3						?			
	Large Polychæta.	A.	0,3	1,3	0,5	0,3					0,5					
10.	Aphrodite aculeata	V.	0,2	5,0	0,6	3,4					6,1					
		A.	60,8	36,3	41,0	53,3	57,8	c. 58,0	57,0	66,8	72,0	26,5	39,5	30,0	37,3	57,3
11.	Vermes var.	V.	3,8	3,1	2,9	4,0	c. 3,9	4,8	3,5	3,5	5,0	1,6	2,0	1,7	1,8	2,3
	Small Gastropod.	A.	10,0	11,8	3,3	73,5	10,8	21,8	18,0	71,0	13,5	17,8	0,8	6,0	4,5	61,8
12.	Philine aperta	V.	0,6	0,5	0,1	2,1	1,0	7,8	0,4	2,9	0,6	0,4	0,1	0,2	0,2	1,7
	Whelk.	A.		0,3			0,5	0,8	2,0		0,3	0,5			0,3	0,5
13.	Buccinum undatum	V.		0,8			?	8,3	27,4		3,9	7,6		1,0	1,0	3,9
	Large Gastropod.	A.	2,8	0,5	0,8	0,3	0,8	1,3	1,5	1,5	1,3	0,8	0,3	1,0		
14.	Nassa reticulata	V.	?	1,0	1,4	0,2	1,1	2,3	4,6	2,1	1,8	1,1	0,9	1,2		
	Small Gastropod.	A.	2,5	8,0	5,8	21,5	7,3	15,8	18,5	40,5	23,0	41,3	14,0	36,0	19,5	36,8
15.	Nassa pygmæa	V.	?	0,6	0,4	1,5	0,6	1,0	1,4	2,1	1,2	2,6	0,7	1,5	0,9	2,4
	Brittle star.	A.	} ?	?	?	?	2,5	1,3	2,0	4,0	4,3	1,5	1,5	1,0	0,3	1,0
16.	Ophioglypha albida	V.					0,6	0,3	0,7	1,0	0,7	0,2	0,5	0,2	×	0,3
	Brittle star.	A.	36,0	40,8	34,0	61,0	37,8	60,3	78,5	55,3	36,3	23,8	13,3	18,0	9,5	5,8
17.	Ophioglypha texturata	V.	12,6	12,9	9,1	16,3	10,5	19,8	21,5	19,7	11,1	8,6	6,0	5,5	3,3	2,9
	Starfish.	A.	0,5	0,5	0,8	0,8	0,3	1,3	1,5		0,5		0,3		0,5	
18.	Asterias rubens	V.	1,9	1,3	11,1	8,2	6,4	11,9	18,0		4,0		2,4		6,1	
	Spatangidæ.	V.	3,5	5,5	5,3	5,5	2,0	1,8	1,5	2,5	2,0	19,3	13,7	7,0	15,8	13,3
19.	Echinocardium cordatum	V.	50,8	82,8	87,8	107,5	33,0	40,6	37,5	c. 50,0	39,2	142,3	145,6	88,0	165,0	172,5
	Spatangidæ juv.	A.	} ?	?	enkelte	?	?	41,5	60,0	30,3	41,3	0,8				2,5
20.	Echinocardium cordatum juv.	V.						29,9	100,0	c.190,0	282,5	0,5				9,8
	Sea-Urchin.	A.	0,8	1,0		2,3	2,0	7,8		0,5		1,0	2,0		2,0	
21.	Echinus miliaris	V.	1,0	1,5		2,1	2,8	8,2		1,3		1,2	2,6		5,4	
	Turnicate.	A.	0,5	5,8	1,8	5,3	0,3	c.105,0		7,0		7,3			2,8	
22.	Ascidiella	V.	2,6	32,5	18,9	19,1	1,4	273,2		36,3		6,9			6,1	

40 samples of 0,1 m. sq. were taken each time, only in April 1913 20, and in August 1915 10.
Some epifauna, especially *Pomatoceros* and *Balanus*, on some stones, are omitted, as also a few larger animal forms. See list p. 27.
× indicates weight under 0,05 gr.

My first idea was that we ought to be able to see, as on land, from spring to autumn, how the stock increased in bulk, partly by growth of existing individuals, partly by the addition of the young in the course of the summer, and that by subtracting the spring stock from that of the autumn, it should be possible to arrive at a kind of minimum for the production. It could not of course be more than a minimal value, as many specimens of most species die off, either by being devoured by the predatory forms, or in some other way, in the course of the year. Some sort of result was in fact also attained by this means (see Report XX) but my attention was soon drawn to other phenomena. Some of these I will here mention, referring also to the two tables on pp. 30 and 31.

In Thisted Bredning, in the spring of 1910, we find that *Abra alba* has entirely disappeared, but in October of the same year, there are 84 individuals per sq. m; from this time, the species decreases in quantity to May 1912. In June 1915, there is enormous stock of *Abra*, with 44 gr. per sq. m. There are thus fluctuations in the stock, extending over more than a year at a time, and not always exhibiting a maximum in autumn. The stock in the different years varies greatly in size, at times falling to zero.

In Nissum Bredning also, we find *Abra* periods, but they do not fall in the same year as those of Thisted Bredning, with their maxima and minima, but seem rather to alternate with them.

It will easily be understood that these periods are so short — abt. $1-1^{1}/_{2}$ years — in the case of small animals with but a short time to live at all, as we have found was the case with *Abra alba* in the Limfjord. Similar periods have also been found to occur with the other small bivalves, such as *Solen*, *Corbula*, *Nucula*, but the periods are as a rule somewhat longer, and the animals often live for several years.

In the case of *Mya truncata*, I was at first entirely unable to comprehend the state of things apparent in Thisted Bredning; there was an enormous stock of *Mya*, up to 250 gr. pr. sq. m. in 1910.[1] The following years, however, showed, that very few young *Mya* appeared, and it was thus a mystery how the stock was renewed; it seemed hardly reasonable to suppose a migration of large *Mya* from different parts of the slope out to the large central area — the slope always quantities of large, and probably also small *Mya;* in 1915, however, quite

---

[1] It has been found that the weight of large *Mya* – i. e. individuals over 3 cm., is not correctly stated in Table V in Report XX. This was the first year we had commenced these investigations, and we did not weigh the entire quantity of *Mya* on board, but determined the weight later on from selected samples. This is not a reliable method, and it was relinquished in the following year. The weight of the large *Mya* for 1909 is thus hardly as given in Table 5, Report XX, viz. 60,01 gr. dry matter pr. 10 sq. m., but greater how much greater we do not know. I regret also, from this very fact, of the great difference between the weight here given and the corresponding value for 1910 l. c. p. 62–63 to have been led to estimate the production of *Mya* too high for Thisted Bredning, whereby the total production of the stock l. c. 68 is also set too high. To find the correct figures must be a task for the future; as mentioned l. c., we were then more concerned to explain a new method than to lay down definite figures.

small *Mya* began to appear in certain quantity at some places out in the Bredning, and in 1916 there were many; they were now somewhat larger, and were found to be distributed over abt. ¹/₃rd of the Bredning, in its western part. This led me to suppose that possibly the stock of *Mya* also is only recruited in certain specially favourable breeding-years, and that such years only occur at considerable intervals. If this were so, then the stock in 1909 and 1910 should presumably also be derived from a single year's breed, or at any rate, chiefly so. And investigation of the average sizes of *Mya* from this year to 1916 (see Table p. 33) also showed that *Mya* over 3 cm. length — the smaller ones were reckoned separately, as being suitable for plaice food — i 1910 only weighed on an average 8,₂ gr. each, but that the average weight gradually increaced to over 20 gr. pr. individual in 1916; in the autumn of 1916 again, it seemed to be on the decline, the large *Mya* gradually dying off. In accordance with this, we find in the table the number of *Mya* for 1910 to 1916 decreasing from over 20 to abt. 10 pr. sq. m. in the last-named year; in other words, it is chiefly the same large *Mya* which we find in the Bredning in 1916 and in 1909—10, but they have become larger and fewer;

Average weight in gr. of Mya truncata in Thisted Bredning, 1909—1916.

Year.	Month.	Mya truncata over 3 cm length pr. sq. m. No.	Mya truncata over 3 cm length pr. sq. m. Rough weight	Mya truncata over 3 cm length Average weight in gr. pr. indiv.	Mya truncata under 3 cm length pr. sq. m. No.	Mya truncata under 3 cm length pr. sq. m. Rough weight.	Mya truncata under 3 cm length Average weight in gr. pr. indiv.	
1909.	Octbr.	13,9?₄	66,0?	5,0?	4,8	c. 18,8	c. 4,0	
1910.	April.	24,3	199,6	8,2	0,2	?	?	
	Octbr.	24,4	250,0	10,2	1,6	?	?	
1911.	May.	17,8	225,0	12,6	0,8	1,5	1,9	
	Octbr.	21,9	294,0	13,4	1,3	1,0	0,8	
1912.	May.	20,2	304,0	15,0	2,2	1,2	0,5	
	Octbr.	21,4	306,0	14,3	1,6	1,1	0,7	
1913.	April.	16,6	270,0	16,3	3,6	0,7	0,2	
	Septbr.	11,2	190,0	17,0	1,4	0,9	0,6	
1914.	April.	16,0	312,0	19,5	1,2	0,9	0,8	Weight in alcohol, Sept.
	Septbr.	6,8	93,9	13,8	2,6	0,5	0,2	
1915.	June.	7,8	192,8	24,7	5,8•	0,9	0,2	Here quite small *Mya* commence to occur in greater number at certain spots
	August.	14,0	268,0	19,1	4,0	2,6	0,7	Only 10 samples taken.
1916.	May.	6,4	140,0	21,9	67,4	20,1	0,3	
	July.	11,4	246,0	21,6	83,6	42,0	0,5	
	Septbr.	10,2	165,1	16,2	48,8	42,0	0,9	

not till the last two years has a new stock grown up, the average weight of these being still only 0,9 gr. pr. individual. The young in 1909 averaged 4 gr., and were thus really fairly large, so that their year of birth must probably have been some two or three years back, i. e. 1906 or 1907.

As already mentioned, I dare not trust to the figures for the large *Mya* in 1909, and they are therefore omitted from the present report.

About 10 years have thus elapsed from the first to the second *Mya* period in Thisted Bredning, which agrees well enough with the longevity of these animals, and with the number of annual rings — which by the way are by no means easy to count — on the shells. The tables on p. 33 show how extremely small was the quantity of young in the intermediate years. In 1915, as mentioned, a new breed of young, somewhat more numerous, begins to appear, but most of them were so small, that many must evidently have passed through the mesh of the sieves; not until 1916 were they so large that the sieves retained them all. The valuation in August 1915 was really not a true valuation; only 10 samples vere taken, and the figures are therefore not of the same value as those from the other valuations, and this we can as a rule see from the figures themselves.

In Nissum Bredning, there are scarcely any *Mya* at all in the valuated area; the species which weighs most out here is *Echinocardium*, which in certain years comes up to over 200 gr. pr. sq m. In 1910 and the spring of 1911, the individuals were few and large, and almost all to be reckoned among the adults; in the autumn of 1911, however, there appeared a quantity of small *Echinocardium*, weighing less than 1 gr. each. The table pg. 31 shows that up to the spring of 1914, it was possible to distinguish between a young group and an adult, but by the autumn of 1914, nearly all were more or less in the adult group, and since then, no young group of any numerical importance has been observed. The stock of this species seems thus likewise to be renewed only in certain favorable breeding-years.

Such renewal of the stock in some few more productive years is known to take place also among oysters, and is probably also a very common phenomenon in our shallower waters; according to the works of Dr. Johan Hjort and his collaborators, the same applies in the case of the stock of herring and cod on the coasts of Norway. And it is doubtless true likewise of many of our Danish fish species, so that it would seem to be a common phenomenon, at any rate in the shallower parts of the sea; as to deeper water, no investigations of this character have as yet been carried out. —

In the Limfjord, then, we have as a rule to deal with a stock which does not grow up from spring to autumn, but covers a longer period in its growth; it is evident, then, that calculation of the whole annual production and that of the different species can only be very roughly carried out from the size of the stock in the course of a single year. Dr. Boysen Jensen has therefore tried another method for the small and numerously represented species, which are easiest to valuate accurately, and are of great importance for the food of plaice and eels in the Limfjord, viz. the small bivalves, small gastropods and polychæta. The production of these forms is partly devoured by fishes, and partly by *Asterias* and the predatory gastropods, though we do not know what portion falls to each. The eels and the other predatory species mentioned, however, also live partly on other food, especially large bivalves and animals from the areas not valuated. The plaice, on the other hand, is, especially in Thisted Bredning, very largely reduced to living on the small animal species referred to, though on the slope, they can also obtain some other food. I am therefore inclined to agree with Dr. Boysen Jensen's argument that it is a good year for the plaice when there is an abundance of these small food animals, and a bad year when they are few, and that an experiment such as that he has made is well worth the trouble, even though we cannot learn therefrom to what degree other animals than the fishes themselves take toll of the lower food-animals.

The closer valuation of the food-animals devoured by plaice and eels was carried out in the following manner on the basis of the table for Thisted Bredning p. 30.

If we have, for instance in the spring of 1911, found by valuation of a locality, a stock a, of say 1000 grown *Abra* or another small bivalve, and in the spring of the following year, 1912, only a remainder a_1 amounting to 200 of these adult *Abra*, each individual therein comprised being now larger, we can see that in the course of the year, 800 *Abra* have disappeared. This, then, is the consumption. The animals thus consumed must be presumed to have reached, before death, a size lying between the average weight of *Abra* in the spring of 1911 and the average weight of *Abra* in the spring of 1912. Calling the total weight of the stocks v and v_1, then the one is $\frac{v}{a}$ and the other $\frac{v_1}{a_1}$ and the consumption, expressed in weight, will then be $(a-a_1) \times \dfrac{\frac{v}{a} + \frac{v_1}{a_1}}{2}$.

The growth increment of the adult individuals in the course of the year is thus equal to the consumption $+$ the remainder of the stock v_1 $\div$ the stock v, and the annual production will be equal to the growth of the adults $+$ a possible newly developed 0 group ($=$ upgrowth).

The consumption will probably often be greater in point of weight than the stock; it will include much of the large stock grown up in the preceding year. Where there is not a large spring stock, and not a large autumn stock in the previous year, then it looks bad for the annual production and for the food of the plaice. *Abra* can, however, even in the period from spring to midsummer, exhibit a considerable upgrowth.

In the course of Boysen Jensen's investigations it has been found that the consumption of those animals which are particularly sought after by the plaice in Thisted Bredning, viz. the small bivalves of various species and the polychæta, amounts on an average, for the years 1910—1915, to 64 gr. rough weight per sq. m fluctuating from 32 to 84 gr. in the different years.

Amount of food consumed by plaice and eel in Thisted Bredning.

With regard to the amount of food consumed by the plaice, all that we know is (Report XX p. 63 ff.) that a plaice in August 1910 (300 gr.) emptied its digestive tract in 8—9 hours during the night, and filled it fairly rapidly again; presumably therefore three time in the course of twentyfour hours; and that the weight of all this food amounts to abt. 30 gr. (rough weight). While the plaice was smaller, from 80—300 gr., (April to August) it ate less in twenty four hours; later, from August to November, evidently more, until it then weighs as a rule 480 gr. We may take the 30 gr. daily for a plaice of this sort as the normal quantity of food, throughout the 240 days in which it feeds, from April to November. It has by then increased 400 gr. since transplantation, and has produced 400 gr. as against a consumption of 7200 gr. food devoured. This means, then, that it takes 18 gr. of food to produce 1 gr. of plaice, or, reckoning with the organic dry matter percentages of 10 % and 25 %, $\frac{18/10}{1/4}$ = 7,2 gr. food to produce 1 gr. plaice, both in organic dry matter.

From 1912—1916, an annual average of 445,000 plaice were transplanted into the Bredning, abt. 70 per hectare, but only abt. 182,000 are brought to market as saleable fish; the quantity consumed by the fishermen themselves is included; altogether, the total of plaice sold has of late years amounted on an average to 68,000 kg. These fish weighed when transplanted abt. 15,000 kg. viz. 8 kg. pr. 100 fish, so that the production is at least 53,000 kg. annually.

This production demands 18 times more food, 14.7 gr. per sq. m., or abt. 1.47 gr. food in weight of dry matter per sq. m.*)

Not all the 182,000 plaice caught attain so great a weight as 480 gr. and it is therefore better here to reckon with the number of kg. produced than with the number of fish; this last would give a consumption of food amouting to abt. 20 gr. per sq. m.

As a matter of fact, however, considerably more than 14.7 gr. rough food per sq. m. is doubtless consumed; many plaice will of course have died off in other ways than by capture, or may be destroyed in the nets; we cannot account for more than half the total number of transplanted plaice as brought to market.

---

*) The area of the Bredning outside the 6 m. limit is 65 million sq. m.

This calculation of the annual consumption of food by the plaice is of course very uncertain, and can only be regarded as the nearest estimate which can at present be formed. The figures here are not throughout the same as, but in some cases better than, those in Report XX p. 65, where such calculations were made for the first time in the case of sea fish.

It will therefore be advantageous to compare them with what we find in the fresh water, where more is known as to the consumption of food by fishes. The carp in particular has been very closely studied in this respect, but the results exhibit considerable variation according to the different conditions, in addition to which, the food of the carp differs very greatly from that of the plaice, so that not much can be gained by any comparison between the two.

W. Cronheim (Bibl. der gesamten Landwirtshaft. Bd. 34, 1907) states that 4 kg. of food in the form of lupine is easily sufficient to produce 1 kg. of flesh in the carp. It is very concentrated food, with a high percentage of organic dry matter, and this will perhaps explain why only 3—4 kg. is required, whereas the plaice consume over 18 kg. of animal food. Cronheim mentions, p 36 l. c, that in good trout hatcheries, 1 kg. of trout can be produced at a cost of 3—4 kg. of food, in this case mostly animal food, as with the plaice, but he expressly adds, that such satisfactory results are only to be attained under the best conditions, so that we may suppose more is usually required.

In this country, C. V. Otterstrøm has carried out experiments with feeding of trout (see Fiskeriberetning for 1911, p. 244—254) and states that in the best season, 4—5 kg. of food was used as a rule in the best season for each kg. of trout produced, the food being fish.

These trout have individually never increased so much in weight as the plaice in Thisted Bredning generally do. It seems therefore, in my opinion, but reasonable that the plaice should use 18 kg. of food to produce 1 kg. of flesh, at the food of the plaice has an organic dry matter content of only abt 10 % whereas the food of the trout, consisting of raw fish, would presumably have a dry matter percentage of abt. 25, so that there would only be abt. 7 kg. of this required to answer to 18 kg. plaice food. The calculation made for the plaice, as regards amount of food, thus does not seem improbable, and in particular, 18 kg. of raw food for the production of 1 kg. rough weight of plaice does not seem too high a figure compared with the food consumption of trout in the good season.

The growth of animals under different conditions varies to such a degree — the plaice, for instance in Nissum Bredning can hardly be said to grow at all — that it is hardly possible to lay down definite figures for the growth in general, or for the consumption of food. To obtain a closer idea as to the food consumption of the plaice in Thisted Bredning, it would be necessary to make further experiments there, and even then the results would probably be found to differ for the different years. It is hardly likely that anything beyond a general outline of the case, as given above, can ever be gained without very detailed investigations, and such do not appear to me necessary for the purpose to which these present researches were directed.

With regard to the consumption of food by the eel in Thisted Bredning, nothing is known, and we must therefore employ the same proportion as for the plaice, abt. 7:1.

From 1912—15, the average annual yield of eels amounted to 48,000 kg. these figures are reckoned as for the production of the entire Bredning, not the middle area alone.

This gives per sq. m. (the whole Bredning)

0,5 gr. eel pr. sq. m. rough weight, or

0,25 gr. weight of dry matter.

For this, the amount of food required will be $7 \times 0{,}25$, or

1,75 gr. fish food dry weight, pr. sq. m.

or abt. 17,5 gr. — rough weight pr. sq. m.

Eels and plaice together thus consume

$14{,}7 + 17{,}5 = 32{,}2$ gr. fish food, rough weight, pr. sq. m.

but many eels leave the Bredning without being captured, and many of the plaice are not brought to market, so that the consumption of fish food is in reality far greater than this. In addition, other fish, such as the blenny, goby, starfish and also gastropods, take toll of the food which should go to feed the plaice and eels.

The quantity of such food available, according to Boysen Jensen's calculations, varied for the years 1910—1915, from 32—84 gr. rough weight per sq. m. and averages abt. 64 gr. per sq. m.; there seems thus to be nothing to spare, especially in bad years, when the eels and plaice alone require 32,2 gr. per sq. m. for they cannot find all the small animals in the sea bottom, and various lower predatory forms also consume a quantity of fish food.

There is thus by no means any unlimited quantity of fish food available in Thisted Bredning. And this I regard as a main result of these valuations in that water.

In the spring of 1917, for various reasons, no plaice were transplanted in the Limfjord, and the customary protective measures for such plaice were therefore suspended. The result of this was, that quantities of small plaice were soon caught; but the few which remained reached in October, November and December an average weight of over $1/2$ kg. apiece. In the summer of 1917, a sudden upgrowth of the food-animals consumed by the plaice had set in. By the winter of 1917—18, practically all the plaice in the Bredning had reached a weight of over $1/2$ kg.; in the years immediately preceding, many were far from reaching this size, and were therefore not taken, so that there was a quantity of small fish remaining at the beginning of 1917. All this I can only take as indicating that the quantity of plaice usually transplanted into the Bredning, 60—70 fish per hectare, is greater than the water can bear as a rule, if the fish are to grow rapidly, and this they must do if new transplantation is to take place each year, after practically all the fish have been captured, at a weight of abt. $1/2$ kg. apiece.

VIII b. Valuation of in the Zostera Region.

The bottom sampler gives good results in most areas of the sea bottom, but it is of very little use where the *Zostera* covers the bottom with its leaves, which are often up to 2 m. long, and set as close together as the stalks in a cornfield. The animals living attached to the leaves or moving between them are as a rule struck off or driven away, so that only few are brought up in the bottom sampler. These *Zostera* areas, however, are extensive, especially in our fjords, and contain a quantity of animal life, so that it was impossible altogether to exclude them from

Fig. 7.

the valuation investigations. I therefore commenced, in 1914, experimenting with hand nets having a bag formed of closely woven stuff, of various sizes and shapes, attached to poles, so that they could be worked by one man in a small rowing boat, the net being drawn for two or four metres through the *Zostera*, and thus bringing up most of the animals from there. With a net $1/2$ m. broad, (see Fig. 7) the area of *Zostera* thus fished would be 1 or 2 sq. m. and we had here, as with the bottom sampler, a definite unit of measurement to go upon.

Numerous test experiments have shown that the quantity of animals captured, e. g. *Rissoa* and various other small forms, really does exhibit a very good degree of uniformity in the results, the difference in quantity between a series of such samples being as a rule less than $1/3$ or $1/4$ (see Figs. 8 and 9).

Fig. 8.

Presuming that the fauna would differ at the different seasons of the year, I arranged these investigations in such a manner that the same localities were investigated several times in the course of the year, sometimes every month, in order to determine the variations. These localities, especially Svendborg Sound, Nyborg Fjord, Holbæk Fjord and the Bay at Nykjøbing in the Limfjord, are all in shallow water: hand nets of this description can only be used for depths to abt.

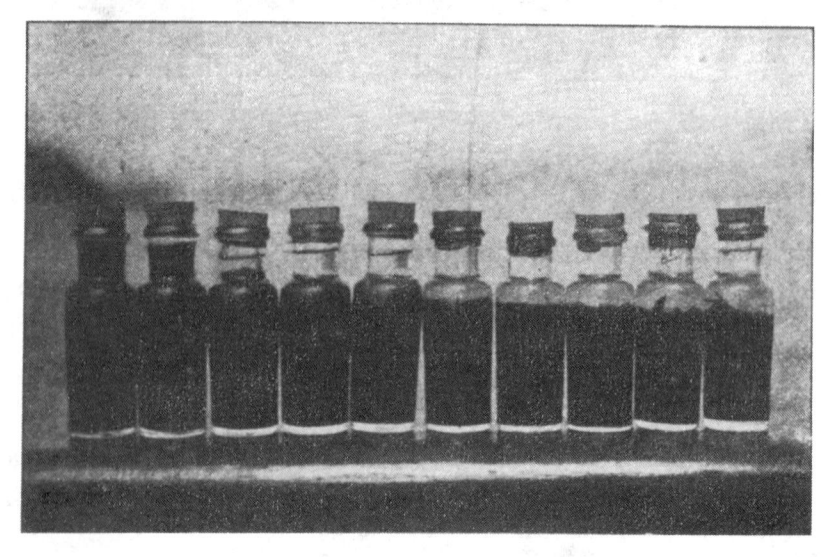

Fig. 9.

5 m. In each locality, a series of stations was taken from abt. $1/_3$ m. depth to as far out as appeared advisable, often as far out as the *Zostera* was to be found at all. In the very shallowest parts, there is no growth of *Zostera*, but other plants (algæ) are found, though they are often absent in the winter. We have now a large quantity of material from these investigations, which has not yet been fully dealt with, and I will therefore here merely give the main outlines of the results arrived at.

What surprised me most was the fact that our two common *Rissoa* species, *R. membranacea* and *R. conspicua*, occurred in highly varying quantities, and exhibited very different appearance, at one and the same station in different seasons; at times only quite a few specimens per. sq. m. were found, at others up to 100,000, and with a weight of nearly 100 gr. It was soon discovered, that these animals live but a single year, and those from the summer of the previous year die off suddenly, to be replaced by a numerous and rapidly growing new generation, which as a rule attains its maximum in number and weight during late summer, and then declines considerably in both respects in course of the winter.

Figs. 8 and 9 show the quantities of *Rissoa* per. sq. m. of *Zostera* growth at the same locality in Svendborg Sound, in 1915, Fig. 8 shows the small quantities found in May; these were all full- grown animals. Fig. 9, again, shows the large quantities of *Rissoa* from the beginning of September. These animals are chiefly young, not yet fully developed individuals, the 0 group, or the young of the came year.

Other gastropds, such as *Cerithium reticulatum*, also occur in summer for a couple of months in some of the hand-net samples, but later disappear. These are, however, not annual animals, as may be seen from a study of their shells the next summer, when we find an old, corroded section and a new growth zone. This species evidently only lives on the *Zostera* for a short summer season, presumably while depositing its eggs on the leaves, and spends the remainder of the year among the roots of the *Zostera*, where the hand net cannot reach. A very few specimens have, however, been found here with the bottom sampler.[*]

In addition to the gastropods, the hand-net samples also contained various crustacea, chiefly *Gammaridæ*, *Isopodæ* and *Mysidæ*, *Asterias rubens*, a few bivalves, polychæta, fishes, etc., in other words all forms living on the *Zostera*. Most, however, were gastropods, and the *Rissoa* has a very large annual production, the approximate amount of which is very easily determined, as all that is found of this species in autumn must be reckoned as production. It is often up to 100 gr. rough weight per sq. m. *Rissoa* is of considerable importance as fish-food, but numbers of them are doubtless devoured by crabs *(Carcinus mænas)* and starfish *(Asterias rubens)*; this latter has its breeding season in the summer, its numerous 0 group will be found living on the *Zostera* in many fjords, where the abundant young *Rissoa* form a favourite article of food for the starfish. Numbers of *Rissoa* also are doubtless

---

[*] See I. Collin: Limfjordens Marine Fauna, p. 77, 1884. This writer is of opinion that *Cerithium* buries itself in the bottom in winter.

destroyed through the falling off of the *Zostera* leaves, which then drift away, especially when the long leaves are changed in the stormy weather of late summer and autumn.

From investigation of stomach content, the crustacea appear to play a more important part as fish food than to the *Rissoa*, despite their comparatively smaller stock; it is possible, however, that the stock is renewed twice a year or more often, and that the annual production is thus really larger than would appear from the size of the stock at any time. Investigations as to this point are at present being carried out by H. Blegvad. The young of *Mytilus*, and on bare spots in the *Zostera* vegetation also older individuals of the same species, are here found at certain localities in enormous quantities; the great bulk of the young doubtless fall a prey to lower animals, among which the starfish, *Asterias rubens*, should once more be mentioned, as the only asterid of this region. The brittle star is not found at all in these localities.

The great struggle for existence which is carried on in our *Zostera* region has evidently a complete counterpart on the coasts of North America, as is seen from two papers in the Bull. U. S. Fish Commission Vol. XIX 1901. One of these papers is by James L. Kellogg, the other by A. D. Mead; they treat of the relation between *Mya arenaria* (the common clam) and the starfishes, including our common *Asterias rubens;* among other points discussed is the question of possibly furthering the production of the one species by measures directed against the other. The clam is a popular article of food in North America, and ought really to be so here as well.

The valuation of the *Zostera* region has raised a number of questions by no means thoroughly dealt with as yet, but the study of which seems to promise a good insight into the economy of our small Danish waters. My intention here has for the present been chiefly to indicate that such investigations are in progress.

As mentioned above, the use of the hand net is for practical reasons restricted to waters of but a few metres' depth; in extending the operations to the deeper parts of the *Zostera* region, other implements will be required. The *Zostera*, it should be noted, can on open coasts, where the water is clearer, and light penetrates farther down, run out to abt. 14 metres depth. The investigations here demanded must be a task for the future, but some preliminary work has already been done.

With regard to fishes and other swiftly-moving forms, such as shrimps, etc. the hand net does not take sufficient quantities to afford material for a valuation. I therefore constructed a fine purse seine, of cheesecloth, to embrace a circle with an area of abt. 70 sq. m. The lower rope of the seine, on the bottom, can be tightened up so as to close in entirely, while the upper rope reaches to the surface of the water; in this manner, the fishes and other swiftly-moving forms of the *Zostera* can be captured. (See Fig. 10).

This seine has been in use for a couple of years at the principal hand-net stations, and the results have shown that it is chiefly the gobies *(Gobius ruthensparri)* and the sticklebacks *(Gasterosteus aculeatus)* which are found in such quantities as to be of importance for a valuation. Both are practically speaking annual forms; that is to say, they might perhaps live more than a single year if they were allowed to do so, but as a matter of fact, the great bulk of them are invariably devoured in the course of the first year. The fifteen-spined stickleback, *(Spinachia vulgaris)*,

Fig. 10.

will, even if left in peace, doubtless die off naturally in the course of abt $1\frac{1}{4}$ years, and the same applies possibly also to *G. ruthensparri*.

These investigations are also in practice restricted to shallow water, of 2—4 metres, and should if possible be replaced by other and more efficacious methods.

The *Zostera* region thus presents very considerable difficulties for valuation investigations, but the work is still of so recent date that future experiments may doubtless be expected to bring improvements. Only a few localities have up to now been valuated as a preliminary study; the operations should if possible be directed towards the comprehension of the entire *Zostera* region.

IX. Fishes of the Danish Waters.

We have in the preceding pages frequently referred, for information as to the quantities of fish for instance in the Kattegat and the Limfjord, to the annual yield recorded in the Fishery Statistics; the actual stock of fish in a water, however, may of course be very different from this. In order to investigate this and other points, I introduced, as far back as 1887, marking experiments with live plaice, the fish being liberated after marking at the spot where they were captured, and later, when retaken by the fishermen, forwarded to the station, so that I could determine not only the percentage of recaptures made in the course of the fishery, but also the growth and migrations of the fish themselves. (See Fiskeri Beretning for 1888—89 — printed 1890 — p. 90—91 and for 1889—90 — printed 1891 — p. 90—91).

A Scottish gunboat came to Copenhagen, as far as I remember, in the year 1888; I demonstrated the method on board, and it was immediately imitated in the Scottish waters, and has since become a general form for investigations of this kind. By this means it has been ascertained that a very great number of all marketable-sized plaice are caught every year in the Kattegat, and doubtless also in the North Sea, more, indeed, than is beneficial to the stock. The same applies in part to the stock of plaice in the Limfjord, which has indeed to be maintained artificially, by restocking portions of the water each year with young plaice from elsewhere.

In order to show that Thisted Bredning contained practically no other plaice than those transplanted from other waters, I applied this marking method in 1895, when abt. 80,000 plaice were transplanted, of which I marked every 7th fish with a hole or two in the fins. On subsequently fishing myself in the Bredning. I found that at least every seventh fish had such holes, and it was thus proved that the water contained practically no other fish than those transplanted there. I mention this, as it is the only method known by which to ascertain the number of fish found in a large area which cannot be drained.*) Unfortunately, the method is not easily applicable in the case of fish which are apt to die as soon as caught, the herring, for instance or the mackerel, but it has nevertheless helped to undermine the prevalent theory that the stock of fish in the sea is inexhaustible; this idea can never be re-established hereafter.

For those who follow the purely practical course of the fisheries also, there are many signs which indicate whether a stock is being overfished or not, if only we can read their meaning. It is becoming more and more generally realised that the Danish fishing industry is very intensive among bottom fish of suitable market value, and we may therefore as a rule regard the figures given in the Fishery Statistics as being fairly near the quantity which the stock of these food fish can afford to lose each year.

---

*) The method laid down by V. Hensen for calculating the number of individuals of a species by counting the pelagic eggs found in the sea is hardly practicable in many waters, but is theoretically of considerable interest.

It would be advantageous also, if we had some means of ascertaining how many fish of all species are to be found on a given area of bottom; i. e. an implement capable of capturing all the fish from the ground on which it worked. I do not, however, know of any such. It is a difficult matter to take fish of large and small sizes in one and the same implement; small-meshed nets, for instance, cannot be drawn rapidly through the water, and with a larger mesh, the small fish will escape. Some fish, again, keep a little above the bottom, while others bury themselves therein, and have almost to be dug up by the implement used, so that it becomes filled with bottom material and lower forms of animal life, at the risk of breaking. Consequently, it is only by knowledge of the fishery statistics, of the practical fishing industry, and by experimental fishery in person, using different implements, large mesh and small, that we can form any idea as to the bottom fauna as a whole.

I order to give something more than generalities as to the fish-fauna of our waters, I have selected a series of actual hauls made with large- and small-meshed nets, the hauls here noted being such as were, in my opinion, among the most typical for the bottom fish in the various waters from the Baltic to the deepest part of the Skagerak. The number of specimens of each species taken in each case is noted. (See Table p. 46).

In several instances, two hauls made in close proximity one to another are taken together, in order to give a fuller view of the fauna from that locality; and to save making the species lists too long, some few of the less frequently occurring forms have been omitted (see notes p. 47). The Table thus does not show exactly the entire fish fauna of the spot concerned, but only those species taken in the particular hauls given, these being naturally as a rule the most common of all. Consequently, the comparison gives, together with the following observations, a good idea of the bottom fishes most characteristic for each of the waters concerned, which is all I have aimed at in the present case.

All the hauls were made in the deeper parts of the waters outside the plant belts; three of them, in the deepest parts of the Skagerak, were made by Dr. Johan Hjort.

It is immediately noticeable that the common dab and the plaice, accompanied by the flounder, are the most common fishes, and from their size also those which bulk most largely in the hauls from the Baltic to up in the Kattegat, where the water of the latter is not too deep. The long rough dab *(Drepanopsetta)* is not found in the Baltic proper east of Gedser, but in the deepest water from south to north of Funen; it is later met with again in the deepest water of the Kattegat, and goes out into the Skagerak, so that its distribution does not follow that of the three first-named species. For the sake of brevity, I may term the area where the three flatfish are characteristic, the dab area, this fish being there as a rule by far the most numerous and most common. In the Baltic east of Gedser, the numerical proportion between the three is different, the plaice being here apparently the most

Locality	Baltic	Great and little Belt	North of Funen		Kattegat				Skagerak				
Community	d.	b.	Eb.	V.	E.Fil.	B.Ch.		B.S.			Al.P.		
Depth in metres	21.	20–38.	c. 20.	12–14.	25.	45–80.		c. 130.	300-430.	c. 400.	430-515.	520-570.	
Zoarces viviparus	c. 20		c. 10										Viviparous blenny.
Lumpenus lampetriformis	c. 20	7	c. 60										Lumpenus lampetriformis
Cottus scorpius	13	1		8									Sea scorpion.
Pleuronectes flesus	23	8	2										Flounder.
" limanda	143	124	4 c. 10	16									Dab.
" platessa	390	c. 800	635 c. 160	90 45	c. 200	2	7						Plaice.
Drepanopsetta platessoides		55	6	9	28		2	6			2		Long rough dab.
Rhombus maximus		212	7	3			43						Turbot
" laevis	1	11	c. 30		1			4					Brill.
Gadus callarias		17	1	1	3		1						Cod.
" merlangus		2	1	2		3	7	8	3		1		Whiting.
Raja radiata		1	c. 12	1			90	3					Starry ray.
Gadus æglefinus		4	1		2			3					Haddock.
" minutus		5			2	22	1	8					Poor cod.
" esmarkii		1	2			1	4	1	3				Norway pout.
Pleuronectes cynoglossus			2				18						Witch.
Lycodes vahlii						6		31	2	14	56		Lycodes vahlii.
Myxine glutinosa						5	1	11 19	43			2	Hag.
Gadus poutassou								2	3			8	Poutassou.
Gadiculus thöri							c. 10	1	1	c. 10			Gadiculus thöri.
Lycodes sarsii								7 12				5	Lycodes sarsii.
Coryphaenoides rupestris								1				8	Coryphaenoides rupestris.
Careproctus reinhardi								2	8	8	281	56	Careproctus reinhardi.
Sebastes viviparus												6	Sebastes viviparus.
Argentina silus									4	4	14	2	Argentina silus.
Chimaera monstrosa									22	1	28	2	King of the herrings.
Raja lintea										1	2?	3?	Raja lintea.
" fyllae												1	" fyllae.
Spinax niger									1		2	1	Spinax niger.

Dab area.   Haddock area.   Coryphænoides area.

Witch area.

common, but the plaice there are wretchedly small and poor, very unlike the large, rapidly growing plaice of the Kattegat.

In the dab area of the Baltic, we also find some few *Rhombus maximus*, *Cottus*, *Lumpenus*, and *Zoarces*, and though none were taken in the hauls mentioned, I may nevertheless add *Gobius minutus*, cod, *Motella*, sand-eels and lumpsuckers; this is practically all that we find there of bottom fish. In the Belts, we encounter, besides all these, from the dab area, *Trigla gurnardus*, *Rhombus lævis*, *Acanthias vulgaris* and *Raja radiata* and the whiting; often also there are occasional finds of rarer marine fish, which I need not mention here. In the southern Kattegat, we find the same fish, vith *Raja clavata* and several new forms, but *Lumpenus* and *Zoarces* are now rare, though the last occurs again near the coasts. *Gobius minutus*, on the other hand, goes out to at least 30 m. in the Kattegat; we do not know in what quantities it is there found, but it is doubtless fairly numerous.

Notes to Table p. 46.

The Baltic. Hestehoved WNW, abt. 4 naut. miles, 21 m. depth. $^{12}/_5$ 1903. English Trawl.
d. 1 *Cottus bubalis* omitted.

Little Belt and great Belt. W of Skjoldnæs, abt. 38 m. depth, $^1/_4$ 1902. English Trawl.
b. { 3 *Gobius minutus* and 4 *Trigla gurnardus* omitted.
Off Kloverhage, abt. 20 m. depth, $^{20}/_{10}$ 1902. English and Danish Trawl.
Omitted: 45 *Gobius minutus*, 11 *Agonus cataphractus* and 1 *Acanthias vulgaris*.

North of Funen. Agernæs Molle SW by W, Munkebo Bakke SSE, abt. 20 m. depth, $^{22}/_9$ 1902.
Eb. { Eng. Trawl.
1 *Gobius minutus* and 1 *Agonus cataphractus* omitted.
E $^1/_2$ N of Æbelo, Einsidelsborg S $^1/_2$ W, abt. 20 m. depth, $^{12}/_8$ 1903. Eng. Trawl.
Omitted: abt. 60 *Trigla gurnardus* and 6 *Acanthias vulgaris*.

Kattegat. A little S of Trekosten on Flyndergrunden, 12 m. depth, $^5/_6$ 1901. Snurrevaad with otterboards.
V. { 6 *Raja clavata* and 3 *Cyclopterus lumpus* omitted.
Aalborg Bay. Buoy at Randers Fjord WSW, abt. 10 naut. miles, 14 m. depth, $^9/_9$ 1902, Eng. Trawl.
1 *Centronotus gunellus*, 1 *Agonus cataphractus* and 5 *Gobius minutus* omitted.

E. Fil. { Hesselø Light N by W 4 naut. miles. hard bottom, many *Modiola*, 25 m. depth, $^{22}/_9$ 1903. Eng. Trawl.
Omitted: 5 *Solea vulgaris*, 1 *Agonus cataphractus*,1 *Trigla gurnardus*, and 3 *Acanthias vulgaris*.

B. Ch. { E of Kobbergrund Lightship, 65–80 m. depth, $^{16}/_6$ 1903. Eng. Trawl.
3 *Motella cimbria* omitted. (2 Tdr. *Brissopsis, Pandalus borealis*).
4 naut. miles N of Kobbergrund Lightship, 45–75 m. depth, $^{26}/_8$ 1898. Seine with otterboards.
Omitted: 3 *Callionymus maculatus* and 1 *Raja batis*.

Skagerak. N of the Skaw, abt. 130 m. depth, $^{15}/_7$ 1897. Seine with otterboards.
B. S. { 10 naut. miles N of Skaw Lightship, abt. 130 m. depth, $^9/_7$ 1898. Seine with otterboards.
7 *Motella cimbria* and 1 *Trigla gurnardus* omitted.

Al. P. {
1) 58° 10' N, 9° 53' E, 300–430 m. depth, $^9/_9$ 1901. Eng. Trawl. — (80–100 large *Pandalus* and *Pasifaë*.)
2) NNW of Skaw, abt. 400 m. depht, $^{21}/_5$ 1897 and $^{28}/_7$ 1897. Steel wire trawl and seine with otterboards.
3) 58° 14' N, 9° 55' E, 430–515 m. depth, $^9/_9$ 1901. Eng. Trawl. (Large *Pandalus* and *Pasifaë*).
4) 58° 20' N, 9° 50' E, 555 m. depth, $^9/_9$ 1901. Eng. Trawl. – (*Pandalus* and *Pasifaë*).
5) Close in to coast of Norway, abt. 520–570 m. depth, $^{28}/_7$ 1897. Seine with otterboards.

The most frequently occurring fish in the dab area, the above-mentioned flatfishes, are thus the same in the Baltic, the Belt Sea and the southern Kattegat, but *Lumpenus* stops at the Belt Sea. In places, however, we find continually more of the true marine fish as we go farther northward, though they are not sufficiently numerous to alter the general character of the fauna.

Not until the deep eastern part of the Kattegat is reached do we encounter an entirely new fish fauna, where the haddock, accompanied by *Gadus minutus*, *Gadus esmarkii*, *Lycodes vahlii*, *Pleuronectes cynoglossus*, and many other forms, commence to become common, together with many long rough dabs; finally, also, the flounder has disappeared, and the dabs and plaice have considerably decreased in number. Here too, we find quantities of whiting, but this species can, especially in the younger stages, be found almost everywhere in the Kattegat. This area should be called the haddock area; the bottom here is soft, blue clay with *Brissopsis* and *Amphiura Chiajei*. *Myxine glutinosa* is found here in numbers.

The haddock area extends up north of the Skaw, where witch *(Pleuronectes cynoglossus)* and haddock increase in numbers doubtless on the Brissopsis-Ophioglypha sarsii community; *Gadus poutassou* and *Lycodes sarsii* are here added to the remaining species, and the plaice and dab have altogether disappeared. This haddock area extends out to abt. 150 metres' depth or more.

At still greater depths we find only *Pleuronectes cynoglossus*, *Lycodes*, and *Myxine;* where the *Pl. cynoglossus* stops, the three others go on, and quite new species appear, of which I need only mention *Coryphænoides rupestris*. (See also List p. 46.)

Outside the haddock area, then, we may note a particular area for the wich *(Pleuronectes cynoglossus)*, and farthest out another for *Coryphænoides (Macrurus) rupestris*.

These deepest areas have been but little investigated, but in the last, probably no edible fish are found.

In the areas of the haddock and the witch, the great trawl fisheries are carried on for these two fish, with ling, etc.

We have thus from the Baltic to the Skagerak, in the deeper parts of the waters, a series of transitions in the fish fauna from the dab area to the haddock area, and further to that of the witch, and finally that of *Coryphænoides*.

Entirely similar areas are met with in the North sea from the coast out into deep water, with dab, haddock, witch and *Coræphænoides* as character forms. See Hjort and Murray: The Depths of the Ocean, p. 451 ff. 1912.

That such a sketch of the fish fauna really is of importance will best be realised from the fact that a fisherman from the Great Belt does not know a haddock by sight unless he has taken part in the Kattegat fishery; the witch he has probably never seen, and I have had a common sole brought up to me by fishermen from the Great Belt, with an enquiry as to what it was; they themselves did not know it at all.

This gives a good idea as to the frequency of these species in some waters, and their rarity in others, such as is not to be gained merely by studying the lists of species in faunistic works.

We were thus able, by following the waters from the Baltic to the Skagerak, to note four different fish areas in the deeper grounds; in the shallower parts, however, it is another matter. The dab area is found in all these waters near the coasts, i. e. on all our coasts from the Baltic to the Skagerak. The coastal waters have to be dealt with separately, owing to the fact that these tracts, with their vegetation, which is only found in any quantity in shallow water, and their fjords and minor waters, which often run far up into the land, exhibit so different conditions that we here encounter other species of fish, not known from the deeper portions of the dab area.

On the open sandy coasts of the Kattegat, where the vegetation is very scanty, the dab itself is a rare form, but the young of plaice, flounder, turbot, brill, and sole are here found in numbers. *Gobius minutus* breeds here, young herring are at times found in enormous quantities, and are preyed upon by sand-eels and mackerel; in other words, we here find mostly the young of these species which are met with in the dab area proper.

Where the coasts, on the other hand, are protected against the action of the waves, so as to permit a rich growth of *Zostera*, new species of fish are encountered, several *Syngnathidæ*, for instance, *Gobius ruthensparri, Gasterosteus, Spinachia* and eels; on the muddy spots between the *Zostera* plants *Gobius niger* is common, accompanied by *Zoarces*, flounders and sea-scorpions in numbers. Only the three last have been previously noted from the dab area.

This fauna is characteristic for great parts of our protected coastal area, where the *Zostera* abounds; it runs up into all our fjords, and covers almost entirely several of our shallower waters as for instance south of Funen and Sealand, but the *Zostera* and its fauna are never met with beyond 14 metres' depth in our waters. In places where this vegetation is succeeded farther out by a rich growth of algæ, especially on stony bottom, we find *Labroidæ, Liparis, Gunellus, Cyclopterus*; these fish will, however, at times make excursions into the *Zostera* zone.

The above-mentioned *Zostera* fauna may be regarded as associated with the rich epifauna of the *Zostera* region, but belonging to the dab area, viz. to the inner, coastal portion of the same, and the algal fauna, with *Labroidæ*, etc., as associated with the algæ in other parts of the dab area, but the limit between the two is not sharply defined; most of the fishes from this area may also be met with in the other. Only *Syngnathidæ (Syngnathus typhle, Nerophis ophidion* and possibly *Syngnathus rostellatus)* belong decidedly to the *Zostera*, and *Gobius niger* as a rule to the bare patches of mud between the roots of the *Zostera* plants or a little outside the limit of the vegetation, while *Nerophis æquoreus* and *Syngnathus acus* doubtless belong chiefly to the algal region.

It is remarkable to find the eel in its adolescent stages, as the yellow eel, so closely associated with the *Macoma* community as it is, since later on, as silver eel, it leaves this region entirely, but I have never caught a yellow eel at any great distance outside this community. In the shallow waters of the Limfjord, where the *Abra* community covers the middle areas of the Brednings, eels have been found, as also of late years *Gobius niger,* in great numbers on this community, but this is only an exception which proves the rule.

The plant zones are visited at certain seasons by quantities of other fish seeking food there; cod, whiting, mackerel, herring, garfish, salmon, trout are probably the most common, but many others may occur, though they do not belong to the true inhabitants.

In the inner parts of the fjords, where the water at times is only very slightly salt, we often encounter various fresh-water fishes; I will here only mention that in the sound between Møen and Sealand, pike and perch play the same part as cod elsewhere; the salinity here is so low that not even the common starfish *(Asterias rubens)* can exist.

It is thus only roughly that we can find any relation between the occurrence of the fish and the distribution of the different animal communities on the sea bottom; it is much the same as with the birds on land; they can fly wherever they please, within the limits of our small country, but we still find the lark keeping to particular parts of the moorland tracts, and the snipe to others, so that there is a certain regularity of distribution in the main features, not easily discernible save by the practised observer. The fishermen often have some sort of knowledge corresponding to this, as regards the denizens of the sea, though the facts are here far less easy to discover, where everything is hidden beneath the water. It is only in the nets that some small quantity of the animal life at the bottom is brought to light, and it is from this alone that it is possible to form an idea of the actual conditions.

The view of the fish fauna given above is of course in entire accordance with the statements in the Fishery Statistics, but it should be borne in mind that the statistics regard each water as a whole, without making distinction, as we have done, between deeper and shallower parts of the same; the accordance therefore, is only to be expected in the principal features, not in all details. Several fish, such as for instance the dab and whiting, are only partially included in the statistics, owing to their small market value; many others are all thrown overboard again.

As regards the herring, the Danish statistics for 1915, including the Skagerak and the North Sea, show a yield of $3,8$ million „Ol" (1 Ol = 80 fish). Taking the weight of the ol as 5 kg., this gives 19 million kg. Sprats, however, are also included in the figures for herring, so that it is difficult to get at the exact weight of the ol; possibly it may be only 3—4 kg. which would then make abt. 15 million kg. or abt. 11 million kg. Most of the herring taken in Danish waters are from the Belt Sea.

Of cod, abt. 14 million kg. were taken, these being from all waters, from Bornholm to the West coast of Jutland.

Plaice amounted to 16 million kg. of which 10 million in the North Sea, and 3 in the Kattegat.

The eel fishery gave a yield of 4 million kg. of which pactically none were taken in the North Sea or the Skagerak, but abt. 1,6 million in the Belt Sea, and 0,9 million in the Limfjord. The remainder from other coasts. Both silver eels and yellow eels are here included. The Limfjord thus gives about $1/4$th of the entire catch of eels; we should, however, expect more from the other parts, since all the eels from the Baltic, for instance, move through the Sound and the Belt unless previously captured, before reaching our coasts. In the Sound and the Belts, we find the greatest fishery of silver eels, with abt. 0,7 million kg. to each, reckoning the Sound as running far to the south. The silver eel fishery of the Limfjord alone is 0,3 million kg.

Of flounder, 1,5 million kg. were taken, mostly from the Belt Sea and the Kattegat.

The witch *(Pleuronectes cynoglossus)* gave in all only 0,7 million kg. nearly all from the Skagerak.

Haddock amounted to 3,7 million kg. in all, almost entirely from the North Sea, only a small part from the Kattegat and Skagerak.

Other species are so poorly represented that they need not be mentioned here.

For the dab, the statistics note a yield of abt. 1 million kg. but, as mentioned above, this fish is only occasionally included.

Altogether, this gives us abt. 42,000 tons of bottom fish, and abt. 15,000 tons of pelagic forms; together with others, abt. 60,000 tons taken by Danish fishermen off the coasts of Denmark in 1915.

The importance of these figures will be realised on comparing them for instance with the yield of Danish agricultural produce, as for instance our exports of eggs. These amount to abt. 20 million score annually, at abt. 1,2 kg. pr. score, or 24,000 tons of eggs; i. e. close on half the weight of our entire fishery yield, but representing a far greater value.

It should here, however, be observed that the production of eggs involves considerable expence, partly in food directly bought for the fowls, partly by the fact that the fowls themselves feed on other agricultural products, in addition to which, there is the cost of the poultry yard, attendance etc. The fish, on the other hand, cost us only the amount disbursed in connection with their capture, so that the net profit of the fishery is doubtless proportionately greater than that of the poultry yards.

In addition, it must be remembered that many other nations also take part in the fishery here, especially in the North Sea and the Skagerak, and these waters are only partially worked by Danish vessels — i. e. not the whole area. We have, however, already seen (Section VIII a) that even in our best waters, as for instance Thisted Bredning, only a small quantity of food-fish per hectare is annually caught, abt. 10—20 kg. per hectare, or less than the poorest-yielding carp ponds; the best run up to several hundred kg. per hectare.

It is the local occurrence of fish in great shoals at certain seasons which has given rise to the incorrect ideas as to the wealth of fish in the sea; on an average, there is only a small quantity of fish per unit of area in the sea, taken as a whole.

X. The Fishing Industry, its past and future.

A century ago, too little fishing was done in our fjords, the Belts, and the Kattegat, not to speak of the North Sea and the Skagerak. There were not enough fishermen. Now, the reverse is the case; one is inclined to fear that, in the waters inside the Skaw at any rate, the fishery is too intensive, so that there is risk of impoverishing the stock thereby, and numerous attempts have been made to hit upon some means of preventing this. There was at one time an idea that the fishing implements destroyed the eggs of the fish, and their food-animals also, on the sea floor, and again and again the use of seines has been prohibited from this point of view; the knowledge acquired in respect of fresh water conditions was applied to the sea, before the true state of the case in the latter had been ascertained. Later on it was found that most marine food fishes do not lay their eggs on the bottom at all, as do the fresh-water fishes, but that the eggs instead float in the water itself, where they cannot be harmed by the fishing implements. And the former idea was consequently abandoned. No real facts have ever been brought to light showing that the animals which furnish food for the fish are destroyed to any serious degree by implements working on the bottom. There are, it is true, exceptions, but as a rule, the small bivalves on which the fish feed pass through the meshes of the nets, and doubtless continue their existence unimpaired thereafter, upon the sea floor. The polychæta, perhaps, may suffer somewhat, but they keep as a rule buried deep down in the bottom itself. Vegetation is mostly found on rocky bottom and in shallow water, where it will hardly be damaged to any great extent by the seines. The animals chiefly taken by the seines, apart from the fishes themselves, are the larger forms, such as large gastropods and starfishes; i. e. predatory forms, and also various useless creatures, such as sponges, *Actiniæ, Hydroida*, etc., and a matter of fact, the taking of such forms would be rather beneficial to the stock of real fish-food, especially if they were brought to land, or at any rate destroyed, instead of being thrown back to the water more or less unharmed. In America, the starfishes caught are boiled on board, in order to kill them and prevent their causing further damage.

Attempts have also been made to combat the activities of those marine animals which interfere with the fishery, such as seals, for instance, or cormorants, which take the fish from the fishermens nets, or the starfishes, *(Asterias)* and certain gastropods *(Buccinum, Nassa)* which work havoc among the fish already caught; or again, the crabs, which do considerable damage to the nets, and useless forms such as the sea-scorpion *(Cottus)* which by their numbers alone occasion much inconvenience. A system of rewards for destruction of seals and birds of prey has doubtless accomplished something in this direction, but in the case of the lower animals, no result seems to have been attained as yet. Various methods have often been suggested whereby the harmful animals might be dealt with (see Report XIX) but it is an endless struggle, and a costly one, as long as no profitable use can be found for the creatures destroyed, so as to pay at any rate part of the expense of their destruction.

In the oyster basins at Ørodde, where no predatory forms were present, a very great production of fish-food was observed, and along the coasts in shallow water, we often find an enormous growth of young bivalves of various species, there being here also no predatory forms in any quantity. It seems, then, that something, at any rate, in this direction might be gained by the keeping down the predatory animals, if it could be done.

Hatching of fish-eggs in the sea — the eggs of cod, plaice etc. — was for some years regarded as a means of improving the stock, and in certain countries, considerable sums have been, and are still, disbursed for experiments in this direction. This idea is likewise borrowed from the fresh-water fisheries, where the point is of great practical importance. In our waters, the artificial hatching of trout appears to have had no slight effect upon the size of the stock and thus of the yield, but there is a difference of principle between improving the stock of a certain species of fish by hatching, accompanied by protection of the breeding grounds, when the latter are all situated in narrowly restricted fresh-water areas, (where practically all breeding can be kept down, since almost all spawning fish can be caught) and improving the stock of a marine species, covering a great area, and with widely extended spawning grounds whereon they breed in great numbers, as the quantity of pelagic eggs sufficiently shows. It is only by the study of these eggs, and the quantities in which they occur, that we have been able to realise properly the magnitude of the phenomena we have to deal with, and the results have forced us to admit that there can be no real comparison with the conditions prevailing in fresh water. The tiny newly hatched young of sea fishes are, moreover, when liberated in the sea, exposed to so many perils that only a small fraction of the number can probably ever survive to be of use as food. All attempts at keeping and feeding them until they have attained a larger size before liberation have proved impracticable. The latest Norwegian experiments with lobsters seem to point in the same direction. If, however, it should be found that these difficulties can after all be surmounted, then the matter would have to be taken under consideration anew.

I have already on a previous occasion (Report IV, 1893) pointed out, in the case of the plaice in the Kattegat, that in order to maintain a profitable fishery, it is not so important to increase the number of small plaice, as it is to let those which have reached a certain size grow larger, so as to attain a really considerable market value, and that hatching is therefore here of no avail. In the case of species producing so many thousands of young as do most marine forms, the most vital point is that there shall be favourable years, where great quantities of the young produced really develope i. e. live and later on grow up into full-sized fish. This theory I have called the theory of growth, as opposed to the general theory of propagation. And the valuation investigations with lower forms, as carried out from the Station, have further convinced me that favourable conditions for the young during their period of growth are of more importance than the number of the newly hatched larvæ. In some years, the growth of the young of many species seems partially to fail; in other years, it may be very successful, though why, we do not know. This is hardly dependent to any great degree upon the number of newly hatched young, as there will doubtless be sufficient as a rule, save in a quite exceptionally bad year for the species, when of course nothing can be done in any way.

This view, that the small fishes — i. e. those up to one or two years old — produced by Nature, should be protected until they have reached a size at which they are of value to man, has led to new endeavours in the direction of prohibitive size limits for fish brought to market. The intervention of authority is here evidently of great importance, as it has been proved that many fishing implements do destroy quantities of such half-grown fish to little or no purpose. This line of action will doubtless enable us to do much for various species of fish; I do not, however, here intend to go into details of the question.

A similar idea has led to the transplantation of two- or three-year old plaice from overstocked waters to other localities where few or no plaice are found, as for instance in the Limfjord. (See Section IV, conclusion).

We have, however, already seen that marine fishes, which in our waters are all carnivorous animals, will never give us any great production for a given area of the sea-floor. It is therefore natural to see if we cannot find other useful animals, which are more peaceable by nature, and thus able to thrive in dense masses on the bottom, living directly upon the vegetation and its products. In other words, we ought, wherever possible, to aim at a transformation of the production in the waters, just as has been done on land in agricultural countries.

Among such directly valuable animal forms, feeding on vegetable matter, and living close together on the bottom, we naturally think of the oyster and the common edible mussel, both of which are already popular as articles of food in this country; there are, however, also other bivalves, such as the clam *(Mya arenaria)* which might likewise be introduced as food here; it is largely eaten in North America. And other forms of the same type might also be mentioned.

Merely to fish for these animals, however, and then sell what is caught, would not be enough. I will here restrict myself to the question of oysters and mussels; these would almost undoubtedly soon be used up if fished for casually, as with most species of fish which are confined to local areas. It would be necessary first of all to set about a rational cultivation, so that the product might be improved, — by being rendered more uniform in quality, for instance, — the quantity increased, and new areas, where they are not at present found in any numbers, be formed. I need not here go into details regarding rational cultivation of these forms; the methods are known in other countries. In both cases, what is generally done is to transform the character of the sea-bottom, rendering it suitable for the growth of an epifauna of oysters or mussels respectively, and this is done very simply by setting ouy foreign bodies — shells for the oysters and poles for the mussels — when the animals will come of themselves, provided the locality be properly chosen. Both methods have been tried of late years in this country, at various places, and it is to be hoped that the experiments may prove successful, leading to a general adoption of the methods, which are based upon a new and correct principle, viz. the production of peaceable, and therefore numerous, useful animals by altering the character of the sea bottom.

I have inserted an illustration here, Fig. 11, showing a mussel pole with large edible mussels, the pole having been set out only $2^1/_2$ years before, by way of experiment, in the Limfjord. In August 1915, there were 42 kg. of mussels on it, of which abt 20 kg. were large enough for human consumption, being indeed of first rate quality.

Fig. 11. Mussel pole from the Limfjord, with 42 kg of live Mussel attached.

These poles can of course be placed fairly close together, only some few metres apart, whereas in the case of fish, we obtain only a few grammes per 1 sq. m.

It is not in all parts of our waters that such rapidly growing mussels are produced as in the Limfjord; possibly, therefore, it may be found that actual cultivation of them will only prove successful here and there. We have, however, in the Belts, so enormous quantities of wild mussels, that it would seem they might at any rate be utilised as food for domestic animals, as they can be captured in large quantities very cheaply, and should then be serviceable, for instance, either crushed in the shell as poultry food, or boiled and extracted from the shells for a like purpose, possibly as food for other animals. Fowls eat them eagerly, and produce good eggs, with no disagreeable taste, as we have found from experiments carried out by the Biological Station itself.

It would be far more economical to utilise the mussels for such purposes than to let them remain in the sea and die there, either of old age, or devoured by crabs and starfish; it is doubtless only a very small proportion of them which serves as nourishment for useful animals, and a sufficient quantity would certainly remain both for this and for maintenance of the stock.

As regards the oyster, we know that there are years when the temperature of the water in the Limfjord falls so low during the breeding season in summer, that only a very few of the young develope into the pelagic stage, i. e. tiny swimming organisms, while still fewer manage to attach themselves to solid bodies and grow up into mature oysters. In warmer summers, on the other hand, the pelagic young are taken in quantities at every haul among the plankton in the Limfjord, and we then find later on, as a rule, great numbers of them in the subsequent stages attached to suitable objects. Nevertheless, the mortality from the pelagic stage to that of the attached oyster, and thence to the mature condition, must be enormous, as the quantity of young produced by a single oyster may be taken at about one million, and even if possibly not all the individuals — though they are hermaphroditic — breed every good year, the hundred million oysters or thereabout in the Limfjord will yet in a favourable season produce so many young that if all grew up, there would be no room for water in the Limfjord, the entire volume would be occupied by oysters.

I stated above, that the oysters in the Limfjord numbered about a hundred million. This figure is not a haphazard guess, but is based on a series of diving investigations on some of the banks in 1907, in order to ascertain how closely the oysters congregate thereon, and what number could be estimated as living in the entire area of the fjord. It was then calculated, that the oysters over 7 cm. length amounted in all to abt. 90 million. (See Report XVII, for 1907, issued 1908.)

A hundred million, or let us say about 90 million oysters over 7 cm., is of course a high figure, yet the oysters were by no means densely packed on the banks. The best spots had only one or two individuals pr. sq. m., and in most places there were far fewer, for instance one for every 15—20 sq. m. It is therefore altogether incorrect to speak of „Østersbanker" (oyster beds; literally, „banks" or „mounds"), as if they represented places where the oysters lay heaped up together; the idea was probably derived from seeing the dredges come up full of oysters, dead and alive, without considering the great area covered to produce such a haul.

The appearance of a typical oyster bed has been described by H. Blegvad („I Dykkerdragt". Naturens Værksted, 3' Hefte 1916, p. 65—77.)

There is thus room on the oyster beds for a far greater number of full-grown oysters than are there found, and there is doubtless food enough for many more, since they live on the organic particles in the fine detritus ooze. Why, then, are there not more of them?

Up to 1910, the oyster fishery had for years yielded only abt. 1 to $1^1/_2$ million oysters a year, but the yield has since been raised to 4—6 millions annually, following on new terms of contract. Up to 1910, then, at any rate, the oyster had been practically protected for a series of years. (See Report XV.)

Before going further into this question, it will be necessary to say something about the age of the oyster generally. On the accompanying pages will be found

— 57 —

Oysters.

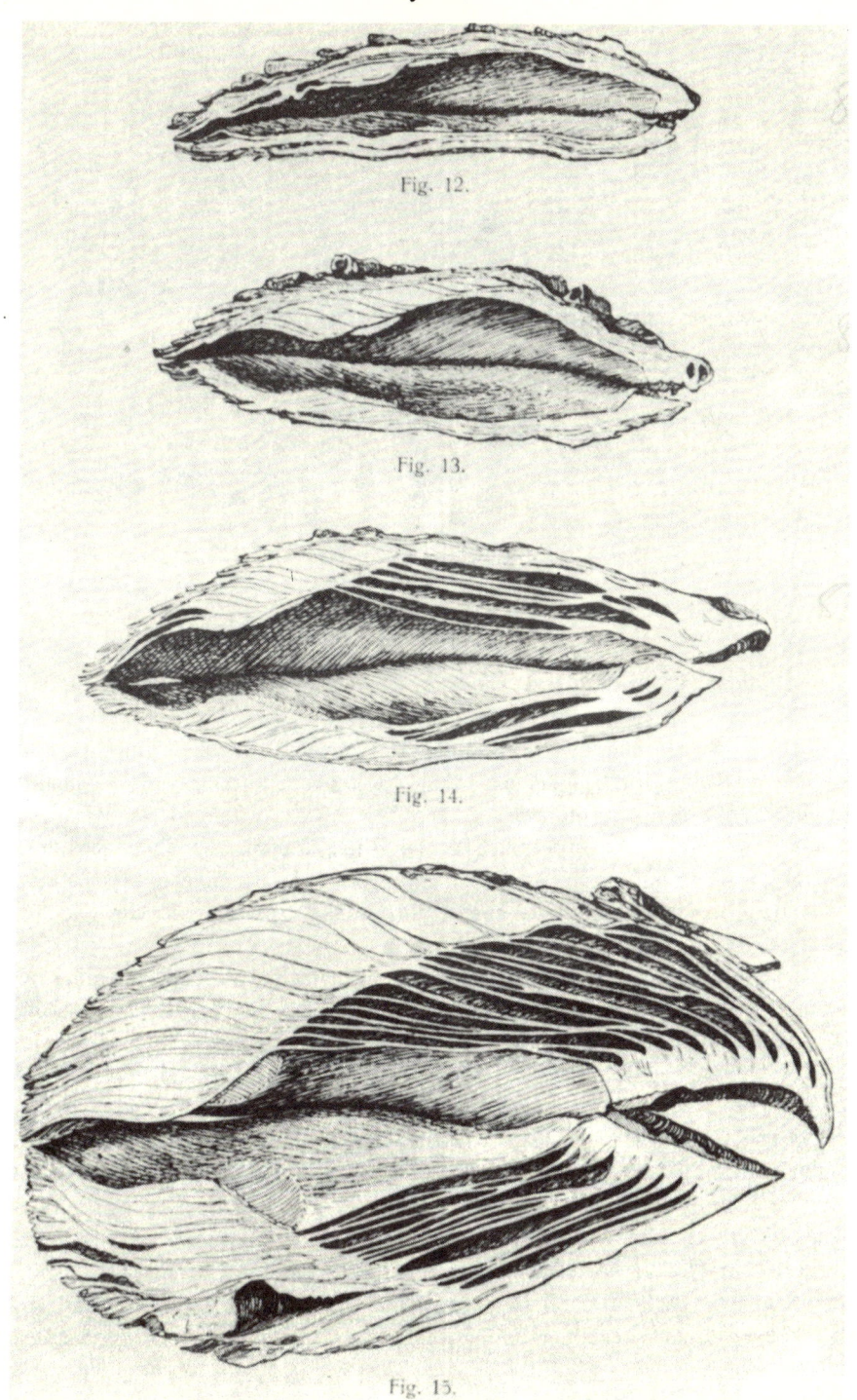

Fig. 12.

Fig. 13.

Fig. 14.

Fig. 15.

Fig. 16.

five illustrations, in natural size, showing sections of oysters, as sawn through in the shell.

Fig. 12 is an oyster over 7 cm. long, and perfectly eatable, though somewhat thin; it is probably abt. 3 years old. Fig. 13 shows a thicker one, presumably at least three years old. For the practised observer, it is possible to estimate roughly the age from the outer appearance of the shell, though the differences are not apparent in these figures. This specimen is of a size suitable for the table, and has solid — i. e. not hollow — shells; this is the case with the three following ones, Figs. 14, 15 and 16. These hollows often contain malodorous water extremely disagreeable to the smell and taste, and which flows out over the oyster when the shell is opened, and the knife breaks through the thin calcareous lamellæ within which the liquid is inclosed. These lamellæ are formed, apparently, one each year, from the soft body of the oyster itself; Fig. 14 has five or six of then, and this specimen must thus have lived for five or six years in addition to the time when its shells were solid   probably abt. 3—5 years — which would make the total age abt. 10 years. I do not know of any definite rule as to when the oyster forms its first lamella; it probably varies a good deal according to the conditions of growth, but young individuals rarely have any such. Oysters such as that shown in Fig. 14 are eatable enough, due precautions being taken, and the water is not always malodorous; those of the types shown in Fig. 15 and 16, on the other hand, are never sent to market, partly because they cost too much in transport, owing to the weight of the shells, up to $1/2$ kg., and partly because the oysters themselves are rarely in good condition; they are therefore generally thrown away. In Fig. 15, abt. 18 lamellæ can be counted, and the specimen is presumably over 20 years old; the oldest layers of the shell seem to have disappeared entirely. In Fig. 16, abt. 31 lamellæ can be discerned, and I should not be surprised if this specimen were abt. 40 years old. These old oysters were taken alive in the dredge, from the Limfjord, some few years back. Such specimens as these, and a quantity of empty shells from dead ones, were some of the results of the protection period, but even in the best beds, the bottom was not found covered with live oysters, as it may be seen to be in pictures of the American oyster beds.

It is difficult to say precisely how old the oysters generally brought to market are, as they grow differently in different places. We may presume that as a rule, they are from abt. 5 to 10 years old. Taking the average as 7 years, then the stock should be able to stand the loss of abt. 15 millions yearly, always provided that the number of oysters over abt. 7 cm. dying in other ways was not too great, and provided, also, that the stock each year was augmented by a sufficient quantity of young capable of further developement. With regard to the degree of mortality among grown-up oysters, very little is known; in the Limfjord, they do not appear to have many enemies among the other animals, but dead *Zostera* will doubtless suffocate them throughout considerable areas. In the winter of 1916–17, during the long period when the fjord was covered with ice, nearly half the stock of oysters died off on many beds, though this perhaps was rather due to lack of oxygen than to cold. As to the production of young each year, we know that it is by no means every year that great quantities of young appear, and especially not on the oyster beds. It is thus impossible to calculate how many million oysters the stock can afford to lose annually; we must find this out by experience. If, however, measures were taken to turn the young produced in good breeding years to better account, by procuring them better conditions for developement, then the fjord would be able to yield a greater quantity of oysters than is at present the case.

Oysters live attached to objects on the sea bottom, at any rate in the younger stages, and are therefore to be regarded as epifauna, like the mussels. Consequently, it is possible to improve the character of the bottom, by covering it with suitable objects, shells, and the like so that the young can attach themselves in greater numbers than they now do.

It has been found that the young oysters find difficulty in attaching themselves to objects which have lain to long – i. e. some weeks or months – on the bottom, as such objects will, in summer, which is the oysters' breeding season, soon become covered with a slippery film consisting partly of living, partly of dead plant growth. (See also W. K. Brooks: "The Oyster". Baltimore 1905.) Young oysters are therefore most frequently found to attach themselves to the newly formed edges of mussel or oyster shells, i. e. the parts of most recent growth, or on the under side of shells, where the light has perhaps been excluded, so that no living vegetation is found, and where detritus is less liable to accumulate; or again, on shells which have been turned over, so that the parts formerly buried in the mud or sand, and therefore devoid of slippery film, have recently come uppermost. Finally also, they attach themselves to shells recently thrown out into the fjord.

The improvement of an oyster bed is often accompanied by the cleaning of the bottom by dredging, so that all large objects, with the useless or detrimental creatures attached thereto, such as sea-anemones, starfish, crabs, etc. can be brought on shore and dried in the sun, whereby the animals rot off, and the shells are then clean and ready to be set out again in the breeding season. This method aims then at increasing the number of oysters, and was arrived at through numerous practical experiments in North America (W. K. Brooks) and is largely followed there; it is now being tried in the Limfjord. Such treatment of the bottom, through somewhat expensive, has proved profitable, despite the fact that the price of oysters in North America is low.

The question raised in the foregoing, on p. 56: Why are there not more oysters per sq. m. in the Limfjord than there are? must thus be answered by stating, as the principal reason, that in good breeding years, the young have difficulty in finding sufficient clean objects to which thay can attach themselves.

There are of course many other factors which tend to keep the stock of oysters within certain limits, but this one, at least, can be altered by human agency, and endeavours in this direction have elsewhere been found successful.

In other parts of the world, it is often the actual quantity of the young originally produced which is insufficient; in such case, the method followed is to import a further supply, the small stages, in size from that of a halfpenny to that of a penny, being transported from the richer breeding grounds to others where conditions are suitable for their growth and nourishment.

To put it briefly, we must in each particular case discover the critical point or points in the life of the stock, before it is possible to take effective measures towards its improvement. —

Should we thereafter come to consider what might be done with other animals belonging to the fauna of the level bottom, and living as a rule buried in the bottom itself, then we have for instance the clam *(Mya arenaria)*, the cockle *(Cardium edule)*, or even the smaller bivalves *(Abra, Corbula, Solen* etc.) which serve as food for the plaice and eels, and which might be worth our attention. Here, however, we should doubtless have to proceed on other lines than in the case of the oyster, and our first aim would be to ascertain what it is that keeps their numbers down to the present level. On this head, our knowledge at present is but vague, but the study of the question should be directed towards finding out whether the cause lies in climatic conditions which cannot be altered, or in purely local circumstances, such as the depredations of other animal forms, etc. whereby the numbers of useful species become reduced, and where effective measures might be taken for protection, as is often done with oyster cultivation, in shallow waters, by removing predatory forms which would otherwise devour the young already on the spot.

If, for instance, it should be found that the brittle star *(Ophioglypha)* in the Limfjord, devours most of the young bivalves which should go to feed the plaice, thus greatly reducing the yield, then it would be necessary to combat these predatory creatures; I do not, however, know for certain that they are responsible; it might for instance be the common starfish *(Asterias)* or the larger gastropods *(Buccinum* and *Nassa)* which do most damage; possibly again, quite other conditions militate against the developement of these food-forms. Obviously, the loss caused by the destruction of numbers of plaice in the nets by the enemies mentioned, can be guarded against by further restocking to a corresponding extent each year (se Report XIX 1911, p. 19); if on the other hand, it is the fish-food itself which is destroyed to a considerable extent by the predatory forms — and there is not too much of it at any time — then the case is more serious. As to how the campaign against these depredators should be carried out, this is a point which I do not propose to deal with further at present; if, however, the animals referred to really do contrive to reduce the production of plaice and eels in the fjord to a fraction of what it might be, then the question will certainly deserve the closest investigation in the future. Of *Ophioglypha* alone, there are in Thisted Bredning often over 100 specimens per sq. m. and if the species has any influence at all upon the young of the fish-food species, then it would surely seem that the influence of such numbers must be enormous in the long run.

Another line of thought leads in the direction of endeavouring to improve the fishery by seeking for other food species than those found in our waters, and which might possibly subsist on the vegetation, or at any rate profit more by the existing stores of animal food than do our own. Foreign species of fish have been already been introduced into Denmark, especially *Salmonidæ*, though these fishes are not herbivorous. Here again, I do not propose to treat the question at any length, having no positive suggestion to offer.

For the sake of completeness however, it should be noted that a direct exploitation of the vegetation, if not for actual human consumption, then at any rate for some other profitable purpose, would be the best solution of all. Se K. Rørdam: Kemisk Undersøgelse af Bændeltangen fra danske Farvande (Den kgl. Veterinær- og Landbohøjskoles Aarsskrift 1917, p. 109—145) where it is stated that "it would most certainly be worth while to institute practical experiments" in the feeding of domestic animals with *Zostera*. The plant can also be used, *inter alia*, for production of explosives of the guncotton type.

Obviously, if the sea bottom and its products should be subjected to proper cultivation, it would be necessary to make suitable provisions in the fishery legislation, the main principle of which at present is, that all Danish subjects have equal right to fishery wheresoever in Danish waters.

As long as our export of eggs alone, (abt. 20 million score, or 24,000 tons a year,) represents a value considerably beyond that of the entire fishing industry of the country under normal circumstances, it will be plain to all that our waters yield but an insignificant amount when compared with the products of the land.

It is not only Danish fisheries, however, to which the above applies; much the same is doubtless the case with marine fisheries everywhere; it is only in fresh water, or similar restricted areas, that a higher production per unit of area is achieved. As long as we have not progressed farther in improving the character of the sea fisheries, all that can be done is to extend the areas fished as far as possible; the Skagerak and the North Sea, for instance, are still only to a comparatively slight degree exploited by Danish fishermen. If this be done, and we furthermore endeavour to profit to the utmost by our smaller waters, little more can be attained at present in the procuring of raw material; as to the treatment of fish after capture, however, much could be said. I will not go further into this question here, but will just mention that a new method of refrigeration, by which the fish are frozen in brine of very low temperature instead of in cold air, offers many possibilities in the direction of enhancing the value of fishery products. Altogether, it is impossible to foresee what improvements may be made by science and invention in the future.

HISTORY OF ECOLOGY
An Arno Press Collection

Abbe, Cleveland. **A First Report on the Relations Between Climates and Crops.** 1905

Adams, Charles C. **Guide to the Study of Animal Ecology.** 1913

American Plant Ecology, 1897-1917. 1977

Browne, Charles A[lbert]. **A Source Book of Agricultural Chemistry.** 1944

Buffon, [Georges-Louis Leclerc]. **Selections from Natural History, General and Particular, 1780-1785.** Two volumes. 1977

Chapman, Royal N. **Animal Ecology.** 1931

Clements, Frederic E[dward], John E. Weaver and Herbert C. Hanson. **Plant Competition.** 1929

Clements, Frederic Edward. **Research Methods in Ecology.** 1905

Conard, Henry S. **The Background of Plant Ecology.** 1951

Derham, W[illiam]. **Physico-Theology.** 1716

Drude, Oscar. **Handbuch der Pflanzengeographie.** 1890

Early Marine Ecology. 1977

Ecological Investigations of Stephen Alfred Forbes. 1977

Ecological Phytogeography in the Nineteenth Century. 1977

Ecological Studies on Insect Parasitism. 1977

Espinas, Alfred [Victor]. **Des Sociétés Animales.** 1878

Fernow, B[ernhard] E., M. W. Harrington, Cleveland Abbe and George E. Curtis. **Forest Influences.** 1893

Forbes, Edw[ard] and Robert Godwin-Austen. **The Natural History of the European Seas.** 1859

Forbush, Edward H[owe] and Charles H. Fernald. **The Gypsy Moth.** 1896

Forel, F[rançois] A[lphonse]. **La Faune Profonde Des Lacs Suisses.** 1884

Forel, F[rançois] A[lphonse]. **Handbuch der Seenkunde.** 1901

Henfrey, Arthur. **The Vegetation of Europe, Its Conditions and Causes.** 1852

Herrick, Francis Hobart. **Natural History of the American Lobster.** 1911

History of American Ecology. 1977

Howard, L[eland] O[ssian] and W[illiam] F. Fiske. **The Importation into the United States of the Parasites of the Gipsy Moth and the Brown-Tail Moth.** 1911

Humboldt, Al[exander von] and A[imé] Bonpland. **Essai sur la Géographie des Plantes.** 1807

Johnstone, James. **Conditions of Life in the Sea.** 1908

Judd, Sylvester D. **Birds of a Maryland Farm.** 1902

Kofoid, C[harles] A. **The Plankton of the Illinois River, 1894-1899.** 1903

Leeuwenhoek, Antony van. **The Select Works of Antony van Leeuwenhoek.** 1798-99/1807

Limnology in Wisconsin. 1977

Linnaeus, Carl. **Miscellaneous Tracts Relating to Natural History, Husbandry and Physick.** 1762

Linnaeus, Carl. **Select Dissertations from the Amoenitates Academicae.** 1781

Meyen, F[ranz] J[ulius] F. **Outlines of the Geography of Plants.** 1846

Mills, Harlow B. **A Century of Biological Research.** 1958

Müller, Hermann. **The Fertilisation of Flowers.** 1883

Murray, John. Selections from *Report on the Scientific Results of the Voyage of H.M.S. Challenger During the Years 1872-76.* 1895

Murray, John and Laurence Pullar. **Bathymetrical Survey of the Scottish Fresh-Water Lochs.** Volume one. 1910

Packard, A[lpheus] S. **The Cave Fauna of North America.** 1888

Pearl, Raymond. **The Biology of Population Growth.** 1925

Phytopathological Classics of the Eighteenth Century. 1977

Phytopathological Classics of the Nineteenth Century. 1977

Pound, Roscoe and Frederic E. Clements. **The Phytogeography of Nebraska.** 1900

Raunkiaer, Christen. **The Life Forms of Plants and Statistical Plant Geography.** 1934

Ray, John. **The Wisdom of God Manifested in the Works of the Creation.** 1717

Réaumur, René Antoine Ferchault de. **The Natural History of Ants.** 1926

Semper, Karl. **Animal Life As Affected by the Natural Conditions of Existence.** 1881

Shelford, Victor E. **Animal Communities in Temperate America.** 1937

Warming Eug[enius]. **Oecology of Plants.** 1909

Watson, Hewett Cottrell. Selections from *Cybele Britannica.* 1847/1859

Whetzel, Herbert Hice. **An Outline of the History of Phytopathology.** 1918

Whittaker, Robert H. **Classification of Natural Communities.** 1962